COALBED METHANE:
SCIENTIFIC, ENVIRONMENTAL AND ECONOMIC EVALUATION

Coalbed Methane: Scientific, Environmental and Economic Evaluation

Edited by

Maria Mastalerz
Indiana University,
Bloomington, U.S.A.

Miryam Glikson
The University of Queensland,
Brisbane, Australia

and

Suzanne D. Golding
The University of Queensland,
Brisbane, Australia

KLUWER ACADEMIC PUBLISHERS
DORDRECHT / BOSTON / LONDON

A C.I.P. Catalogue record for this book is available from the Library of Congress.

ISBN 0-7923-5698-5

Published by Kluwer Academic Publishers,
P.O. Box 17, 3300 AA Dordrecht, The Netherlands.

Sold and distributed in North, Central and South America
by Kluwer Academic Publishers,
101 Philip Drive, Norwell, MA 02061, U.S.A.

In all other countries, sold and distributed
by Kluwer Academic Publishers,
P.O. Box 322, 3300 AH Dordrecht, The Netherlands.

Printed on acid-free paper

Printed in the Netherlands.

Table of Contents

COAL-SOURCED LIQUID HYDROCARBONS: GENERATION, ACCUMULATION

NOTES / SHORT PAPERS

PREFACE

Coalbed gas has been considered a hazard since the early 19th century when the first mine gas explosions occurred in the United States in 1810 and France in 1845. In eastern Australia methane–related mine disasters occurred late in the 19th century with hundreds of lives lost in New South Wales, and as recently as 1995 in Queensland's Bowen Basin. Ventilation and gas drainage technologies are now in practice. However, coalbed methane recently is becoming more recognized as a potential source of energy; rather than emitting this gas to the atmosphere during drainage of gassy mines it can be captured and utilized. Both economic and environmental concerns have sparked this impetus to capture coalbed methane.

The number of methane utilization projects has increased in the United States in recent years as a result, to a large extent, of development in technology in methane recovery from coal seams. Between 1994 and 1997, the number of mines in Alabama, Colorado, Ohio, Pennsylvania, Virginia, and West Virginia recovering and utilizing methane increased from 10 to 17. The Environmental Protection Agency estimates that close to 49 billion cubic feet (Bcf) of methane was recovered in 1996, meaning that this amount was not released into the atmosphere. It is estimated that in the same year total emissions of methane equaled 45.7 Bcf. Other coal mines are being investigated at present, many of which appear to be promising for the development of cost-effective gas recovery.

Methane from coal is increasingly becoming a substantial contributor to the United States' natural gas resource base, production having increased at least fivefold since 1990. Although numerous basins have considerable coalbed gas potential, more than 90 percent of the gas produced comes from two basins: the San Juan and the Warrior, where the current development represents only a fraction of the estimated 675Tcf (19.1 Tm^3) of U.S. coalbed methane resources. The San Juan Basin is the most productive in the world, and accounts for about 80 percent of the entire U.S. coalbed gas production. In 1992 alone this basin produced 44 Bcf (12.6 Bm^3) of gas, with cumulative production through the end of 1994 exceeding 1 Tcf (28 Bm^3).

This special publication on coalbed gas is multidisciplinary in scope and covers a wide range of aspects from exploration through production, resource calculations, emissions, and regulatory processes. One part of this volume deals also with qualifying and quantifying oil generation from coal. Furthermore, quality and quantity of oil and gas generation from coals as a function of heating mode is reported. New models for oil and gas generation in coal seams by convective heat transfer through hydrothermal circulation in basins are presented in this volume for the first time.

We acknowledge the following reviewers for their efforts to improve the papers included in this volume: John Comer, John Rupp, Colin Ward, Chris Boreham, Stewart Gillies, Julian Baker, and Sue Golding.

M. Mastalerz and M. Glikson

COAL SEAM GAS IN QUEENSLAND - FROM THERE TO WHERE?

S.G. SCOTT
Department of Mines and Energy
GPO Box 194 Brisbane, Queensland, Australia

1. Introduction

Coal seam gas exploration has had a long history in Queensland. While most people think that coal seam gas exploration is a relatively new industry in Queensland, exploration started in May 1976 with the granting of Authorities to Prospect 226P, 231P and 233P to Houston Oil & Minerals of Australia Inc. Since then over 200 wells have been drilled and nearly $200,000,000 has been spent exploring for and appraising coal seam gas. All this effort has resulted in the development of three areas for coal seam gas production; the Moura Mine, the Dawson Valley Project and Fairview with many more possible.

2. Coal Seam Gas Exploration History

Queensland did not lag too far behind the United States in the rush to attempt to exploit coal seam gas. In May 1976 Authorities to Prospect (ATPs) 226P, 231P and 233P were granted to Houston Oil & Minerals of Australia Inc over the central and eastern Bowen Basin (Figure 1). In October 1976, the first well, HOM Kinma 1 was drilled. This was followed closely by HOM Carra 1 and two conventional petroleum wells, HOM Shotover 1 and HOM Moura 1 were re-entered and completed as coal seam gas wells. Production testing was intermittent due to mechanical problems and the wells were eventually abandoned and the ATPs surrendered.

It was another four years before the next attempt at exploiting coal seam gas with the granting of ATP 273P to BHP Petroleum Pty Ltd (BHP) in the central Bowen Basin (Figure 2). The main aim of the work was to degasify the working seam at BHP's coal mining division's Leichhardt Colliery. Four wells (HPP Gemini 1 - 4) were drilled and fracture stimulated, with production testing lasting between 9 and 17 months. Production rates were considered uneconomic and the wells were abandoned and the ATP surrendered.

1

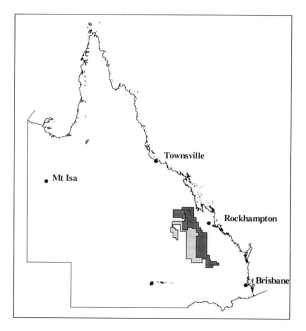

Figure 1: Houston Oil & Minerals coal seam gas Authorities to Prospect

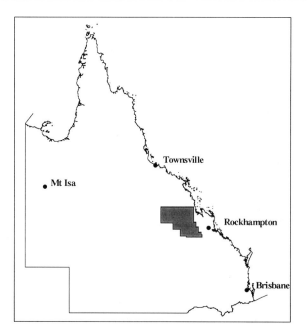

Figure 2: BHP's coal seam gas Authority to Prospect

Five years later, in 1985, ATP 352P was granted to Median Oil NL and Baraline Pty Ltd. ATP 352 was located in the northern Bowen Basin. A year late ATP 352P was conditionally surrendered and ATP 364P was granted (Figure 3). The granting of this ATP was the catalyst that sparked a surge in coal seam gas exploration, which has not yet ceased. Between 1986 and 1996 171 wells were drilled, 156 of these in the Bowen Basin (Figure 4).

In November 1995 the first petroleum leases (PL 90, 91 and 92) were granted for the production of coal seam gas. These PLs, referred to as Fairview, are located near Injune in the southwest Bowen Basin and were granted to Tri-Star Petroleum Company. The granting of these leases was closely followed by the granting of a petroleum lease (PL 94) and a pipeline licence (PPL 26) to a Conoco Australia Pty Ltd. PL 94, referred to as the Dawson Valley Project, is located between Moura and Theodore in the southeast Bowen Basin. PPL 26 connects the Dawson Valley Project with the PG& E Gas Transmission Pty Ltd's Wallumbilla to Gladstone gas pipeline. In late 1996 PL 101, located north of Wandoan in the southeast Bowen Basin, was granted to Pacific Oil & Gas Pty Ltd (Figure 5).

3. Present Day Status

Presently there are 11 ATPs in the Bowen, Surat/Bowen and Eromanga/Galilee Basins where the titleholders have indicated their main exploration target is coal seam gas. Another three ATPs in the Surat/Bowen Basins have indicated that they are targeting both conventional and coal seam gas (Figure 6). In total these ATPs cover 80,400km^2. The major players are Oil Company of Australia Ltd (ATPs 467P, 525P, 564P, 602P, PLs 94 and 101), Tri-Star Petroleum Company (ATPs 526P, 584P, 592P, 606P, 623P, PLs 90, 91 and 92) BHP Petroleum Pty Ltd (ATP 364P), Santos Ltd (ATP 378P) and Galilee Energy Pty Ltd (ATP 529P). On 1 April 1998 Oil Company of Australia Ltd concluded a deal in which they bought out Conoco Australia Pty Ltd to become the titleholder of all the ATPs previously held by Conoco Australia Pty Ltd.

Exploration and production is also being carried out at a number of coal mines in the Bowen Basin. The most successful at the present is the BHP Mitsui Coal's Moura Mine where horizontal wells have been drilled into the seam left when mining has ceased in the open-cuts. There wells have been drilled for over 1,000m to drain coal seam gas from the seams from the Baralaba Coal Measures.

4. Coal Seam Gas Production

Prior to the granting of the Fairview PLs, only production testing had occurred in Queensland for coal seam gas. During this time, companies would test the production rates of their wells by flaring the gas produced. This allowed the companies to

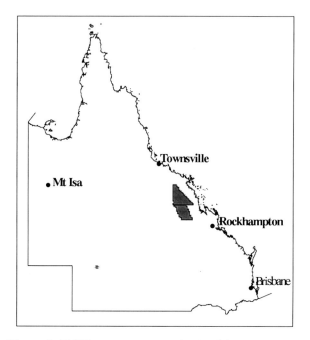

Figure 3: NQE's coal seam gas Authorities to Prospect

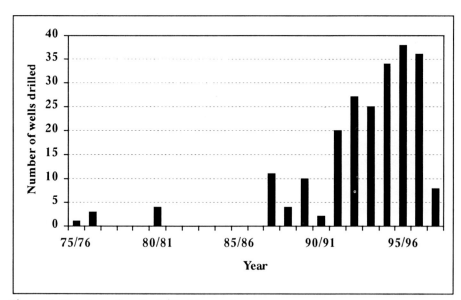

Figure 4: Number of wells drilled for coal seam gas

determine whether the rate of the well would be commercial when the rate had settled down. Many wells initial rates were very encouraging but over time decreased to a level where they would have been uneconomic. There are a number of reasons why this occurred;

- if the well was unstimulated, the natural fracturing (cleats) within the coal was not extensive enough to access enough coal volume to provide enough gas

- if the well has been stimulated, then either;
  - the stimulation was not effective, the fractures being more horizontal the vertical or
  - the stimulation was effective but deposition of minerals within the fractures restricted gas flow.

The companies also had to demonstrate that commercial production rates could be sustained over long periods. As shown in Figure 7 the total Queensland production rate, which includes commercial production and production testing, has increased substantially since 1996.

The first commercial production was achieved from the Dawson Valley project (PL 94) which started production in late 1996 into PPL 30 to Gladstone. Approximately 4TJ/day is produced from Dawson Valley from two major areas in PL 94; Moura and Dawson River. Each of these areas has 12 wells which feed into a dehydrator and then a compressor before being transported via PPL 26 to PPL 30.

In early February 1998 Tri-Star Petroleum Company began from the Fairview area via a lateral extension of PPL 30. Approximately 3 TJ/day is produced from Fairview from 11 wells.

Production testing is also happening at Peat (PL 101). This testing is presently not continuous but is being used by the present operator of PL 101, Oil Company of Australia Ltd, to attempt to understand the nature and conditions of the reservoir at Peat.

Production also comes from BHP Mitsui Coal's Moura Mine, east of Moura, in the southeast Bowen Basin. Commercial production began in mid 1996 with the gas being delivered along a twenty kilometre pipeline to PG&E Gas Transmission Pty Ltd's Wallumbilla to Rockhampton/Gladstone pipeline. Because this operation is being carried out in a coal mining lease, the operation comes under the control of the Queensland *Mineral Resources Act 1989* and not the Queensland *Petroleum Act 1923* as all the other coal seam gas operations do.

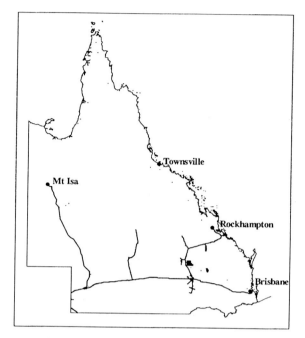

Figure 5: Current coal seam gas Petroleum Leases and pipelines in Queensland

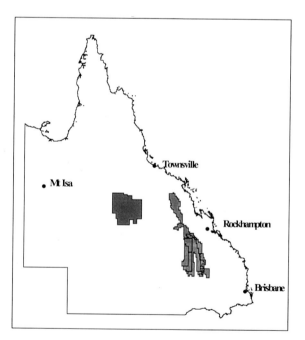

Figure 6: Current coal seam gas Authorities to Prospect in Queensland

5. Coal Seam Gas Resources/Reserves

Coal seam gas resources and reserves and an oft confused idea. While there is no formal definition for resources and reserves in coal seam methane the most accepted definitions are;

- resources are all of the potentially useable gas in a defined area, and are based on points of observation and extrapolations from those points.

- reserves are those parts of the resource for which sufficient information is available to enable detailed or conceptual production planning and for which such planning has been undertaken.

Currently there are no defined reserves for coal seam methane in Queensland. A number of authors have estimated the inferred resources of coal seam methane in Queensland by basin. The Bowen Basin is viewed as having the greatest potential as a coal seam methane province and resources for the Bowen Basin have been estimated between 124,000 and 192,000 PJ (Warren Centre for Advanced Engineering and The Earth Resources Foundation, 1994 and Decker, Reeves & White, 1990).

Conventional gas reserves in Queensland, as at 30 June 1995, are 2,267 PJ (Scott, Suchocki & Kozak, 1997).

6. Gas Demand And Supply

Gas demand for Queensland and the entire eastern Australia region is anticipated to grow substantially over the next 30 years (Figures 8 and 9) (Australian Gas Association, 1997). While demand will increase, the supply from traditional conventional sources like the Bowen, Surat and Cooper Basins in Queensland and the Gippsland Basin in offshore Victoria will decline and not be able to meet anticipated demand. This decline when demand can not be met by supply is anticipated between 2000 and 2008 (Australian Gas Association, 1997).

Increased conventional exploration in the Eromanga and Cooper Basins, the Surat and Bowen Basins and the proposed introduction of gas from Papua New Guinea via a new pipeline to Gladstone may go someway to delaying the shortfall. If coal seam gas is not included as a factor then the shortfall will happen sooner than later. The lack of reserve data for coal seam gas makes it difficult to factor coal seam gas into the supply and demand curve.

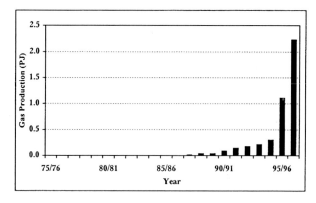

Figure 7: Coal seam gas production in Queensland

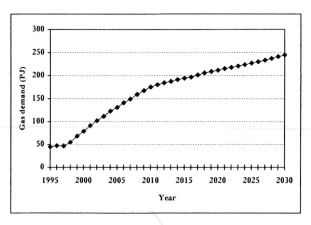

Figure 8: Gas demand in Queensland

(Australian Gas Association, 1997)

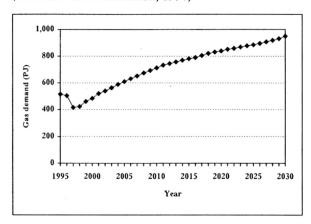

Figure 9: Gas demand in eastern Australia

(Australian Gas Association, 1997)

7. Coal Seam Gas Verses Conventional Gas

Coal seam methane enjoys a number of factors over conventional gas sources in Queensland.

- Coal seam gas resources are substantially greater than conventional reserves.

- On the whole, coal seam gas is closer to the east coast and population centres than conventional sources making transportation cost of the product much less.

- Coal seam gas wells tend to be shallower (ranging between 500 and 1,000m) compared to conventional gas wells (1,000 and 2,500m).

- Coal seam gas areas tend to be closer to settled areas, making development easier and less expensive (transportation costs of materials, services etc should be less).

Coal seam gas also has a number of mitigating factors;

- Lack of proven reserves.

- Perceived difficulty in extracting coal seam gas.

- High cost of wells (drilling and stimulating) and large number of wells to achieve gas volumes.

The one telling factor that makes coal seam gas a very attractive supply option for future markets is the very large inferred resources when compared to the estimates for undiscovered conventional reserves. These undiscovered conventional reserves range from 2,500 and 4,800 PJ (Australian Gas Association, 1997). The Australian Gas Association's best case scenario is that there will be a shortfall by 2008, even with an increase in production from the Gippsland Basin, additional reserves of 2,500 PJ found in other eastern Australian Basins but with only 50 PJ of coal seam gas production per year. Clearly an increase in the production of coal seam gas could delay the shortfall.

8. Conclusion

Coal seam gas exploration has had a long history in Queensland. There is no doubt that there are very substantial resources of coal seam gas in basins relatively close to Queensland's east coast and major population and industrial centres. The forecast demand for gas in Queensland and the rest of eastern Australia will increase dramatically over the next thirty years. Conventional gas reserves in eastern Australia will not be able to meet this demand but coal seam gas may well be the resource to fuel this demand. Research into recovery technologies that will turn these substantial resources into substantial reserves must be a high priority.

9. References

AUSTRALIAN GAS ASSOCIATION, 1997: *Gas supply and demand study 1997. A fourth report to the participants on the findings of a study into Australia's natural gas supply and demand to the year 2030.* Participants report. The Australian Gas Association. Canberra. ACT.

DECKER, D., REEVES, S. & WHITE, J., 1990: A basin exploration strategy for the development ofcoalbed natural gas in the Bowen Basin. *IN, Bowen Basin Symposium 1990*, Geological Society of Australia, 81 – 84.

SCOTT, S.G., SUCHOCKI, V. & KOZAK, A., 1997: Petroleum operations, production and reserves – Annual Report 1995/96. Queensland Geological Record 1997/4.

WARREN CENTRE FOR ADVANCED ENGINEERING AND THE EARTH RESOURCES FOUNDATION (WITHIN THE UNIVERSITY OF SYDNEY), 1994: *Coal Bed Methane Extraction: A proposal for a field trial in the Southern Sydney Basin*, NSW, January.

DEVELOPING A NEW COAL SEAM GAS REGIME FOR QUEENSLAND

Stephen G. Matheson
Queensland Department of Mines and Energy
GPO Box 194
Brisbane 4001
Australia

Acknowledgements

This paper is a modification of a paper published in the Queensland Government Mining Journal (January, 1998, p.30-34) and also draws extensively on the Government position paper 'A new coal seam gas regime for Queensland' published by the Department of Mines and Energy in November 1997. The assistance of Steven Scott in providing information for the overview section (including Figures 1&2) is also acknowledged.

1. Introduction

Coal seam gas is an important energy commodity. While commercial production of coal seam gas is in its infancy as an industry in Queensland, the State has vast resources of coal (over 35 billion tonnes) and much of that coal has potential for containing significant quantities of coal seam gas (largely methane). New policy and legislation is required in order to ensure appropriate development of those resources and at the same time ensure that the already established and economically important coal mining industry is not jeopardised. This paper sets out the proposed new regime for administering exploration and production titles for coal explorers and miners under the *Mineral Resources Act 1989* (Qld) and coal seam gas explorers and developers under the *Petroleum Act 1923* (Qld).

1.1 OVERVIEW OF THE COAL SEAM GAS INDUSTRY

Coal seam gas exploration in Queensland began in earnest, in 1986 with the grant of Authority to Prospect (ATP) 364P over the northern Bowen Basin. Since the grant of ATP 364P, more than 35 ATPs have been issued and over $200 million has been spent exploring for and testing coal seam gas. This has included the drilling of more than 200 wells.

Presently there are 11 ATPs in the Bowen, Surat/Bowen and Eromanga/Galilee Basins where the titleholders have indicated their main exploration target is coal seam gas. Another three ATPs in the Surat/Bowen Basins have indicated that they are targeting both conventional and coal seam gas (Figure 1). In total these ATPs cover 80,400km². The major players are Oil Company of Australia Ltd (ATPs 467P, 525P, 564P, 602P, PLs 94 and 101), Tri-Star Petroleum Company (ATPs 526P, 584P, 592P, 606P, 623P, PLs 90, 91 and 92) BHP Petroleum Pty Ltd (ATP 364P), Santos Ltd (ATP 378P) and Galilee Energy Pty Ltd (ATP 529P).

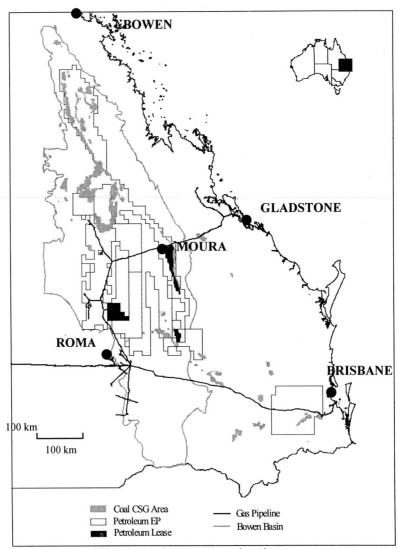

Figure 1 Coal seam gas exploration areas

On 1 April 1998 Oil Company of Australia Ltd concluded a deal in which they bought out Conoco Australia Pty Ltd to become the titleholder of all the petroleum titles previously held by Conoco Australia Pty Ltd.

Exploration and production of coal seam gas is also being undertaken at the BHP Mitsui Coal Moura mine where horizontal wells have been drilled into coal seams exposed in the highwalls of open cut pits where mining has ceased

The first commercial production in Queensland commenced production in late 1996 from the Dawson Valley project (PL 94) into PPL 30 to Gladstone (Figure 2). Approximately 4TJ/day is produced from Dawson Valley from two major fields in PL 94 (Moura and Dawson River). Each of these fields has 12 wells which feed into a dehydrator and then a compressor before being delivered via a 47 km pipeline (PPL 26) into the Wallumbilla to Gladstone pipeline (PPL 30). The project has recently been acquired from Conoco by the Oil Company of Australia.

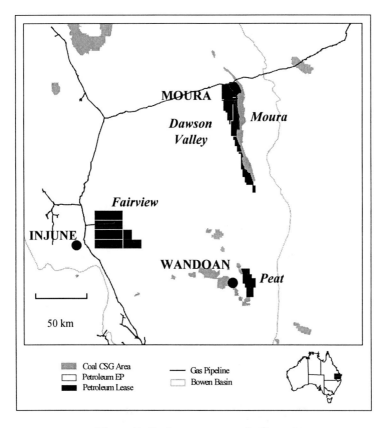

Figure 2 Coal seam gas production areas

In early February 1998, Tri-Star Petroleum Company began production from the Fairview area, some 40km north east of Injune. Approximately 2 TJ/day is produced from 11 wells. A 25 km pipeline has been constructed to join the Tri-Star field to the Wallumbilla to Gladstone pipeline.

Production testing is also occurring at the Peat Deposit, some 20 km east of Wandoan. This testing is currently not continuous but is helping the operator of PL 101, Oil Company of Australia Ltd, gain a greater understanding of the nature and conditions of the reservoir at Peat.

Production also occurs at BHP Mitsui Coal's Moura Mine, east of Moura, in the southeast Bowen Basin. Commercial production began in mid 1996 with the gas being delivered along a 20 km pipeline to Wallumbilla to the Gladstone pipeline. BHP are currently producing 1.5 TJ/day

1.2   LEGISLATIVE RIGHTS TO ACCESS AND UTILISE THE GAS

In Australia all minerals (with only a few isolated historical exceptions) belong to the Crown, and the administration of mineral property rights to allow private companies and individuals to explore and mine minerals is undertaken by the relevant State Government.

The development of a coal seam gas industry in Queensland has not been assisted by the situation whereby rights to explore and develop coal seam gas (CSG) have potentially been available to title holders over the same ground under both the *Mineral Resources Act 1989* (Qld) and the *Petroleum Act 1923* (Qld).

Currently there is little direct conflict between the two industries because coal mining in Queensland has largely been undertaken by open cut methods. Whilst underground mining is becoming more prevalent, it is still only being conducted at relatively shallow depths (less than 300m) compared with other parts of the world, including New South Wales. However, the problem is a significant one because there are overlapping coal and petroleum tenures covering most of the coal measures throughout Queensland.

The Queensland Department of Mines and Energy (DME) recognised the need therefore to amend the administrative and legislative regime under which coal seam gas explorers and producers and coal miners investigate and exploit coal seam gas resources in Queensland. A number of discussion papers, policy frameworks and administrative procedures have been considered and reviewed over the last three years in an effort to develop a new regime suitable to both parties. This has been complicated by having two industries with opposing viewpoints who had failed to provide any conciliatory position on which a new framework could be developed; the lack of consistent and well defined policy; and legal questions surrounding the interpretation of the relevant Acts.

2.   Recent Policy Process

2.1   29 NOVEMBER 1996 POSITION PAPER

During 1996 the Department of Mines and Energy resolved to address this issue and developed proposals for a new administrative and legislative regime for the exploration and development of coal seam gas. A position paper which outlined the new regime was presented to key stakeholder groups on 29 November 1996. Copies of the position paper (essentially a Government Green Paper) were also provided at that time to all key stakeholders likely to be affected by the proposals.

The response to the Department's position paper was limited as only eleven individual stakeholder submissions were received. Overall comment on the paper from both the coal and CSG industries has been positive, with general agreement with the principles although there were concerns with the amount of discretion and lack of certainty provided by the regime. In addition to the eleven individual responses, the Queensland Mining Council (QMC) initiated the formation of a joint CSG industry/coal industry working group (and acted as secretariat) to prepare a joint industry response to the Department's position paper. The joint response represents a major step forward in resolving the issues and the development of a new cooperative regime between the two industries.

2.2   DEVELOPMENT OF A FINAL POSITION PAPER - DECEMBER 1997

The DME worked to bring together all the comments from the stakeholders and develop a final position paper. This was undertaken in close consultation with the QMC taskforce and has proved to be a very successful approach to resolving a difficult issue.

On 12 December 1997 the final position paper was released (or Government White paper) and presented a new legislative and administrative regime that outlines the rights to coal seam gas and provides for appropriate interaction between the coal seam gas and coal industries.

While not intended as a discussion document, stakeholders were given until 20 February 1998 to provide any comment on the position paper. One stakeholder (Oil Company of Australia) provided comments, which are being considered.

3.   The Proposed New Regime

The new regime has a number of clear objectives, relating to the resolution of rights to extract coal seam gas and access to conduct exploration and mining. The regime does not address issues relating to the *Coal Mining Act 1925* (Qld) nor does it address issues relating to the environment or landholders which will be addressed in the current review of the *Petroleum Act 1923* (Qld).

3.1 GUIDING PRINCIPLES

The proposed regime is based on a number of key guiding principles. The guiding principles are:

- Both coal and coal seam gas are valuable resources, and the extraction and utilisation of coal seam gas should be encouraged while ensuring potentially mineable identified coal resources are protected from conflicting resource utilisation.

- The rights to extract and utilise coal seam gas need to be clarified and certainty provided to both industries.

- For safety reasons a coal miner must have overriding control of the management of coal seam gas being released into a working mine.

- The effect on the existing rights of current tenure holders is to be minimised and existing production facilities and investments must be safeguarded where new exploration or production facilities are being considered for approval.

- Recognition of simultaneous development or access is to be encouraged, but potential impacts of either industry on the other should be minimised, and procedures for dispute resolution are to be simple.

3.2 KEY OBJECTIVES

- To clarify rights to coal seam gas under the *Mineral Resources Act 1989* (Qld) and *the Petroleum Act 1923* (Qld).

- To protect existing operations and investments.

- To permit surface access so concurrent exploration and development operations can occur.

- To ensure safety in existing and future coal mining operations.

3.3 KEY FEATURES

1. Rights to CSG will be provided under both the *Mineral Resources Act 1989 (Qld)* and the *Petroleum Act 1923 (Qld)* and will be secured by application for production title, subject to certain conditions and in line with the principles and objectives outlined above.

2. No distinction will be made between CSG and other petroleum exploration production tenements.

3.   All existing petroleum and coal mining leases and lease applications (as of 28 October 1996) will be given CSG rights to the 'centre of the earth'.

4.   In regard to future lease applications, primary areas of interest for the coal seam gas producer and the coal developer will be identified and rules of priority and access will apply in each of the areas.

5.   To avoid surface access conflict problems the following restrictions on the grant of production titles will apply:

   • where there are any conflicting titles (coal and petroleum) an application for a production title will be strata titled (ie. the depth limits of the 'coal CSG areas' apply).

   • where there is no other tenure present an application for a production title will go to the 'centre of the earth'.

6.   'Coal CSG areas' will be considered priority areas for coal miners. 'Coal CSG areas' will be areas of identified, measured and indicated coal resources to specified depth limits in accordance with the mineability of the coal.

7.   Within the surface boundaries of a 'coal CSG area':

   • CSG rights (along with coal rights) will attach to mining leases (MLs) down to specified depths. The exception will be existing MLs and mining lease applications (MLAs) where CSG rights (and coal rights) will extend to the centre of the earth.

   • Applications for petroleum leases (PLs) may be granted below the depth limit of a 'coal CSG area' i.e. strata title.

   • The holder of a prior Mineral Development Licence (MDL) or a ML will be obliged to provide surface access on reasonable terms to an applicant for an underlying PL. The reverse will apply in regard to a prior PL and an overlying ML or MDL.

8.   Outside the boundaries of a 'coal CSG area', priority will be given to CSG operators:

   • Where there is no conflicting prior title, a PL or ML will be granted to the 'centre of the earth'.

   • Where there is a pre-existing PL, an ML or MDL applicant will be required to obtain the approval of the PL holder prior to grant. Where the prior title is an ATP, the holder will be invited to apply for a PL and thereby establish priority production title rights.

- Where there is a pre-existing ML or MDL, the holder will be obliged to provide access on reasonable terms to the applicant for an underlying PL.

9.   Both within and without the boundaries of a 'coal CSG area':

- Where overlapping titles occur, the CSG rights of the original holder will be protected, but surface access will have to be negotiated between the two parties. An arbitration process will be set in place where agreement is not reached.

- The prior title holder will be entitled to some form of compensation from the subsequent ML, MDL or PL applicant for the dislocation caused by admitting that person to the surface of the land. The extent and scope of such compensation have yet to be finalised. Compensation may be determined by the Mining Warden in the absence of an agreement.

- The range of deductible expenses used to arrive at the value for royalty assessment (for both petroleum and coal) may be expanded to include the compensation costs such as those incurred by the subsequent title applicant in gaining surface access (this would be up to approved limits).

- Measures to encourage the dual utilisation of resources will be developed. For instance exemption from royalty (for defined time periods or up to maximum limits) may be provided for CSG which is produced as a by-product of coal mining and used on the mining lease - for example, for on-site power generation.

10.   Both ML and PL holders (where overlapping coal and CSG titles occur) will be required to have an approved Production Development Plan before commencement of production of CSG. This will include details of proposed operational activities, measures to reduce impacts on other resources or operations and planned production.

4.0   Some Areas of Interest

4.1   COAL SEAM GAS AREAS

Areas of primary interest to the coal miner will be defined as 'coal CSG areas' and will define areas where the primary resource utilisation will be coal. In these areas, priority of production title will be provided to the coal miner, in terms of CSG rights and surface access. Thus rights to CSG within 'coal CSG areas' will not be available to a petroleum title holder but will lie exclusively with the coal miner (or Minister if not covered by a MDL or ML). Holders of a mining lease for coal that is deemed as a 'coal CSG area' will have rights to utilise all CSG to the depth limit of the 'coal CSG area'.

Coal CSG areas will be determined in terms of a number of criteria:

- The level of geological data available and status of coal resources (e.g. are the resources of Indicated status in accordance with the Australian Code for Reporting of Identified Coal Resources and Reserves?).

- Any plans for the mining of the deposit, or if none are available the plans for mining similar deposits, or the number and history of operating mines in a similar and adjacent geological environment.

- The depth to which current or immediate future mining methods might be applicable to the geological setting. For the principal coal bearing provinces in Queensland this will be:
  Bowen/Ipswich/Laura Basins   400m depth exclusion
  Surat/Moreton/Tarong/Galilee Basins 150m depth exclusion
  Other Basins   Upon application

- The areal extent on the surface of the identified coal resources to these depths plus a defined buffer zone.

- A 'coal CSG area' will be considered as a rhomboid, that is from the definition on the surface it shall extend vertically to the appropriate depth as defined above.

The 'coal CSG areas' that will apply as of 28 October 1996, will include:

- all Mining Leases for coal (granted or applications)
- all Mineral Development Licences for coal (granted or applications)
- some identified coal deposits that lie outside of these areas (ie. may lie within an EPC or have no current tenure).
- the Theiss Peabody Mitsui (TPM) coal field area.

The blanket inclusion of all MDLs and all MLs for coal is justified because, these tenures must contain proven resources, they are known to not contain resources at depths greater than the current criteria and they already include buffers for infrastructure and access. All other areas have been determined by DME in accordance with the above criteria.

It should be noted that under transitional arrangements, all the pre-existing MLs (ML applications and granted MLs pre 28 October 1996) will have coal seam gas rights to the "centre of the earth". There are a total of 206 in this category and 79 additional 'coal CSG areas' that would have strata title CSG rights. New areas will be able to be added to the register as they are identified.

4.2   USE OF STRATA TITLE

Outside of 'coal CSG areas', utilisation and commercial production of coal seam gas will be determined by a "first in first served" policy where priority is provided by the application date of the production title, with a first right of refusal to any prior petroleum title holder.

Production titles in all cases will however be strata titled to the appropriate depths of primary interest. The position paper proposes that where no form of title exists at all (coal or petroleum) then the production title could be granted to the centre of the earth. Consideration is now been given to the suggestion that all production titles be strata titled if there is any possibility for future potential production titles. The depths of the production title would be made in accordance with their proposed usage.

4.3   OTHER MEASURES

A number of measures are being considered in terms of the Government's role in fostering an environment where there is open communication to allow commercial outcomes that would benefit both coal and coal seam gas industries.

For example, measures are being considered to encourage parties to negotiate compensation agreements, where there is simultaneous production by overlying and different operators, rather than proceed through the formal determination processes.

Measures are also being considered to encourage further exploration and development activity in the industry, particularly where production of two resources such as coal seam gas and coal from the one production title occurs, and the CSG reserves may otherwise be flared, vented or disposed of.

Whilst these measures require further detailed investigation, the Government intends to encourage the best economic use of these two resources in areas where they co-exist.

5. Next Steps

Drafting instructions for the various legislative amendments that will be required to the *Mineral Resources Act 1989* (Qld) and the *Petroleum Act 1923* (Qld) are currently being prepared. Development of any associated policies and technical guidelines is also planned and the relevant Cabinet and legislative formulation processes would then be followed. Department of Mines and Energy staff will consult closely with the QMC taskforce and other stakeholders as this work is being undertaken, to ensure that the intent of the final position paper is reflected in the draft legislation.

It is planned to have an Authority to Prepare a Bill to Cabinet by July 1998, dependant on the concurrent review and rewriting of the *Petroleum Act 1923* (Qld). However given the work involved, Parliament's current legislative program, and 13 June 1998 State election, legislative amendments are unlikely to be enacted until late 1998.

6. Conclusion

Coal seam gas is potentially a major supplier of gas for Queensland industry. It is geographically well placed to supply gas at competitive prices to new industrial developments on the east coast of Australia. Coal seam gas producers should benefit from the more accessible and competitive market being developed in the national gas market. However, commercial viability of coal seam gas will ultimately depend on the economics of extraction of the gas. This is particularly dependent on gaining a better understanding of the dynamics of the coal seam gas environment and on developing cost effective techniques of extraction.

The Queensland Government has attempted to remove perceived impediments to the rational development of coal seam gas resources by developing a legislative framework that has struck a balance between ensuring that resources will not be sterilised and providing certainty to explorers and developers. The use of coal seam gas areas will define the areas that are of prime interest to the coal miner. The use of strata title and surface access provisions will ensure that the coal seam gas explorers can have title to coal seam gas resources and also access those resources where conflicting titles may occur.

Now that the agreement of key representatives of the two industry groups has been reached, the path has been laid to allow legislative and administrative amendments to be made that should provide security and certainty for future coal seam gas and coal development.

The new regime addresses the main concerns expressed by stakeholders:- the regime has been simplified, security of title has been defined, ministerial discretion has been reduced, there are not two types of petroleum, and mechanisms to allow operations in areas of overlap have been clarified.

The Fairview Coal Seam Gasfield, Comet Ridge, Queensland Australia

S.G. SCOTT
Department of Mines and Energy
GPO Box 194 Brisbane, Queensland, Australia

1. Introduction

The Fairview coal seam gasfield is located in central eastern Queensland, approximately 30km northeast of the town of Injune which is approximately 475km northwest of Brisbane (Figure 1). Fairview was discovered in August 1994 with the drilling of TPC Fairview 1. Since August 1994, 18 more wells have been drilled for a total of 14,455.93m with an average depth of 803.1m. The gasfield is covered by three granted Petroleum Leases (PLs) 90, 91 and 92 with eastern and southern extensions covered by two PL applications (PL 99 and 100). The five PLs and the surrounding Authority to Prospect (ATP) 526P are owned and operated by Tri-Star Petroleum Company.

2. Structure

Fairview is located on the Comet Ridge, a prominent basement high that separates the Denison Trough from the much larger and thicker Taroom Trough (Figure 2). The Denison Trough is composed of a series of grabens and half grabens formed during the Bowen Basin's Early Permian rifting phase (Murray, 1990). The Taroom Trough also formed by rifting due to the possible presence of mantle diapirs.

The Comet Ridge is part of the Clermont Stable Block, one of the main tectonic units of the Bowen Basin (Dickins & Malone, 1973). Rocks from the Denison Trough, on the west, and the Taroom Trough to the east, thin and onlap onto the Comet Ridge.

3. Stratigraphy

The target formation within the Fairview area is the Late Permian Bandanna Formation (Figure 3). The Bandanna Formation was defined by Hill (1957) who divided the unit into an upper sandstone and coal section and a lower shale section. In 1967 Power restricted the name to the upper sandstone and coal section and in 1969 Mollan & others named the lower section the Black Alley Shale. The Bandanna Formation and its equivalents, the Rangal Coal Measures in the north and central Bowen Basin, the

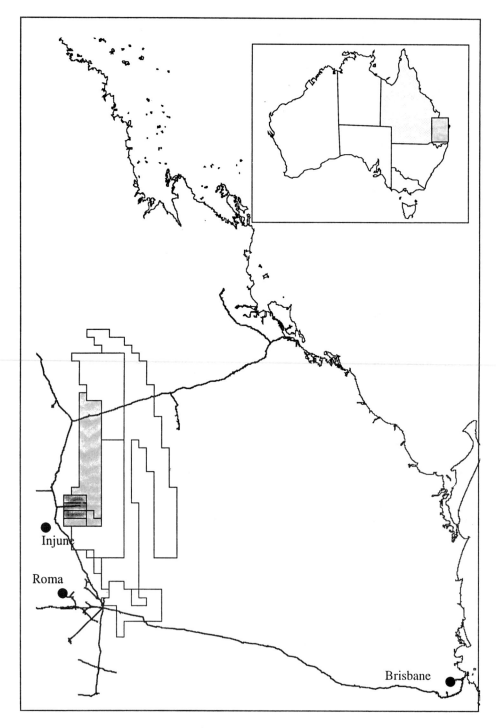

Figure 1: Fairview coal seam gas project location

Baralaba Coal Measures in the eastern basin and the Burunga Formation in the southeast are the most widely spread coal bearing unit in the Bowen Basin.

The Bandanna Formation is composed of grey to dark grey mudstone and siltstone, grey, white to cream sandstone and bituminous coal (Green & others, 1997). The mudstone and siltstone commonly have disseminated carbonaceous laminae. The sandstone is fairly well sorted, lithic labile and also has disseminated carbonaceous laminae. Coals from the Rangal and Baralaba Coal Measures in the north, central and southeast Bowen Basin are extensively mined for steaming and coking coal with 22 million tonnes mined in 1996/97.

The Bandanna Formation conformably overlies the Black Alley Shale with the boundary between the two being gradational (Green & others, 1997). Overlying the Bandanna Formation is the Late Permian to Early Triassic Rewan Group. This relationship is conformable over most of the Taroom Trough but is unconformable on the margins and on the Roma Shelf (Green & others, 1997).

At Fairview, the top of the Bandanna Formation has been taken at the top of the upper coal seam. This follows the usage suggested by Jensen (1975). At this location two well-developed coal seams are found in the Bandanna Formation.

The upper seam has a relatively consistent thickness across the field, ranging between 4.9 and 6.4m. Gamma, density and sonic logs run in the wells all show a consistently bright upper coal seam. In TPC Fairview 11, situated on the eastern flank of the field, this upper seam is beginning to split and in TPC Fairview 15, on the northwestern side, the seam has split into three plies; 7.3m of coal in 24.7m of strata. In TPC Fairview 8, which is located 12 km southwest of the field, the upper seam is also split into two plies of 3.7 and 4.0m.

The lower seam is thinner but is more consistent in thickness, ranging between 3.0 and 3.7m. This seam also splits, in TPC Fairview 2, on the southern flank, into an upper ply of 0.9m and a lower ply of 1.5m. In TPC Fairview 8 the lower seam is considerably thinner (1.5m) and much shalier. If this splitting and deterioration is consistent in the south of the Fairview area, then this seam will not contribute any coal seam gas resources.

4. Exploration History

Exploration for coal seam gas began in the Comet Ridge area in 1989 when Authority to Prospect (ATP) 419P was granted to Tri-Star Petroleum Company. Two years later, Tri-Star applied for ATP 490P, which covered an area to the west of ATP 419P. Not long after the granting of ATP 490P, the two ATPs were conditionally surrendered and re-issued as a single ATP; ATP 526P.

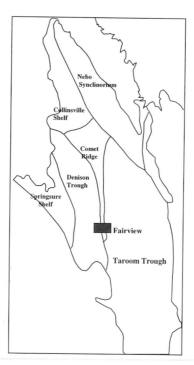

Figure 2: Bowen Basin structural units

Moolayember Formation
Snake Creek Mudstone Member
Showgrounds Sandstone
Clematis Group
Rewan Group
Bandanna Formation
Black Alley Shale
Tinowan Formation
Muggleton Formation
? non deposition ?
Combarngo Volcanics

Figure 3: Stratigraphic units - Fairview

Tri-Star's initial drilling interest on the Comet Ridge was a prospect in the eastern portion of the neighbouring ATP, ATP 337P held by Santos Petroleum Operations Pty Ltd and Oil Company of Australia Ltd. Tri-Star entered into the deal to farm-in to the block by funding the drilling of SPO Comet Ridge 1. This well was spudded in November 1993 but was plugged and abandoned.

In August 1994, TPC Fairview 1 was spudded in ATP 526P. This well reached a depth of 1,340.82m before terminating in basement. The drilling of TPC Fairview 2 to 15 during the next eleven months followed this well. Four more wells, TPC Fairview 16 to 18A, were drilled between November 1996 to February 1997 (Figure 4). Total meterage drilled at Fairview is 14,455.9m with an average depth of 803.1m.

In January 1995 Tri-Star announced the discovery of coal seam gas at Fairview with the TPC Fairview 4 well flowing a over 1TJ a day. This was followed in November 1995 with the granting of three petroleum leases (PLs 90, 91 and 92) over Fairview. In April 1996 Tri-Star applied for a further 2 leases (PLs 99 and 100) over southern and eastern extensions to the Fairview Field. PLs 99 and 100 have not been granted because of complications caused by the Australian High Court decision regarding Native Title, which was handed down on 23 December 1996.

PG& E completed an extension of its' Wallumbilla to Gladstone pipeline, Pipeline Licence (PPL) 30, to Fairview on 1 November 1997 and on 8th February 1998 Tri-Star began commercial production from the Fairview coal seam gasfield with daily production of 2TJ.

5. Completion Methods

Of the 19 wells drilled at Fairview only TPC Fairview 1 and 2 have been drilled to below the Bandanna Formation. In all wells, once the total depth was reached, the well was conditioned then underwent high pressure cavitation for up to 10 days. In this method, a mixture of air, air and water and water was forced into the well bore under high pressure for extended periods. The pressure was then suddenly released causing the coal seams to collapse into the well bore. The coal fragments were then removed from the well or allowed to settle into a sump below the floor of the seam.

This treatment of the wells was designed to do a number of things. The seams were fractured and any natural cleats within the seam opened up to improve the permeability of the coal. The well diameter was also increased by this treatment which resulted in a larger area of the seam being exposed to the well bore and so increasing the desorption rate of the coal.

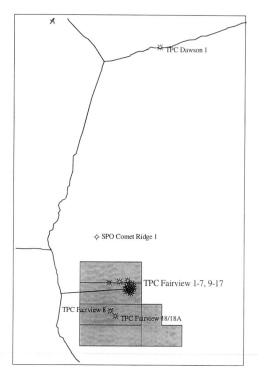

Figure 4: Tri-Star well locations

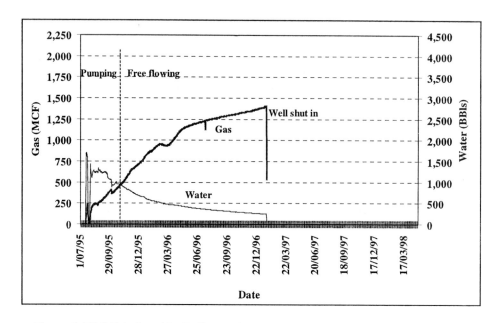

Figure 5: TPC Fairview 13 - Daily gas & water Production

6. Resources

There has been very little analytical work carried out at Fairview. Cores have been cut in only one well, TPC Fairview 2. These cores were cut between 720.2 and 726.6m and 940.0 and 949.1m. The first core was cut to sample the upper coal seam for gas analyses and gas volume studies, the second to study the sandstone of a potential conventional reservoir. Because these cores were cut with a conventional oilfields' coring unit, very little of the coal core was recovered. The results from the analytical work have not been publicly released.

Samples of the upper seam in TPC Fairview 2 and from cuttings of the upper seam in TPC Fairview 1 were used to determine vitrinite reflectance. The mean maximum vitrinite reflectance for coal from the upper seam in TPC Fairview 1 was 0.89% and for the upper seam in TPC Fairview 2, 0.85%. It should be noted that due to the method of coring, very little vitrinite was collected in the recovered core from TPC Fairview 2. Most of the brighter, more brittle coal was flushed out of the core barrel during drilling leaving the harder, more massive exinite. With this in mind, the maturity of the coal at Fairview should be higher than what should be expected from a coal with a RV_{max} between 0.85 and 0.89%.

Most of the testing at Fairview has been conducted in-situ, with most wells undergoing extensive production testing. TPC Fairview 4 and 6 exhibited initial flow rates of over 1TJ a day and very little produced water. Even now after nearly four years these wells, when producing, deliver next to no water. Another factor with these wells is that they can be suspended for relatively long periods and then allowed to flow without apparent deterioration.

Other wells at Fairview when testing began produced very little gas and quite large volumes of water but over time have become very good gas producers with a large drop in water production. A number of these wells have produced production curves that follow classic coal seam gas graphs; small initial gas production, large water production then a steady increase in gas production with a corresponding decrease in water production with eventually the well becoming free flowing (Figure 5).

The wells that are the best producers are situated in the central to southern part of the field. While most are situated towards the centre of the field and so benefiting from the interference effect of surrounding wells, this does not explain why these wells are such good producers. In fact two wells are situated on the edge of the field. These wells do not have the thickest coal intersections either but are the highest wells in the field. This would seem to point to structure having a very large governing effect on deliverability at Fairview. One final factor is the lack of secondary mineralisation within the coal cleats. This lack of mineralisation allows gas to more easily access the well bore after stimulation.

No gas volume measurements have been reported from the coals at Fairview. Taking into account the maturity of the coal and the depth of burial then a range between 400 and 500 SCF/ton (10 -12m³/tonne) is not extreme. Again there has been no published coal seam gas reserves for the Fairview field but to determine an approximate in-situ resource in PLs 90, 91, and 92 a number of factors can be used in the formula;

$$GIP = A \times h \times Gc \times C$$

where:

	At Fairview
GIP = gas in place,	
A = well drainage area	$675 \times 10^6 \, m^2$
h = net coal thickness	5.6m
Gc = methane content	$10m^3$/tonne
C = coal density	1.45

If these figures are used, an in-situ resource of approximately 1,700PJ is calculated. It should be noted that only the upper seam was used, it was considered that the upper seam was present across all three PLs and an average gas volume of $10m^3$/tonne was used based on the coal's maturity. It should be remembered that in-situ resources for the northern Bowen Basin are estimated as being at least 124,000PJ (Warren Centre for Advanced Engineering and The Earth Resources Foundation, 1994).

Conventional gas resources are also inferred to be present at Fairview in the Peawaddy and Freitag Formations, underlaying the Bandanna Formation. Log interpretations from TPC Fairview 1 indicated that the formations below the Bandanna Formation were gas saturated but in most cases these units have a low permeability.

7. Conclusion

The Fairview coal seam gas field is only the second coal seam gas field to have began commercial gas production from vertically drilled wells in Queensland. It is currently producing more than 2TJ daily from seams in the Late Permian Bandanna Formation. The coal seams at Fairview apparently have very little in-seam mineralisation and this combined with the fact that the seams are thick, are mature and the field is located on a broad anticline are major contributing factors influencing the very good flow rates from many of the wells. These factors combine to make Fairview a very promising coal seam gas field and the Comet Ridge area of the Bowen Basin, potentially Queensland's best coal seam gas region.

8. References

DICKINS, J.M. & MALONE, E.J., 1973: Geology of the Bowen Basin, Queensland. *Bureau of Mineral Resources, Geology & Geophysics, Bulletin* **130**.

GREEN, P.M., CARMICHAEL, D.C., BRAIN, T.J., MURRAY, C.G., McKELLAR, J.L., BEESTON, J.W. & GRAY, A.R.G., 1997: Lithostratigraphic units in the Bowen and Surat Basins, Queensland. *In* Green, P.M. (Editor) The Surat and Bowen Basins, south-east Queensland. *Queensland Minerals and Energy Series*, Queensland Department of Mines and Energy, 41-108.

HILL, D., 1957: Explanatory Notes on the Springsure 4-Mile Geological Sheet. *Bureau of Mineral Resources, Geology and Geophysics, Australia, Note Series* No. 5.

JENSEN, A.R., 1975, Permo-Triassic stratigraphy and sedimentation in the Bowen Basin, Queensland. *Bureau of Mineral Resources, Geology and Geophysics, Australia,* Bulletin **154**.

MOLLAN, R.G., DICKINS, J.M., EXON, N.F. & KIRKEGAARD, A.G., 1969: Geology of the Springsure 1:250 000 Sheet area, Queensland. *Bureau of Mineral Resources, Geology and Geophysics, Australia, Report* **123**.

MURRAY, C.G., 1990: Tectonic evolution and metallogenesis of the Bowen Basin. *In* Beeston, J.W. (Compiler): *Bowen Basin Symposium 1990 Proceedings.* Geological Society of Australia, Queensland Division, Brisbane, 201-212.

POWER, P.E., 1967: Geology and hydrocarbons, Denison Trough, Australia. *American Association of Petroleum Geologists Bulletin,* **51**, 1320-1345.

WARREN CENTRE FOR ADVANCED ENGINEERING AND THE EARTH RESOURCES FOUNDATION (WITHIN THE UNIVERSITY OF SYDNEY), 1994: *Coal Bed Methane Extraction: A proposal for a field trial in the Southern Sydney Basin,* NSW, January.

COST BENEFIT ANALYSIS OF COALBED METHANE RECOVERY ACTIVITIES IN AUSTRALIA AND NEW ZEALAND- IMPLICATIONS FOR COMMERCIAL PROJECTS AND GOVERNMENT POLICY

Paul Massarotto, FIPENZ, SPE, Tripenta Energy Solutions, New Zealand

Abstract

This paper presents drilling, completion and stimulation costs for Australia and New Zealand coalbed methane (CBM) wells and compares them with reference international costs from the USA, Canada and China to deduce the local "premium factor" faced by the fledgling Australian and New Zealand CBM industry. CBM exploration and development economics in the Queensland fiscal environment are then analysed with a 100-well designed and costed prototype model. It is concluded that CBM economics are sub-marginal and costs must come down to make the industry attractive to potential investors. As well, investment tax credits and royalty "holidays" are investigated in the context of government policy. They are recommended as efficient fiscal tools to "kick start" a vibrant CBM industry, thereby achieving economies of scale and favourable social return.

Introduction

Wellbore stimulation and reservoir enhancement technologies for increasing CBM production and ultimate recoveries have been mainly borrowed from the oil and gas industry, covering both conventional and unconventional projects. Costs for these technologies in oil and gas are fairly high, unless accompanied by great economies of scale and competitive pressures, and/or by tailor-made low cost adaptations, through R&D advances. Furthermore, many oil and gas techniques have yet to be tried in CBM applications.

In Australia and New Zealand, costs for drilling, completion and stimulation of wells have been fairly high, mainly reflecting the small size of the service industries supporting the oil & gas and CBM projects. There have been relatively few wells drilled specifically for CBM exploration and production in Australia (some 50) and fewer yet in New Zealand (12); cost data is available for only a portion of these. These high costs also reflect the exploratory and R&D nature of these CBM programmes. Future costs, should a reasonably-sized industry emerge, would be expected to compare to mature USA CBM regions (San Juan, Black Warrior and Powder River Basins) and, as the technology matures fully, to the very-low cost shallow gas fields of Alberta,

Canada, where over 1000 wells are drilled annually. Historical cost data on the these regions are presented, classified by depth of wells and translated from year of expenditure to Australia 1997 dollars. Cost data include wells in remote, pilot and commercial operations.

Main stimulation techniques applied in Australia and New Zealand have been hydraulic fracturing and, to a lesser extent, open-hole cavitation. Results on productivity increases have been inconsistent, while costs have been fairly high. Other stimulation and enhancement techniques are possible, to increase both productivity and gas reserves. These are presented for consideration and trial by operators of future schemes. These technologies will require an R&D programme to transfer their potential to CBM recovery in a cost-effective way.

In order to evaluate the current economic status of the Australian CBM industry, the present study contains a preliminary design of a prototype production and cost model, including exploration and development, based on a 100-well project. Completed well costs have been reduced by 25% from the Australian trend line, to reflect advances industry needs to make to reduce costs to international standards. Yet, even with these savings, project economics are only marginally attractive. Government policy action, via fiscal incentives by both the Commonwealth and State governments, are examined as to their impact, singly and jointly, on project economics.

Reference International Drilling, Completion and Stimulation Costs

It was surmised at the outset that the costs in the mature and very large CBM field operations in the USA, and in the shallow gas fields of Alberta, Canada would be lower than the very limited initiatives in Australia and particularly New Zealand. Factors that affect costs are the scale of operations, the competitiveness of the service industry, the remoteness of operations from supply bases, the maturity of the technology and the combined complexity of the technology and field conditions (ie deep, high pressure wells will cost more to drill and more to frac).

-POWDER RIVER BASIN

Hobbs in 1990 [ref. 1]published some figures on the drilling and completion costs of shallow CBM wells in the eastern margin of the Powder River Basin. They related to the Ft. Union formation, at depths of up to 243m. Hobbs made no mention of wellbore stimulation and pumping costs. His quoted figure of $US60,000 per well was escalated at 3% per year and converted at year-end 1997 exchange rates. Holland in 1994 [2] revisited the Powder River Basin, particularly the Marquis field development, and referred to investments of $US120,000 per well, but after it had been bought with Section 29 tax credits. Because of this, the Hobbs data are considered more representative of actual costs. The equivalent 1997 Australian cost is $A104,000.

-BLACK WARRIOR BASIN

Hobbs and Holland reported in 1997 [3] on anticipated costs for a 250-well development project in the Cedar Cove Coalbed Methane field of the Black Warrior Basin. In this, they also included cost data for core holes and experimental test wells. Their 1991 estimate gave an average of $260,000 per development well, including pumping equipment, for wells at depths of 915m. They did not mention if their data included stimulation costs; because of the following information, the author deduced that their data probably excluded stimulation costs. Contacts with industry consultants in 1998 yielded information on costs for a 25-well delineation programme in the Cedar Cove area, amounting to $US351,000 per well in the 1991 time frame. These wells included experimental stimulations and were scattered over a wide field area; costs would therefore be higher than a development programme. The Hobbs and Holland data was translated to an updated 1997 equivalent cost of $A438,000 per well.

-WASHINGTON STATE

A company called Epic Resources applied for resource development consents for a small prototype CBM recovery project to be built in 1998 [Kahill and Sinclair 4]. Total depth was expected to be at about 600m. Drilling costs were detailed to include coring, logging and drill-stem testing, totalling $US135,000. Completion costs included a 14-day well test and a small hydro frac, on top of the usual casing, cementing and perforating services. Sub-total for completion was $US240,000. The total costs reflect the remote location and are equivalent to $A640,000 in1998.

-CANADIAN SHALLOW GAS WELLS

There have been few reports published of wells drilled specifically for CBM production in Canada. However, Sinclair and Cranstone [5] mentioned that, between 1978 and 1997, there were some 137 wells drilled either solely as CBM wells or in conjunction with regular exploration or as recompletions of suspended oil and gas wells. However, the province of Alberta is rich in shallow sandstone and carbonate rock formations that yield low pressure gas across extensive areas. Over 1000 wells per year are drilled to exploit these reserves, some being infill wells. Because of the scale of this industry, there are many service companies and contractors supplying the operating companies. There is a strongly competitive market with numerous field centres for quick, low cost access to services, maintenance and equipment. Furthermore, the competitive environment drives innovation, so that new, low cost solutions and tools quickly fill the market needs.

The Petroleum Services Association of Canada publishes an annual well cost study [6] for different areas and depths of formations. A typical result is reproduced in Table 1 for the Hatton area (Medicine Hat formation) at 550m depth, to show the detail that goes into these reports and highlight the efficiencies achieved (eg use of coiled tubing units to reduce completion costs). The equivalent Australian dollar cost for drilling, completing and stimulating these wells is $A120,000. With an installed bottom-hole pump, costs would increase to $A170,000. Other examples for different depths and more remote areas are given in Table 2.

Table 1

TYPICAL CANADIAN SHALLOW GAS WELL
DRILLING AND COMPLETION COST BREAKDOWN *

Field: Hatton   **Formation: Medicine Hat (sweet gas)**   **Depth: 550m**

	Activity	**Est. Cost($C 1997)**
1.	Drilling contract	16,800
2.	Road and Site preparation	1,200
3.	Rig Trans. & Misc. Transport	5,000
4.	Drilling Fluids	2,800
5.	Open Hole Logging	0
6.	Drill Stem testing	0
7.	Cement and Cementing Services	8,300
8.	Casings and Attachments	13,300
9.	Drill Bits	3,200
10.	Other Equipment and Services	5,800
11.	Land, Engineering, Supv. and Admin.	11,400
	Drilling Subtotal	**67,500**
12.	Service Rig	6,500
13.	Trucking and Misc. Trans.	500
14.	Cased Hole Logging and Perforating	2,500
15.	Tubing and Attachments	5,500
16.	Stimulation and Treatments	10,500
17.	Pumping Eq't	0
18.	Wellhead	2,500
19.	Other Equip't and Services	1,000
20.	Engineering, Supv. and Admin.	2,900
	Completion Subtotal	**31,900**
	Completed Production Well(w/ 10% contingency)	**109,300**

*Extracted from the "1997 Well Cost Study" of the Petroleum Services Association of Canada

Table 2

**DEPTH AND REGIONAL COST VARIATIONS
CANADIAN SHALLOW GAS WELLS**

Drilling, Completion and Stimulation Costs ($C 000's)
Region, Formation and Depth

	Hatton Med.Hat 550m	Athabasca Nisku 680m	Pelican McMurray 690m	Stettler Mannville 1330m
1.Total Drill and Case	67.5	163.1	105.3	218.5
2.Completion*	31.3	50.5	55.3	118.4
3.Stimulation	10.5	5	0	0
Total D,C&S	109.3	218.6	160.6	337

* Includes tubing, pump, other completions and contingencies

-CHINESE WELL COSTS

Recent literature contains cost information prepared by Wang Xingjin [7] on costs for a potential project in the Chinese interior for 400m depth wells. Cost of development wells is quoted at RMB1.1 million per well; converted to Australian currency, this yields a D&C cost of $A185,000. Interestingly, this cost is on trend with USA costs for equivalent depths - see Figure 1.

The New Zealand Experience

There have been a dozen core holes drilled exclusively for evaluating coal seams for methane presence. M. Cave has reported on the New Zealand experience at this same International Coalseam Conference [8]. Detailed costs were made available to the author [9] for two notable well tests for CBM production, including massive hydraulic fracture stimulations, as part of the Southgas Coalseam project in 1995. Well depths were 472m and 570m. Total drilling costs, adjusted for exclusion of fishing-related costs, amounted to $NZ422,000 and $NZ 489,000, respectively.

Completion costs, including the massive frac jobs but excluding extended well testing time, totalled some $NZ664,000 per well.

Updating to an Australian 1997 base yields total well costs of $A960,000 and $A1,018,000, respectively. Large mobilisation/demobilisation costs to a remote area of New Zealand contributed a large portion of these unusually high costs.

Costs for core holes in the Westgas project were more in line with international trends, amounting to $NZ36,000 , $NZ50,000 and $NZ102,000, respectively for 300m, 437m and 502m wells. These wells were drilled with local HQ coring rigs used for coal mine planning, confirming that an existing service industry can keep costs down.

Australian CBM Well Costs

The Australian data was derived from personal contact with the major operating companies with experimental CBM recovery projects in Queensland and with a major coal producer for the cost of Goaf wells. Projects referenced herein include the Peat, Moura and Dawson River CBM areas of the Bowen Basin. Well depth ranges for each are, respectively, 900 to 1100m, 700 to 800m and 1000m. Total well costs vary considerably due to the range of wellbore stimulation techniques used.

-DRILLING AND COMPLETION

Many wells had drill stem tests (DST's) and some extended initial flow tests performed, masking the cost of future development wells. Excluding these tests, drilling and completion costs for a five well group averaging 1000m in depth amounted to an average of $A405,000, including tubing and pump. Pumping equipment used for de-watering wells included Moyno progressive cavity pumps, powered by rods driven by a generator-fed electric motor, and conventional beam-driven, sucker rod bottom hole pumps . Cost variation with depth for this group of five wells averaged $A480 per meter. Two other groups of wells, one at the same 1000m and the other at 750m, gave a similar cost-depth trend line, but lower overall costs due to less costly stimulation programmes.

Typical drilling and completion costs for experimental CBM wells, derived from the above clusters of wells in the Bowen Basin averaged to a well depth of 900m, are presented in Table 3.

Table 3

**TYPICAL 1997 AUSTRALIAN CBM
DRILLING & COMPLETION COSTS**

Basis: (1) excludes stimulation
     (2) 900 m well - Development Basis - (excludes DST's)

	Est. Costs ($A 000's)
DRILLING	
Civil (road, site & sump)	15
Rig contract (incl. mob/demob)	120
Transportation costs	20
Drilling fluids & bits	15
Logging (open hole & geol.)	15
Cement & services & anc.	25
Casings (surf & prodn) & att'ch	60
Land, eng, survey, supv, & admin.	25
Other services & cont. @ 10%	30
SUBTOTAL	**325**
COMPLETION	
Service rig	15
Tubing (13), wellhead (2)	15
Pump & prime mover	35
Transportation, open hole logging	5
Eng. & admin.	5
SUBTOTAL	**75**
TOTAL D & C	**$400**

-STIMULATIONS

These included water hydraulic fracturing, acid hydraulic fracturing, foam fracs and "dynamic" open hole, or cavitation, completions. Stimulation costs varied from $A250,000 for two-seam fracs in one well, to $400,000 for multiple-seam cavitation treatments or massive, multi-zone frac jobs. Because of the combination of experimental treatments, remote operations and low numbers of completions and a small service industry, costs for these are guesstimated to be some 50% higher than a vibrant industry could provide (basis USA and Canadian frac costs).

-PRODUCTION BENEFITS

Information on production increases is hard to come by due to the inherent commercial sensitivity. Nonetheless, reference to official records and deduction of timing of treatments reveals some limited useful data. Cavitation completions seem to yield a tripling of production rate, utilising a stable period for comparison both before and after the treatment. The available information on nitrified frac jobs indicates roughly a doubling of production rates. Stimulation costs and production benefits are summarised in Table 4.

Table 4

**RECENT AUSTRALIAN & NZ CBM
STIMULATION COSTS & BENEFITS**

AUSTRALIA	Est. Costs ($A 000's)	Prodn Benefit
Water Hydraulic fracs	300	NA
Nitrified HCl Acid fracs w/- proppant (incl. rig time)	400	2:1
Cavitation w/- open hole (incl. rig time)	400	3:1
NEW ZEALAND		
Water hydr. fracs w/- proppant & L. gel	495	nil

-GOAF WELLS & IN-SEAM DRILLING

The cost of wells drilled from surface into the goaf area of coal mines for methane drainage purposes, also varies with depth and completion technique. For open hole, shallow wells of 250m depth, the cost can be as little as $A10,000. For deeper 300m wells encountering reasonable pressure and thus requiring casing to surface, the cost approaches $A100,000.

In-seam wells, used for pre-mining methane drainage, average between $A50 and $A100 per meter, depending on the diameter, length and operational factors.

International Cost Comparison

All of the above data has been assembled and normalised to a cost variation with depth and with region of the world, as summarised in Figure 1. The plotted data sets include some with total well costs (ie including well stimulation treatments) as well as wells with only drilling and completion costs (ie the NZ wells). The inclusion or exclusion of stimulation costs is an important distinction; for example, shallow gas wells in Alberta, usually receive minimal well stimulation treatments of $10,000 to $15,000, while +$200,000 are spent on Australian CBM wells with similar gas reserves per well. The implication is that, in order to achieve economics comparable to shallow gas reservoirs, CBM operators will have to find ways to reduce total well costs, with a focus on well completions.

Fig. 1

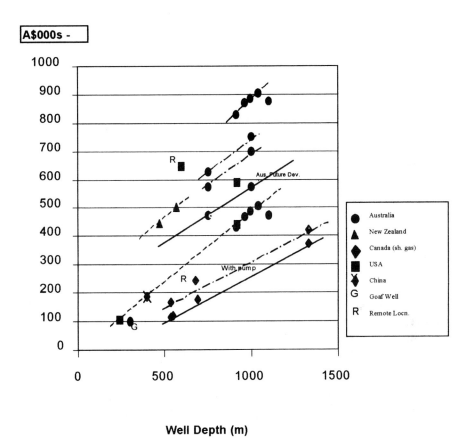

| CBM DRILL & COMPLETION & STIMULATION COSTS |

vs DEPTH & REGIONS

Well Depth (m)

The Australian data is presented in two main categories: costs with and without stimulation. Also, a hypothetical "Australia future development" line is postulated, at costs 25% lower than the current low range of costs. This is based on operators' expectations of more service companies entering the market and being able to achieve economy-of-scale costs for a large number of development wells. They also hope to have solved, with sufficient R&D, the stimulation cost challenge.

The USA data highlights a reasonable well depth-cost correlation with other regions of the world ; the low development costs achievable with large numbers of wells; and low costs generally to reflect access to an active service industry.

Costs are some 16% to 20% lower than the "future Australia" costs. The Washington State project data reflects its remoteness and experimental nature; the remote premium is similar to the NZ premium over Australian costs.

Interestingly, the China data set falls on the USA depth-cost correlation line, implying an anticipation of economy of scale and focused use of technology. Some lower costs in China (eg labor) are probably offset by costs for imported technology and equipment, with associated higher mobilisation/demobilisation costs. Australian costs without stimulation also fall on the USA trend line.

The Canadian data set reveals how low costs can be driven, given the scale of operations, the incentives of competition and time to continuously improve all facets of design and operations. The two lines reflect gas well costs with and without pumps. The sourced data set from the PSAC excluded pump costs on the gas wells (as most shallow gas is relatively dry). Pump and driver costs were added on the second line to reflect the current practice of pumping off CBM wells. However, as mentioned later, one can visualise advanced technology being utilised here to reduce costs to quasi-pumpless operation, by adapting the "gas lift" process to CBM operations. In the long run, the challenge to Australian and New Zealand operators is to aim for the low Canadian cost line for their CBM wells.

A comparison of limited borehole/test hole costs is shown in Figure 2.

Figure 2

COREHOLE - DRILLING & TESTING COSTS

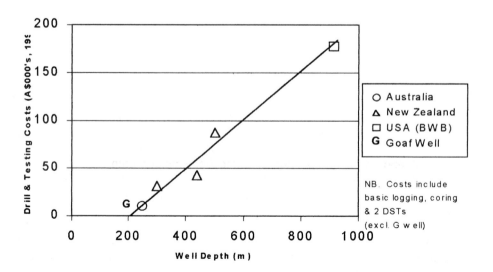

Practised and Potential Stimulation and Enhancement Technologies

These technologies can be grouped into two main categories:
-stimulation of the wellbore and near-wellbore region, increasing short   term production rates;
-enhancement of the demethanation process in the main reservoir flow paths and hence ultimate recovery of CBM.

A summary is presented in Table 5.

Table 5

**PRACTISED & POTENTIAL
STIMULATION & ENHANCEMENT TECHNOLOGIES**

Production Stimulation (Wellbore & Near WB)	Reserves Enhancement (Reservoir Flow)
Dewatering	Dewatering
Progressive cavity pumps	
Sucker rod pumps	
Electric Submersible pumps	
Gas lift mandrels	
Mini Hydraulic Fracturing	Massive Hydraulic Fracturing
Various fluids	
Various or no proppants	
Open Hole "Cavitation"	
Drain Hole Drilling	Horizontal Drilling
&Ultrashort Radius	(Short, Medium and Long Radius)
Under-reaming	CO_2 flooding with Peripheral Pressure Maintenance
Lance Perforating	N2 Peripheral Press.Mtn.
Water Jetting	Microbial Enhanced CBM (MECoM)
Gas " Huff & Puff"	Periph. Water Flood with MECoM
Acidization/Chemical Leaching	

The current practise includes:

●Hydraulic fracturing utilising various fluids and proppants, both mini-fracs for   stimulation and massive fractures for enhancement;
●The novel "cavitation" stimulation process (which many drillers cursed when going through thick coal seams looking for deeper oil), with deep enough effects (30m or more) to be considered enhancement also;
●Dewatering, with various pumping mechanisms for both stimulation and enhancement benefits. Some coal seams, particularly shallow ones at low pore pressures, will not flow gas without significant dewatering beforehand. Other, higher pressure seams will still benefit from some form of artificial lift of the free water to accelerate and maximise methane production rates. The author has not seen any reports of gas lift being attempted in CBM operations, yet it has promise for reducing both capital and operating costs and could level out initial production rates.

In the Bowen Basin, acidising and acid fracs have been trialed to remove the calcite and clay mineralisation in the cleats and larger fractures. Progress is also being made in Queensland to develop cost-effective water-jet drilling techniques and tools which may deliver economic ultra-short radius drainholes as well. The author understands field trials have recently been undertaken.

A simple and potentially cost-effective stimulation technique is the application of the under-reaming drilling process, but with longer extendable arms, on open hole completions.

A potential process to improve the relative permeability to methane involves the injection and immediate production of gaseous fluids , a sort of "huff and puff" operation (borrowing a term from heavy oil steam recovery operations). This needs to be researched and developed.

Production rates decline usually due to a decline in the reservoir pressure. High abandonment pressures in coal seams can lead to loss of 35 % to 50 % of the methane in place. To overcome both of these factors, peripheral injection with CO_2 or N_2 could add substantially to CBM economics. Finally, the Bureau of Economic Geology of Texas   (Andrew Scott) has proposed a microbial injection process (MECoM) to generate more methane than initially in the coal. This could be extended further by co-injecting with peripheral water, should natural hydrodynamic gradients not be sufficiently developed to carry the microbes.

Figure 3

PRODUCTION PROFILES OF PROTOTYPE 100 WELL CBM RECOVERY PROJECT

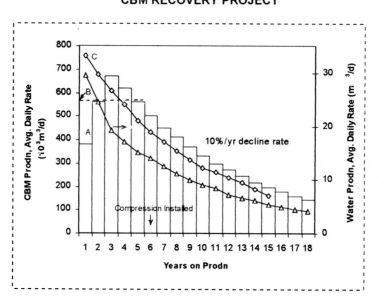

Design Basis for Prototype CBM Project

-RESERVOIR ENGINEERING

In order to report on the economic viability of CBM in the Australian context, one needs a realistic physical model tied to the local geology, engineering practice and operations. Following a review of some of the current projects and geology associated with potential acreage, a reservoir engineering model has been assembled to yield representative average values for the Bowen Basin. These reservoir engineering parameters, both input and derived, are listed in Table 6. The 100 well prototype is designed to yield a first year average production rate of 380 thousand m3 per day, design rate of 560 thousand m3 per day and reserves of 2600 million m3 of methane. First year rates and the decline thereafter were varied upwards for Cases B and C, by incrementally investing in 50 cavitation completions for Case B and a further 50 well frac stimulations for Case C. The resulting production profiles are shown in Figure 3.

Table 6

DESIGN BASIS FOR
PROTOTYPE CBM PROJECT
RESERVOIR ENGINEERING PARAMETERS

Inputs	metric units	(american)
Avg. coal seam depth	600 m	(2000 ft)
Net coal pay	6 m	(20 ft)
Coal methane adsorbed (saturated)	$7 \ m^3/t$	(220 scf/t)
Initial reservoir pressure	7000 kpa	(1000 psig)
Reservoir temp.	32 C	(90 F)
Abandonment reservoir pressure	1400 kpa	(200 psig)
Avg. water saturation	20 %	
Bulk coal porosity	4 %	
SP gravity	1.4	
Bulk coal absolute permeability	10 md	
Initial well rate	$3.8 \times 10^3 \ m^3/d$	(135 mcfd)
Plateau well rate	$6.0 \times 10^3 \ m^3/d$	(210 mcfd)
Gas-water ratio (initial)	$5.6 \ m^3 /10^3 \ m^3$	(1 Bbl/mscf)
Derived		
CBM gas-in-place	$714 \times 10^3 \ m^3/ha$	(10.1 mmcf/ac)
CBM reserves per well	$26 \times 10^6 \ m^3$	(940 mmcf)
(avg. rec. factor - 58%)		
CBM Reserves-Total Project	$2,600 \times 10^6 \ m^3$	(94 bcf)
CBM production rate		
-initial case A	$380 \times 10^3 \ m^3/d$	(13.6 mmcfd)
-initial case B	$560 \times 10^3 \ m^3/d$	(20 mmcfd)
-initial case C	$760 \times 10^3 \ m^3/d$	(27.2 mmcfd)
No. wells - producers	100	
Well spacing	64 ha	(160 ac)
Acreage required	6,400 ha	(16,000 ac)

-CAPITAL COSTS

Costs for the prototype were tied to a scaled exploration effort, some lease acquisition/right-of-entry expenditures and a small R&D programme to optimise well locations and engineering design. These costs are estimated at $A1.6 million and are detailed as follows:

Table 7

EXPLORATION, LEASE ACQ. & R&D COSTS

		Est. Cost ($A m)
◆Exploration		
	Geological studies	0.25
	Core holes & testing (5)	0.40
	Lab tests / desorption, permeability, etc)	
◆Lease acq.		0.6
◆R & D		
	Geol./geochemical	0.15
	Eng./prodn	0.2
TOTAL		**$1.6**

Producing facilities and equipment requirements were tied to the reservoir engineering model, with 64ha per well spacing controlling costs of flowlines, satellite separation and metering stations and gathering lines. The high local transmission line pressure was taken into account for the amount of compression required (1000 HP). Capital costs for this portion were estimated at $A33.9 million, with details given in Table 8.

Table 8

PROTOTYPE FACILITIES & EQUIPMENT ENGINEERING PARAMETERS & COSTS

	Est. Cost ($A m)
Flowlines	10
Gas gathering lines	5
Gas delivery line	0.9
Satellite stations	4
Gas treatment & compression plant	8.2
Field compression (1,000 HP)	1.8
Electric power rectification	1.5
Engineering, Admin & Management @ 10%	2.5
TOTAL F & E	**$33.9**

Well designs were tied to the assumed 600m depth of the coal seam, allowing for a further 10m of sump. Total well costs for Case A (drilling and completion) were nominally reduced 25% form the current Australian trend, to $A300,000, anticipating a breakthrough in technology (see above) and reductions in contractor costs for a large project. Without expensive well stimulations, initial production rates for Case A were kept at a reasonable 140MCF/D per well. Case B is premised on 50 of the wells being subject to cavitation completions at an extra cost of $A140,000 per well. Case C took this well cost further, by some $A15 million, by subjecting the remaining 50 wells to massive hydro fracs at $A300,000 each. The total well costs ended at $A30m, $A37m and $A52m, respectively for Cases A,B and C- details shown in Table 9. Reserves were kept the same in all cases, due to the design plan to install field compression in year 6. Total capital costs for Case A amount to $A65.5 million, or $A25 per 1000 m3 ($A0.71/MCF).

Table 9

PROTOTYPE WELL DESIGN & COSTS

	Est. Cost ($m A)
1.Production wells (610 m TD) (Drilling & completion; excl. fracs & special completions)	30
2.Cavitation completions (50) (14 days @ $10k/d)	7
3.Hydraulic fracturing (50) (@ $300k)	15
CASE A TOTAL D & C:	$30.0
CASE B TOTAL D & C & S: (1+2)	$37.0 m
CASE C TOTAL D & C & S: (1+2+3)	$52.0 m

-OPERATING COSTS

Operating costs are categorised as fixed, variable and well workovers, the latter not often mentioned but usually a necessity with bottomhole pump operations. Provision has been made for electrification of the field, with associated maximum demand and energy charges. Electric-drive compressors have been assumed, however there may be some advantages to burn project methane to drive the compressors. Some nominal costs have been allowed for water treatment and disposal, though the practice in remote Queensland is for individual well evaporation ponds. Well workovers have been scheduled to start on the third year, with a three year cycle thereafter, at $10,000 per well. Total operating costs in year 1 amount to $A2.9 million per year. The unit operating costs work out to $0.40 per MCF. Details to these costs are contained in Table 10.

Table 10

PROTOTYPE OPERATING COSTS

	Est. Cost ($mA/yr)
Production - fixed costs - Electr. Power (max. demand) - Admin, Eng & Mgt (4) - Operations (8)	0.9
Production - variable costs - Compression @ $6/10 3 m3 - Water disposal @ 0.30/m3 - Treatment .... @ $2/10 3 m3	2.0
Well workovers - Production Wells ($10k/yr/ w- 3 yr cycle)	0.33
TOTAL OPER. COSTS	**$2.9m (Yr 1)** $3.4m (Yr 3+)
UNIT OPERATING COST	**$0.40/MCF(Yr 1)** $0.47/mcf (Yr 3+)
ANNUAL COST ESCALATION	3 %/yr

-FINANCIAL AND FISCAL TERMS

From contacts with local operators, gas prices were interpreted to vary with terms and conditions, as usual, yet an average plant gate price was surmised at $A90 per thousand m3 ($A2.50/MCF). Conditions usually require operators to pay for the tie-in to the main transmission line; a cost for a 1 km line has been allowed in the facilities cost estimate. As well, the lines can run at very high pressures (some 8500kpa/1200psig), putting extra cost pressures on operators.
Royalty in Queensland is 10% of the wellhead "net revenue" while the corporate income tax rate is 36%, after allowing for straight line depreciation over 15 years for all development capitalised expenditures. These are $A65.5m, $A72.5m and $A87.5m for Cases A,B and C respectively. The summary of terms is provided in Table 11.

Table 11

ECONOMIC ANALYSIS
PROTOTYPE CBM PROJECT
FINANCIAL AND FISCAL TERMS

Plant gate gas price	$90/10 3 m3 ($2.50/mcf)
Royalty	10%
(on wellhead 'net' revenue)	
Depreciation life	15 yrs
(on tangibles, avg)	
Income tax rate	36.%
Total capital investment	
Case A.	$65.5 m
Case B.	$72.5 m
Case C.	$87.5 m

Economics

The economic parameters resulting from the above design, costs and fiscal terms have been determined for the base case (Case A), for the base case with incentives and for the two accelerated production scenarios (Cases B and C).

-BASE CASE

The Internal Rate of Return (known also by the Discounted Cash Flow Rate of Return) for the base case is a sub-marginal 12% (see Table 12). Consequently, the Net Present Value at 15% is a negative $A8.2 million. Payout (without interest) is a lengthy 5.7 years. Both parameters reflect the high capital costs per unit of gas reserve. The implication is that costs must be reduced even further beyond the nominal 25% assumption in order to attract substantial industry investment. In all practicality, this can only come about over many years, with substantial R&D work and an economy of scale similar to Alberta, Canada where 1000 wells are drilled per year. Yet, the base case still yields a substantial $A43.7 million of combined royalties and income tax, giving a total social return, after adding operating profit, of $A105 million. There is room, in other words, for both State and Commonwealth governments to forgo some of their economic rents to help develop a vibrant and environment-friendly industry.

Table 12

ECONOMICS- BASE CASE

Case A

Payout	7 yrs
Return-On-Investment (P/I)	0.94
Net Present Value (@ 15%)	<$8.2 m>
Internal Rate of Return (DCF-ROR)	12 %
Social Return	$105.4 m
(incl. taxes & royalties of $43.7 m)	

-FISCAL INCENTIVES

In the USA, Congress passed the Section 29 Non-Conventional Fuels Tax Credit as part of the Windfall Profit Tax Act , which created a price incentive to drill for unconventional gas supplies. The response did not happen overnight as it wasn't until the late 1980's that saw the industry successfully drilling CBM wells in the Black Warrior, Powder River and San Juan basins. Previously, the USA had enacted at various times investment tax credits(ITC's), usually 10% to 15 %, to direct developments in the oil and gas and other capital intensive strategic industries.

It is proposed that the Australian Commonwealth enact a similar investment tax credit, at the 15% level for qualifying capitalised expenditures in CBM production projects, to "kick start" the industry not just in Queensland but also in New South Wales. This level of ITC would amount to $A10 million of the $A30 million income tax take, which would not exist without a successful project. The effect on the economics of this prototype would be enough to start attracting serious long term investors to look at the CBM industry, as the IRR would increase to 14.2%. A further incentive boost would put such projects over the 15% hurdle rate most operators use for decision making. This is where State governments can do their share, by establishing royalty incentives.

Royalty incentives have been used successfully in Alberta to further the development of enhanced oil recovery activities, for both conventional oil and synthetic oil derived from the huge oil sands deposits. These were mainly cost sharing/risk sharing formulas, yet Alberta also enacted royalty holidays up to five years for conventional oil and gas well drilling. It is proposed that this royalty holiday approach be established, as it is quite straightforward to implement, with very little overhead required. For the 100 well prototype, a five year royalty holiday increase the IRR from 12% to 13.6%.. The cost to the State would be a reduction in the royalty take from $A13.8 million to $A7.1 million, still a substantial sum for its economic rent, which it would probably never collect without extending more favourable fiscal terms. Results are summarised in Table 13.

The total social return after incentives still amounts to some $A105 million, including $A27.4 million of taxes and royalties, indicating the tremendous benefit even one small project can have on the local economy.

Table 13

ECONOMICS AND POTENTIAL INCENTIVES

Commonwealth Incentive:-
   Invest Tax Credit (@ 15% of cap. Inv.): $9.6 m

      IRR increases to   14.2 %

      NPV increases to   -$2.1 m

State Incentive:-
   Royalty Holiday (@ 5 yrs) $6.7 m

      IRR increases to   13.6 %

      NPV increases to   -$3.8 m

The combination of ITC and royalty holiday yields a prototype IRR of 16.1% and a payout of 4.2 years, suspected to be enough to attract serious investors from around the world.

-PRODUCTION RATE AND INVESTMENT SENSITIVITY

An analysis was done on the effect on project economics by accelerating production rates, as represented by Cases B and C, with their associated higher capital costs. Due to the small and constant reserve base, results showed an unfavourable change in all parameters (see Table 14), as the IRR dropped to 10% and 6.7%, respectively, for Cases B and C.

Table 14

PROTOTYPE CBM PROJECT ECONOMICS
PRODN RATE & INVESTMENT SENSITIVITY

Parameter	Units	CASE A	CASE B	CASE C
Initial Prod. Rate	km3/d	380	560	760
Capital Investment	$Am	65.5	72.5	87.5
Payout	yrs.	5.7	6.3	7.4
IRR	%	12	10	6.7
NPV	$Am	-8.2	-14.3	-26

These poor results highlight the waste of physical plant capacity associated with many CBM projects after the peak initial years. In order to overcome this, the high decline rate of most CBM wells has to be engineered into the overall design of a project so as to achieve load-leveling of all plant and equipment for as many years as possible. This *value-engineering* requires a comprehensive and integrated programme of multi-disciplined thinking at the pre-feasibility and feasibility stages of projects.

Conclusions

1. Total CBM well costs in Australia, including drilling, completion and stimulation, have been between 25% and 40 % higher than the USA total costs for large scale development programmes. If an Australian CBM industry were to be launched in a big way, well costs would be expected to come down by this range.

2. In remote areas and for exploratory-type wells with extended testing, total costs both in Australia and the USA can be 30% to 70% higher than development-type wells near service centres in these regions.

3. Even with a future 25% reduction in CBM drilling and completion well costs, and assuming no major stimulation costs are required, the economic viability of a 100-well prototype CBM project in Queensland's Bowen Basin would be marginal at best. Partly this is due to the high total capital cost per unit of methane reserve ($A0.71 per MCF).

4. Principal causes of this high unit capital cost are believed to be the high services costs from limited contractors, the high compression requirements and the high landed cost for compressors, tubulars, and other processing equipment. Essentially, these all result from a small industry, with little economy of scale.

5. The very limited number of New Zealand wells drilled and tested (2), and their remoteness contributed to high mobilisation/demobilisation costs for rigs and services. Total well costs were some 40% above similar remote costs in Australia and the USA. Operators of remote and exploratory CBM projects in other " wildcat" regions of the world should allow for these in their budgets.

6. Reasonable and effective economic incentives, in the form of investment tax credits and royalty holidays, can increase the profitability of future CBM projects to above a risked threshold. This would "kick-start" an industry to generate economies of scale and advances in R&D in but a few years. Government incentives would be paid by industry CBM revenues, leaving a valuable net social return to the local economy ($A105 million for a 100 well operation).

Acknowledgments

The author wishes to thank members of the Australian and New Zealand CBM industry for sharing some of their operational information and the challenges and opportunities they see ahead. In particular, I would like to thank Wendell Beavers of Conoco Australia, Bruce Robertson of Shell Coal Australia, Ross Nausmann and Martin Riley of Oil Co. of Australia, and Dr. Murry Cave and the directors of Coal Seam Gas Limited, of New Zealand. For Washington State information, I wish to acknowledge the assistance of Allan Kahill and Ken Sinclair. Thanks are also due to Drs. Victor Rudolph and Sue Golding at the University of Queensland for help and encouragement to prepare and present this paper. A final note of thanks to Carl Svoboda of Northstar Energy Corporation of Alberta for assistance in sourcing Canadian information and Dr. Miryam Glikson for editorial review.

References

1 Hobbs, G.W., 1990: " Coalbed Methane Developments in the Powder", Western Oil World, August, P.17.

2 Holland, J.R. and Kimmons, J.W., 1995: " Characteristics and Economics of Coalbed Methane Production From the Ft. Union Formation, Powder River Basin", Intergas '95, Tuscaloosa, P. 115

3 Hobbs, G.W., Holland, J.R., Winkler, R.O., 1997: " Updated Production and Economic Model for Cedar Cove Coalbed Methane Field", International Coalbed Methane Symposium, Tuscaloosa, P. 75.

4 Kahill, A. and Sinclair, K., 1998 : Washington State CBM project by Epic Resources; personal contact.

5 Sinclair, K.G. and Cranstone, J.R., 1997: " Canadian Coalbed Methane: The Birth of an Industry", International Coalbed Methane Symposium, Tuscaloosa, P.115.

6 Petroleum Services Association of Canada, 1998: " PSAC 1997 Well Cost Study", prepared by Winterhawk (1997) Ltd.

7 Wang, X. and Zuo, W., 1997: " Method Study on Economic Evaluation for Coalbed Methane Development", International Coalbed Methane Symposium, Tuscaloosa, P.59.

8 Cave, M.P.,1998: " Natural Gas from Coal Seams: Investigations in New Zealand", International Conference on Coal Seam Gas and Oil, Brisbane, March.

9 Coal Seam Gas Limited, of New Zealand, 1998: personal contact

THE USE OF MONTE CARLO ANALYSIS TO EVALUATE PROSPECTIVE COALBED METHANE PROPERTIES

Michael D. Zuber
S. A. Holditch & Associates, Inc.
1310 Commerce Drive
Park Ridge 1
Pittsburgh, PA 15275-1011 USA

1. Abstract

The exploration of coalbed methane resources is underway in many countries throughout the world. The ability to accurately predict and/or estimate expected production from a prospective coalbed methane play is imperative for evaluating the potential for economic success of proposed developmental projects. However, to accurately predict coalbed methane production requires the use of a reservoir simulator and generally requires knowledge of numerous reservoir properties. It is usually difficult to reliably predict key reservoir properties based on the datasets which are available for exploratory-type coalbed projects. We have found that a probabilistic approach is useful for estimating the expected range of producibility for a prospective coalbed methane area, and for quantifying the risk associated with finding commercial production in prospective areas. The probabilistic approach provides a forecast of the expected distribution of reserves that would be expected to be realized from large-scale development of a particular area.

This paper describes a probabilistic approach for evaluating prospective coalbed methane projects which combines (1) coalbed methane reservoir simulation to predict coalbed well production, and (2) Monte Carlo simulation analysis to determine the distribution of expected reserves for a prospective area. This methodology allows us to predict, in a probabilistic manner, the expected average productivity and distribution of reserves for a prospective coalbed methane project. This paper explains in detail the approach used and presents a case study to show how this method is applied to a developmental coalbed methane project. The methodology presented in this paper can be applied to any new or emerging coalbed methane development project to assist in quantification of the economic viability of the project. It also provides a basis for risking investments in prospective coalbed methane projects.

2. Introduction

There are numerous areas of prospective coalbed methane development throughout the world. Outside the United States, most exploration is in areas with little previous development. This includes countries such as China, Australia, India, Eastern Europe, and Canada. In many of these areas, the coalbed methane resources are in the early stages of development. There are no coalbed projects with long productive histories. Therefore, the knowledge base regarding the long-term productivity potential of these resources is small.

Commercial coalbed methane development usually requires a large-scale development. This means development of projects with many wells (10 to hundreds) and a large acreage position. Therefore, a large number of wells in a large infrastructure for gathering, transportation, etc. must be developed for commercial production. This means that a large initial investment is required to initiate commercial coalbed methane production. This is especially true in areas that have had little historic conventional oil or gas production.

Because a large initial investment is needed to develop commercial coalbed methane production, it is often desirable for companies to team in partnerships or secure outside investors for these projects. To secure investors, it is most often necessary to have an estimate of potential reserves and also to estimate the risk involved in the prospective project.

To accurately predict coalbed methane production requires the use of a reservoir simulator,[1] and generally requires knowledge of numerous reservoir properties such as net thickness, gas content, permeability, etc. It is usually difficult to reliably predict these key parameters based on the datasets which are available for exploratory-type coalbed methane projects. Therefore, it is difficult to predict coalbed methane production prior to the development of a pilot project or initial development wells. The data required for estimating reserves is at a minimum in the early life of a project.

3. Analysis of Coalbed Methane Reserves

Coalbed methane reservoirs are complex reservoirs.[2] Coals are typically characterized as dual porosity systems with adsorbed gas in the matrix.[2] Generally two-phase (gas/water) flow exists in the fracture system and gas production is governed by desorption and diffusion characteristics in the coal matrix system. Because of the complexity of coal reservoirs, reservoir simulators which model the complex coal flow mechanisms are generally required to accurately forecast production.[3] In producing areas, these simulators are calibrated by history matching historic coalbed methane well production.[4] Experience has shown that each commercial production area has its own characteristics, therefore, analogy with existing producing areas is often difficult.

Coalbed methane reservoir simulators require many properties as input for their use.[3] In historic producing areas, models are calibrated by matching historic production. A summary of the data required for coalbed methane simulation is shown in Table 1. For analysis purposes the reservoir properties required for simulation are generally grouped as either *fixed properties* or *match properties*. *Fixed properties* are those which will not be varied in the history matching process. These values will be determined using the best available measurements. *Match properties* are those properties which will be varied in the calibration process when history matching historic production data. These properties are generally determined from history matching. A methodology for estimating reserves for coalbed methane reservoirs has been documented in previous articles.[5]

TABLE 1. Summary of data used for analysis of coalbed methane reserves (Compiled after Sawyer, W. K. *et al.*)

• **Fixed Properties**
- Adsorbed Gas Content
- Drainage Area
- Net Thickness
- Characteristic Sorption Time
- Relative Permeability Curves
- Initial Reservoir Pressure
- Gas & Water Properties
- Well Completion Efficiency
- Well Flowing Pressure
• **Match Properties**
- Permeability
- Porosity (fracture system)
- Initial Gas & Water Saturations
- Relative Permeability

4. Reserves Analysis Methods

Prospective coalbed methane projects exist throughout the world. Many of these projects are currently being evaluated for coalbed methane potential. These projects have the following characteristics.

- Large area for development.

- Regional geology is generally known.

- Data may be available from mine coreholes or conventional well logs.

- There are sparse (or no) permeability and production data.

- Production test wells or pilot data are sometimes available.

To evaluate the potential for these prospective areas, a *deterministic* or *probabilistic* approach to reserves evaluation may be used. A *deterministic* approach is one that results in the "best estimate" of performance based on estimated "average" properties for the study area. A *probabilistic* approach results in a distribution of expected

M. D. ZUBER

performance for the study area.[6] The probabilistic approach incorporates distributions of
key input variables. The results from this approach can be analyzed using statistical
measures to quantify the risk associated with the forecasts for the study area. The
probabilistic approach attempts to capture the distribution of reserves that would be
realized from large-scale development of the study area. By nature, the reserves of a
group of coalbed methane wells in a given area tend to fall on a log-normal distribution.
Fig. 1 shows an example of the distribution of forecast 30-year reserves for a group of
coalbed methane wells. This graph shows a plot of the cumulative frequency of forecast
30-year gas recovery for a group of more than 100 coalbed methane wells in a field in
the Black Warrior Basin. The goal of the probabilistic approach is to define this
reserves distribution for the study area.

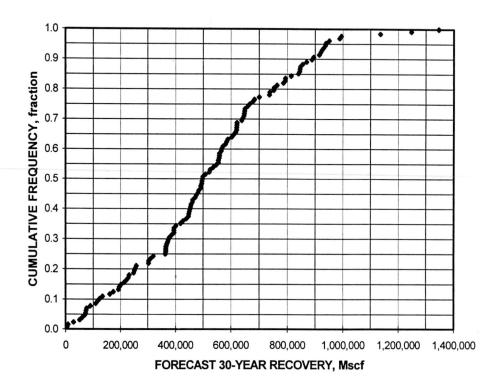

**Fig. 1 – Distribution of forecast 30-year production for individual wells in a coalbed
methane field.**

5. Monte Carlo Analysis Approach

Based on analysis of numerous prospective coalbed methane projects, we believe that a
probabilistic approach is most appropriate for evaluation of these prospect areas. This

approach is consistent with the quantity and quality of data generally available for these types of projects. This approach is also consistent with the type of economic analysis that is required. Typically, companies evaluating coalbed prospective projects wish to attempt to quantify the risk associated with these reserves estimates. This is only possible using a probabilistic reserves approach.

A Monte Carlo analysis approach has been developed and used for evaluation of numerous coalbed methane projects. The general steps used in this approach are described below.

1. Define the area to be evaluated.

2. Gather all available geologic, engineering, and production data for the evaluation area.

3. Develop a "base case" reservoir model using estimated average properties for the development area and/or history matching available production test data.

4. Determine key reservoir properties. These are those properties which are expected to have a significant impact on the outcome (reserves forecast) and are not well measured.

5. Estimate distributions of key reservoir properties.

6. Run a Monte Carlo simulation using the "base case" dataset and incorporating the distributions of key input variables. Typically, we utilize a 1,000 case Monte Carlo analysis using Latin Hypercube Sampling.[7]

7. Analyze the Monte Carlo simulation results using statistical measures. Typically, the 30-year cumulative gas production is our common measure for analysis and comparison.

Care should be taken when selecting a study area. An area of prospective development is normally considered the study area. However, the prospective area may be broken into several study areas which are similar in geologic nature, otherwise graded based on prior geologic analysis. The target coal seams in the study area should be similar in quality and geologic setting. As a general guideline, the range of average depth of the target coals within the study area should be within approximately a 300 meter (1,000 ft) interval. The study area should also be characterized by a similar degree of data quality across the study area. In other words, the degree to which key properties are known should be similar for the entire study area.

In developing a "base case" reservoir model, care should be taken to make the simulation model as simple as possible. This minimizes the simulation time required to conduct the Monte Carlo analysis. We normally use a single-layer, radial simulation model of a single-well. Multiple target coals are generally grouped into a single simulation layer. This is found generally to be a reasonable assumption. In this approach, we use the effective well radius concept to approximate fractured well performance. In calibrating the "base case" model, it is important to use all available production or permeability test data to calibrate the model. This will provide the best possible base case dataset.

One of the key aspects of the probabilistic approach is defining the expected distribution of key reservoir properties. For coalbed methane reservoirs, reservoir properties which have the most impact on the production forecasts tend to be (1) gas content, (2) permeability, (3) net coal thickness, (4) fracture system porosity, and (5) initial gas saturation in the fracture systems. In selecting distributions for the key properties, we normally find it is acceptable to keep the distributions simple. Typically, it is sufficient to define the minimum, maximum, and median expected value for the variable property within the study area. This is shown schematically is Fig. 2. Triangular distributions are generally utilized for gas content, thickness, porosity, and initial gas saturation. A log normal distribution is generally used for permeability.

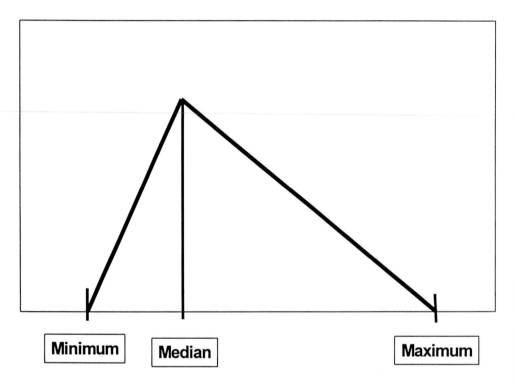

Fig. 2 – Schematic of triangular distribution used for variable reservoir properties.

Once the "base case" reservoir model is determined, and the distributions of the key properties are determined, a Monte Carlo simulation analysis is conducted where a series of simulations are made in which the "base case" dataset is used and values for the key reservoir properties are randomly selected from their distributions. A schematic of the software structure used for this purpose is shown in Fig. 3. COALGAS™, a three-dimensional, coalbed methane reservoir simulator is used for this purpose.[8]

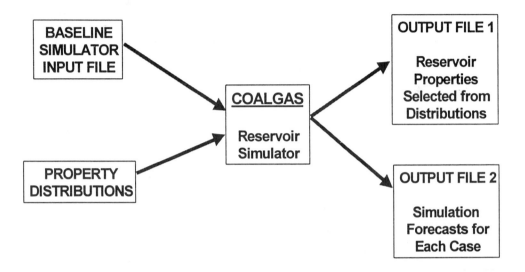

Fig. 3 – Structure of software for Monte Carlo analysis.

6. Example Application

To illustrate the use of the probabilistic approach presented in the previous section, we provide an example of the use of this approach for a prospective coalbed project. The prospect area is a lease of approximately 8,000 acres. Available data for the prospect area include (1) a regional geologic description developed from conventional well data, (2) data from six corehole "data wells" drilled in the prospect area, and (3) data from one production test well. Available data included several desorption gas content tests for the target coal seams and two desorption isotherm tests. Based on the available geologic description, the average depth of the target coals in the study area ranges from 600 m to 750 m. Normal pressure gradient is assumed. The average net coal thickness for the two major target coals is estimated to be 15 m. From the experience gained from the corehole data wells and the one production test well, a "base case" reservoir model was constructed. An average development well spacing of 80 ac/well is assumed. It is expected that the wells will be cased-hole completions and the two target coals will be fracture stimulated.

On the basis of available data for the prospect area, seven key properties were identified. We then estimated expected distributions of these key properties for the prospect area. The distributions of the key reservoir properties are shown in Table 2.

TABLE 2. Example of variable properties

Parameter	Minimum	Most Likely	Maximum	Units
Pressure	933	986	1,093	psia
Thickness	20	50	95	ft
Permeability	3	10	20	md
Porosity	0.002	0.005	0.020	fraction
Ash	4	7	10	percent
Moisture	10	11	12	percent
Gas Content	122	175	225	scf/ton

A Monte Carlo simulation analysis consisting of 1,000 single-well reservoir simulations was generated using the properties shown in Table 2. The reservoir properties randomly selected from the variable reservoir distributions and the output 5, 10, and 30-year gas production forecasts were stored in output files. A distribution of the resulting 30-year gas production forecast from this Monte Carlo simulation is shown in Fig. 4. For this example, the median 30-year gas recovery is approximately 750 MMscf/well. From the distribution shown in Fig. 4, we can also determine the range of expected performance to assess the risk associated with the production forecast for this property. From Fig. 4 we determine that there is a 10% probability that the average well production from the prospect area will be less than 500 MMscf/well, and a 90 % probability that the average well production from the prospect area will be less than 1,250 MMscf/well. This assumes that the prospect area is developed entirely using the average well spacing of 80 ac/well assumed in this analysis.

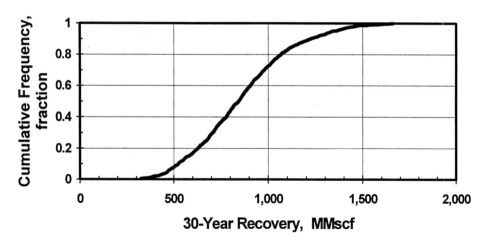

Fig. 4 – Distribution of forecast 30-year gas production from Monte Carlo analysis.

Using the results from the Monte Carlo simulation analysis, individual well production profiles can be generated for use in economic analyses. It is often useful to generate production profiles for the median case (50% probability) and for a high-side and low-side case. Fig. 5 shows production forecasts generated for the median case (50% probability), the low-side (25% probability), and high-side (75% probability). These forecasts were generated by determining the combinations of parameters from the distributions of variable properties used to generate the production forecast shown in Table 2 at the various levels of cumulative frequency.

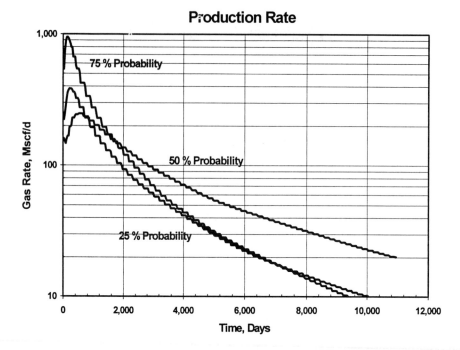

Fig. 5 – Single well forecasts for median, 25% and 75% probability cases from the reserves distribution shown in Fig. 4.

It is often desirable to determine the sensitivity of the key variable input properties to the resulting production forecasts. This helps to define which parameters have the most impact on the production forecast. This sensitivity study is accomplished by using the "base case" reservoir properties and the average or most-likely values of the key reservoir properties. A series of parametric simulations is then conducted in which a similar *percentage* change in each key reservoir property is made. For instance, a simulation run may be made where a 10% increase and a 10% decrease in reservoir permeability are used as inputs, with all other parameters staying constant. In this manner, we can test the sensitivity to key properties on the resulting production forecast. Fig. 6 shows the results of such a sensitivity analysis for the example used in this paper. As can be seen in Fig. 6, the reservoir properties which have the most impact on the production forecast for a similar percentage change in their values are the reservoir thickness and gas content.

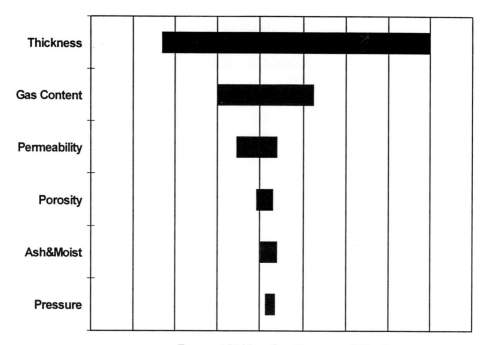

Forecast 30-Year Gas Recovery, MMscf

Fig. 6 – Example of parameter sensitivity analysis results.

7. Conclusions

Based on the work presented in this paper, we have drawn the following conclusions.

1. A probabilistic approach using a coalbed methane reservoir simulator and Monte Carlo analysis is the recommended method to evaluate prospective coalbed methane projects.

2. This approach incorporates the distribution of key variables across the property evaluated, and provides results in a statistically meaningful format.

3. The utility of the Monte Carlo analysis results are dependent upon the ability of the "base case" model to accurately represent the production character of the prospect area.

4. The Monte Carlo analysis may need to be run at several different well spacings to determine the optimal commercial spacing.

8. References

1. Zuber, M. D. and Kuuskraa, V. A.: "A Reservoir Simulator-Based Methodology for Calculating Reserves of Coalbed Methane Wells," Proc., Coalbed Methane Symposium, Tuscaloosa (1989).

2. Gray, I.: "Reservoir Engineering in Coal Seams: Part I – The Physical Process of Gas Storage and Movement in Coal Seams," *SPE Reservoir Engineering* (February 1987) pp. 28-34.

3. Zuber, M. D. and Semmelbeck, M. E.: "Simulators Analyze Coalbed Methane," *The American Oil and Gas Reporter* (September 1990), 40-44.

4. Sawyer, W. K. *et al.*: "Using Reservoir Simulation and Field Data to Define Mechanisms Controlling Coalbed Methane Production," paper 8763, presented at the 1987 International Coalbed Methane Symposium, The University of Alabama/Tuscaloosa (November 16-19, 1987) pp. 295-307.

5. Zuber, M. D.: "Basic Reservoir Engineering for Coal," *A Guide to Coalbed Methane Reservoir Engineering*, Gas Research Institute (1996), 3.1-3.32.

6. Megill, R. E.: *An Introduction to Risk Analysis,* Petroleum Publishing Company, Tulsa (1977), 110-136.

7. Crout, L. E.: *Improving the Computing Efficiency of Risk Simulation Techniques,* MS Thesis, Texas A&M University, College Station, TX (1985).

8. COALGAS: A Comprehensive Coalbed Methane Simulator, User's Manual, Version 3.1, S. A. Holditch & Associates, Inc. College Station, TX, January 1990.

DEFINING COALBED METHANE EXPLORATION FAIRWAYS: AN EXAMPLE FROM THE PICEANCE BASIN, ROCKY MOUNTAIN FORELAND, WESTERN UNITED STATES

ROGER TYLER and ANDREW R. SCOTT
Bureau of Economic Geology
The University of Texas at Austin
Austin, Texas, USA
W. R. KAISER
Consultant, Austin, Texas, USA

1. Abstract

A basin-scale coalbed methane producibility and exploration model has been developed on the basis of research performed in the San Juan, Sand Wash, Greater Green Rivers, and Piceance Basins of the Rocky Mountain Foreland and reconnaissance studies of several other producing and prospective coal basins in the United States and worldwide. The producibility model indicates that depositional setting and coal distribution, coal rank, gas content, permeability, hydrodynamics, and tectonic/structural setting are controls critical to coalbed methane production. However, knowledge of a basin's geologic and hydrologic characteristics will not facilitate conclusions about coalbed methane producibility because it is the interplay among geologic and hydrologic controls on production and their spatial relation that govern producibility. High producibility requires that the geologic and hydrologic controls be synergistically combined. That synergism is absent in the marginally producing, hydrocarbon-overpressured Piceance Basin. As predicted from the coalbed methane producibility model, significant coalbed methane production (greater than 1 MMcf/d [28 Mm^3/d]) may be precluded in many parts of the hydrocarbon-overpressured Piceance Basin by the absence of coalbed reservoir continuity, high permeability, and dynamic ground-water flow. The best potential for coalbed methane production may lie in conventional and compartmentalized traps basinward of where outcrop and subsurface coals are in good reservoir and hydraulic communication and/or in areas of vertical flow potential and fracture-enhanced permeability. In the low-permeability, hydrocarbon-overpressured Piceance Basin, exploration and development of migrated conventionally and hydrodynamically trapped gases, in-situ-generated secondary biogenic gases, and solution gases will be required to achieve high coalbed methane production.

2. Introduction

Methane from coal beds is a potentially important source of natural gas worldwide, but to date successful exploitation of coalbed methane resources has been limited to the United States. Triggered by success in the United States, exploration for coalbed methane has begun in coal-rich areas of the United Kingdom, eastern and western Europe, China, South Africa, Zimbabwe, and Australia. What has not been widely recognized about U.S. experience is that although coalbed methane resources in some basins have been successfully exploited, other basins with seemingly similar geologic and hydrologic attributes have proven to be disappointing methane producers.

The traditional view of gas production from coal reservoirs is inadequate to explain the contrasts in coalbed methane producibility among coal basins. In the traditional view, coal gases are generated in situ during coalification and are sorbed on the coal's large internal surface area. Sorption is pressure dependent and is promoted by increasing pressure. Gas production is achieved by reducing reservoir pressure through depressuring (dewatering), which liberates the gases from the coal surface. Gases are then diffused to the cleat system for subsequent flow to the wellbore. This traditional view is oversimplified because it fails to recognize that additional sources of gas—beyond that generated during coalification—are required to achieve high gas content following basinal uplift and cooling. Migrated conventionally trapped and hydrodynamically trapped gases, in-situ-generated secondary biogenic gases, and solution gases are required to achieve high gas contents and/or fully gas-saturated coals for consequent high productivity (Scott and others, 1994a).

Delineating the presence and origin of these additional sources of gas requires the application of the coalbed methane producibility model and an understanding of the interplay among tectonic and structural setting, depositional setting and coal distribution, coal rank, gas content, permeability, and hydrodynamics (Kaiser and others, 1995) (fig. 1). This coalbed methane producibility model identifies critical geologic and hydrologic controls on production and synergistically integrates them to show how they interact for commercial gas production (Kaiser and others 1994a and b). As shown in earlier studies of the San Juan, Sand Wash, Greater Green River, and Piceance Basins (Ayers and others, 1991; Kaiser and others, 1993; Tyler and others, 1994; Tyler and others, 1995; Tyler and others, 1996) (fig. 2), a synergistic interplay among these controls determines high coalbed methane producibility (fig. 1). Our basin-scale model for coalbed methane producibility has evolved out of a comparison of the prolific San Juan Basin and the marginally producing Sand Wash, Greater Green River, and Piceance Basins (Tyler and others, 1997; Scott and Tyler, 1998) (fig. 2). The comparison is apt because these basins share similar geologic and hydrologic attributes but in dissimilar combinations. Before development efforts, the Sand Wash, Greater Green River, and Piceance Basins were viewed as being potentially highly productive; however, subsequent drilling and production efforts showed them to be poor to moderate producers. To date, and with currently available technology, the Sand Wash Basin coals have yielded large volumes of water and little gas, and the Piceance Basin coals, little water and limited gas. Refinement of the model in the Piceance Basin was considered critical because it is a low-permeability, hydrocarbon-overpressured basin, as are many basins throughout the world (Tyler and others, 1997).

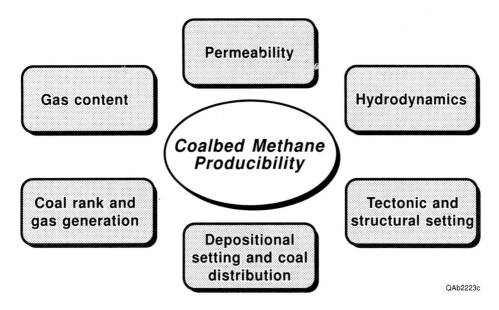

Figure 1. Geologic and hydrologic controls on the producibility of coal gas (Kaiser and others, 1994a). Synergistic interplay among controls and their spatial relations governs producibility.

3. Coalbed Methane Producibility Model

To delineate the presence, origin, and extent of coalbed methane producibility in the Piceance Basin requires an understanding of the interplay among tectonic and structural setting, depositional systems and coal distribution, coal rank, gas content, hydrodynamics, and permeability (fig. 1). Structural and depositional setting control the distribution of coal resources and determine their location with respect to the thermally mature parts of a basin. Structural and depositional setting also determine the orientation of coal resources relative to ground-water flow direction, which may be orthogonal or parallel to structural grain or up or down the coal-rank gradient. Knowledge of depositional framework enables prediction of coalbed thickness, geometry, and continuity, especially when data are sparse.

Permeability of a coal bed is determined by its fracture (cleat) system, which in turn is largely controlled by the tectonic or structural setting. Coals are commonly aquifers with permeabilities that may be orders of magnitude larger than those of associated sandstones; they may also be conduits for gas migration. Permeability that is too high can be as detrimental to economic production as very low permeability because of high water production (Scott, 1996). Basinward flow of ground water requires recharge of laterally continuous permeable coals at the elevated, structurally defined basin margins.

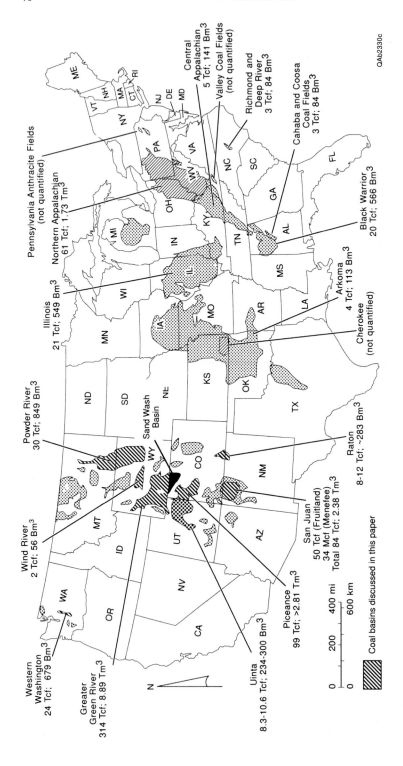

Figure 2. Coal basins and coalbed methane resources of the United States. Total coalbed methane resources are estimated to exceed 19.1 Tm³ (675 Tcf) (Scott and others, 1994b). The model of the controls on coalbed methane producibility is based on comprehensive geologic and hydrologic studies of the San Juan, Sand Wash (Greater Green River), and Piceance Basins and on reconnaissance studies of Raton, Powder River, Uinta, Wind River, and Western Washington Basins in the western United States.

Coal rank is a direct indicator of gas generation but not of gas content (Scott, 1996). Maximum generation of thermogenic gas occurs at ranks of medium-volatile to low-volatile bituminous (R_o of 1.3 to 1.8%). Although gas content generally increases with rank, it is not determined by rank alone but reflects permeability contrasts and hydrodynamics, as they influence conventional trapping of gas, reservoir pressure, generation of secondary biogenic gases, and long-distance migration of gas. Thus, the ability of a coal reservoir to store gas depends on more than rank alone. Because of the interplay among the controls on production, gas content in coals can be much higher than that predicted on the basis of rank alone.

Synergistically, controls of the coalbed gas producibility model that can be used to predict high-productivity fairways include (1) thick, laterally continuous coals of high thermal maturity, (2) basinward flow of ground water through high-rank coals down the coal-rank gradient toward no-flow boundaries such as structural hingelines, faults, facies changes, and discharge areas that are oriented perpendicular to the regional flow direction, (3) generation of secondary biogenic gas, and (4) conventional and hydrodynamic trapping of gas along no-flow boundaries (Kaiser and others, 1994a and b; Scott and others, 1994a and b) (fig. 3).

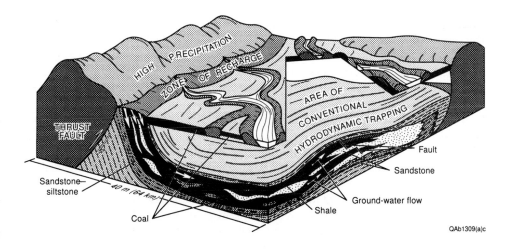

QAb1309(a)c

Figure 3. Conceptual model for high coalbed methane producibility. The essential elements of the coalbed methane producibility model, developed from the San Juan, Sand Wash, and Piceance Basin studies, are (1) thick, laterally continuous coals of high thermal maturity, (2) basinward flow of ground water through high-rank coals down the coal-rank gradient toward no-flow boundaries such as structural hingelines, faults, facies changes, and discharge areas that are oriented perpendicular to the region flow direction, (3) generation of secondary biogenic methane, and (4) conventional and hydrodynamic trapping of gas along no-flow boundaries (Tyler and others, 1996).

4. Geologic and Hydrologic Controls Critical to Coalbed Methane Producibility in the Piceance Basin

The following section is adapted from the Tyler and others (1997) paper, which reviews the key geologic and hydrologic controls on coalbed methane producibility in the low-permeability, hydrocarbon-overpressured Piceance Basin. An application of the producibility model and the synergistic interplay (or lack thereof) of key geologic and hydrologic factors used in defining exploration fairways in the Piceance Basin are discussed.

4.1. TECTONIC AND STRUCTURAL SETTING

Delineating areas of conventional trapping and fracture-enhanced permeability is very important in low-permeability, hydrocarbon-overpressured basins such as the Piceance Basin. In the structurally complex Piceance Basin, Upper Cretaceous Williams Fork Formation coals are the prime target for coalbed methane exploration. Greater structural complexity in the Williams Fork Formation reflects folds and faults above thrust detachment zones in the Mancos Shale; the most productive coalbed methane wells are on structural anticlines. The Piceance Basin is bounded by uplifts that formed in the Rocky Mountain Foreland during the Laramide Orogeny, approximately 72 to 40 mya (Dickinson and Snyder, 1978; Ogden, 1980; Tyler and others, 1996) (fig. 4). The onset of the Laramide Orogeny is recorded by an unconformity at the top of the Upper Cretaceous, which may represent the removal of thousands of feet of Upper Cretaceous rocks (Johnson and Nuccio, 1986). Subsidence continued through the Paleocene to nearly the end of the Eocene, during which time as much as 12,000 ft (3,600 m) of Paleocene and Eocene sediments was deposited. During late Eocene to early Oligocene time, a second erosional surface developed across the basin. This resulted in the removal of little or no section in the basin center but as much as 2,000 to 3,000 ft (610 to 910 m) along the upturned margins of the basin (Johnson and Nuccio, 1986). At the end of the Laramide Orogeny, about 40 mya, basin subsidence ceased; little structural movement or sedimentation occurred in the basin until the Colorado River system began to cut deep canyons, about 10 mya. Significant topographic relief, developed during the Miocene to Recent, is an important factor in coalbed methane development (Kaiser and Scott, 1996). Many exploration and production wells are restricted to lower elevations or canyons because of the excessive drilling depths required to reach the coal-bearing intervals.

Higher production in the Piceance Basin also is attributed to fracture-enhanced permeability associated with tectonic flexure or differential compaction or both. Regional mapping in the Piceance Basin of fracture (cleat) sets in coal beds, an indicator of permeability pathways for gas and water to the well head, shows that face-cleat strikes are oriented normal to the basin-fold axis and to the Grand Hogback thrust front and parallel to the maximum horizontal compressive paleostresses (Tremain and Tyler, 1996) (figs. 4 and 5). Upper Cretaceous face cleats correlate with prominent joints in clastic rocks that strike east-northeast in the southern Piceance Basin and west-northwest in the northern basin. East-northeast-trending face-cleat domains in the southeast part of the basin are oblique to current maximum stress directions, whereas face-cleat domains in the northern half of the basin are parallel to current maximum stress directions. Therefore, face cleats in the northern half of the basin may provide more permeable pathways for gas production.

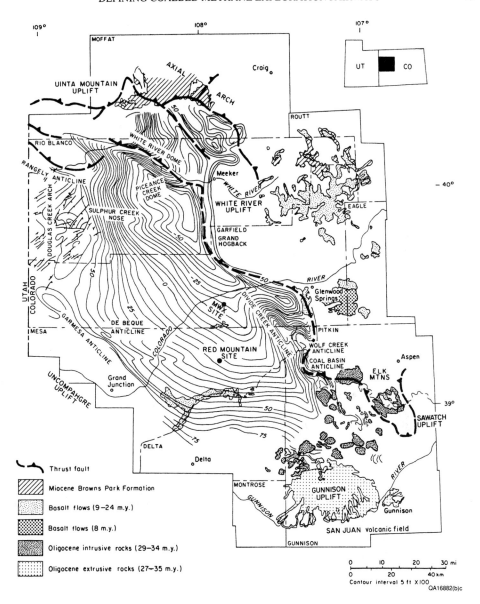

Figure 4. Structure map, contoured on top of the Rollins–Trout Creek Sandstone Members, Piceance Basin. Contour interval 500 ft (152 m). Modified from Johnson and Nuccio (1986).

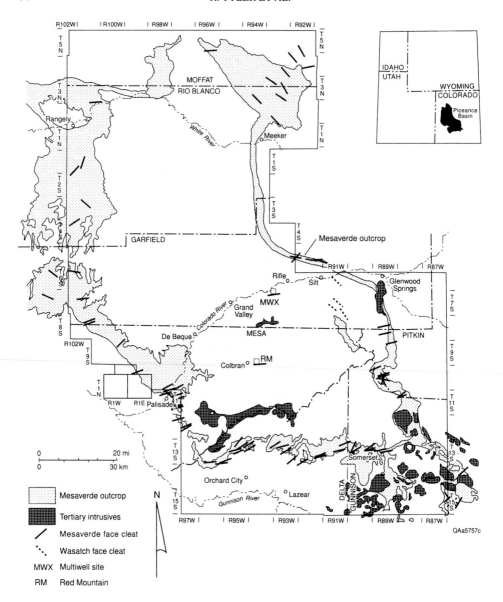

Figure 5. Face-cleat strikes in the Mesaverde Group and Wasatch Formation coal beds, Piceance Basin (Tremain and Tyler, 1996).

4.2 DEPOSITIONAL SETTING AND COAL DISTRIBUTION

The Piceance Basin has a single major coalbed methane target, the Cameo-Wheeler-Fairfield coal zone (Williams Fork Formation, Mesaverde Group), which ranges in thickness from 91 to 183 m (300 to 600 ft) and lies at an average depth of approximately 1,800 m (~6,000 ft). The most continuous and thickest coal beds (individual seams from 6 to 11 m [20 to 35 ft] thick) formed in coastal plain environments landward (westward) of the progradational strandplain/delta plain deposits of the Rollins–Trout Creek Sandstone, in a depositional setting almost identical to that of the San Juan and Sand Wash Basins (McMurray and Tyler, 1996). Net coal thickness of the Williams Fork Formation averages 80 to 150 ft (24 to 45 m) and is greatest in a north-south belt, parallel to depositional strike (fig. 6).

The Williams Fork Formation can be divided into several genetic depositional sequences (fig. 7). Genetic unit 1 is a progradational/aggradational couplet that extended coal-bearing coastal plain deposits beyond the present-day basin margin. Genetic units 2 and 3 are clastic wedges displaying a similar arrangement of depositional systems. Three depositional systems are recognized in each genetic unit. A north- to northeastward-oriented linear shoreline system dominated the easternmost part of the basin and was backed landward by a coastal plain system, which in turn graded westward into an alluvial plain system. Coal beds pinch out against and/or override the shoreline sandstone to the east, and their ultimate lateral extent is limited by the final shoreline position beyond which marine conditions prevail.

Continuity of the Williams Fork coals is variable. Some individual seams, particularly in genetic unit 1, are correlatable for up to 48 km (30 mi) in the southeastern half of the basin (McMurry and Tyler, 1996). Other seams could be correlated only when grouped within coal packages. Coals that reach outcrop are reduced in number and total thickness, and, consequently, their ability to receive and transmit ground-water recharge basinward is reduced. Limited recharge may have implications for the producibility of coal gas. In the absence of dynamic ground-water flow, less gas is available for dissolution and for sweeping basinward for eventual resorption and conventional trapping along potential no-flow boundaries. Thus, high coalbed methane productivity is precluded; conventional and compartmentalized traps therefore have the greatest potential for coalbed methane production in the Piceance Basin.

4.3 COAL RANK

Coals in the Piceance Basin have reached the thermal maturity level required to generate significant quantities of methane. Coal rank in the Cameo-Wheeler-Fairfield coal group changes abruptly over relatively short distances, ranging from high-volatile C bituminous at the northeast, west, and south margins of the basin to semianthracite in the eastern part of the basin, along its synclinal axes (Scott, 1996) (fig. 8). Coal rank approaches the semianthracite rank in the deepest part of the basin, according to vitrinite-reflectance profiles. Coals adjacent to Tertiary intrusives have reached the anthracite stage, and some graphite has been reported locally (Collins, 1976). Coal rank generally parallels the structure of the Rollins-Trout Creek Sandstones over much of the basin but cuts across structure in the north part of the basin (figs. 4 and 8) (Johnson and Nuccio, 1986).

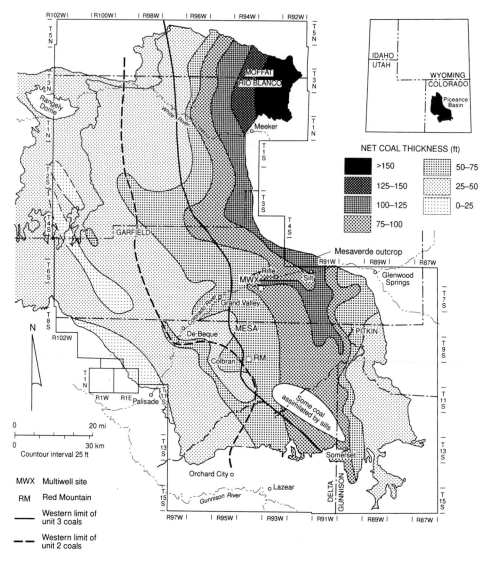

Figure 6. Net coal thickness map of genetic units 1 through 3, Williams Fork Formation. North-northwest to south-southeast net coal thickness trends in the eastern Piceance Basin occur above thick, north-south-oriented progradational shoreline (strandplain/delta) systems (McMurry and Tyler, 1996).

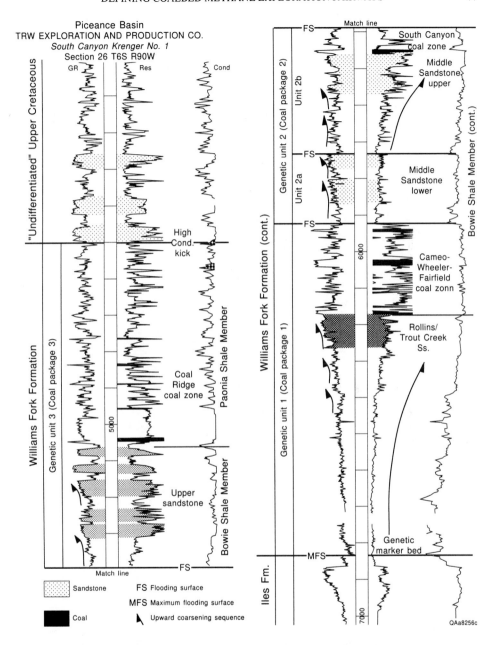

Figure 7. Genetic stratigraphy and type log of the upper Mesaverde Group in the southeastern Piceance Basin. Coal beds are identified on accompanying density logs. Surfaces bounding genetic units are defined by regionally extensive, low-resistivity shale marker beds, which define flooding surfaces (McMurry and Tyler, 1996).

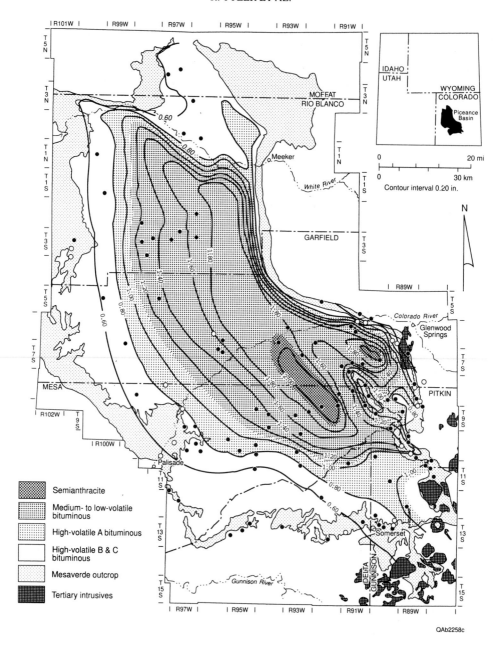

Figure 8. Coal-rank map of the Cameo-Wheeler-Fairfield coal group from measured and calculated data. Map is based on data from previous studies (Tremain, 1982; Tremain and Toomey, 1983; Bostick and Freeman, 1984; and Scott, 1996).

High-volatile B bituminous coals are located on the crests of the Divide Creek and Wolf Creek Anticlines in the east part of the basin (Scott, 1996) (fig. 4). The presence of lower-rank coals on these anticlines suggests that these structures either formed before or syntectonically with the main stage of coalification and are related to Laramide Orogenic events rather than to Tertiary volcanic activity (Johnson and Nuccio, 1986). Coal rank changes abruptly off the crest of the anticlines, increasing from high-volatile B bituminous to low-volatile bituminous over approximately 5 mi (8 km). Medium-volatile bituminous to high-volatile B bituminous coals are located on the Coal Basin Anticline, suggesting that these coals may have been buried deeper and were subsequently uplifted after the main coalification stage. The presence of quartz monzonite below the Coal Basin Anticline suggests that the anticline may have formed in response to the emplacement of an Oligocene pluton (Johnson and Nuccio, 1986) (fig. 4).

4.4   HYDRODYNAMICS AND PERMEABILITY

The Williams Fork formation is the major coal-bearing hydrostratigraphic unit in the Mesaverde Group and is confined below by the Mancos Shale and only partially confined or unconfined above by the undifferentiated Upper Cretaceous (Kaiser and Scott, 1996). Geopressure and hydropressure are both present in the basin; regional hydrocarbon overpressure is dominant in the central part of the basin (fig. 9). Permeability in the Piceance Basin is generally very low, and sandstones in the basin have permeability values in the microdarcy range (Kukal and others, 1992). Permeability in Williams Fork coals, determined from production testing and laboratory in situ stress measurements taken on core samples, is also very low—ranging from nanodarcies to around 1 md, with most estimates being in the microdarcy range (Logan and Mavor, 1995; Reinecke and othres, 1991). Although most of the coal permeability estimates were from within the hydrocarbon-overpressured part of the basin, low-permeability values are also reported in the hydropressured parts of the basin (for example, at Red Mountain).

Very low permeability and extensive hydrocarbon overpressure indicate that meteoric recharge, and, hence, hydropressure, is limited to the basin margins and that long-distance migration of ground water is controlled by fault systems (Kaiser and Scott, 1996). Along the Divide Creek Anticline in the southeastern part of the basin, active recharge and northwestward ground-water flow occur in association with a northwest-trending thrust fault system. Aquifer confinement basinward results in artesian overpressure. Recharge is limited along the eastern and northeastern margins of the basin because the Grand Hogback and Danforth Hills thrust faults offset Williams Fork coals, separating outcrops and restricting subsurface to meteoric flow basinward (Kaiser and Scott, 1996; Scott, 1996). Although dynamic flow is limited in the northern part of the basin, a potentiometric mound and large vertical pressure gradients (fig. 9) in the White River field indicate potential for convergent, upward flow.

4.5   GAS CONTENT

Ash-free gas-content values for Mesaverde coals for all samples range from less than 1 to more than 25.6 m³/t (<1 to >540 scf/ton) but are generally less than 12.5 m³/t (<400 scf/ton)

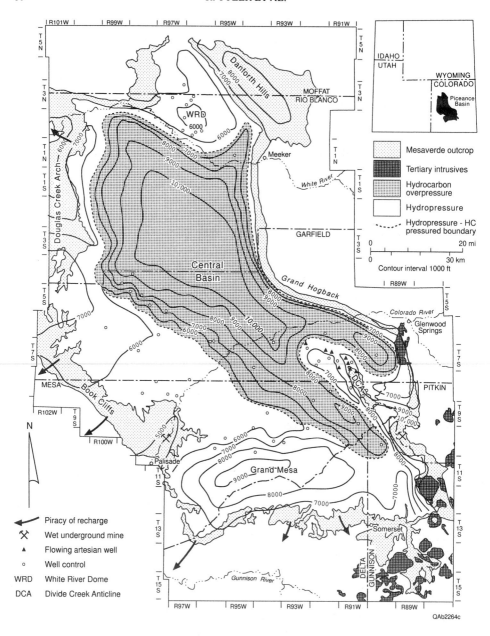

Figure 9. Williams Fork potentiometric-surface map, Piceance Basin. The surface is dominated by a series of mounds and ridges. Only the ridge on the Divide Creek Anticline slopes upward to outcrop to indicate recharge along the anticline. Piracy of recharge along the south, southwest, and west margins of the basin limits is diverted away from the deep basin (Kaiser and Scott, 1996).

(average 7.6m^3/t [243 scf/ton]) (fig. 10). Profiles of gas content versus depth show a gradual increase in gas content with increasing burial depth and coal rank, similar to gas-content profiles in other western basins (Scott and Ambrose, 1992). Overall, there is a gradual increase in gas content with increasing coal rank. However, the unusually high gas content values in lower rank (high-volatile B bituminous) coals in the White River field suggest migration and conventional trapping of coal gases (Scott, 1996). Additionally, relatively high gas content values (3.1 to 9.4 m^3/t [100 to 300 scf/ton]) are found in shallow coals (<457 m [<1,500 ft]), suggesting that generation of secondary biogenic and/or migration and conventional trapping of thermogenic gases may have occurred along basin margins.

However, the gas-content profiles in the Piceance Basin are unusual because there is less scatter in the gas-content data than in other basins (fig.10). The progressive increase of gas content with increasing rank and depth over much of the basin is consistent with the presence of predominantly thermogenic coal gases over much of the basin. The very low permeability in the basin limits meteoric circulation and, therefore, the amount of secondary biogenic gas generation that would increase gas contents. Additionally, gas content values are often increased or decreased in a dynamic flow system, resulting in additional scatter on gas-content profiles. Therefore, the gas-content profile of the low-permeability Piceance Basin probably reflects the progressive increase of gas content with rank because of thermogenic gas generation more accurately than do gas-content profiles in other basins.

5. Defining Exploration Fairways in the Piceance Basin

Coalbed methane exploration fairways were identified using our basin-scale coalbed methane producibility model. Currently available technology and the absence of dynamic groundwater flow in the southern Piceance Basin preclude significant coalbed gas production (fig. 11). Predictably, the parts of the basin with the best potential for coalbed methane exploration fairways are those with fracture-enhanced permeability and/or those basinward of where outcrop and subsurface coals are in good hydraulic communication. Good communication is required for consequent generation of secondary biogenic gas, advective gathering and transport of gas, and subsequent basinward resorption and conventional trapping, which promote fully gas-saturated coals and high productivity. Hydrocarbon overpressure is the dominant pressure regime in the Piceance Basin and represents an area of limited coalbed gas potential because of low permeability, as evidenced by marginal coalbed gas production in Grand Valley/Parachute/Divide Creek fields (figs. 5 and 12). However, the transition between hydropressure and hydrocarbon overpressure in western Grand Valley field may be an area of modest coalbed gas potential, as may be the Buzzard Creek area to the southwest along the pressure transition (fig. 12). Likewise, the same type of transition northwest of the Divide Creek Anticline is prospective. Moreover, coalbed methane production is favored throughout the Piceance Basin in areas of upward flow associated with the flow barriers along the pressure transition zone. Convergent flow and conventional structural trapping favor coalbed methane accumulation on the White River Dome (fig. 9).

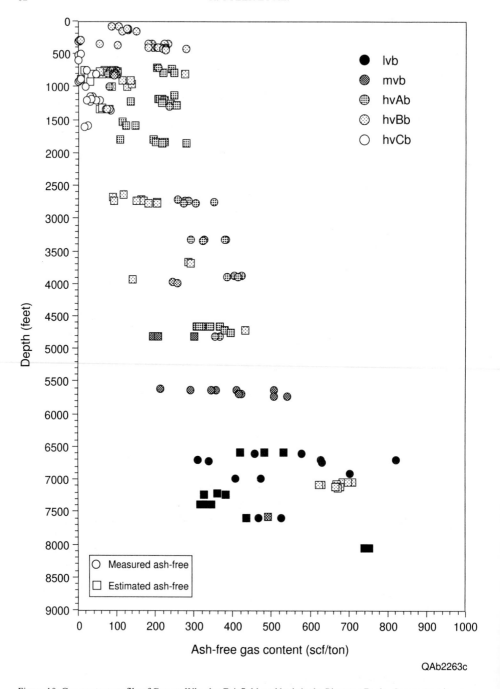

Figure 10. Gas-content profile of Cameo-Wheeler-Fairfield coal beds in the Piceance Basin. Gas content increases with increasing depth and coal rank. Unusually high gas contents in high-volatile B bituminous coals at 7,100 ft (2,165 m) suggest that gas migration and convential trapping occurred in the northern Piceance Basin (Scott, 1996).

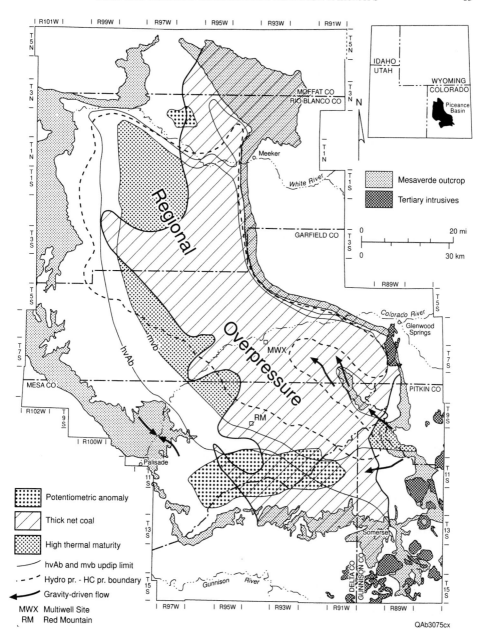

Figure 11. Geologic and hydrologic characterization of the Williams Fork Formation, Piceance Basin. Hydrocarbon overpressure dominates the basin and corresponds to the distribution of high-rank coal. Gravity-driven flow is mainly from the southeast margins of the basin; overpressure and low permeability limit its basinward extent (Tyler and others, 1996).

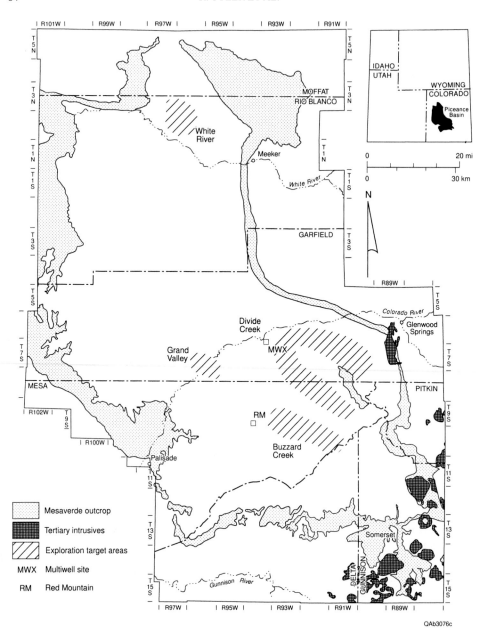

Figure 12. Exploration fairways and target areas, Piceance Basin. Target areas are defined mainly by the transition zone from hydropressure to hydrocarbon overpressure and thicker net coal, except for the White River Uplift, which is structurally controlled (Tyler and others, 1996).

5.1   COAL AND COALBED METHANE RESOURCES

Coal and coalbed methane resources were calculated from digitized structure, topographic, and net-coal-thickness maps on a 9.1-km^2 (3.5-mi^2) grid, using plots of gas content versus depth, density, and coal volume as described by Scott (Scott, 1996). Total subsurface coal resources in the Piceance Basin are estimated to be 289 billion tons, whereas coalbed methane resources are approximately 3.4 Tm^3 (99 Tcf), although estimates range between 2.8 and 4.8 Tm^3 (80-136 Tcf), depending on the calculation method used (Scott and others, 1996a and b).

6. Acknowledgments

This research was funded by the Gas Research Institute under contract no. 5091-214-2261. Several operators, including Barrett Resources Corporation; Oxy USA, Inc.; Mobil; Chevron; Anadarko; AA Production, Inc.; Snyder Oil; and several other operators within the Piceance Basin supplied data that greatly enhanced the quality of the study; these contributions are gratefully acknowledged. This paper benefited from assistance from Carol Tremain, Colorado Geological Survey. The computing staff of the Bureau provided guidance in data processing and computer-assisted mapping. Drafting was by the graphics staff of the Bureau under the direction of Joel Lardon. Word processing was by Susan Lloyd. Publication was authorized by the Director, Bureau of Economic Geology, The University of Texas at Austin.

7. References

Ayers, W. B., Jr., Kaiser, W. R., Laubach, S. E., Ambrose, W. A., Baumgardner, R. W., Jr., Scott, A.,R., Tyler, Roger, Yeh, Joseph, Hawkins, G. J., Swartz, T. E., Schultz-Ela, D. D., Zellers, S. D., Tremain, C. M., and Whitehead N. H., III, 1991: "Geologic and Hydrologic Controls on the Occurrence and Producibility of Coalbed Methane, Fruitland Formation, San Juan Basin": The University of Texas at Austin, Bureau of Economic Geology, topical report prepared for the Gas Research Institute (GRI-91/0072), 314 p.

Bostick, N. H., and Freeman, V. L., 1984: "Tests of Vitrinite Reflectance and Paleotemperature Models at the Multiwell Experiment Site, Piceance Creek Basin, Colorado, in Spencer, C. W., and Keighin, C. W., eds., Geological Studies in Support of the U.S. Department of Energy Multiwell Experiment, Garfield County, Colorado: U.S. Geological Survey Open-File Report 84-757, p. 110–120.

Collins, B. A., 1976: "Coal Deposits of the Carbondale, Grand Hogback, and Southern Danforth Hills Coal Fields, Eastern Piceance Basin, Colorado": Quarterly of the Colorado School of Mines, V. 71, No. 1, 138 p.

Dickinson, W. R., and Snyder, W. S., 1978: "Plate Tectonics of the Laramide Orogeny," in Matthews Vincent, III, ed., Laramide Folding Associated with Basement Block Faulting in the Western United States: Geological Society of America Memoir 151, p. 355–366.

Johnson, R. C., and Nuccio, V. F., 1986, "Structural and Thermal History of the Piceance Creek Basin, Western Colorado, in Relation to Hydrocarbon Occurrence in the Mesaverde Group," in Spencer, C. W., and Mast, R. F., eds., Geology of Tight Gas Reservoirs: American Association of Petroleum Geologists Studies in Geology 24, p. 165–205.

Kaiser, W. R., Hamilton, D. S., Scott, A. R., and Tyler, Roger, 1994a: "Geological and Hydrological Controls on the Producibility of Coalbed Methane": Journal of the Geological Society of London, V. 151, p. 417–420.

Kaiser, W., R., and Scott, A. R., 1996: "Hydrologic Setting of the Williams Fork Formation, Piceance Basin, Colorado, *in* Tyler, Roger, Scott, A. R., Kaiser, W. R., Nance, H. S., McMurry, R. G., Tremain, C. M., and Mavor, M. J., eds., Geologic and Hydrologic Controls Critical to Coalbed Methane Producibility and Resource Assessment: Williams Fork Formation, Piceance Basin, Northwest Colorado: The University of Texas at Austin, Bureau of Economic Geology, topical report prepared for the Gas Research Institute, GRI-95/0532, p. 252–268.

Kaiser, W. R., Scott, A. R., Hamilton, D. S., Tyler, Roger, McMurry, R. G., Zhou, Naijiang, and Tremain, C. M., 1993: Geologic and Hydrologic Controls on Coalbed Methane: Sand Wash Basin, Colorado and Wyoming: The University of Texas at Austin, Bureau of Economic Geology, topical report prepared for the Gas Research Institute under contract no. 5091-214-2261 (GRI-92/0420), 151 p.

Kaiser, W. R., Scott, A. R., Hamilton, D. S., Tyler, Roger, McMurry, R. G., Zhou, Naijiang, and Tremain, C. M., 1994b: Geologic and Hydrologic Controls on Coalbed Methane: Sand Wash Basin, Colorado and Wyoming: The University of Texas at Austin, Bureau of Economic Geology Report of Investigations No. 220, and Colorado Geologic Survey, Department of Natural Resources, Resource Series 30, 151 p.

Kaiser, W. R., Scott, A. R., and Roger, Tyler, 1995: "Geology and Hydrogeology of Coalbed Methane Producibility in the United States: Analogs for the World": Intergas Unconventional Gas Symposium, May 15–16, Tuscaloosa, Alabama, Intergas '95 Short Course, 516 p.

Kukal, G. C., Price, E. H., Hill, R. E., and Monson, E. R., 1992: "Results of Field Verification Tests in the Tight Mesaverde Group: Piceance Basin, Colorado": CER Corporation: U.S. Department of Energy, DOE/MC/24120-312 (DE93000201), 153 p.

Logan, T. L., and Mavor, M. J., 1995: "Western Cretaceous Coal Seam Project Final Report": topical report prepared for the Gas Research Institute, GRI-94/0089, 155 p.

McMurry, R. G., and Tyler, Roger, 1996: "Genetic Stratigraphy and Coal Deposition of Progradational Sequences in the Upper Cretaceous Williams Fork Formation, Mesaverde Group, Piceance Basin, Colorado," *in* Tyler, Roger, Scott, A. R., Kaiser, W. R., Nance, H. S., McMurry, R. G., Tremain, C. M., and Mavor, M. J., Geologic and Hydrologic Controls Critical to Coalbed Methane Producibility and Resource Assessment: Williams Fork Formation, Piceance Basin, Northwest Colorado: The University of Texas at Austin, Bureau of Economic Geology, topical report prepared for the Gas Research Institute under contract no. 5091-214-2261, GRI-95/0532, p. 112–148.

Nuccio, V. F., and Johnson, R. C., 1983: "Preliminary Thermal Maturity Map of the Cameo-Fairfield or Equivalent Coal Zone in the Piceance Creek Basin, Colorado": U.S. Geological Survey Miscellaneous Investigations Series Map MF-1575, scale 1:253,440.

Reinecke, K. M., Rice, D. D., and Johnson, R. C., 1991: "Characteristics and Development of Fluvial Sandstone and Coalbed Reservoirs of Upper Cretaceous Mesaverde Group, Grand Valley Field, Colorado," *in* Schwochow, S. D., Murray, D. K., and Fahy, M. F., eds., Coalbed Methane of Western North America: Rocky Mountain Association of Geologists, p. 209–225.

Scott, A. R., 1996: "Coalification, Cleat Development, Coal Gas Composition and Origins, and Gas Content of Williams Fork Coals in the Piceance Basin, Colorado," *in* Tyler, Roger, Scott, A. R., Kaiser, W. R., Nance, H. S., McMurry, R. G., Tremain, C. M., and Mavor, M. J., eds., Geologic and Hydrologic Controls Critical to Coalbed Methane Producibility and Resource Assessment: Williams Fork Formation, Piceance Basin, Northwest Colorado: The University of Texas at Austin, Bureau of Economic Geology, topical report prepared for the Gas Research Institute under contract no. 5091-214-2261, GRI-95/0532, p. 220–251.

Scott, A. R., and Ambrose, W. A., 1992: "Thermal Maturity and Coalbed Methane Potential of the Greater Green River, Piceance, Powder River, and Raton Basins (abs.)," *in* "Calgary: American Association of Petroleum Geologists 1992 annual convention official program": American Association of Petroleum Geologists, p. 116.

Scott, A. R., Kaiser, W. R., and Ayers, W. B., Jr., 1994a: "Thermogenic and Secondary Biogenic Gases, San Juan Basin, Colorado and New Mexico—Implications for Coalbed Gas Producibility: American Association of Petroleum Geologists Bulletin, V. 78, No. 8, p. 1186–1209.

Scott, A. R., Kaiser, W. R., and Tyler, Roger, 1996a, Development and Evaluation of a Basin-Scale Coalbed Methane Producibility Model; San Juan, Sand Wash, and Piceance Basins, Rocky Mountain Foreland," *in* Tyler, Roger, Scott, A. R., Kaiser, W. R., Nance, H. S., McMurry, R. G., Tremain, C. M., and Mavor, M. J., eds., Geologic and Hydrologic Controls Critical to Coalbed Methane Producibility and Resource Assessment: Williams Fork Formation, Piceance Basin, Northwest Colorado: The University of Texas at Austin, Bureau of Economic Geology, topical report prepared for the Gas Research Institute, GRI-95/0532, p. 310–407.

Scott, A. R., and Tyler, Roger, 1998, Geologic and hydrologic controls critical to coalbed methane production and resource assessment - The United States experience: Analogs useful for Australian coal basins: The Texas Bureau of Economic Geology, Coalbed Methane Workshop, in conjunction with the University of Queensland, Australia, and Indiana University, United States of America, International Conference on Coal Seam Gas and Oil (CSGO), Brisbane, Australia, 668 p.

Scott, A. R., Tyler, Roger, Hamilton, D. S., and Naijiang, Zhou, 1994b: "Coal and Coal Gas Resources of the Greater Green River Basin: Application of an Improved Approach to Resource Estimation," in Sonneburg, S. A. (compiler), Rocky Mountain Association of Geologists and Colorado Oil and Gas Association, First Biennial Conference, Natural Gas in the Western United States, Lakewood, Colorado, October 17–18, 1994, unpaginated (4 p).

Scott, A. R., Tyler, Roger, Kaiser, W. R., McMurry, R. G., Nance, H. S., and Tremain, C. M., 1996b: "Coal and Coalbed Methane Resources and Production in the Piceance Basin, Colorado," in Tyler, Roger, Scott, A. R., Kaiser, W. R., Nance, H. S., McMurry, R. G., Tremain, C. M., and Mavor, M. J., eds., Geologic and Hydrologic Controls Critical to Coalbed Methane Producibility and Resource Assessment: Williams Fork Formation, Piceance Basin, Northwest Colorado: The University of Texas at Austin, Bureau of Economic Geology, topical report prepared for the Gas Research Institute under contract no. 5091-214-2261, GRI-95/0532, p. 269–285.

Tremain, C. M., 1982: "Coalbed Methane Potential of the Piceance Basin, Colorado": Colorado Geological Survey Open-File Report 82-1, 49 p., 5 sheets.

Tremain, C. M., and Toomey, J., 1983: "Coalbed Methane Desorption Data": Colorado Geological Survey Open-File Report 81-4, 514 p.

Tremain, C. M., and Tyler, Roger, 1996, "Cleat, Fracture, and Stress Patterns in the Piceance Basin, Colorado: Controls on Coal Permeability and Coalbed Methane Producibility," in Tyler, Roger, Scott, A. R., Kaiser, W. R., Nance, H. S., McMurry, R. G., Tremain, C. M., and M.J. Mavor, eds., Geologic and Hydrologic Controls Critical to Coalbed Methane Producibility and Resource Assessment: Williams Fork Formation, Piceance Basin, northwest Colorado: The University of Texas at Austin, Bureau of Economic Geology, topical report prepared for the Gas Research Institute under contract no. 5091-214-2261, GRI-95/0532, p. 33–81.

Tweto, Ogden, 1980: "Summary of Laramide Orogeny in Colorado," in Kent, H. C., and Porter, K. W., eds., Colorado Geology: Rocky Mountain Association of Geologists, p. 131–132.

Tyler, R., Kaiser, W. R , Scott, A. R., Hamilton, D. S., and Ambrose, W. A., 1995: Geologic and Hydrologic Assessment of Natural Gas from Coal: Greater Green River, Piceance, Powder River, and Raton Basins, Western United States: The University of Texas at Austin, Bureau of Economic Geology Report of Investigations No. 228, 219 p.

Tyler, Roger, Kaiser, W. R., Scott, A. R., Hamilton, D. S., McMurry, R. G., and Naijiang, Zhou, 1994: "Geologic and Hydrologic Assessment of Natural Gas from Coal Seams in the Mesaverde Group and Fort Union Formation, Greater Green River Basin, Wyoming and Colorado": The University of Texas at Austin, Bureau of Economic Geology, topical report prepared for the Gas Research Institute, GRI-93/0320, 120 p.

Tyler, Roger, Scott, A. R., and Kaiser, W. R., 1997, Defining coalbed gas exploration fairways in low permeability, hydrocarbon overpressured basins: an example from the Piceance Basin, northwest Colorado, in Proceedings, International Coalbed Methane symposium: Tuscaloosa, The University of Alabama, p. 527–539.

Tyler, Roger, Scott, A. R., Kaiser, W. R., and McMurry, R. G., 1997, The application of a coalbed methane producibility model in defining coalbed methane exploration fairways and sweet spots: examples from the San Juan, Sand Wash, and Piceance Basins: The University of Texas at Austin, Bureau of Economic Geology, Report of Investigations No. 244, 59 p.

Tyler, Roger, Scott, A. R., Kaiser, W. R., Nance, H. S., McMurry, R. G., Tremain, C. M., and Mavor, M. J., 1996: "Geologic and Hydrologic Controls Critical to Coalbed Methane Producibility and Resource Assessment: Williams Fork Formation, Piceance Basin, Northwest Colorado": The University of Texas at Austin, Bureau of Economic Geology, topical report prepared for the Gas Research Institute, GRI-95/0532, 398 p.

IMPROVING COAL GAS RECOVERY WITH MICROBIALLY ENHANCED COALBED METHANE

ANDREW R. SCOTT
Bureau of Economic Geology
The University of Texas at Austin
Austin, Texas, USA

1. Abstract

Microbially enhanced coalbed methane (MECoM) imitates and enhances the natural process of secondary biogenic gas generation in coal beds that occurs in coal basins worldwide. MECoM involves the introduction of anaerobic bacterial consortia, which consists of hydrolyzers, acetogens and methanogens, and/or nutrients into coalbed methane wells. Coalbed methane production may increase through generation of additional methane, removal of pore-plugging coal waxes, and permeability enhancement as cleat-aperture size increases during biogasification. The amount of coal gas potentially generated by MECoM is large. If only one-hundredth of 1% (1/10,000) of U.S. (lower 48) coal resources were converted into methane using MECoM, gas resources would increase by 23 Tcf, or approximately 16% of current lower 48 nonassociated reserves. However, coal surface area and biogasification reaction rates in the subsurface may potentially limit gas generation, indicating that permeability enhancement may be the most significant benefit of MECoM. Additional research, including microbial sampling of deeply buried bituminous coals to identify genetically unique bacterial consortia, is required to fully evaluate MECoM and determine if the process will improve coalbed methane producibility. Successful implementation of MECoM requires an integrated approach towards understanding the geologic, hydrologic, organic and inorganic geochemistry, microbiological, and engineering factors that may limit MECoM in the subsurface. If economically feasible, MECoM can generate methane in coal beds that currently have limited coalbed methane potential, and thereby provide cheap, environmentally clean energy for many parts of the world.

2. Introduction

"Combustible air" has been known since ancient times, but widespread interest and the scientific foundation for studying biological gas started with the experiments of Alessandro Volta in 1776 who collected "combustible air" by poking his cane in the bottom of marshy sediments (Wolfe, 1993). Although the association of combustible gas with water was initially made by Volta, the microbiological origin of the gas was not discovered until nearly a century later, and the recognition of methanogens as a genetically separate group of bacteria (Archaebacteria) until the late 1970's (Fox and others, 1977; Woese and Fox, 1977). Since the simple experiments of Volta, methanogenic bacteria have been found in a wide variety of anaerobic environments ranging from acidic bogs to deep ocean sediments, from freshwater environments to hypersaline lakes, and from tundra to hot thermal vents over temperatures ranging from 36° to 230°F (2° to 110°C), (Winfrey, 1984; Stettler, 1986; Goodwin and Zeikus, 1987; Kelly and Deming, 1988; Zinder 1993). More recently, anaerobic bacterial consortia were recovered from sidewall cores taken from overpressured (0.51 to 0.56 psi/ft; 11.5 to 12.6 kPa/m) rocks at depths ranging from 8,700 to 9,180 ft (2,650 to 2,798 m), at temperatures between 149° to 167°F (65 to 75°C) (Balkwill and others, 1994), suggesting that bacteria are probably more ubiquitous in sedimentary basins than previously recognized.

Research on Microbially Enhanced Oil Recovery (MEOR) in the late 1940's indicated that anaerobic sulfate-reducing bacteria indigenous to oil reservoirs could displace oil under the appropriate conditions (Jack, 1983). Most MEOR processes involve the injection of carbohydrates and nutrients into wells to stimulate naturally occurring or introduced organisms (Sperl and Sperl, 1993). Enhanced oil recovery is achieved when carbon dioxide and other gases evolved during fermentation repressurize the reservoir and dissolve into the oil to decrease viscosity and surfactants, produced during microbial growth, lower the interfacial tension between oil and water (Jack, 1983).

The acronym for Microbially Enhanced Coalbed Methane (MECoM) is derived from MEOR and applies to the processes by which enhanced coalbed methane recovery is achieved through the introduction of bacterial consortia and nutrients into coal beds. Improved coalbed methane productivity may be accomplished through the generation of additional methane, removal of pore-plugging coalbed waxes and paraffins, and an increase in permeability as coal is removed from cleat surfaces. The objectives of this paper are to briefly review previous work on coal biogasification and microbial processes and to describe possible limitations and benefits of MECoM.

3. Coal Biogasification

After the ability of fungi to degrade coal solids was first recognized in the early 1980's, a wide range of organisms have shown the ability to promote coal solubility (Cohen and Gabriele, 1982). Much of the earlier research focused on using bacteria for the precombustion removal of organic and inorganic sulfur from coal (Olson and others, 1986), whereas subsequent research included the biological conversion of coal into methane (Barik and others, 1988; Barik and others, 1991a, b; ARCTECH, 1990, 1993). Nearly all previous studies on coal biogasification have focused on using bacterial consortia to convert coal into methane, carbon dioxide, or other organic molecules in surface bioreactors. ARCTECH (1990) proposed a two-stage process during which pulverized coal is first solubilized and transferred to a bioreactor where the coal is converted into smaller organic compounds by bacterial consortia. These organic precursors are then transferred to a second bioreactor containing methanogens. Conversion of lignite by bacterial consortia resulted in a direct conversion efficiency of coal carbon into methane ranging from 35 to 50% (ARCTECH, 1990). Preliminary cost estimates for methane production from bioreactors ranges from $2.69 to $9.78USD/Mcf (Barik and others, 1991a).

Lignites and/or subbituminous coals are used in substrates in nearly all coal biogasification experiments, whereas Paleozoic-age bituminous coals are used less frequently. Most bituminous coals are believed to be resistant to biogasification because microbial activity generally decreases with decreasing oxygen content (fig. 1), suggesting that most bituminous coals cannot be biodegraded (ARCTECH, 1990). However, very little, if any, biogasification research has been performed on hydrogen-rich Cretaceous and Tertiary bituminous coals. One bituminous coal in figure 1 has the same biosolubility as lower rank, higher oxygen content coals, suggesting that bacterial degradation of higher rank coals is possible. The highly bioreactive nature of a Texas lignite sample was attributed to a relatively high concentration of sporinite (Barik and others, 1991a). Subsequent research indicates that biogasification of individual macerals in the lignite occurs more rapidly than whole coal biogasification because hydrogen-rich liptinitic macerals, such as sporinite, generally have higher biogasification rates than other macerals. Therefore, additional biogasification research on hydrogen-rich bituminous coals is warranted.

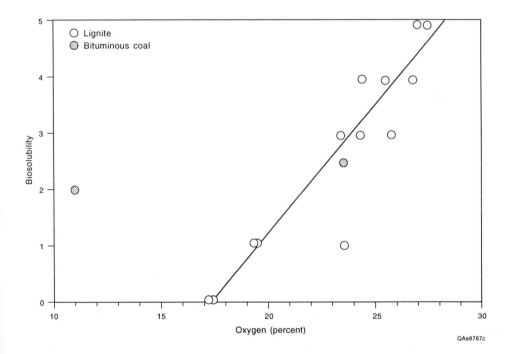

Figure 1. Relation between biosolubility and rank. Biosolubility is ranked from minimum (0) to maximum (5) based on the discoloration of the aqueous medium during the culture period. Biosolubility generally decreases with decreasing oxygen content (increasing rank) although some higher rank coals have higher biosolubility, suggesting that bitiminous coals may be metabolized by bacteria. Modified from Quigley and others (1989).

Although microorganisms have been isolated from coal-associated sources such as coal outcrops, coal slurry pits, and run-off from coal piles (ARCTECH, 1990; Roberts and Crawford, 1991; Scott and others, 1991a), the presence of viable organisms in coal samples has been debated. Vorres (1990) reported the presence of bacteria in Argonne coal samples kept under anaerobic conditions, but Polman and others (1993) examined three coal standards and found no viable bacteria. Polman and others (1993), however, used inoculation levels that were 25 to 1,000 times less than the initial study, suggesting that viable organisms are present but at low concentrations. Microbial consortia capable of generating methane from coal has been isolated from a bituminous coal in an abandoned mine (Volkwein, 1990), providing additional evidence that bituminous coals can be metabolized. Hydrogeologic, gas compositional, and isotopic data from coalbed methane wells in the Upper Cretaceous Fruitland Formation of the San Juan Basin (Kaiser and others, 1991; Scott and Kaiser, 1991; Scott and others, 1991b; Scott and others, 1994) also suggest that microbial activity in bituminous (and lower rank) coals is probably more widespread than previously thought.

Bacteria for biogasification experiments are generally derived from the gastrointestinal tracts of ovines and bovines, leaf litter, chicken waste, termite guts, sewage sludge, and/or animal wastes, and then acclimated to coal samples (ARCTECH, 1990, 1993). The adaptation of some cultures to specific coal samples resulted in a 200 to 300% increase in methane production in laboratory cultures after only 4 mo (ARCTECH, 1990). The relatively short doubling times for bacterial growth, coupled with genetic enhancement through natural selection, suggest that bacteria in coal beds will genetically evolve over a relatively short time span to more efficiently metabolize coal, and that bacterial consortia isolated from coal beds are probably unique genetically.

4. METHANOGENESIS

Aerobic bacteria utilize oxygen as an electron acceptor and are considered to be obligate aerobes if only oxygen can be used as an electron acceptor. Bacteria that utilize oxygen as an electron acceptor when it is available, but are capable of using organic compounds as electron acceptors (fermentation), are called facultative anaerobes (Chapelle, 1993). Obligate anaerobes grow only in the absence of oxygen. Methanogenic bacteria have the reputation of being the strictest of anaerobes, requiring an oxidation/reduction potential of <-0.3 V, which corresponds to 10^{-56} mole per liter of oxygen (Hungate, 1967, in Zinder, 1993).

Microbial degradation of organic compounds often requires the metabolic interaction of microbial consortia, during which individual bacterial species feed off of the products of another species. Three functionally different trophic groups of bacteria (fig. 2) are required to convert organic matter to methane (Winfrey, 1984; McInerney and Beaty, 1988; Thiele and others, 1988): (1) hydrolytic fermentative bacteria, (2) syntrophic acetogenic bacteria, and (3) methanogenic bacteria.

Fermentative bacteria initially hydrolyze complex organic compounds to acetate, longer chained fatty acids, carbon dioxide, hydrogen, NH_4^+, and HS^-. These bacteria are mostly obligate anaerobes, although facultative anaerobes may be present as well. Syntrophic, hydrogen-producing (proton-reducing) acetogenic bacteria reduce intermediary metabolites into acetate, carbon dioxide, and hydrogen. Hydrogen-utilizing acetogenic bacteria

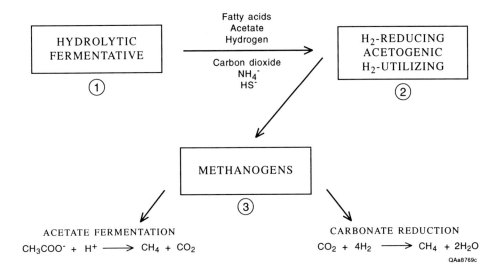

Figure 2. Generalized flow diagram for anaerobic decomposition of organic matter and generation of methane. Modified from Winfrey (1984). Hydrolytic, hydrogen-reducing, acetogenic, and hydrogen-utilizing bacteria provide the organic compounds metabolized by methanogens.

demethoxylate low molecular weight ligneous compounds and ferment some hydroxylated aromatic compounds (McInerney and Beaty, 1988). Methanogenic bacteria are dependent on hydrogen produced by the other bacteria to reduce carbon dioxide (or bicarbonate) to methane. The hydrogen partial pressure in methanogenic environments is low, ranging between 10^{-5} to 10^{-4} atm (1 and 10 Pa) (Thauer, 1990) because of the great capacity methanogens have for hydrogen. If the methanogenic populations become stressed, hydrogen increases, making reactions thermodynamically unfavorable for the degradation of organic compounds by hydrogen-producing acetogens (McInerney and Beaty, 1988). Therefore, syntrophic-interspecies hydrogen transfer regulates metabolic efficiency and growth yield of acetogenic and methanogenic bacteria.

Aerobic bacteria are capable of metabolizing a wide variety of organic substrates. One species of Pseudomonas was shown to grow on at least 127 different organic molecules and possibly more; the researchers simply ran out of ideas as to what organic compounds to feed it (Chapelle, 1993). Methanogens, however, are generally limited to hydrogen, carbon dioxide, and simple organic compounds, most of which contain only one atom, including formate, carbon monoxide, methanol, acetate, methylated amines, short-chained alcohols, and methyl mercaptan (Winfrey, 1984; Zinder, 1993). Although one methanogenic species can use up to seven substrates, many other methanogens are highly specialized, using only one or two substrates. This reinforces the importance of other bacteria to provide suitable organic compounds for the methanogens to metabolize.

Carbonate reduction and acetate fermentation are the metabolic pathways for methane generation in natural environments (fig. 3a). Carbonate reduction occurs primarily in, but is not restricted to, marine waters containing sulfate, whereas acetate fermentation is the dominant process by which methane is produced in fresh water (Whiticar and others, 1986). During carbonate reduction, 75% of the hydrogen atoms in the methane are derived from formation water and not hydrogen gas, as suggested by equation 1 (Whiticar and others, 1986). Furthermore, bicarbonate ions, in addition to carbon dioxide, are involved in metabolic processes. Finally, carbon dioxide is not necessarily consumed in an overall reaction involving multiple bacterial species. Carbon dioxide/bicarbonate is used in intermediary steps such that the net chemical reaction involving two or more bacterial species may result in a net gain of carbon dioxide (bicarbonate ions) rather than the removal of carbon dioxide from the system (Zinder, 1993)(table 1). This explains why coal gasps in the San Juan Basin contain high concentrations (>10%) of isotopically heavy carbon dioxide, whereas the black warrior basin contains < 1% of isotopically heavy carbon dioxide.

Acetate fermentation, which involves the reduction of acetate by methanogenic bacteria, is a different process from the creation of acetate by syntrophic acetogenic bacteria discussed previously. Unlike carbonate reduction, the hydrogen in methane generated through acetate fermentation is derived primarily from the methyl group, and only 25% of the hydrogen is obtained from the formation water (Whiticar and others, 1986). During acetate fermentation, the fermentation of acetate, through reduction of the methyl group to methane, is coupled with the oxidation of the carbonyl group to carbon dioxide (fig. 3b).

(a)   (b)

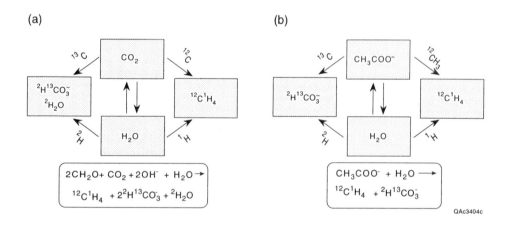

Figure 3. Bacteria metabolic pathways. Although carbonate reduction occurs primarily in marine waters containing sulfate, several lines of evidence suggest that carbonate reduction is probably the preferred metabolic pathway in coal beds. Bacteria derive nearly all of the hydrogen from formation water to produce methane, resulting in methane depleted in ^{13}C and ^2H (a). Acetate fermentation may also occur in some coal beds, but may not be the dominant metabolic process in the subsurface; three of the hydrogens in the methane are derived from the organic matter and only one hydrogen comes from the formation water (b). Both metabolic process generate isotopically light methane (^{12}C and ^1H) and isotopically heavy bicarbonate (^{13}C and ^2H).

TABLE 1. Methanogenesis involves several types of bacteria working in a syntrophic relationship. Methanogenesis is occurring by carbonate reduction in both examples below; bicarbonate (or carbon dioxide) is being consumed to produce methane. However, the sum of reactions involving multiple bacterial reactions consumes bicarbonate in Example 1 and generates bicarbonate in Example 2. Therefore, secondary biogenic gases may be carbon dioxide rich or carbon dioxide poor, depending on the bacterial consortia present in the subsurface. Equations from Zinder (1993).

Example 1:

"S" Organism	$2 \text{ Ethanol} + 2H_2O \rightarrow 2 \text{ Acetate} + 2H^+ + 4H_2$
Methanogen	$4H_2 + HCO_3^- + H^+ \rightarrow CH_4 + 3H_2O$
SUM	$2 \text{ Ethanol} + 2HCO_3^- \rightarrow 2 \text{ Acetate} + CH_4 + H_2O$

Example 2:

S. Wolinii	$4 \text{ Propionate} + 12H_2O \rightarrow 4 \text{ Acetate} + 4 HCO_3^- + H^+ + 12H_2$
Methanogen	$12H_2 + 3HCO_3^- + 3H^+ \rightarrow 3CH_4 + 9H_2O$
SUM	$4 \text{ Propionate} + 3H_2O \rightarrow \text{Acetate} + HCO_3^- + H^+ + 3CH_4$

A variety of factors, including types of bacteria species, relative abundance of organic substrates available for methanogenesis, and the partial pressure of hydrogen, determine which metabolic pathway is more favorable for methane generation (Winfrey, 1984). The processes of carbonate reduction and acetate fermentation can be distinguished using isotopic data. Methane δD values for biogenic methane derived from carbonate reduction and acetate fermentation range from -250 to -170 ‰ and -400 to -250%, respectively (Whiticar and others, 1986). Methane δD values can provide information concerning bacterial metabolic pathways and gas origin but may be unreliable where gas mixing has occurred. Because most of the hydrogen required for carbonate reduction is derived from formation water, hydrogen fractionation during carbonate reduction produces methane depleted in ^{13}C and deuterium and formation waters that become progressively enriched in ^{13}C and deuterium.

5. MICROBIALLY ENHANCED COALBED METHANE

Primary biogenic methane that is generated shortly after deposition during early coalification stages is probably not retained by peat in large quantities because of low pressures and the high moisture content of peat. Secondary biogenic gases are generated after burial, coalification, and subsequent uplift and erosion along basin margins (Scott, 1993; Scott and others, 1994). Data from the San Juan Basin in Colorado and New Mexico provide the best evidence for secondary biogenic gas generation. Overpressure in the north part of the basin is artesian in origin and probably developed during the middle Pliocene (Kaiser and others, 1991); a carbon-14-age date of 31,176 yr (Mavor and others, 1991) is additional evidence for geologically young formation waters. Evidence for the generation of secondary biogenic gases when bacteria were transported into permeable coal beds during meteoric recharge include methane, carbon dioxide, and bicarbonate isotopic values, abrupt change in coal-gas chemistry, and the "biodegraded" n-alkane pattern of some coal-extract samples in the northern San Juan Basin (Kaiser and others, 1991; Scott and Kaiser, 1991; Scott and others, 1994).

MECoM imitates and enhances the natural processes of secondary biogenic gas generation through the introduction of nutrient and/or trace elements and/or anaerobic bacterial consortia into coalbed methane wells to stimulate coalbed methane production. Anaerobic bacterial consortia can probably be collected from produced formation waters and/or fresh sidewall and whole core samples; pressurized cores present the best opportunity to collect viable consortia. Although methanogens will not grow or generate methane in the presence of even trace quantities of oxygen, they can remain viable in the presence of oxygen for as much as 24 hr by forming multicellular lumps (Kiener and Leisinger, 1983). Anoxic/reducing microenvironments in an oxygenated system can potentially extend viability longer. This oxygen tolerance of methanogenic bacteria has important implications in the recovery of microbes from subsurface cores and water samples.

The relatively quick adaptation of bacteria to local environmental conditions suggests that consortia collected from basins, or possibly individual coal beds, may be genetically unique. Once collected, these bacteria can be grown in laboratory cultures to evaluate and determine factors enhancing and/or limiting the conversion of coal into methane. Genetic engineering of bacteria may also enhance biogasification of coal. Recombinant DNA techniques in the early 1970's were used to engineer aliphatic (n-alkane) hydrocarbon-degrading bacteria (Chakrabarty and others, 1973; Chakrabarty, 1972; Chakrabarty and others, 1978). With the progressive development of genetic engineering technology, biologists are now capable of genetically engineering microorganisms to have abilities beyond their "normal" capacities (Bayer, 1991). Special bacterial consortia adapted over time in the laboratory to efficiently metabolize coal and/or genetically engineered bacteria along with appropriate nutrients when injected into coal beds may enhance coalbed methane production. However, there remain many unknowns about the technical aspects of microbially processing coal in the subsurface and whether the technology is economically feasible.

5.1 POSSIBLE BENEFITS OF MECoM

The amount of biogenic gas that could be generated from coal using MECoM is staggering. If one weight percent of coal is converted into methane using bacterial consortia, gas content would increase by 488 scf/ton (15.2 cm^3/g). For a 32-ft (10-m)-thick, ash-free coal bed having a density of 1,659 tons/acre-ft (1.22 g/cc) and covering 10 acres (40,468 m^2), over 265,700 Mcf (7,400 m^3) of methane would be produced. At \$2.00 USD/Mcf this comes to over \$531,000 (USD). If only one-hundredth of 1% (1/10,000) of U.S. coal resources were converted into methane using MECoM, gas reserves would increase by 23 Tcf (Scott, 1995; Scott and Kaiser, 1995), or approximately 10% of current reserves. MECoM could therefore have a significant impact on U.S. gas reserves.

The quantity of gas generated during MECoM ultimately depends on bacterial access to the coal and the surface area of coal exposed to the bacteria. Permeability in coal beds is restricted to an interconnected fracture or cleat system in the coal because the coal matrix is impermeable (fig. 4). Face cleats develop first, are more continuous, and usually have the higher permeability, depending on in situ stresses. Butt cleats form an abutting relation with the face cleats and are less continuous, resulting in lower permeability. Face-cleat spacing is highly variable, ranging from <0.05 cleats per inch (0.02 cleats/cm) in lignites (W. R. Kaiser, personal communication, 1994), to >100 cleats per inch (40 cleats/cm) in vitrain bands (Close, 1993; Laubach and others, 1998). Cleat spacing is highly variable and

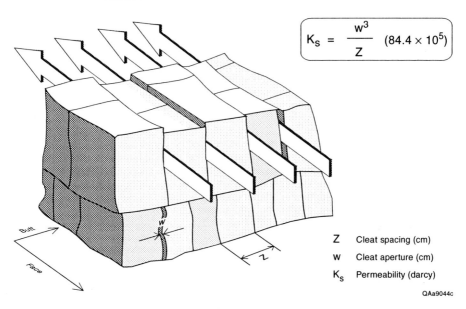

$$K_s = \frac{w^3}{Z} \, (84.4 \times 10^5)$$

Z Cleat spacing (cm)

w Cleat aperture (cm)

K_s Permeability (darcy)

QAa9044c

Figure 4. Permeability is restricted to the face and butt cleats in coal beds. The face cleats are the first formed and tend to be more continuous than the butt cleats that form an abutting relationship with the face cleats. Permeability in coals and carbonate rocks is similar because the matrix between the cleats (fractures) is impermeable. Therefore, permeability can be estimated using the equation of Lucia (1983), which relates permeability to aperture size and the number of fractures.

depends on maceral type and ash content, although cleat spacing generally increases with increasing coal rank through high-volatile bituminous coal (Law, 1993).

Fracture permeability in coal beds ranges from microdarcys to darcys. Direct permeability measurements on coal are difficult or impossible to obtain, but indirect measurements suggest that permeability in many producing coalbed methane wells generally ranges between 1 and 100 md. The relation among fracture (or cleat) spacing, permeability, and fracture aperture was developed by Lucia (1983) for carbonate rocks. The equation relating these three variables can be used to estimate cleat-aperture size in the subsurface because both fractured carbonate and coal reservoirs are characterized by a fracture network separated by an impermeable matrix.

Assuming a cleat spacing of 10 cleats/inch (3.9 cleats/cm) and permeabilities between 1 and 100 md, cleat-aperture width in the subsurface probably ranges from 3 to 15 microns (fig. 5). If the permeability range (1 to 100 md) remains the same, but cleat spacing is decreased to 1 cleat/inch (0.4 cleat/cm), cleat-aperture size would range between 7 and 30 microns. These estimated ranges of cleat apertures are well within the size range of many bacteria (1 to 3 microns in diameter), and are much larger than ultramicrobacteria, which have diameters of <0.3 microns (Torella and Morita, 1981; Zinder, 1993). Therefore, the smallness of bacteria indicates that bacteria can be injected into low-permeability coal reservoirs and can be expected to access the coal through a cleat aperture <1 micron wide.

In addition to generating additional methane, MECoM may increase coalbed methane production by increasing permeability. Because bacteria are capable of accessing the coal

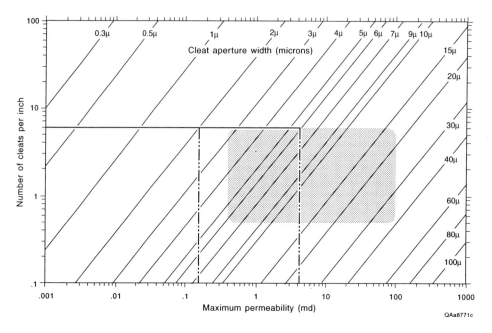

Figure 5. Relation among face-cleat spacing, permeability, and face cleat (Scott, 1995). Cleat tortuosity will decrease fluid permeability, suggesting that cleat-aperture size in the subsurface may be larger than the apertures indicated on this figure.

only through the cleat system, cleat-aperture size will increase as the bacteria convert coal into methane (fig. 6). For example, coal having a cleat spacing of 6 cleats/inch (2.4 cleats/cm) and a cleat aperture of 2 microns has an estimated permeability of 0.16 md (fig. 5). If bacteria remove 2 microns from each side of the cleat, the cleat aperture increases to 6 microns and the permeability to 4.31 md (figs. 5 and 6). Because permeability increases with the cube of cleat-aperture size, permeability enhancement is more pronounced in low-permeability reservoirs having narrow cleat apertures than in higher permeability reservoirs, assuming the same volume of coal is removed in each reservoir (fig. 7). For example, if the cleat width in a 1-md reservoir is increased by 1 micron, permeability will double, whereas a 1-micron increase in cleat-aperture size in a 1,000-md reservoir will have negligible effect on permeability (fig. 7).

5.1.1. *Well Remediation*

Chemically wet gases, condensate, and waxes are generated from hydrogen-rich coals during coalification and are produced from coalbed methane wells. These organic compounds can be metabolized by bacteria and may represent an important food source for bituminous coals. Significant quantities of these compounds can be generated from high-volatile hydrogen-rich coals. Excessive wax production in parts of the San Juan Basin has resulted in some coalbed methane wells being temporarily shut in. Coal is a complex organic

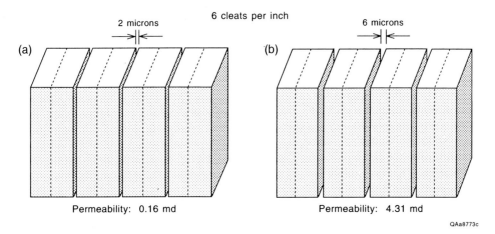

Figure 6. Changes in face-cleat-aperture size during microbially enhance coalbed methane (MECoM). Removal of coal from cleat surfaces during MECoM will increase cleat aperture, resulting in permeability enhancement (Scott, 1995).

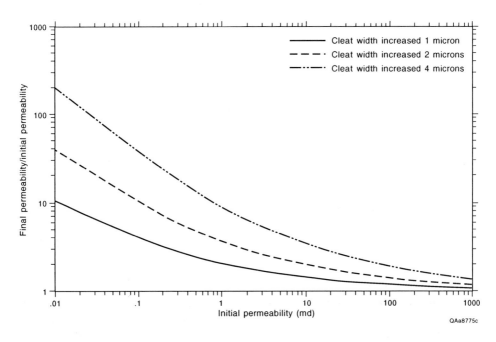

Figure 7. Potential permeability enhancement during MECoM assuming 6 cleats/inch. Permeability enhancement may be more important to coalbed methane producibility than the total amount of methane produced. The ratio of final permeability (after MECoM) to initial permeability (before MECoM) is much greater in low-permeability coal reservoirs (permeability <1 md).

substance, and organic solvents injected into coalbed methane wells to remove coal wax have lowered permeability and damaged the reservoir rather than enhanced producibility. Coal gases and organic compounds are readily sorbed onto the coal surface and may result in a decrease in relative permeability (Penny and Conway, 1993) or coal swelling, and/or decrease sorption capacity, depending on the organic solvent and coal properties.

Waxes from coalbed methane wells in the San Juan Basin are composed predominantly of long-chained n-alkanes (Vorkink and Lee, 1993), suggesting that they may be susceptible to biodegradation. Although the lighter weight n-alkanes ($<C_{20}$) are preferentially removed first during biodegradation, heavier weight n-alkanes (C_{15} to C_{35}) are biodegraded before branched alkanes (Hunt, 1979; Connan, 1984). Bacteria capable of metabolizing pore-occluding n-alkanes and waxes in coal beds to generate biogenic methane may also modify the coal-surface structure to increase methane adsorptive capacity and total gas content. When bacteria metabolize and remove wet gases, n-alkanes, and other pore-plugging organic compounds on the coal surface during biogasification, the methane generated during MECoM may be readily sorbed onto the sorption sites previously occupied by the pore-plugging organic compounds. In a sense, bacteria can be thought of as nature's soxhlet extractor for coal beds because they are capable of removing soluble organic compounds from the coal. Unlike organic solvents, bacterial consortia may remove organic compounds without damaging the reservoir.

5.2 POSSIBLE LIMITATIONS OF MICROBIALLY ENHANCED COALBED METHANE

5.2.1 *Biodegradation of Organic Compounds*

Methanogenic bacteria are just one component of complex trophic consortia (consortia having specific nutritional levels) required to convert coal into methane under anaerobic conditions. Much research has been performed on the aerobic degradation of hydrocarbons in oil reservoirs and organic compounds during bioremediation (Hunt, 1979; Connan, 1984; Rittman and others, 1992). Anaerobic degradation of hydrocarbons and organic compounds occurs naturally, and biodegradation of many oils in deeper reservoirs probably occurred under anaerobic conditions rather than aerobic; even methane oxidation occurs under aerobic and anaerobic conditions (Zhender and Brock, 1980; Coleman and others, 1981; Whiticar and Faber, 1986).

Most oil samples are composed predominantly of organic compounds that are relatively easily biodegradable, whereas coals are composed of an interconnected network of hydroaromatic and cycloalkanes, which are less easily biodegradable. The resistance of lignin and lignocellulose to biodegradation in anoxic environments results in the accumulation of peat and eventually coal upon maturation. However, recent studies indicate that lignin, lignocellulose, and other complex organic compounds thought to be representative of the coal structure can be biodegraded under anaerobic conditions (Colberg and Young, 1985; Benner and others, 1984; Crawford and others, 1990; Davison and others, 1990). Therefore, the complexity and nature of the coal structure should not be used to dismiss the possibility of MECoM to improve coalbed methane producibility. However, if waste products toxic to bacteria (generated during coal biogasification) are not removed from the system, MECoM recovery could be reduced or inhibited.

5.2.2. *Temperature*

Methanogenic bacteria are found at temperatures ranging from 36° to 230°F (2° to 110°C) (Winfrey, 1984; Stettler, 1986; Kelly and Deming, 1988; Zinder, 1993), but the optimum temperature range for biogenic gas generation is 95° to 113°F (35° to 45°C) (Zeikus and Winfrey, 1976). Muralidharan and others (1991) suggested that higher microbial reaction rates may be achieved at higher temperatures. How long the bacteria present at the time of deposition can survive during burial and thermal maturation remains unknown at this time. At very high temperatures and/or in the absence of food sources, or without appropriate nutrients over extended periods of time, bacteria will probably not survive. However, bacteria may remain viable, though dormant during burial, and when environmental conditions become more favorable during uplift and erosion, growth is reinitiated (Clayton, 1992). Therefore, the full effects of temperature on in situ biogasification of coal remain uncertain but are an important consideration for MECoM.

5.2.3. *Cleat Surface Area*

Another important factor affecting MECoM is the surface area available for bioconversion of coal into methane. A large coal surface area available for coal biogasification will result in higher methane yields during MECoM. An evaluation of the total surface area based on cleat spacing can therefore be used to estimate the amount of gas generated during MECoM. A bituminous ash-free coal having a density of 1.22 g/cc (1,659 tons/acre-ft) will occupy a volume of 26.26 ft³ (0.744 m³) equivalent to a cube 35.67 inches (0.906 m) per side. Therefore, each face cleat in a ton of coal will have a surface area equal to 17.67 ft² (1.641 m²), and the total cleat surface area per ton of coal will increase linearly with decreasing cleat spacing (fig. 8). A 32.8-ft-thick (10-m), ash-free bituminous coal covering 10 acres (40,468 m²) will contain 544,226 tons (493,714 t) of coal (fig. 9). Assuming a cleat spacing of 6 cleats/inch (2.4 cleats/cm), the total face-cleat surface area would be 2.06×10^9 ft² (1.91×10^8 m²), whereas a cleat spacing of 10 cleats/inch (3.9 cleats/cm) would have a total face-cleat surface area of 3.43×10^9 ft² (3.18×10^8 m²). If the cleat-aperture size of the butt cleats is large enough for bacterial access, then the total cleat surface areas calculated above would be doubled, assuming similar cleat spacing.

The amount of gas generated during MECoM is proportional to the surface area and the amount of coal removed from either side of the cleat during biogasification. A previous example demonstrated that permeability in a bituminous coal having a cleat spacing of 6 cleats/inch (2.4 cleats/cm) and an initial cleat aperture of 2 microns would increase from 0.16 to 4.31 md if bacteria remove 2 microns from either side of the cleat (4 microns total). The amount of methane generated during MECoM can be estimated using cleat surface area and the volume of coal converted into methane.

One ton of ash-free bituminous coal occupies a volume of 26.25 ft³ (0.744 m³), equivalent to a cube 35.67 inches (90.60 cm) on a side. Therefore, a total of 214 cleats are in 1 ton of coal and the total surface area is 351 m² (fig. 8). Assuming 2 microns of coal are removed from both sides of each cleat, the total volume of coal would be 42.9 inches³ (7.026×10^{-4} m³), or 0.095% of the total weight of the coal. The total yield for a one-weight-percent conversion of coal into methane is estimated to be 488 scf/ton (15.3 cm³/g), indicating that the total methane generation from 0.095% conversion is between 46 and 92 scf/ton (1.4 and 2.9 cm³/g) (table 2). The conversion of 10 microns per cleat of coal into

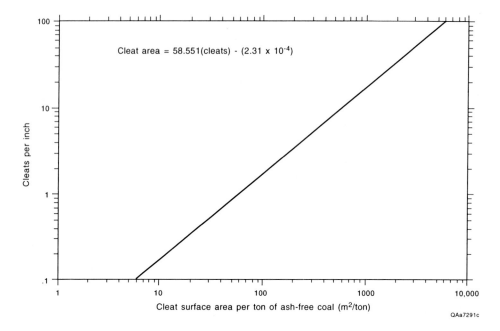

Figure 8. Relation between cleat spacing and cleat surface area per ton of ash-free bituminous coal. Face-cleat and butt-cleat spacing may differ slightly, and the total butt-cleat surface area may be partially inaccessible to bacteria. Equations in Scott and others (1995) can be used to correct cleat surface area (coal volume) for weight percent ash.

methane in a coal with a cleat spacing of 10 cleats/inch (3.9 cleats/cm) would yield between 192 and 384 scf/ton (6.0 and 12.0 cm^3/g). However, the preceding examples assume that bioconversion of coal generates only methane and no other gases. Therefore, the actual amount of methane generated per ton of coal during MECoM could be significantly less than the amount indicated in table 2.

5.2.4   Reaction Rates

Another major factor affecting MECoM economics is the rate at which bacterial consortia will convert coal into methane. If the conversion rate is too slow, the process will be uneconomic regardless of the quantity of gas generated or how much permeability is increased. Unfortunately, insufficient data are available to fully evaluate laboratory reaction rates in the context of the subsurface environment and to extrapolate those reaction rates right now.

Laboratory experiments on the direct conversion of lignite by a bacterial consortia resulted in a direct conversion efficiency of coal carbon into methane ranging from 35 to 50%, and found that methane generation began soon after the coals were inoculated with bacteria and increased steadily for 2 to 3 wk (ARCTECH, 1990). One experiment using bacterial consortia especially adapted to converting coal into methane achieved a peak methane-generation rate of 30.5 cm^3/g of coal/d (978 scf/ton/d) after only 7 d; methane production subsequently declined after 21 d (ARCTECH, 1993) (fig. 10). However, coal

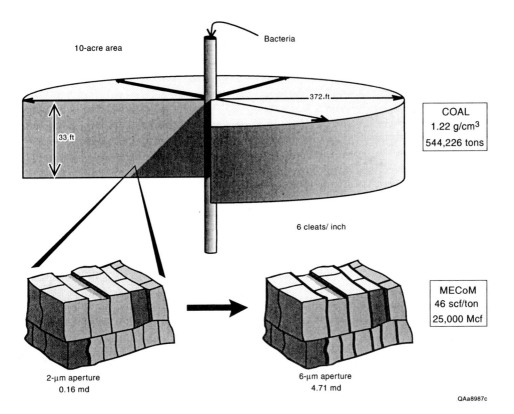

Figure 9. Volumetrics of MECoM. Assuming that bacteria injected into a 33-ft-thick (10-m) coal bed though a well bore will be transported and/or migrate only 322 ft (113 m), the total volume of coal affected by the bacteria will be 544,226 tons (246,815 t). If the bacteria generate an average of 46 scf/ton (1.4 cm^3/g), then approximately 25,000 Mcf (696 m^3) of bacterial coal gas is added to the existing gases already sorbed onto the coal. Note that in the MECoM process, permeability will be increased over one order of magnitude.

TABLE 2. Estimated quatitites of methane generated during MECoM. Calculations assume that methane is the only gas generated during biogasification. Other assumptions are discussed in text.

Available surface area	scf/ton[b]	scf/ton/cleat/micron[c]
Face cleat only	46.13	1.92216
Face cleat and partial butt cleat[a]	69.19	2.88325
Face and butt cleat	92.26	3.84433

[a] assumes butt cleats are partially accessible to bacteria, therefore having only half of the surface area of the face cleat.
[b] estimated based on 6 cleats/inch and 4 microns of coal per each cleat converted into methane
[c] "micron" refers to how many microns of coal in each cleat are converted into methane; for the example discussed in the text: (1.992216)(4 microns) (6 cleats) = 48.128 scf/ton

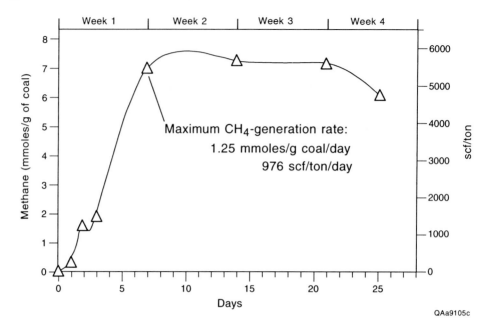

Figure 10. Laboratory experimental results for coal biogasification experiments (modified from ARCTECH, 1993). Gas generation increases for 7 d, stabilizes, then declines after 21 d. The maximum gas generation rate is 976 scf/ton/d (1.25 mmoles/g coal/d) and over 5,500 scf/ton (171 cm³/g) are generated over 7 d.

biogasification experiments are performed on crushed coal samples having a large net surface area and, therefore, much higher reaction rates. Additionally, laboratory experiments are performed under controlled conditions, whereas bacterial consortia utilized in MECoM would have to perform under more adverse conditions.

Assuming that reaction rates in the subsurface are controlled primarily by total surface area, subsurface reaction rates can be estimated from laboratory reaction rates with known surface areas. The generation of gas during MECoM in the subsurface is roughly proportional to the total surface area of the coal in the subsurface that is available for biodegradation relative to the coal surface areas used during the laboratory experiments (fig. 11). Based on the sieve size used during the laboratory experiments, the volume (Vc) and surface areas (Ac) of a single cube of coal are estimated to be 8.5×10^{-8} cm³ and 5.8×10^{-5} cm², respectively (fig. 11). From the volume of the coal cubes, there are approximately 8.7×10^{12} cubes in 1 ton of coal and the total surface area of all the cubes could be 50,700 m². The total surface area of 1 ton of coal in the subsurface containing 6 cleats per inch is 351 m². Therefore, the amount of gas generated from MECoM over 7 d (figs. 10 and 11) is 38 scf/ton (1.2 cm³/g) or 20,681 Mcf (576 m³) for a 32.8-ft-thick (10-m), ash-free bituminous coal covering 10 acre (40,468 m²).

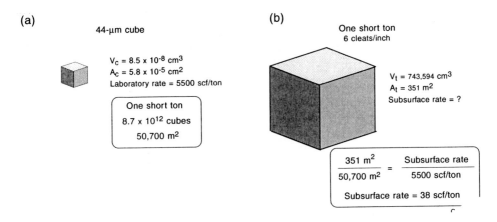

Figure 11. Extrapolation of laboratory results to the subsurface. Assuming that MECoM gas generation rate is directly proportional to the surface area available for bacterial activity, the subsurface reaction rate can be estimated. For a large volume of coal (fig. 9), the total gas generated over 7 d is estimated to be 20.7 MMcf (644 Mm³).

If the laboratory reaction rates discussed above are reduced to only 0.98 scf/ton/d (0.03 cm³/g of coal/d), the additional gas generated by MECoM could be significant in a large volume of coal. For example, a 32.8-ft (10-m) coal covering an area of 10 acre (40,468 m²) with a total coal tonnage of 544,226 tons (493,714 t) of coal would produce an additional 533 Mcf/d for a total of 195 MMcf in 1 yr from the face-cleat surfaces only (fig. 11); at $2.00USD/Mcf, this comes to an additional $390,000 USD. However, additional bacteria and/or nutrients would probably have to be reinjected periodically to maintain high gas generation rates. Therefore, bacterial consortia would probably not generate gas at high rates consistently over long periods of time without reinoculation, particularly if toxic waste products accumulated.

Bacteria are capable of generating large quantities of methane in relatively short time periods under natural conditions, sometimes with dramatic results. During construction of the Buckman Bridge outside of Jacksonville, Florida, water from the St. Johns River was used to fill hollow bridge pilings (Dunker and Rabat, 1993). The prestressed 2- × 2-ft (61- × 61-cm) concrete pilings with cardboard-lined hollow centers supported piers of the 3.1-mi (5-km) bridge. Anaerobic bacteria, derived from raw sewage dumped into the St. Johns River, metabolized the cardboard liners and generated enough methane to rupture the reinforced concrete pilings and destabilize part of the bridge (Dunker and Rabat, 1993). Divers later found cracks ranging in size from hairline to 10 inches (25.4 cm) wide in 32 pilings; the remaining pilings were vented to prevent additional explosions. Although laboratory data and unusual events such as the Buckman Bridge suggest that coal biogasification rates may be adequate for MECoM, additional research using bacterial consortia derived from bituminous coals and/or consortia genetically adapted to bituminous coals is required to fully evaluate the economic feasibility of MECoM.

6.0   MECoM Summary

MECoM may improve coalbed methane production through the generation of additional methane, removal of pore-plugging coal waxes, and/or permeability enhancement as cleat aperture is enlarged during biogasification. Removal of coal waxes and enlargement of cleat apertures as bacteria convert coal into methane will generate additional methane and increase permeability. Preliminary calculations suggest that permeability enhancement will be more pronounced in low-permeability coal reservoirs than in high-permeability ones and may prove to be the greatest benefit of MECoM.

The genetic development or discovery of bacteria consortia capable of efficiently converting coal into methane, available coal surface area, and biogasification reaction rates, are factors that will impact MECoM development. Previous studies, using genetically adapted bacterial cultures, indicate that 35 to 50% of the coal can be converted to methane in the laboratory, which is equivalent to 17,080 to 24,000 scf/ton (534 to 763 cm^3/g). However, calculations of coal surface area available for biogasification on the basis of cleat spacing indicate that the quantity of methane generated in the subsurface will probably be substantially less. Insufficient data currently exist to extrapolate laboratory biogasification reaction rates to the subsurface, but the results of these previous studies are encouraging.

Research including microbial sampling of deeply buried bituminous coals is required to fully evaluate MECoM and to determine if the process will improve coalbed methane producibility and recovery. In addition to microbiological considerations, a detailed understanding of hydrogeologic factors affecting coalbed methane producibility is critical for the successful implementation of MECoM. Coalbed methane producibility is determined by the complex interplay of coal distribution, coal rank, gas content, permeability, groundwater flow, and depositional and structural setting (Kaiser and others, 1994), which may vary from one part of a basin (or subbasin) to another.

7. Conclusions

(1) MECoM imitates and enhances the natural process of secondary biogenic gas generation in coal beds that occur in basins worldwide through the introduction of anaerobic bacterial consortia into coalbed methane wells.

(2) If only one-hundredth of 1% (1/10,000) of U.S. (lower 48) coal resources were converted into methane using MECoM, gas resources would increase by 23 Tcf, or approximately 16% of current lower 48 nonassociated reserves.

(3) MECoM benefits include the generation of additional coal gas, removal of pore-plugging waxes, and permeability enhancement as cleat apertures are increased during biogasification.

(4) Limitations of MECoM include the total cleat surface area (which is a function of cleat spacing) accessible to bacteria in the subsurface, biogasification reaction rates, very high temperatures (>230°F; 110°C), and the possible generation and/or removal of toxic waste products associated with microbial activity.

(5) Permeability enhancement during MECoM may prove to be more beneficial to coalbed methane producibility than additional gas generation. Permeability enhancement would be more important in low-permeability coal reservoirs than in higher permeability reservoirs.

(6) A detailed understanding of hydrogeologic factors affecting coalbed methane producibility in addition to microbiological considerations is critical for the successful implementation of MECoM.

8. Acknowledgments

The concepts discussed evolved from coalbed methane research funded by the Gas Research Institute under contract numbers 5087-214-1544 and 5091-214-2261. Critical reviews of the manuscript by W. R. Kaiser, Ronald G. McMurry, Muhammad Razi, and Tucker F. Hentz improved the manuscript. Word processing was by Susan Lloyd and layout was by Jamie H. Coggin. Editing was by Amanda R. Masterson and Nina Redmond, under the supervision of Susie Doenges. Drafting was provided by the cartographic staff of the Bureau of Economic Geology under the direction of Joel Lardon. This publication was authorized by the Director, Bureau of Economic Geology, The University of Texas at Austin.

9.0   References Cited

ARCTECH, Inc., 1990, Biological gasification of coals: final report prepared for the Department of Energy under contract no. DE-AC21-87MC23285: DOE/MC/23285-2878 (DE90015337), 130 p.

ARCTECH, Inc., 1993, Biogasification of low-rank coals: final report prepared for the Electric Power Research Institute, EPRI TR-101572, variously paginated.

Balkwill, D., and others, 1994, DOE seeks origin of deep subsurface bacteria: Transactions, American Geophysical Union, v. 75, no. 34, p. 385,395–385,396.

Barik, S., Isbister, J., Hawley, B., Forgacs, T., and Reed, L., 1988, Microbial conversion of coals to clean fuel forms: preprints of papers presented at the 196th American Chemical Society National Meeting, September, v. 33, no. 4, p. 548–553.

Barik, S., Isbister, R., and Berrill, P., 1991a, Coal biogasification research and process development: Proceedings, Second International Symposium on the Biological Processing of Coal: Palo Alto, California, Electric Power Research Institute, EPRI GS-7482, p. 6-23–6-38.

Barik, S., Tiemans, K., and Isbister, J., 1991b, Investigations of metabolites and optimization of bacterial culture for coal biogasification: final report prepared for the Gas Research Institute under contract no. 5088-221-1756 (GRI-91/0051), 67 p.

Bayer, P. E., 1991, Gene transfer in the environment and coal bioprocessing with genetically engineered microorganisms: Proceedings, Second International Symposium on the Biological Processing of Coal: Palo Alto, California, Electric Power Research Institute, EPRI GS-7482, p. 2-17–2-32.

Benner, R., Maccubbin, A. E., and Hodson, R. E., 1984, Anaerobic biodegradation of the lignin and polysaccharide components of lignocellulose and synthetic lignin by sediment microflora: Applied and Environmental Microbiology, p. 998–1004.

Chakrabarty, A. M., 1972, Genetic basis of the biodegradation of salicylate in Pseudomonas: Pseudomonas: Journal of Bacteriology, v. 112, p. 815–823.

Chakrabarty, A. M., Chou, G., and Gunsalus, I. C., 1973, Genetic regulation of octane dissimulation plasmid in Pseudomonas: Proceedings, National Academy of Sciences, v. 70, p. 1137–1140.

Chakrabarty, A. M., Friello, D. A., and Bopp, L. H., 1978, Transposition of plasmid DNA segments specifying hydrocarbon degradation and their expression in various microorganisms: Proceedings, National Academy of Sciences, v. 75, p. 3109–3112.

Chapelle, F. H., 1993, Ground-water microbiology and geochemistry: New York, John Wiley & Sons, 424 p.

Clayton, J. L., 1992, Role of microbial processes in petroleum systems: the petroleum system—status of research and methods: U.S. Geological Survey Bulletin 2007, p. 16–19.

Close, J. C., 1993, Natural fractures in coal, in Law, B. E., and Rice, D. D., eds., Hydrocarbons from coal: American Association of Petroleum Geologists Studies in Geology no. 38, p. 119–132.

Cohen, M. S., and Gabriele, P. D., 1982, Degradation of coal by the fungi Polyporus versicolor and Poria monticoloa: Applied Environmental Microbiology, v. 44, p. 23–27.

Colberg, P. J., and Young., L. Y., 1985, Aromatic and volatile acid intermediates observed during anaerobic metabolism of lignin-derived oligomers: February, Applied and Environmental Microbiology, p. 350–358.

Coleman, D. D., Risatti, J. B., and Schoell, M., 1981, Fractionation of carbon and hydrogen isotopes by methane oxidizing bacteria: Geochimica et Cosmochimica Acta, v. 45, p. 1033–1037.

Connan, Jacques, 1984, Biodegradation of crude oils in reservoirs: Advances in Organic Geochemistry, 1984, v. 1, p. 300–335.

Crawford, D. L., Gupta, R. K., Deobald, L. A., and Roberts, D. J., 1990, Biotransformations of coal and coal substructure model compounds by bacteria under aerobic and anaerobic conditions: Proceedings, First International Symposium on the Biological Processing of Coal: Palo Alto, California, Electric Power Research Institute, EPRI GS-6970, p. 4-29–4-43.

Davison, B. H., Nicklaus, D. M., Misra, A., Lewis, S. N., and Faison, B. D., 1990, Utilization of microbially solubilized coal: Applied Biochemistry and Biotechnology, v. 24/25, p. 447–456.

Dunker, K. F., and Rabat, B. G., 1993, Why America's bridges are crumbling: Scientific American, p. 68.

Fox, G. E., Magrum, M. W., Balch, W. E., Wolfe, R. S., and Woese, C. R., 1977, Classification of methanogenic bacteria by [16]S ribosomal RNA characterization: Proceedings of the National Academy of Science, USA, v. 74, p. 4537–4541.

Goodwin, S., and Zeikus, J. G., 1987, Ecophysiological adaptations of anaerobic bacteria to low pH: analysis of anaerobic degradation in acidic peat bog sediments: Applied Environmental Microbiology, v. 53, p. 57–64.

Hunt, J. M., 1979, Petroleum geochemistry and geology: San Francisco, W. H. Freeman, 617 p.

Jack, T. R., 1983, Enhanced oil recovery by microbial action, in Yen, T. F., Kawahara, F. K., and Hertzberg, R., eds., Chemical and geochemical aspects of fossil energy extraction: Ann Arbor, Ann Arbor Science, 266 p.

Kaiser, W. R., Swartz, T. E., and Hawkins, G. J., 1991, Hydrology of the Fruitland Formation, San Juan Basin, in Ayers, W. B., Jr., eds., Geologic controls on the producibility of coalbed methane, Fruitland Formation, San Juan Basin: Gas Research Institute Topical Report GRI-91/0072, p. 195–241.

Kaiser, W. R., Hamilton, D. S., Scott, A. R., Tyler Roger, and Finley, R. J., 1994, Geologic and hydrologic controls on the producibility of coalbed methane: Journal of the Geological Society of London, v. 151, p. 417–420.

Kelley, R. M., and Deming, J. W., 1988, Extremely thermophilic Archaebacteria: biological and engineering consideration: Biotechnological Progress, v. 4, p. 47–62.

Kiener, A., and Leisinger, T., 1983, Oxygen sensitivity of methanogenic bacteria: Systematic and Applied Microbiology, v. 4, p. 305–312.

Laubach, S. E., Marrett, R. A., Olson, J. E., and Scott, A. R., 1998, Characteristics and origins of coal cleat: a review: International Journal of Coal Geology, v. 35, p. 175–207.

Law, B. E., 1993, The relation between coal rank and cleat spacing: implications for the prediction of permeability in coal: Proceedings, 1993 International Coalbed Methane Symposium, v. I, p. 435–442.

Lucia, F. J., 1983, Petrophysical parameters estimated from visual descriptions of carbonate rocks: a field classification of carbonate pore space: Journal of Petroleum Technology, March, p. 629–637.

Mavor, M. J., Close, J. C., and Pratt, T. J., 1991, Western Cretaceous coal seam project summary of the Completion Optimization and Assessment Laboratory (COAL) site: Gas Research Institute Topical Report, GRI 91/0377, variously paginated.

McInerney, M. J., and Beaty, P. S., 1988, Anaerobic community structure from a nonequilibrium thermodynamic perspective: Canadian Journal of Microbiology, v. 34, p. 487–493.

Muralidharan, V., Hirsh, I. S., Peeples, T. L., and Kelley, R. M., 1991, Mixed cultures of high temperature bacteria: prospects for bioprocessing of fossil fuels: Proceedings, Second International Symposium on the Biological Processing of Coal: Palo Alto, California, Electric Power Research Institute, EPRI GS-7482, p. 6-51–6-67.

Olson, G. J., Brickman, F. E., and Iverson, W. P., 1986, Processing of coal with microorganisms: final report prepared for the Electric Power Research Institute, EPRI AP-4472, variously paginated.

Penny, G., S., and Conway, M. W., 1993, Coordinated studies in support of hydraulic fracturing of coalbed methane: Gas Research Institute Annual Report, GRI 93/0125, variously paginated.

Polman, K. J., Breckenridge, C. R., and Delezene-Briggs, K. M., 1993, Microbiological analysis of Argonne Premium coal samples: Energy and Fuels, v. 7, p. 380–383.

Quigley, D. R., Ward, B., Crawford, D. L., Hatcher, H. J., and Dugan, P. R., 1989, Evidence that microbially produced alkaline materials are involved in coal biosolubilization: Applied Biochemistry and Biotechnology, v. 20/21, p. 753–763.

Rittman, B. E., Valocchi, A. J., Seagren, E., Ray, C., Wrenn, B., and Gallagher, J. R., 1992, A critical review of bioremediation: topical report prepared for the Gas Research Institute, (GRI-92/0322), 152 p.

Roberts, M. A., and Crawford, D. L., 1991, Peroxidase and esterase production in coal depolymerizing Pseudomona strain DLC-62: Proceedings, Second International Symposium on the Biological Processing of Coal: Palo Alto, California, Electric Power Research Institute, EPRI GS-6970, p. 4-5 to 4-22.

Scott, A. R., 1993, Composition and origin of coalbed gases from selected basins in the United States: Proceedings, 1993 International Coalbed Methane Symposium, v. I, p. 207–216.

Scott, A. R., 1995, Limitations and possible benefits of microbially enhanced coalbed methane: Proceedings, International Unconventional Gas Symposium, University of Alabama, Tuscaloosa, p. 423-432

Scott, A. R., and Kaiser, W. R., 1991, Relation between basin hydrology and Fruitland gas composition, San Juan Basin, Colorado and New Mexico: Quarterly Review of Methane from Coal Seams Technology, v. 9, no. 1, p. 10–18.

Scott, A. R., and Kaiser, W. R., 1995, Microbially enhanced coalbed methane: limitations and possible benefits (abs.): American Association of Petroleum Geologists Program, v. 4., p. 86A.

Scott, C. D., Faison, B. D., Woodward, C. A., and Brunson, R. R., 1991a, Anaerobic solubilization of coal by microorganisms and isolated enzymes: Proceedings, Second International Symposium on the Biological Processing of Coal: Palo Alto, California, Electric Power Research Institute, EPRI GS-7482, p. 4-29–4-46.

Scott, A. R., Kaiser, W. R., and Ayers, W. B., Jr., 1991b, Composition, distribution and origin of Fruitland and Pictured Cliffs Sandstone Gases, San Juan Basin, Colorado and New Mexico, in Schwochow, J., ed., Coalbed methane of western North America: Rocky Mountain Association of Geologists Guidebook, p. 93–108.

Scott, A. R., Kaiser, W. R., and Ayers, W. B., Jr., 1994, Thermogenic and secondary biogenic gases San Juan Basin, Colorado and New Mexico-implications for coalbed gas producibility: American Association of Petroleum Geologists Bulletin, v. 78, no. 8, p. 1186–1209.

Scott, A. R., Zhou, Naijiang, and Levine, J. R., 1995, An improved approach to estimating coal and coal gas resources: example from the Sand Wash Basin: American Association of Petroleum Geologists Bulletin, v. 79, p. 1320–1336.

Sperl, P. L., and Sperl, G. T., 1993, New microorganisms and processes for MEOR: final report prepared for the U.S. Department of Energy, DOE/BC/14663-11, 46 p.

Stettler, K. O., 1986, Diversity of extremely thermophilic Archaebacteria, in Brock, T. D., ed., Thermophiles: general, molecular, and applied microbiology: New York, John Wiley & Sons.

Thauer, R. K., 1990, Energy metabolism of methanogenic bacteria: Biochimica et Biophysica Acta, v. 1018, p. 256–259.

Thiele, J. H., Chartrain, M., and Zeikus, J. G., 1988, Control of interspecies electron flow during anaerobic digestion: role of floc formation in syntrophic methanogenesis: Applied and Environmental Microbiology, v. 54, no. 1, p. 10–19.

Torella, F., and Morita, R. Y., 1981, Microcultural study of bacterial size changes and microcolony and ultramicrocolony formation by heterotrophic bacteria in seawater: Applied and Environmental Microbiology, p. 518–527.

Volkwein, J., 1990, Preliminary evidence microbial gasification of Pittsburgh seam coal: Proceedings, First International Symposium on the Biological Processing of Coal: Palo Alto, California, Electric Power Research Institute, EPRI GS-6970, p. P-37–P-43.

Vorkink, W. P., and Lee, M. L., 1993, Appraisal of heavy hydrocarbons in coal seam gas reservoirs: Gas Research Institute Annual Report, GRI 92/0501, 31 p.

Vorres, K. S., 1990, The Argonne premium coal sample program: Energy & Fuels, v. 4, p. 420–426.

Whiticar, M. J., and Faber, E., 1986, Methane oxidation in sediment and water column environments—isotopic evidence: Advances in Organic Geochemistry, v. 10, p. 759–768.

Whiticar, M. J., Faber, E., and Schoell, M., 1986, Biogenic methane formation in marine and freshwater environments: CO_2 reduction vs. acetate fermentation-isotopic evidence: Geochimica et Cosmochimica Acta, v. 50, p. 693–709.

Winfrey, M. R., 1984, Microbial production of methane, in Atlas, R. M., ed., Petroleum Microbiology: New York, Macmillan, Chapter 5, p. 153–220.

Woese, C. R., and Fox, G. E., 1977, Phylogenetic structure of the prokaryotic domain: the primary kingdoms: Proceedings of the National Academy of Science, USA, v. 74, p. 5088–5090.

Wolfe, R. S., 1993, An historical overview of methanogenesis, in Ferry, J. G., ed., Methanogenesis: New York, Chapman & Hall, p. 1–32.

Zeikus, J. G., and Winfrey, M. R., 1976, Temperature limitations of methanogenesis in aquatic sediments: Applied and Environmental Microbiology, v. 31, p. 99–107.

Zhender, A. J. B., and Brock, T. D., 1980, Anaerobic methane oxidation: occurrence and ecology: Applied and Environmental Microbiology, v. 39, p. 194–204.

Zinder, S. H., 1993, Physiological ecology of methanogens, in Ferry, J. G., ed., Methanogenesis: New York, Chapman & Hall, p. 128–206.

COALBED METHANE EXPLORATION IN STRUCTURALLY COMPLEX TERRAIN
A Balance Between Tectonics And Hydrogeology

F.M. Dawson
Geological Survey of Canada – Calgary, Alberta, Canada
3303 33 St. N.W., Calgary, Alberta, T2L 2A7

Abstract

Coalbed methane exploration in the Western Canada Sedimentary Basin has focused mainly on the foothills and mountain regions of Alberta and British Columbia because of the numerous thick coal seams of suitable rank for thermogenic gas generation. Cumulative coal thicknesses in excess of 50 m and gas contents ranging from 12 to 22 cc/g yield in situ resources estimates of more than 1.2 x109 m^3 (42x109 Bcf) per sq. km. These regions are structurally complex and present unique challenges in defining and exploring for the optimum CBM exploration target.

In order to delineate potential reservoir traps, it is necessary to understand the style of deformation and the main structural features where permeability enhancement may occur. The axial regions of anticlines and synclines appear to have the greatest potential. Here, a tensional rather than compressional stress regime exists and the natural fracture system of the coal, (cleating) could be enhanced, resulting in better reservoir permeability. In anticlines the coal gas is trapped in the crest of the structure, where "free" or easily desorbed gas lies in the enhanced natural fracture system, above what would be traditionally defined as the gas/water contact. In synclinal structures, the overpressuring of the reservoir from the adjacent structural limbs may lead to enhanced gas storage within the open fracture system.

In addition to structural complexity, topography and local hydrogeology influence the depth of the reservoir and subsequent reservoir pressure and gas retention within the coal seam(s). Gas desorption experiments completed in the Mist Mountain Formation of southeast British Columbia have demonstrated that in most cases, coals must lie below the regional water table, or be a sufficient distance from subcrop, to ensure that the adsorbed methane is retained within the seam. In synclinal structures, the dipping limbs can provide the hydrological recharge and hydrostatic head to allow overpressured reservoir conditions to exist in the axis of the fold. The regional migration of gas downdip via the coal seam aquifer can also result in the formation of biogenic methane that may enhance the resource potential.

2. Introduction

The coalbed gas potential of the Western Canada Sedimentary Basin (WCSB) is estimated to be approximately four times the known conventional natural gas reserves, yet only sporadic limited production has been achieved. In the late 1970's to mid 1990, exploration efforts were concentrated in both the plains region where the coal-bearing strata are essentially flat lying and in the foothills and mountain regions, where tectonics have created structurally complex terrain. The plains region is characterized by thin to thick coals of lower rank with, for the most part, lower gas contents, (2-10 cc/g). The foothills and mountain regions contain thick coal sequences with medium to high gas contents (10-20 cc/g).

In the structurally complex region of the foothills and mountains, regional horizontal stress gradients appear to be higher as a result of the tectonic folding and overthrusting. This paper examines the key geological parameters that influence the producibility potential of coal gas in this region. Areas where horizontal stress gradients are reduced and where local hydrology has allowed gas retention, appear to be the most prospective. Examples from both Canada and Australia of coal gas reservoirs in structurally complex terrain are presented to illustrate the significance of structure and hydrogeology on reservoir producibility.

3. Regional Stress

The Western Canada Sedimentary Basin (WCSB) is bounded on the west by the fold and thrust belt of the Canadian Rocky Mountains and on the east by the subcrop edge of the Pre-Cambrian age Canadian Shield. Horizontal stress generally increases towards the west and the stress gradients parallel the edge of the fold and thrust bell (Fig. 1). The magnitude of the stress increases from 16 kPa/m near the Alberta border to greater than 20 kPa/m near the edge of the Rocky Mountains (Bell, 1996). Exploration projects have been completed but no wells have been effectively stimulated and no economic coalbed gas fields have been developed within the disturbed belt of the Rocky Mountains.

In structurally complex regions, effective stress values are widely variable, dependent on the location with respect to the major and minor geological structures. The coal gas reservoirs with the greatest potential will be located in areas where horizontal stress is reduced and effective stress is minimized. In the fold and thrust belts of both the WSCB and the Bowen Basin the coals commonly attain the greatest thickness and achieve a rank suitable for thermogenic gas generation. The potential reservoirs in these regions must be examined in view of the high horizontal stresses that are present. In order to minimize the impact of the high horizontal stress gradients, areas of stress relaxation, such as the axes of anticlinal and synclinal structures, represent the most prospective exploration targets (Fig. 2). In contrast, the limbs of the structures would tend to exhibit areas of maximum compressional forces and high horizontal stress. Stress test measurements conducted for regions of the Bowen Basin confirm this hypothesis (Enever, pers. comm., 1998).

Similarities exist between the WSCB foreland basin and the Bowen Basin in Queensland Australia. In the Bowen Basin, the New England Fold Belt represents the eastern edge of the foreland basin similar to the Rocky Mountains, and in the west a thin clastic sequence lapping onto the basement rocks of the Clermont Stable Block, similar to the Canadian Shield. Figure 2 presents a schematic diagram of the main structural elements of the Bowen Basin. Coalbed methane exploration projects in both the northern and southern Bowen Basin have encountered high horizontal stress gradients and in most cases, limited coalbed methane production has been achieved (Enever and Hennig, 1997). The most productive coal gas fields in the Bowen Basin are the Peat field, located south of Moura on the southeast edge of the basin, and the Fairview field, located on the western edge of the southern Bowen Basin, near the Comet Ridge.

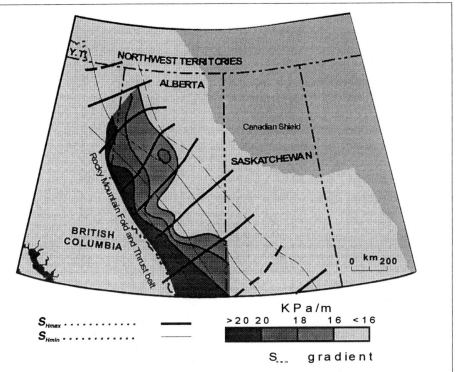

Figure 1:   Schematic map of Western Canada illustrating the increase in horizontal
             stress from east to west, roughly paralleling the Laramide age orogenic front
             of the Rocky Mountains, (after Bell, 1996).

4. Permeability Enhancement

Reservoir producibility is controlled primarily by permeability. In the WSCB and Bowen Basin, regional permeability is commonly low, averaging less than 5 mD and often less than 0.1 mD (Dawson, unpublished, 1997). In the tectonically disturbed regions, where folding and faulting is prevalent, the coals are commonly sheared and much of the cleat fabric is destroyed. Where the cleat is preserved, high horizontal stresses tend to close the cleat system. The axes of the

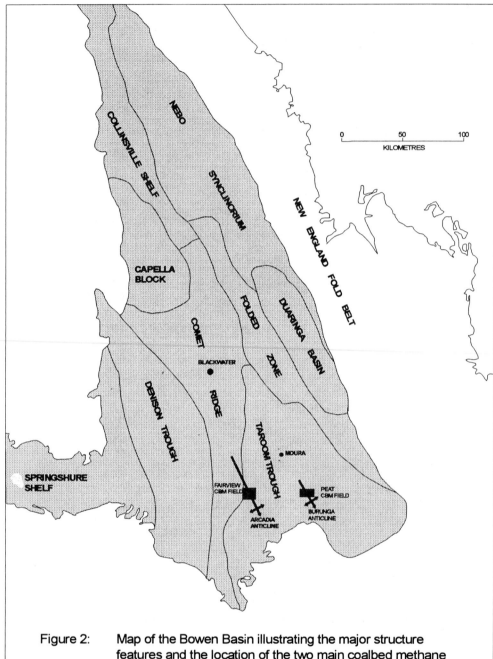

Figure 2:   Map of the Bowen Basin illustrating the major structure
              features and the location of the two main coalbed methane
              fields, Peat and Fairview.

anticline and syncline structures however represent areas where the coals lie within a zone of extension where the cleat system will be open. In these areas, one would expect to have higher permeabilities and hence higher potential gas production. Each of these potential reservoir trap types is examined in detail with examples from both the WSCB and Bowen Basin provided.

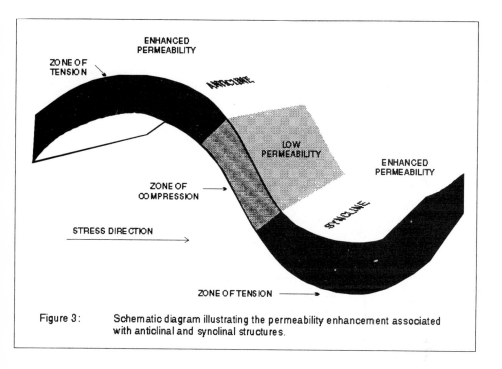

Figure 3 :   Schematic diagram illustrating the permeability enhancement associated with anticlinal and synclinal structures.

4.1 ANTICLINAL STRUCTURES

The axial regions of anticlinal folds present attractive CBM exploration targets in a similar manner to conventional hydrocarbon traps. In the crest of the fold, extension fractures commonly exist and the cleat system may be more developed as well as being open, leading to enhanced reservoir permeability. Horizontal stress gradients would be minimized in the axis of the fold, while the limbs would display higher stress gradients due to the compressional forces associated with the folding. As with conventional hydrocarbons, the coal gas would have the potential to migrate up dip on the limbs, producing a conventional gas cap. In this gas cap, "free gas", residing within the fracture system could yield high gas flow rates with minimal water in the early stages of the wellbore production. The enhanced fracturing of the coal-bearing strata in the axial region of the fold can lead however, to some geological complications. If the fracturing is too extensive, the integrity of the seal rock above the coal seam could be compromised. Minor faulting can also result in compartmentalization of the reservoir thus complicating the production volumes and adequate drainage of the reservoir. Figure 4 presents a schematic illustration of an anticline reservoir type.

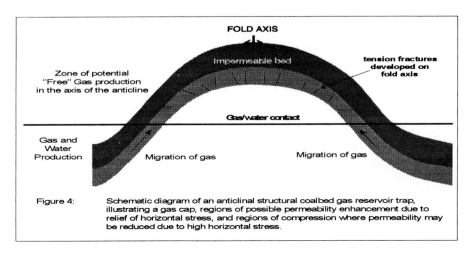

Figure 4:   Schematic diagram of an anticlinal structural coalbed gas reservoir trap,
             illustrating a gas cap, regions of possible permeability enhancement due to
             relief of horizontal stress, and regions of compression where permeability may
             be reduced due to high horizontal stress.

A number of examples of anticlinal coalbed methane reservoirs have been drilled or are in the
process of being developed in both Canada and Australia. Table 1 summarizes the key geological
elements of these examples.

Table 1

Geological Parameter	Peat Structure Australia	Fairview Structure Australia	Flatbed Structure Canada
Depth of Coal Measures	700-850 m	650-750 m	400-500 m
# of Seams	4-10	2	8
Cumulative Thickness	Up to 35 m	9 m	15-20 m
Gas Content	10-14 cc/g	10-12 cc/g	16-21 cc/g
Gas Resource	0.45 Bcm/sq. mile	0.22 Bcm/sq mile	0.64 Bcm/sq mile
Fold Axis Width	20-30 km	4-8 km	4-8 km
Dip of Limbs	Moderate (15-20°)	Gentle (<15°)	Moderate (10-20°)
Structural Complexity	Minor faulting and erosional unconformity	Minor faulting, minimal compartmentalization	Unknown, structure is located in hanging wall of thrust fault

4.2 SYNCLINAL STRUCTURES

Synclinal structures, like the anticlines, have the potential for increased permeability in the axis of the fold due to extension. Cleat development could be enhanced and the cleat system may be open. No gas cap would be present and the migration of water down-dip of the limbs of the structure would lead to perhaps higher than normal water flows from wells that are drilled in the axial region. Although the down-dip flow of water may be seen as a disadvantage with respect to borehole operations, the water has the potential to enhance the reservoir pressure and the gas storage within the coal seams. In Canada high gas contents have been reported at shallow depths, where the hydraulic head associated with the limbs of the structure has resulted in overpressured conditions in the axis. In the Mount Allen syncline in the Canadian Rocky Mountains, gas contents as high as 25 cc/g have been reported at depths as shallow as 100 m (Algas Resources Limited unpublished data). The development of biogenic gas, associated with the meteoric recharge of the reservoir can also lead to higher than predicted gas contents (Fording River Coal, unpublished data). Figure 5 presents a schematic diagram of a typical synclinal structural reservoir. A number of Canadian examples are presented in Table 2 that illustrate the geological parameters associated with synclinal coalbed gas reservoirs.

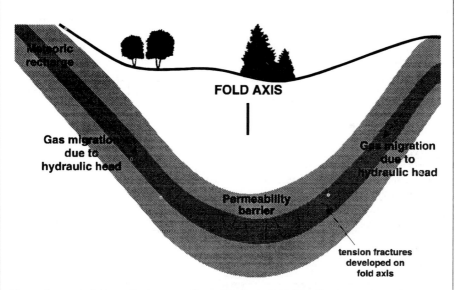

Figure 5:   Schematic diagram of a synclinal structural coalbed gas reservoir trap, illustrating structure axis where hydraulic overpressuring may be present, regions of possible permeability enhancement due to relief of horizontal stress, and regions of compression where permeability may be reduced due to high horizontal stress.

Table 2

Geological Parameter	Alexander Creek Syncline	Elk Valley Syncline	Mount Allen Syncline
Depth of Coal Measures	350 – 700 m	450 – 750 m	100 – 250 m
# of Seams	10-20	10-20	10-20
Cumulative Coal Thickness	15-40 m	15-40 m	10-20 m
Gas Content	12-15 cc/g	12-15 cc/g	10-25 cc/g
Gas Resource	0.64 Bcm/sq mile	0.64 Bcm/sq mile	0.60 Bcm/sq mile
Fold Axis Width	2 km	3 km	2 km
Dip of Limbs	35-50°	35-50°	35-70°
Structural Complexity	Minor faulting on the structural limbs	Minor faulting and some shearing of coals on structural limbs	Coal highly sheared and fractured due to folding and faulting
Artesian Conditions	Wellbores flowing water and gas in axis of structure	Wellbores flowing water and gas in axis of structure	Wellbores flowing water and gas in axis of structure

5. Hydrogeology and Topography

In the Canadian Rocky Mountains, regional topography is highly variable ranging from high mountain peaks to incised valley bottoms. Much of the topography is controlled by the underlying geological structural framework. In general, the anticlinal structures occur in the hanging wall of thrust faults and tend to form the mountain peaks. The synclines tend to lie within the valley bottoms (Fig. 6). The structures tend to plunge along strike.

In order for the adsorbed coal gas to be retained within the potential coal reservoirs, sufficient

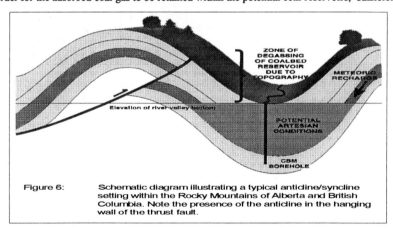

Figure 6:   Schematic diagram illustrating a typical anticline/syncline setting within the Rocky Mountains of Alberta and British Columbia. Note the presence of the anticline in the hanging wall of the thrust fault.

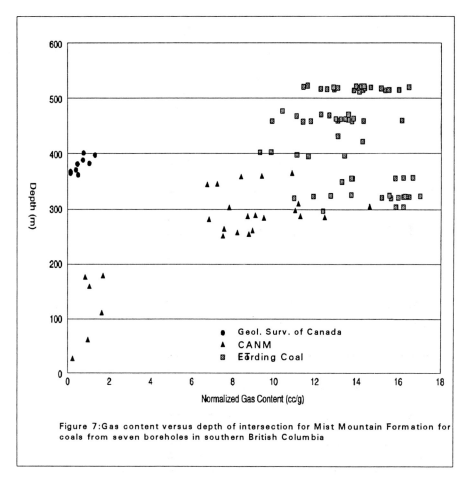

Figure 7:Gas content versus depth of intersection for Mist Mountain Formation for coals from seven boreholes in southern British Columbia

pressure must be present within the seam. This pressure is commonly produced by regional and local water tables. In most cases, in order for the gas to be retained in significant quantities, the coal beds must lie below the topographic elevation of the local water table. Studies conducted by Dawson and Clow (1992) have demonstrated the effect of a local water table on gas retention in coals of the Mist Mountain Formation in southeast British Columbia. In Figures 7 and 8, measured gas contents from canister tests are plotted versus depth of penetration and elevation. In Figure 7, there is a general trend of increasing gas content with depth. There are a number of anomalous data points where gas contents are extremely low (less than 2 cc/g) even though the coals were intersected at depths greater than 350 m. These coal samples were collected from wellbores that were located at the top of a mountain peak approximately 900 m above the valley floor. The same data are plotted against elevation of intersection (Fig. 8) and it can be seen that at an elevation of approximately 1600 m above sea level, gas contents increase dramatically from 1-2 cc/g to greater than 6 cc/g. This elevation coincides with the elevation of the Fording River (local water table) in the Fording River valley. It is interpreted that the presence of water within the formation presents a pressure gradient and permeability barrier preventing the adsorbed gas from desorbing to the atmosphere.

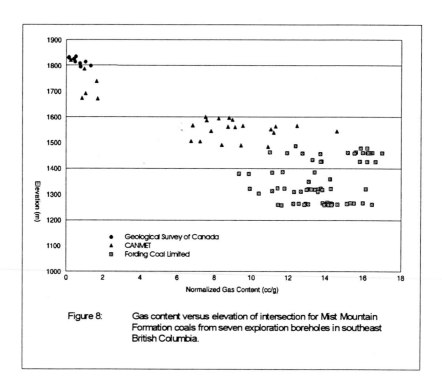

Figure 8:   Gas content versus elevation of intersection for Mist Mountain
             Formation coals from seven exploration boreholes in southeast
             British Columbia.

5.2 HYDROLOGIC RECHARGE

In addition to water flow being an indicator of permeability and providing reservoir pressure, it may also allow the migration of gas to potential trap sites. Scott (1995), demonstrated the importance of gas migration in relation to regional aquifer flows within the fairway zone of the San Juan Basin. Aquifer flow within the coal seam allows gas to migrate and trap against more impermeable areas resulting in overpressuring and elevated gas contents. In addition, the influx of meteoric water can lead to the development of biogenic methane that may enhance the reservoir's overall gas resource potential. In the Elk Valley syncline structure in southeast British Columbia, gas and water analyses indicate that as much as 30% of the methane gas present within the coals may be from biogenic sources (Harrison, unpublished data). This gas enrichment has resulted in high gas contents (> 15 cc/g) at depths as shallow as 250 m.

The negative component of hydrologic recharge is the ability of the mobile aquifer to strip the potential coal reservoir of adsorbed gas over time. In the Fernie Basin of southeast British Columbia, several exploration boreholes intersected coals with anomalously low gas contents (Dawson, unpublished data). Examination of the core revealed that the coals were extremely fractured with abundant calcite deposition within the cleat and fracture faces. The wells had intersected a number of bedding plane thrust faults within the coals and the gas had effectively been flushed from the seam and migrated to outcrop near the valley bottom by passage of formation waters.

6. Conclusions

The development of economically productive coalbed methane gas resources outside of the prolific fields of the Warrior and San Juan basins in United States has been limited by permeability. Many CBM exploration wells have been drilled in areas where high horizontal stress gradients have resulted in low reservoir permeabilities. In the CBM fields such as Peat and Fairview, where productive gas volumes have been attained, it appears that the permeability enhancement that has occurred is the result of structural overprinting in the form of anticlinal structures. Similarly, in a number of synclinal structures in Canada and China that were drilled for CBM, encouraging results have been obtained although no commercial production has been developed to date. Structural trap examples from Canada and Australia indicate that permeability enhancement is required to achieve economic production from most CBM fields. This overprinting along with the local hydrological conditions of the potential reservoir are critical to the producibility of the CBM reservoir in structurally complex terrain.

7. References

Bell, J.S. (1996) In situ stresses in Sedimentary Rocks (part 2): applications of stress measurement, Geoscience Canada, 23,. 135-153.

Dawson, F.M. and Clow, J.T. (1992) Coalbed methane research in southeast British Columbia, in The Canadian Coal and Coalbed Methane Geoscience Forum, February 2-5, 1992, Parksville British Columbia, 57-72.

Enever, J.R.E. and Hennig, A. (1997) The Relationship Between Permeability and Effective Stress for Australia Coals and Its Implications with Respect to Coalbed Methane Exploration and Reservoir Modelling, in Proceedings of International Coalbed Methane Symposium, May 12-17, 1997, Tuscaloosa Alabama, 13-22.

Scott, A.R. (1995) Limitations and Benefits of Microbially Enhanced Coalbed Methane in Proceedings of International Unconventional Gas Symposium, May 14-20, 1995, Tuscaloosa, Alabama, 423-432.

COALBED METHANE EXPLORATION RESULTS OF THE LIULIN PERMIT IN CHINA

ZUO WENQI (1), WANG XINGJIN (1), IAN WANG (2), ZHANG WENHUI(1)

(1) North China Petroleum Bureau, China National Star Petroleum Company, Zhengzhou City, Henan Province
(2) Lowell Petroleum N.L, Melbourne, Australia

Abstract

Since 1994, North China Petroleum Bureau has cooperated with Lowell Petroleum N.L to exploit coalbed methane in the Liulin Permit which is 218 square kilometers in area. The Permit is structurally situated in the southern limb of the Lishi structural nose which is in the middle of the eastern margin of Ordos Basin. Main coal seams are Numbers 4 and 5 of the Shanxi Formation, Lower Permian and Number 8 of the Taiyuan Formation, Upper Carboniferous.

Three wells have been completed and operations of coring, gas content determination, geophysical logging and well testing have been performed. The exploration results have proved that Seam 4、 5 and 8 are the main potential seams of the Permit. In the eastern part of the Permit, three target seams have similar thickness of 3 m to 5 m, and the formation pressure is relatively low. In the west part of the Permit, Seam 8 is dominant among all seams with thickness of 8-10 m, and the formation pressure is 1.2 Mpa higher than normal pressure. The gas content of the Permit is high and varies from 10 to 15 m^3/ton. Well testing and reservoir simulation of Seam 8 yield a rather high permeability, which is comparable to one of the main production seams in the San Juan Basin. On the whole, the reservoir characters of the Permit show that Liulin Permit possesses high producibility.

Based on exploration results, coalbed methane resources of Liulin Permit was calculated and the value is 24.26 billion m^3.

1. Introduction

Huawell's Liulin Coalbed Methane Permit is situated in Liulin County, Shanxi Province (Fig. 1), which occupies about 218 Km^2. The Permit was registered by Huawell Coalbed Methane Company, a joint venture for exploring and developing coalbed methane in China, between Lowell Petroleum (China) Pty. Ltd. from Australia

and North China Petroleum Bureau. Huawell is the first Coalbed Methane Joint
Venture in China, which were ratified by Chinese Government. According to the
Contract, Huawell will explore and evaluate the coalbed methane in the Permit to
identify commercial development potential. So far, the project has been implemented
for more than three years, and plentiful exploration results have been obtained. Based
on these results, a synthetical study, reservoir simulation and economic evaluation for
the Permit were conducted. It is clearly proved that the Permit is the most promising
area for developing coalbed methane commercially in China. Therefore, Lowell and
North China Petroleum Bureau have decided to implement a pilot development plan,
and are preparing a development Program in Liulin.

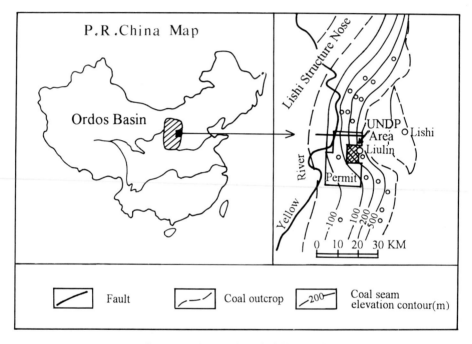

Figure 1. The location of Liulin Permit
(after Zhang Wenhui et al., 1995).

2. General Geology

2.1. STRUCTURE

Liulin Permit is structurally situated in the southern limb of Lishi Structure Nose in
the eastern margin of the Ordos Basin. In the north of the Permit , the structure dips to
the west and gradually dips to the southwest and the south. Strata dip very gently
between 3 and 8°. hence, the seams occur at higher depths towards the southwest.
Liulin Permit is structurally simple and was developed a series of minor folds along
the hinge of the nose fold. At the north boundary of the Permit, a fault zone occurs

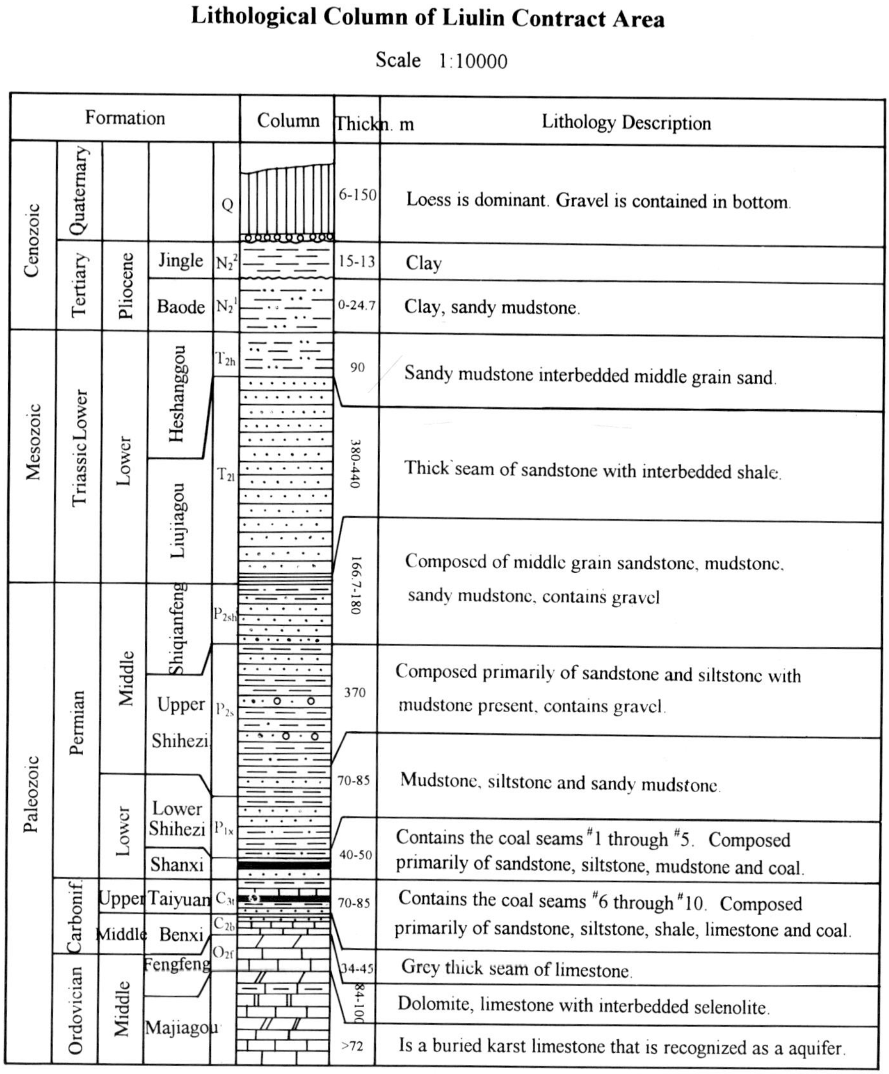

Lithological Column of Liulin Contract Area

Scale 1:10000

Formation				Column	Thickn. m	Lithology Description
Cenozoic	Quaternary			Q	6-150	Loess is dominant. Gravel is contained in bottom.
	Tertiary	Pliocene	Jingle	N_2^2	15-13	Clay
			Baode	N_2^1	0-24.7	Clay, sandy mudstone.
Mesozoic	Triassic Lower	Lower	Heshanggou	T_{2h}	90	Sandy mudstone interbedded middle grain sand.
			Liujiagou	T_{2l}	380-440	Thick seam of sandstone with interbedded shale.
Paleozoic	Permian	Middle	Shiqianfeng	P_{2sh}	166.7-180	Composed of middle grain sandstone, mudstone, sandy mudstone, contains gravel
			Upper Shihezi	P_{2s}	370	Composed primarily of sandstone and siltstone with mudstone present, contains gravel.
		Lower	Lower Shihezi	P_{1x}	70-85	Mudstone, siltstone and sandy mudstone.
			Shanxi		40-50	Contains the coal seams #1 through #5. Composed primarily of sandstone, siltstone, mudstone and coal.
	Carbonif.	Upper	Taiyuan	C_{3t}	70-85	Contains the coal seams #6 through #10. Composed primarily of sandstone, siltstone, shale, limestone and coal.
		Middle	Benxi	C_{2b}		
	Ordovician	Middle	Fengfeng	O_{2f}	34-45	Grey thick seam of limestone.
			Majiagou		84-100	Dolomite, limestone with interbedded selenolite.
					>72	Is a buried karst limestone that is recognized as a aquifer.

Figure 2. The lithological column of Liulin Coalbed Methane Permit (modified from Wang Xingjin, 1994).

across the Permit, which is composed of normal faults and formed a grabben with a width of about 500m (Zhao Shu et al., 1995).

2.2. STRATIGRAPHY

Stratigraphic framework of Ordovician, Carboniferous, Permian, Triassic, and Cenozoic were established in the Permit. Cenozoic sediments is prevail through the permit, and the rest can only be observed along the gully and road. Ordovician in this area is represented by sedimentation of carbonate rocks, limestone intercalated with dolomite limestone and gypsum. The thickness of Middle and lower Ordovician sediments varies from 400 to 650 meters, and Upper Ordovician was denuded out in the Permit. An erosion at zone occurs at top of the Ordovician and now represents a regional aquifer. The Carboniferous-Permian is a series of littoral-delta sediments with a thickness of 400-800 meters. Six groups can be recognized from bottom to top of this section: Middle Carboniferous Benxi Formation, Upper Carboniferous Taiyuan Formation, Lower Permian Shanxi Formation and Lower Shihezi Formation, Upper Permian Upper Shihezi Formation and Shiqianfeng Formation (Fig. 2). Coal seams in Taiyuan and Shanxi Formations are the major target seams for coalbed methane exploration in the Permit. The Taiyuan Formation with an average thickness of 90 meters consists of limestone, sandy mudstone and coal seams. Limestone occurs in the middle of the formation, and forms another key aquifer. The Shanxi Formation averages about 60 meters, and is composed of dark mudstone interbedded with sandstone and coal seams. The Triassic in this area is represented mainly a set of purple and grey-purple fluvial sandy mudstone sediments which were distributed in the west of the area and termed Liujiagou Formation with a thickness of 410 meters. The Cenozoic is represented by Pliocene and Pleistocene sediments with the gravel deposited at the bottom, loess and silty loess on top. The thickness varies from 0 to 150 meters (Zhang Wenhui and Wang Xingjin, 1997).

Altogether 17 coal seams were developed in the Shanxi Formation and the Taiyuan Formation. Seam 4, 5 in the Shanxi formation and Seam 8 in the Taiyuan Formation are major target seams and thick enough for developing coalbed methane. The thickness of Seams 4 and 5 varies between 1 meter an 4 meters in the Permit. The distribution of Seam 8 varies, averaging of 3.1 meters. Seam 8 thins towards the southeast from the northwest. In the northwest part of the permit, the thickness of Seam 8 reaches 9 meters.

3. Exploration Results

Coal exploration is very intense in the Permit, where about 72 wells were drilled for coal. Therefore, coal structure, stratigraphy, and thickness are relatively well known (Zhang Wenhui et al., 1995). In the early 1990's, North China Petroleum Bureau began to implement a pilot development project for coalbed methane an with an assistance of United Nations Development Program (UNDP). Seven wells have been drilled and plentiful reservoir data have been obtained (Mei Shixin et al., 1997). The data from UNDP Area can be used to evaluate the development potential of Liulin Permit CBM because it is close to the center of our Permit and geology properties in both areas are similar. However, it is necessary to drill pilot development wells in the Permit and collect enough data on thickness, gas content, permeability and formation pressure for calculating CBM reserve and recoverability of CBM. So far, three wells have been completed, two of which are located in the southwest of the Permit and one is located in the northeast of the Permit. A set of data on coal distribution, gas content, permeability and formation pressure have been obtained.

3.1. COAL THICKNESS

Interesting from coalbed methane potential viewpoint coal seams occur in the Lower Permian Shanxi Formation (P_{1s}) and the Upper Carboniferous Taiyuan Formation(C_{3t}) Fig. 2). About 10 of 17 seams, of which 6 seams in the Shanxi Formation and 11 seams in the Taiyuan Formation, are over 0.5 meter in thickness. Seams 4, 5, and 8 are main target as CBM reservoirs. In addition, Seams 2, 3, 6, 9 and 10 have consistent thickness over the Permit, and may be regarded as target seams.

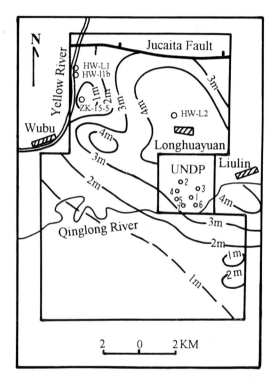

Seam 4 occurs in the lower part ofthe Shanxi Formation, and occurs in most of the area with an average thickness of 3.16 meters (Fig. 3). It is mostly developed with a thickness of 4.85 meters around Longhuayuan Village and pinches out to the southwest. The area with thickness less than 1 meter is only 1.5 square kilometers around ZK-15-5 well.

Figure 3. Thickness Isopach Map of Seam 4, Liulin Permit (Wang Xingjin et al., 1997).

Therefore, Seam 4 is a thick seam with consistent deposition throughout the whole Permit.

Seam 5 occurs in the base of the Shanxi Formation. In the Permit, it has an average thickness of 2.59 meters. In the northern part it thickens from NW to SE, whereas in the southern part it thins toward the south. Seam 5 is missed in an area of 4.5 square kilometers around HW-L1 well (Fig. 4). Similar to Seam 4, Seam 5 is most developed around Longhuayuan Village with a thickness of 4.94 meters.

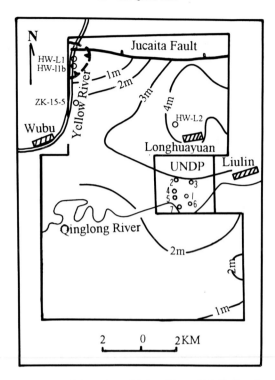

Figure 4. Thickness Isopach Map of Seam 5, Liulin Permit
(Wang Xingjin et al., 1997).

Seam 8 occurs in the lower-middle of the Taiyuan Formation, and distributes in most of the Permit with an average thickness of 3.07 meters. The thickness increases towards the northwest and reaches 9 meters at the west boundary (Fig. 5). The continuity of the distribution is stopped by the Jucaita Fault near the north boundary of the permit. The structure of Seam 8 distribution is similar to that of Seam 5 except it is 50 meters deeper.

In summary, Seams 4, 5 and 8 occur in the whole permit area. The thickness of Seam 8 increases towards the west from 4 meters to 9 meters. In contrast, Seams 4 and 5 is thinning towards the west from more than 3 meters to less than 1 meter. Therefore, the west of the Permit is dominated by Seam 8. Seams in the southeast are slightly thinner than the northeast in accumulated thickness, which is 3-9 meters commonly, as each seam in the southeast is thinner than that in the northeast. The Permit is characterized by thick multiple seams, favorable to form coalbed methane reservoir. Distribution of each main seam varies over the Permit.

3.2. GAS CONTENT

During the coal exploration period in the Permit, 43 coal samples from Seam 4, 5, 8, 9 and 10 in 15 wells have been analyzed for gas content by desorption method. The data of gas contents in main seams are over 10 m^3/ton, and gas content increases with

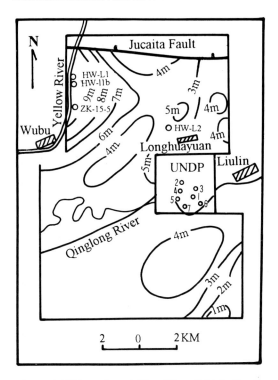

Figure 5. Thickness Isopach Map of Seam 8, Liulin Permit
(Wang Xingjin et al., 1997).

TABLE 1. Gas content obtained from Permit wells and UNDP wells (Wang Xingjin, 1997)

Seam No.	Gas Content (m³/ton)									
	Permit			UNDP Area						
	HW-11	HW-11b	HW-12	1	2	3	4	5	6	7
4	9.8	9.53	9.14	14.8	14.7		13.1	14	11	11
5			8.73	11.8	7.8		12.8	11.6	8.2	10
8	14.38	14	18	13.3	14.8	12	16.2	16.7	11	12
9	16	15.9	15							
10	16.3	16	14.8							

depth. The gas contents of Seams 8, 9, and 10 are about 15 m³/ton (TABLE 1). These values may be lower than the actual gas contents because some deficiencies of technology for testing and coring (Zhang Wenhui and Wang Xingjin, 1997). In the UNDP Area, North China Bureau conducted gas content determination for 7 wells by the USBM (US Mine Bureau of Mine) Method. Samples of Seams 4, 5 and 8 buried at 350-400 meters yielded a gas content variation between 7.8 m³/ton and 22.4 m³/ton, which is higher than the result by the desorption method used. These gas content values of the UNDP Area can be helpful for the permit. Fifty samples of main seams from three wells drilled by Huawell were determined for gas content using the USBM

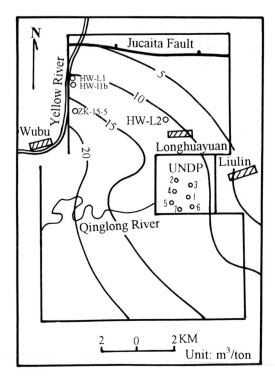

Figure 6. Gas Content Contour Map of Seam 4, Liulin Permit
(Wang Xingjin, 1997).

Method. The gas content of Seam 8 in HW-L1 Well is 14.38 m³/ton and 18.15 m³/ton
in HW-L2 well. Gas content contour maps for the Permit were prepared based on
these data from three sources mentioned above to analyze distribution patterns of gas
content in the Permit (Fig. 6, 7, 8).

a. Gas content of three main seams varies between 5 and 20 m³/ton, with an average of
10 m³/ton. Depthwise, the gas content of Seam 8 is the highest, seam 5 is the lowest
and Seam 4 is between Seam 8 and Seam 5.

b. Horizontally, gas content tends to be higher towards southwest and south, following
the direction of strata dipping.

c. Gas content of each seam declines towards the Jucaita Fault. The reason is that
extensional movement of the fault caused formation pressure to drop and gas was lost.

d. Gas content changes significantly in a single seam. An inhomogeneity was caused

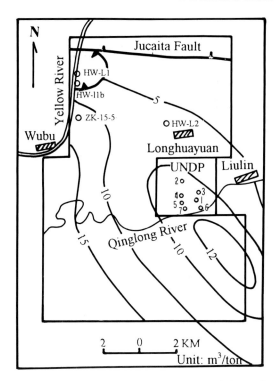

Figure 7. Gas Content Contour Map of Seam 5, Liulin Permit
(Wang Xingjin, 1997).

by the heterogeneity of maceral types and ash rate. For example, the gas content of Seam 8 in HW-11b well decreases with the increase of ash content (Fig. 9).

3.3. PERMEABILITY AND FORMATION PRESSURE

In order to determine the permeability and formation pressure in Liulin Permit, the transient well testing was conducted for Seam 4 and Seam 8 in HW-L1 Well and HW-L2 Well. Though Seam 5 is a main target seam, it was not tested as it is close to seam 4 and in the same hydraulic system, and may be represented by Seam 4 in permeability and formation pressure.

The results of transient testing proved that Seam 8 possesses high permeability with a range of 13.72 to 24.81 md (TABLE 2). The permeability of Seam 4 may not be reliable with a value of 0.0106 md, because of short injection time and ineffective detective radius. With regard to the properties of Seam 4 in maceral type, pore structure, cleat pattern, burial depth and hydraulic condition, the Permit is similar to the UNDP Area. The permeability in the UNDP Area ranges from 7 to 10 md. Therefore, the permeability in the Permit is expected to be in the same range. 7 wells

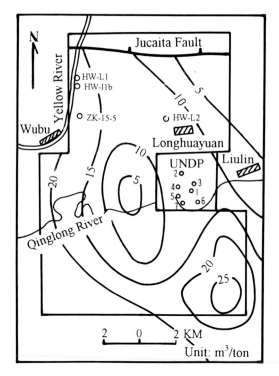

Figure 8. Gas Content Contour Map of Seam 8, Liulin Permit
(Wang Xingjin, 1997).

in the UNDP Area have produced for 2 years and efficient dynamic data have been achieved (Fig 10). The permeability simulated based on these data is 10 scale points higher than one derived from transient testing, which may represent the real nature of the reservoir. For example, the permeability tested of Seam 8 in No. 5 Well is 12.5 md, but the permeability simulated based on actual production (Fig. 10) is 24 md. Hence, permeability of Seams 4, 5 and 8 are expected to be range from 15 to 20 md.

The formation pressure of Seams 4 and 8 derived from transient testing in HW-L1 Well are 6.55 Mpa and 8 Mpa respectively. Seams 4 and 8 are buried to a depth of 584.8-587 m and 659.45-668.5 m with pressure factors of 1.14 and 1.19 respectively, which shows that formations in the west of the Permit are overpressured. The pressure of Seams 4 and 8 in HW-L2 Well are 3.97 Mpa and 4.09 Mpa, which are buried to a

The Relationship of gas Content and Coal rank

HW-11b Well

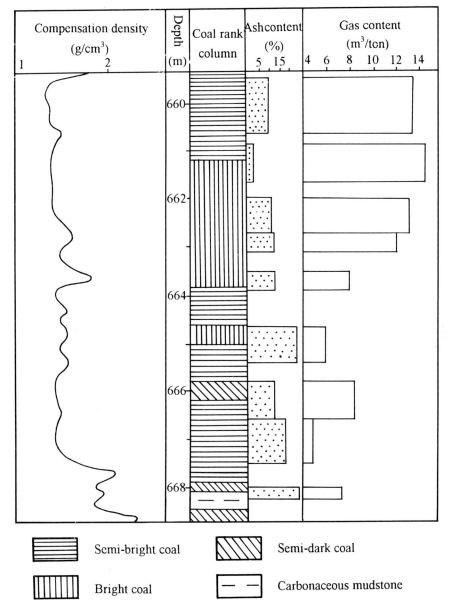

Semi-bright coal   Semi-dark coal

Bright coal   Carbonaceous mudstone

Figure 9. The gas content decreases with the increases of coal rank and ash content of coal samples from HW-11b well (Zhang Wenhui, 1996)

TABLE 2 The Permeability data obtained from transient testing (Wang Xingjin, 1997)

Well	Permeability (md)		
NO.	Seam 4	Seam 5	Seam 8
HW-l1	0.0106	10	13.72
HW-l2	5.00		24.8
1 (UNDP)	1.98		1.12
2 (UNDP	0.01	0.04	8.86
3 (UNDP)	0.12	0.15	0.29
4 (UNDP)		12.72	0.93
5 (UNDP)			12.5
6 (UNDP			
7 (UNDP)		6.7	11.6

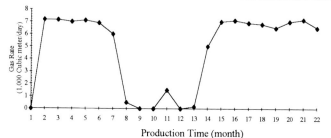

Figure. 10 The gas production profile for Seam 8 of No. 5 Well in UNDP Area (Wang Xingjin, 1996)

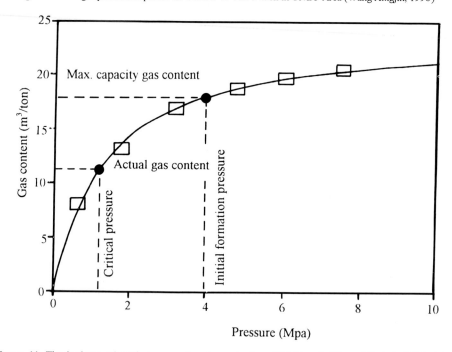

Pressure (Mpa)

Figure 11. The isotherm adsorption curve of coal sample from HW-l2 well (Zhang Wenhui, 1996). The saturation (critical desorption pressure/initial formation pressure) represents the difficulty degree of gas desorption from coal. The higher the saturation is, the easier the desorption from coal seam is.

depth of 606.4-610 m and 618.7-621.8 m with pressure factors of 0.666 Mpa and 0.672 Mpa. That shows that the formation in east of the Permit is under normal pressure (Zhang Wenhui and Wang Xingjin, 1997).

Matching gas content and isotherm adsorption data (Fig 11), analysis formulates a gas saturation (equal to critical pressure/formation pressure) corresponding to the pressure of Seam 4 in the west of the Permit 68%, Seam 5 74%, Seam 8 75.9%. In the east of the Permit, the saturation of Seam 4 is 83% and Seam 8 is 85.79%. That indicates the pressures should fall more excessively in the west than that in the east before desorption of gas. In contrast, the pressure drop of the west is easier than that in east as it is overpressured in the west and it is underpressured in the east. It is expected that the western reservoir has more energy to degas.

4. Resource evaluation

Based on the exploration results, coalbed resources were calculated. volumetric method (equation (1)) (Modified after King, 1993) were employed to calculate the coalbed methane resources of Liulin Permit.

$$G = Ah\rho C \qquad (1)$$

G	gas in place	m^3
A	drainage area	m^2
h	net thickness of coal seam	m
ρ	pure coal density	ton/m^3
C	gas content	m^3/ton

Considering the heterogeneity of reservoir in the permit, the permit was divided into 218 cells according to Fig. 12, each of which is 1 square km in area and regarded as homogeneous subject. Before calculation of the resources, all data should be distributed to each cell. Following, the resources of each cell is calculated according to equation (1). The total resources of permit is equal to the sum of resources of all cells of the permit (Fig 13). As a result, the total resources of three main seams (Seams 4, 5 and 8) in the whole Permit is $242.59 \times 10^8 m^3$.

Conclusions

1. There are 10 seams in total which are thicker than 0.5 m, of which Seams 4, 5 and 8 are the thickest ones with stable distribution and the average accumulated thickness is 14 m. Seam 4, 5 and 8 are main reservoir seams in the Permit. That caused high gas generating capacity and developed cleat which is favorable to reserving migration.

2. Gas content is high in the Permit, commonly from 5 to 20m^3/ton. The average gas content of the Permit is over 10m^3/ton.

13.9	13.9	13.4	12	10.6	9	7.3	5.4	3.5	2	1.5
14.6	15	14.6	12.9	11.4	9.8	8	5.9	3.8	2	1.5
15	16.6	15.7	14.2	12.2	10.6	8.8	6.8	4.4	2	1
15.3	17.3	16	15	12.7	11.4	9.9	8	5.6	2.6	2
14.6	15.3	15	14.3	13.6	12.8	11.2	9.8	7	4.8	3
12.7	13.3	13.5	14.4	13.2	13.1	12.7	11.6	9.4	7.3	5.7
11.2	11.5	11.8	11.9	12.2	12.8	13.2	13	11.5	10	8.5
9.5	9.6	9.9	10.2	10.9	11.9	13.3	13.7	12.8	11.5	10.4

8.2	8.7	7.9	7.8	7.8	8.1	9.18	11.2
7.1	6.9	6.5	6.1	5.7	5.7	7	10.1
6.3	5.3	4.9	4.8	4.3	3.3	4.5	8.9
4.9	4.6	4.2	4.5	3.8	3.4	4.6	7.6

Unit: m^3/t

4.4	4.1	3.9	4.2	4.0	4.6	4.9	6.9	9.1	10.8	12	11	9	10
4.1	3.8	4.6	3.8	3.9	4.4	5.5	6.8	8.1	10.5	11	10	12	16
5.0	4.1	4.1	4.2	4.5	5.0	5.6	7	8.5	10.3	12.5	16.0	18	19
4.4	4.3	4.4	4.6	5.0	5.0	6.0	7.2	8.6	10.1	11.8	13.7	20.4	20
4.6	4.6	4.8	5.0	4.7	5.5	6.4	7.5	8.8	10.2	12.0	13.8	17.1	21.5
4.8	4.9	5.2	4.8	5.2	6.0	6.9	7.9	9.2	12.1	14.3	17.1	18.0	16
5.1	4.6	4.9	5.0	5.7	6.4	7.3	8.3	9.6	12.5	14.5	17	17	14

Fig. 12. Gas Content Distribution Map of Seam 8, Liulin Permit (Distributions of other data like this) (Wang Xingjin, 1998)

	172.3	160	122.7	87.8	71.6	65.6	57.5	46.4	31	12.2	7.6		
	2057	195.1	159.6	107.8	94.7	83.8	67.6	58.1	37.2	17.3	8.1		
	215.9	222.6	164.3	171.2	123.8	114.5	101.9	84.3	64	27.3	9.1		
	221.0	249.4	232	202.4	173.6	160	131.1	104.6	92.3	43.2	19.6		
	223.5	237.6	229	248.8	232.8	195.5	169.6	143.0	104.2	59.3	43.3		
	240.8	248.8	236.2	198.9	199.0	209.6	199.9	183.4	147.8	106.5	77.6		
	227.9	250.8	252.3	198.8	181.7	197.1	228.1	217.2	217	99.6	131.1		
	187.7	189.1	221.2	191.5	181.8	210.7	169.2	228.2	210.8	173	147.1		
106.8	112.7	122.5	136.7	136.3	137.5	102.8	207.7						
83.5	86.8	122.5	90.3	94.3	98	119.1	166.7	Unit: $10^6 m^3$					
74	72	71.9	51.1	74.1	71.3	81.5	122						
61.7	61.6	63.2	66.4	59.8	61.4	76.7	103.8						
51.4	50.7	49	50.6	33.5	51.7	65.6	86.2	113	138.7	151	149.5	137.3	147.6
42.3	46.9	46.9	33.3	50.3	54.8	65.2	76.4	98.5	126.4	138.1	134	141.7	135.5
47.3	43.3	48.2	49.3	46.6	56.7	62.7	74.8	94	115.9	134.4	165.3	173.8	120.2
44.7	49.7	50.9	51.3	53.8	54	60.9	70.1	85	102.2	118.2	133	149.2	129.4
45.4	50.6	51.8	53.7	52.8	57.8	65.3	71.4	83.4	97.8	106.4	122.7	109.6	118.4
50.2	49.3	48.7	51.8	54.7	61	66.7	77.5	90.5	108.4	121.6	128.7	94.7	84.7
52.4	46.2	51.2	49.6	56	61.1	68.9	76.8	86.2	104.1	114.8	104.3	56.11	64.1

Fig. 13 Resources of all cells in Liulin Permit. (Wang Xingjin, 1998)

3. The permeability simulated is commonly over 20 md commonly. With regard to permeability, the permit is one of the best areas in China. Formation pressure in the west of the Permit is over normal pressure which offers gas an efficient power for desorption.

4. Coalbed methane resources of the Permit are 24.26 billion m^3.

In conclusion, Liulin Permit is a very promising area in terms of resources.

References

1. King, G. R. (1993) Material-Balance Techniques for Coal-Seam and Devonian Shale Gas Reservoirs with Limited Water Influx, *SPE Reservoir Engineering (Feb. 1993)*, 67-72.

2. Mei Shixin et al. (1997) Exploration for Deep Coalbed Methane, Coal Program Report of UNDP Project, No. UNDP/CPR/91/214, 41-47.

3. Wang Xingjin (1996) Analysis on Production of No. 5 well in UNDP Area, Reserch Report on UNDP project. 50-60.

4. Wang Xingjin and Zuo Wenqi (1997) Method Study on Economic Evaluation for Coalbed Methane Development, *Proceedings of 1997 International Coalbed Methane Symposium*, Tuscaloosa, Alabama, USA. 59-64.

5. Wang Xingjin et al. (1998) A Study on the Method of Coalbed Methane Reserve Calculation, Natural Gas Industry Vol. 18 No. 4, 24-27.

6. Zhang Wenhui et al. (1995) A Study on Coalbed Methane Geology Characterizing and Area Selecting. The Report of the Eighth-five Year National Research Project. N0. 85-102-11-01, 143-144.

7. Zhang Wenhui and Wang Xingjin (1997) Geological Report of Initial Exploration On Coalbed Methane In Liulin Permit, 50-120.

8. Zhao Shu, Zuo Wenqi, Zhang Wenhui, Wang Xingjin (1995) Sino-Australia Coalbed Methane Exploration Project in Huawell's Liulin Contract Area, *United Nations International Conference on Coalbed Methane development and Utilization(A)*, 170-175.

RESIDUAL GAS CONTENT OF COAL IN THE LIGHT OF OBSERVATIONS FROM THE UPPER SILESIAN COAL BASIN, POLAND

I. GRZYBEK

State Mining Authority
Poniatowskiego 31
40-956 Katowice, Poland

Through centuries coalbed methane (CBM) was only treated as a hazard for coal mining operations. Fortunately, this point of view has changed and now the methane is also perceived as a valuable source of energy, if captured and utilized, or as an impedence for the Earth's climate, if released to the atmosphere. These three faces of CBM force mining engineers, geologists and environmental officers to assess the amounts of methane captured in coal, available for recovery and capable for venting to the air. But anybody who wants to make any reliable assessment of CBM resources or describe rules the amount of gas release from coal depends on needs to take into account a lot of gassy characteristics of coal.

Residual gas content is, among the others, one of the key such a characteristics. To understand what does the term 'residual gas content' mean the processes of CBM retention in and release from coal should be described first. As it is commonly known, in coal-bearing formations methane is contained in both coal seams and surrounding barren rocks. In the barren rocks and fractures of coal it is collected as a free gas. During its recovery the gas is released dynamically, with a rate that is proportional to the gradient of pressure between borehole wall and unaffected rock mass. At the other hand, methane in coal matrix is sorbed on the internal surface of coal substance. Release of that methane from coal is somewhat different. According to Grzybek et al. (1997) 'at the early dynamic stage its release is the same as from barren rocks, [but] when the pressure gradient approaches close to zero, the velocity of methane release depends on rate of diffusion through coal (Mazzsi 1992). Usually, dynamic release is very quick, while the diffusion is very slow.' It was estimated, that copmlete diffusion of methane may take from ten of hours (Gawraczyński & Borowski 1986) to many years (Smith & Sloss 1992), adequately from crushed and in-situ coals.

In the light of the above description it is obvious, that certain part of methane cannot be released from coal seams during recovery of gas or will remain even in just extracted coal. This part is called residual gas content. For scientific unmistakability, residual gas content has been defined as the volume of methane standarized on a tonne of coal and retained in coal sample when release of methane stops to be dynamical and starts to progress under diffusion laws (Grzybek et al. 1994). It is accepted, that the moment is when the pressure in coal equalized to the atmospheric pressure.

Direct methods for testing of residual gas content, such as U.S. Bureau of Mines (USBM) method for example (see: Yee et al. 1993), are not common in Polish praxis. So, the analytical method for determining it has been developed. However, the method was incorrectly used long time and its results were not comparable one to another and to results of direct measurements. Reasons of such situation were formerly found (Grzybek 1997a) and the new procedure of analytical method was proposed. But correctness of the procedure has not been fully confirmed up to date. To confirm it the procedure was employed for a few CBM deposits in Poland. The results from each of they were compared one to another and, where possible, to the results found by direct measurements of residual gas content. Outcome of the comparison is presented in the paper, on the background of assumptions of the analytical method. The new procedure of the method is also presented there, as well as methods for measuring gassy parameters of coal used by the analytical method.

1. Procedures for Determining Gassy Parameters of Coal

During exploration of coal-bearing formations and exploitation of coal only two gassy parameters of coal are commonly measured in Poland. They are total gas content and intensity of gas desorption. Other parameters are measured rarely and mainly for scientific purposes. Total gas content (G; cu m/Mg) defines volume of methane, and sometimes also of other hydrocarbons, contained in the mass unit of pure coal substance (i.e. ash and water free). Intensity of desorption, expressed by desorption ratio (ΔP_2; mm H_2O), fixes the amount of gas released from coal sample during specified interval of time. The desorption ratio shows the amount indirectly by means of overpressure equivalent to volume of gas desorbed. Polish procedures for determining the two parameters were previously presented internationally (e.g. Grzybek 1993 and 1997b, Hampton III & Schwochow 1994, Kobiela et al. 1992). But for better understanding of the analytical method there is a need to remind them.

According to Grzybek (1997b) Polish methods for testing total gas content can be divided into the three following groups:

1. The first group consists of several methods, called together 'side-wall methods' or 'hermetic containers methods'. They are based on coal samples from side-walls of

underground workings, collected in one piece into hermetic containers immediately after getting coal seam, and then degassed at laboratory. Laboratory procedures for sample degassification were differentiated in the past (for details see: Kobiela et al. 1992), but now only one of them is in use. This procedure was described in Grzybek (1993) in the following words: 'At laboratory [coal] sample is pulverized by steel balls put into container before sampling, through a 120 minutes shakering them in a vibrator. After pulverization the sample is degassed by vacuous degassification system connected to the container (...). The gas [liberated] from the sample is measured and examined to establish its volume and composition (...). Next, the sample degassificated is weighted and analyzed for ash, water [and volatile matter content of coal]. Then, basing on these examinations and measurements, the so-called 'substantial gas content' (G_p, cu m/Mg) is calculated. The G_p is volume of methane released after coal sampling, counted for 1 Mg of [pure coal substance]. To determine the total gas content its volume lost [during] sampling must be assessed and added to G_p value:

$$G = G_p + G_l \qquad (1)$$

where G_l is lost gas' content. The lost gas content (cu m/Mg) is assessed using an equation in the following form:

$$G_l = 0{,}33 G_p \qquad (2)$$

2. The second group consist of only one method, termed 'two-phase vacuum degassing method'. It is based on core samples from surface boreholes drilled to unmined coal seams and uses two-phase vacuum procedure for samples degassification. During the first phase, a coresample is degassed through 3—4 hours under conditions of 1.333×10^{-3} MPa underpressure. Then, for the second phase small part (150 g) of the sample is put into vacuum mill, pulverized and totally degassed. Next steps of laboratory procedure are the same as for the described method of the first group.

3. The third group contains two similar methods, named 'direct-hole method' and 'modified direct-hole method'. The former of them was employed up to 1983. Currently only the further one is in use. The modified direct-hole method tests cuttings of coal taken from the last 20 centimeters of the four meters long, horizontal borehole drilled from underground workings to parts of coal seams unaffected by mining. Time of coal sampling is equal to 2 minutes, starting from drilling the sampled interval of borehole and ending when vacuum container is closed. Laboratory procedure for this method is the same as for the method of the first group, except of lost gas calculations. The modified direct-hole method defines lost gas content as a sum of lost of free and desorbed methane. The lost-free gas content (G_{lf}, cu m/Mg) is determined on the basis of average volume of coal macropores saturated by gas, assumed to be equal to 0,050 cu m/Mg, and measurements of partial pressure of methane in-situ (p, bar). The lost-desorbed gas content (G_{ld}, cu m/Mg) is evaluated basing on intensity of desorption measured. For determining the above lost gas contents the following empirical formulae are used:

$$G_{ld} = 5{,}45 \cdot 10^{-3} \Delta P_2 \qquad (3)$$

and

$$G_{lf} = 4{,}971 \cdot 10^{-3} \, p \cdot e^{1.73 \cdot 10^{-3} p} \qquad (4)$$

Results of gas content measurements made by different methods described were long time assumed comparable one to another. However analysis, drawn mainly on the basis of the results of previous works by several authors, shows that (Grzybek 1997b):

1. Core samples for testing by two-phase vacuum degassing method are gathered from not degassed coals, in opposition to the side-wall samples, partially degassed before sampling (9 % in average). Thus, use of the same formula to calculate lost gas content by both of the methods is unproper.
2. Results of gas content measurements by side-wall method and two-phase vacuum degassing method are comparable, if lost gas content for the further method is assessed multiplying substantial gas content by the 0.11—0.16 factor, dependent on coal properties.

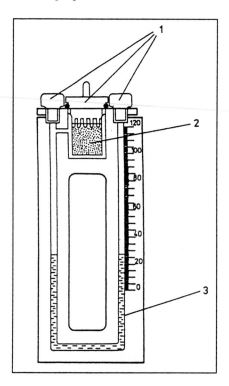

3. The average value of a lot of measurements made by modified direct-hole method is close to the average of measurements made for the same coals by both side-wall and two-phase vacuum degassing methods, of course if the lost gas content has been calculated correctly.
4. Outcomes of measurements made by modified direct-hole method are very similar to results of measurements by U.S. Bureau of Mines method.

Contemporaneously to measurements of total gas content, desorption ratio is also tested very often. The tests are made using manometric, fluidal desorbometer of the DMC-2 type (Fig. 1). It consist of U-shaped glass tube scaled in milimmeters, a box for collecting coal cuttings and hermetic stoppers for plugging the tube and the box. The glass tube is partially infilled by coloured water, changes of level of which may be read out in milimeters from the scaled part of the tube. Coal samples for

Figure 1. The manometric, fluidal desorbometer of the DMC-2 type; 1 - stoppers, 2 - box for coal cuttings, 3 - glass tube, partially infilled by coloured water (lined); acc. to Grzybek (1993).

desorbometric examinations are collected either from underground holes drilled for total gas content measurements by direct-hole method, or from a crushed piece of corehole tested by two-phase vacuum degassing method. Particles of coal sample should be $0.5-1.0 \times 10^{-3}$ m in size and have to be gathered into desorbometer during 2 minutes after start of drilling the sampled interval of underground borehole or after the moment, when corehole reached surface. On gathering of coal sample, gas desorbing from coal causes change in water level in the tube. Magnitude of the change should be read out after the next two minutes. The magnitude recalculated for 3 g of pure coal substance creates the value of desorption ratio.

Finishing the description of methods for determining total gas content and desorption ratio it is important to notice the above mentioned parameters are often measured simultaneously for the same coal sample. Thus, the results of such simultaneous measurements create a pair (ΔP_2, G). The whole population of such pairs from any deposit given is used by the analytical method.

2. Basic Assumptions of Analytical Method for Determining Residual Gas Content [1]

Values of total gas content and desorption ratio depend on a lot of agents. But from this work point of view, from among the others internal (pore) pressure of coal is the most important one. The pressure because it states sorption capacity of coal (i.e. total gas content for saturated coals), as well as initial pressure of desorption, intensity of desorption depends on (Borowski 1975, Borowski & Gawraczyński 1986, Rogowska 1990). Then, the values can be defined as functions of pore pressure (p_p, MPa):

$$G = f(p_p) \tag{5}$$

$$\Delta P_2 = g(p_p) \tag{6}$$

Consequently, after Borowski and others (Borowski 1975, Borowski & Gawraczyński 1986, Borowski & Matuszewski 1978) it is possible to assume that total gas content and desorption ratio are in proportion one to another (comp.: Fig. 2). Of course, with reference to physical system of coal and gas the assumption is not fully precise, first of all because:

- Not all of coal samples are fully saturated by methane;
- Methane sorption and desorption on coal have different courses (Fig. 3);
- Measurements of total gas content are only made for methane, in opposition to tests of desorption ratio doing for real gas mixtures contained in coal, the composition of which strongly influences the progress of desorption (comp. Greaves et al. 1993).

---

[1] after (Grzybek 1997a), changed.

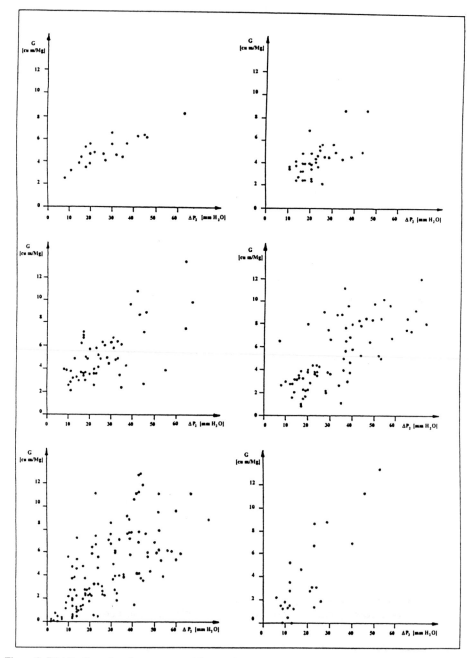

Figure 2. Distribution of total gas content (G, cu m/Mg) versus desorption ratio (ΔP_2, mm H_2O) for coals of different rank, expressed by volatile matter content (V^{daf}, %), observed in area I; acc. to Grzybek (1997a). Results of total gas content found by two-phase vacuum degassing method are presented without adequate recalculation.

Moreover, there are also important the additional, following agents: inacuracies of mensurations of desorption ratio and total gas content, different temperatures of particular tests of desorption ratio and - for corehole samples - differentiated times of uncontrolled desorption before the sample putting into desorbometer. All of the above reasons, and other ones too, cause the proportion between total gas content and desorption ratio is not a function, but only creates a regression equation, typically in the following form:

$$G = a_0 + a_1 \Delta P_2 \qquad (7)$$

where a_0 & a_1 are coefficients of the equation found using the least square method.

At the same time, analysis of desorption process enables to state, that for ΔP_2 equal to zero, what means there is no dynamic desorption, pore pressure of coal is not greater than the atmospheric pressure. In such case total gas content measured simultaneously is in general typical for unknown value of pore pressure (p_x, MPa) meeting the inequality:

$$0 \le p_x \le p_{atm} \qquad (8)$$

where p_{atm} (MPa) is atmospheric pressure.
So, it is obvious, that the (ΔP_2, G) pairs containing value of ΔP_2 equal to zero cannot be used for calculating the regression equation (7).

At the other hand, if pore pressure of coal is equal to the atmospheric pressure, then desorption ratio is equal to zero and equation (7) can be simplified as follows:

$$G = a_0 \qquad (9)$$

In such case desorption of methane from coal may proceed by diffusion only. Consequently the assumption is allowed that average residual gas content (G_r, cu m/Mg) of any coal seam given in respect of figures is equal to value of a_0 coefficient of the regression equation calculated for the seam:

$$G_r = a_0 \qquad (10)$$

The assumption is essential for the analytical method for determining residual gas content.

3. New Procedure of the Analytical Method for Determining Residual Gas Content

Residual methane content was formerly calculated on the basis of the whole population of (ΔP_2, G) pairs found for any deposit or basin given. But such a population grew from year to year. As a consequence of those facts values of residual gas content calculated for the same deposit, but based on the pairs of ΔP_2 and G measurements from different periods of time were not comparable one to another (Table 1). It has been recognized the

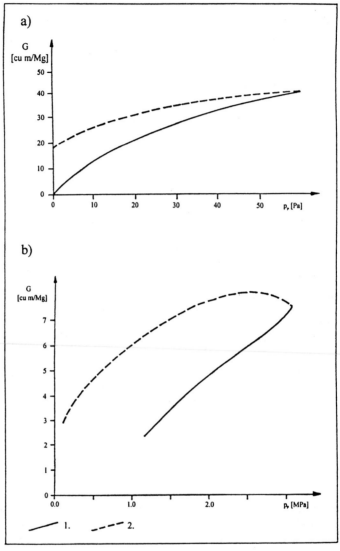

Figure 3. Examples of various progresses of methane sorption (1) and desorption (2) from coals from Hungary coal basins; a - acc. to Radnai & Bakai-Papp (1993), b - acc. to Nodzeński (1990).

situation was because a rank of coal had not been taken into account (Grzybek 1997a). On the ground of this finding a new procedure for calculating residual gas content by the analytical method has been proposed (Grzybek 1997b). In general the procedure consist in:

(1) Dividing the population of $(\Delta P_2, G)$ pairs into subpopulations, correspondingly to volatile matter content of coal samples the pairs of measurements been made for. The division should be twice done in different ways to avoid further calculations influencing by the choice of volatile matter content limits, subpopulations been defined through.

(2) Calculating regression equation and residual methane content separately for each subpopulation, accordingly to the assumptions described in the chapter above.

(3) Calculating average value of volatile matter content of coal samples from each subpopulation.

(4) Finding any relation between average values of volatile matter content and residual gas contents calculated for the subpopulations. [2]

TABLE 1. A comparison of values of residual gas content found for the same coal deposits from the south-west part of the Upper Silesian Coal Basin, Poland, basing on measurements of total gas content and desorption ratio from the years 1977—1985 (after: Jaworek et al. 1986) and 1981—1991 (after Kandora & Grzybek 1992). The rank of coal has not been taken into account during the calculating values presented.

Coal deposit name	Residual gas content, after: (cu m/Mg)	
	Jaworek et al. (1986)	Kandora & Grzybek (1992)
Anna	1,26	2,05
Borynia	1,74	0,75
Jastrzębie	2,56	2,30
Pniówek	1,58	1,51

4. Programme of Research and the Results

The new procedure of analytical method was first tested on the ground of data from a single CBM deposit (for details see: Grzybek 1997a). However results of the test were very good, they were unsufficient to confirm comparability of otcomes of the procedure using in different deposits. To confirm it the procedure was employed for three areas (deposits) from the north (area I), south-west (area II) and south-east (area III) parts of the Upper Silesian Coal Basin, Poland (Fig. 4). Measurements of total gas content had there been made by the two-phase vacuum degassing method (areas I & III) and modified direct-hole method (area II). So, results of they found by the two-phase method were first recalculated, to make them comparable to results of measurements by the

---

[2] It is important to notice here, that the average value of volatile matter content (V_i^{daf}, %) and residual gas content (G_n, cu m/Mg) create a new pair (V_i^{daf}, G_n) for each subpopulation. Then, all of such pairs found in the whole deposit establish series of pairs, which can be treated by statistic methods.

modified direct-hole and U.S. Bureau of Mines methods. Next, calculations were performed separately for subpopulations characterized by difference in volatile matter content of coal samples no greater than 4 %. The calculations were made in two series, for which different divisions into subpopulations were established through displacement by 2% the limits of volatile matter content, subpopulations been defined through.

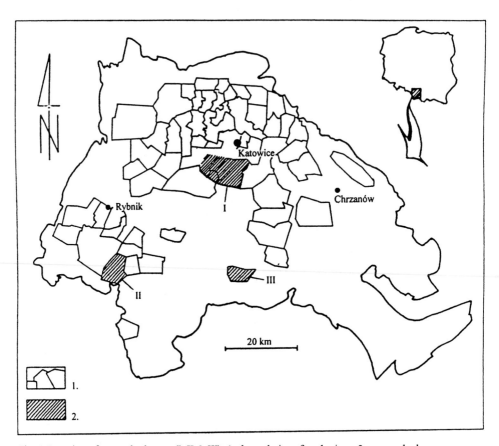

Fig. 4. Location of researched areas (I, II & III); 1 - boundaries of coal mines, 2 - researched area.

Results of the calculations are presented in Table 2 and Figs. 5 & 6. They show residual gas content dependency on volatile matter content of coal. In general, residual gas content rises sharply up, if volatile matter content decreases down to 28 %, and next is more or less stable for the volatile matter content in the range from 14 to 28 %. In detail, for two of analized deposits, i.e. the north (I) and the south-west (II) ones, the dependency is very similar and can be described either by regression equation in the form:

$$G_{r_i} = 5.620 - 0.149\, V_i^{daf} \tag{11}$$

TABLE 2. Results of calculations of residual gas content for two series of the calculations made by the new procedure of analytical method for different areas of the Upper Silesian Coal Basin, Poland

V^{daf} [%]	Area I				Area II				Area III			
	n	r	G_{ri} [cu m/Mg]	V_i^{daf} [%]	n	r	G_{ri} [cu m/Mg]	V_i^{daf} [%]	n	r	G_{ri} [cu m/Mg]	V_i^{daf} [%]
				1st series of calculations								
14.01—18.00	10	0.962	2.362	16.74	18	0.721	2.271	16.81	—	—	—	—
18.01—22.00	30	0.649	2.488	20.16	135	0.667	1.726	20.45	—	—	—	—
22.01—26.00	44	0.474	2.705	24.30	295	0.194	—	—	—	—	—	—
26.01—30.00	63	0.752	1.440	28.12	95	0.682	1.814	27.04	7	0.743	3.660	28.36
30.01—34.00	110	0.691	1.046	32.22	14	0.695	1.332	31.72	34	0.738	2.469	32.51
34.01—38.00	72	0.686	0.239	35.55	1	—	—	—	62	0.608	0.928	35.74
				2nd series of calculations								
16.01—20.00	21	0.846	2.656	18.01	59	0.686	18.75	2.001	—	—	—	—
20.01—24.00	38	0.621	1.693	21.87	210	0.667	22.23	1.834	—	—	—	—
24.01—28.00	57	0.696	2.171	26.16	258	0.736	25.55	1.687	3	—	—	—
28.01—32.00	75	0.717	1.515	30.17	21	0.825	29.72	0.753	13	0.701	30.39	2.319
32.01—36.00	118	0.680	0.649	33.89	6	—	—	—	63	0.693	34.18	1.594

Denotations:

n - number of (ΔP_2, G) pairs, r - correlation coefficient, V_i^{daf} - average value of volatile matter content, G_{ri} - average value of residual methane content.

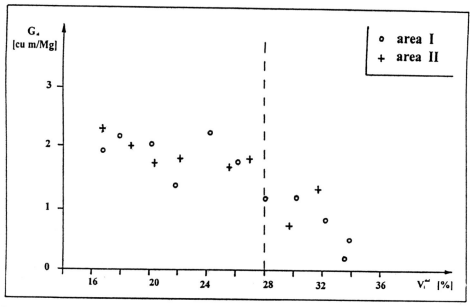

Figure 5. Distribution of residual gas content (G_{ri}, cu m/Mg) versus average volatile matter content (V_i^{daf}, %) for areas I & II.

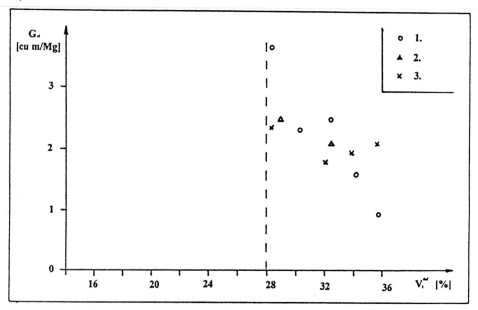

Figure 6. Distribution of residual gas content (G_{ri}, cu m/Mg) versus average volatile matter content (V_i^{daf}, %) for area III, and comparison of residual gas contents calculated (1) and measured directly (2 - by USBM method, 3 - by sorption izotherm method).

— for V_i^{daf} from 28 to 36 % (correlation coefficient $r = -0.840$);
or by the average:

$$G_{ri} = 1.899 \qquad (12)$$

— for V_i^{daf} between 14 and 28 % (standard deviation $\delta = 0.252$).
For the last analysed area, i.e. the south-east (III) one, the dependency between V_i^{daf} and G_{ri} has been found only in the following regression equation form:

$$G_{ri} = 12.864 - 0.331 V_i^{daf} \qquad (13)$$

for V_i^{daf} between 28 and 38 % ($r = -0.949$). Data for more coalificated coals were absent there.

Additionaly results of the calculations performed for area III were compared to outcomes of residual gas content measurements made by either direct USBM method or laboratory, desorption isotherm method. For comparison 25 outcomes of the USBM method and 4 of the isotherm method were used, which were averaged first for the same ranges of volatile matter content, as for division into subpopulations described above. The results, shown in Fig. 6, are found to be similar to the findings of analytical method calculations.

5. Conclusions

The results of the research presented show, that residual gas content is strongly dependent on coalification of coal. The dependency can be described either by a linear regression equation — for coals ranked in the range $V_i^{daf} = 28$ to $V_i^{daf} = 38$ %, or by the average value of residual gas content — for coals ranked in the range $14 \leq V_i^{daf} \leq 28$ %. However, there has been found difference between values of residual gas content of the same rank, but from various areas (I & II versus III), what enables to point the coalification of coal is not only the agent gas desorption from coal depends on.

At the other hand, consistence of calculations results both for different areas (I & II) and with direct measurements for the same area (III) shows the new procedure of analytical method correct. So, the procedure may be used for industrial praxis.

6. References

Borowski, J. 1975. Nowe metody określania metanonośności pokładów węglowych. *Przegląd Górniczy*, **10**, 399—406.

Borowski, J. & Gawraczyński, Z. 1986. Zwiększenie dokładności określeń metanonośności pokładów węgla metodą desorbometryczną. in *Metody rozpoznawania zagrożeń metanowych w kopalniach węgla kamiennego, Katowice, czerwiec 1986*. Katowickie Gwarectwo Węglowe — Zarząd Oddziału SITG w Katowicach, Conference Proceedings, pp. 163—177.

Borowski, J. & Matuszewski, J. 1978. Kierunki udoskonalenia systemu określania gazonośności pokładów węgla. *Przegląd Górniczy*, **1**, 13—19.

Gawraczyński, Z. & Borowski, J. 1986. Określenie zależności między metanonośnością ociosową i pokładową. in *Metody rozpoznawania zagrożeń metanowych w kopalniach węgla kamiennego, Katowice, czerwiec 1986*. Katowickie Gwarectwo Węglowe — Zarząd Oddziału SITG w Katowicach, Conference Proceedings, pp. 151—162.

Greaves, K.H., Owen, L.B., McLennan, J.D. & Olszewski, A. 1993. Multicomponent gas sorption-desorption behavior of coal. in *Proceedings of the 1993 International Coalbed Methane Symposium, Birmingham, May 17—21 1993*, **I**, pp. 197—205.

Grzybek, I. 1993. The Polish methods of coalbed methane content testing and its reserves estimating. in *Proceedings of the 1993 International Coalbed Methane Symposium, Birmingham, Alabama, USA, May 17—21 1993*, **I**, pp. 61—68.

Grzybek, I. 1997a. Nowa procedura obliczeń metanonośności resztkowej węgla metodą analityczną w świetle wyników badań w centralnej części Zagłębia Górnośląskiego. *Zeszyty Naukowe Politechniki Śląskiej, Górnictwo*, **235**, 71—87.

Grzybek, I. 1997b. *The New Polish Method for Estimating Methane Resource of Coalbed Methane*. Poster for the European Coal Conference'97, Izmir, Turkey, May 5—10 1997, (in preparation for publication).

Grzybek, I., Gawlik, L., Suwała, W. & Kuzak R. 1994. Wstępne wyniki implementacji analitycznej metody określania metanonośności resztkowej w GZW. in *Węgiel kamienny - własności, akumulacja, uwalnianie i poszukiwanie gazów kopalnianych. Kraków, 11—12 października 1994*. Akademia Górniczo-Hutnicza, Kraków, Workshop Proceedings, pp. 7—9.

Grzybek, I., Gawlik, L., Suwała, W. & Kuzak R. 1997. Method for estimating methane emissions from Polish coal mining. in Gayer, R. & Pešek, J. (eds), *European Coal Geology and Technology*. Geological Society Special Publication, **125**, pp. 425—434.

Hampton III, G.L. & Schwochow, S. 1994. Gas content analysis of coals: A comparison of the U.S. Bureau of Mines direct method with the Polish "Barbara" Experimental Mine method. in *Proceedings of the Silesian International Conference on Coalbed Methane Utilization, Katowice, Poland, October 5—7 1994*, **I**.

Jaworek, J., Kandora, P. & Michalik, H. 1986. Modyfikacja desorbometrycznej ruchowej metody określania metanonośności pokładów węgla w dostosowaniu do kopalń ROW. in *Metody rozpoznawania zagrożeń metanowych w kopalniach węgla kamiennego, Katowice, czerwiec 1986*. Katowickie Gwarectwo Węglowe — Zarząd Oddziału SITG w Katowicach, Conference Proceedings, pp. 119—134.

Kandora, P. & Grzybek, I. 1992. *On the Criteria of the Possibilities of Balancing and Exploitation of Coalbed Methane*. Proceedings of the United Nations Economic Comission for Europe Workshop on the Recovery and End-Use of Coal-Bed Methane, Katowice, Poland, March 16—21 1992. Report A—9.

Kobiela, Z., Modrzejewski, Z., Simka, A., Sobala, E. & Wrona, B. 1992. *The Methods of Sampling, Laboratory Methods of Determining Gas Content in Coal Seams and Methods of Predicting Methane Emission in the Mine Workings Used in Poland*. Proceedings of the United Nations Economic Comission for Europe Workshop on the

Recovery and End-Use of Coal-Bed Methane, Katowice, Poland, March 16—21 1992. Report B—4.

Mazzsi, D. 1992: *Cavity Stress Relief Method for Recovering Methane from Coal Seam.* United Nations Economic Comission for Europe Workshop on the Recovery and End-Use of Coal-Bed Methane, Katowice, Poland, March 16—21 1992.

Nodzeński, A. 1990. *Wysokociśnieniowa desorpcja dwutlenku węgla z węgli kamiennych w aspekcie procesu uwalniania gazu z pokładu węglowego.* Zeszyty Naukowe AGH, Chemia, 17, Kraków, Akademia Górniczo-Hutnicza.

Radnai, S. & Bakai-Papp K. 1993. Coalbed methane exploitation possibilities in Hungary. in *Proceedings of the 1993 International Coalbed Methane Symposium, Birmingham, Alabama, USA, May 17—21 1993*, **II**, pp. 421—427.

Rogowska, J. 1990. Analiza wskaźnikowa szybkości desorpcji CO_2 i CH_4 z węgli wyrzutowych jako kryterium zagrożenia wyrzutowego. in Litwiniszyn, J. (ed), *Górotwór jako ośrodek wielofazowy. Wyrzuty skalno-gazowe*, **II**, Instytut Mechaniki Górotworu Polskiej Akademii Nauk — Akademia Górniczo-Hutnicza, Kraków, pp. 573—593.

Smith, I.M. & Sloss, L.L. 1992. *Methane Emissions from Coal*. Perspectives. IEA Coal Research.

Yee, D., Seidle, J.P. & Hanson, W.B. 1993. Gas sorption on coal. in Law, B.E. & Rice, D.D. (eds), *Hydrocarbons from coal*. American Associacion of Petroleum Geologists, Tulsa, Oklahoma, USA, 1993, pp. 203—218.

COAL COMPOSITION AND MODE OF MATURATION, A DETERMINING FACTOR IN QUANTIFYING HYDROCARBON SPECIES GENERATED

M.GLIKSON*, C.J.BOREHAM** & D.S.THIEDE*
*Department of Earth Sciences, University of Queensland,Australia
**Australian Geological Survey Organisation, Canberra, Australia

Abstract

Products from various temperatures and heating rates of vitrinite and extractable bitumen from coals of different rank were studied by pyrolysis gas chromatography - mass spectrometry (py-gc-ms) and flash pyrolysis, and compared to 'naturally' matured Bowen basin coals. Generation temperatures and quantities of hydrocarbon species from vitrinite and bitumen were shown to be dependant on initial rank of the coal as well, as H/C of vitrinite. A significant amount of bitumen formation characterises rapid heating of vitrinite as established by py-gc-ms, as well as microscopy of residues. Similarly significant bitumen input is noted for Bowen Basin coals, supporting maturation by rapid heating . Using flash pyrolysis, bitumen has been confirmed to be a major source of methane. The light oil during pyrolysis is readily expelled from the coal leaving behind bitumen in vitrinite micro-cleats and char (inertinite) cavities. The ratio of solvent extractable to non-extractable bitumen may be used as indicator of methane generation in these coals.

Comparison of open system pyrolysis with confined pyrolysis under varied pressure showed accelaration of organic maturation at presures up to 250 bar, and retardation at higher pressure. Very low pressure of 2-3 bar showed peak bitumen generation at the same tempertaure as in open system. On the other hand, light oil and methane peaked at higher temperature with significantly higher yields that in open system due to bitumen cracking.

TEM observations of vitrinite and inertinite (chars) in the coals studied, as well as their residues from pyrolysis experiments highlight differences in microporosity between the two macerals. Microporosity in chars explains diffusion of hydrocarbons through char walls as well as storage of bitumen and gases within the large internal volume of closely spaced micro and nano-pores.

The conversion of vitrinite to char has been followed experimentally by heating vitrinite from 300 to 800 degrees C.

The gradual development of mosaic texture was recorded by TEM in residues from a range of temperatures. A 3- dimensional reconstruction of mosaic textures facilitates understanding of processes associated with hydrocarbon adsorption and diffusion.

Introduction

Previous studies of coalseam gas and generally hydrocarbons sourced from coal in the Bowen Basin were based on conventional thermal maturation by step-wise subsidence and burial of the coal, with heat source in the basement, and heat transfer predominantly by conduction. The present study is part of an extensive integrated multidisciplinary study of the Permian coals of the Bowen Basin, Queensland, Australia (Fig.1) using a range of techniques and methods to elucidate their thermal history and composition of the hysrocarbons generated. These studies provide evidence that the coals in the central and northern parts of the basin matured to their present rank as a result of rapid heating by hydrothermal processes rather than through basin subsidence and depth of burial (Glikson *et al*. 1995; Golding *et al*. 1996; Golding *et al*.,this volume). The Bowen Basin is a back-arc extensional foreland basin that contains up to 6500 m of Permo-Triassic mostly terrigenous sediments and volcanics. Intense rifting and magmatism accompanied the breakup of the Gondwana supercontinent in the late Mesozoic and continued into the Tertiary (Veevers 1989) with a pronounced effect on the existing coal measures. Bowen basin coals have been recorded as reaching their present day maturation during the early Triassic (Mallet *et al*.,1990) when Permian coalseams were buried to about 3000-4000m. at a maximum temperature of 120 degrees C.

Our previous studies of coals throughout the Bowen Basin, documented irregular maturation profiles, local to sub-regional thermal aureoles, as well as irregular coking and extensive mineralisation (Crossley 1994; Barker 1995; Glikson *et al*.. 1995; Golding *et al*.. 1996), suggesting dominant magmatic-hydrothermal effects in the late Triassic and Cretaceous on coal maturation and cleat mineralisation rather than the earlier deep burial effect.

Migrated coal-sourced hydrocarbons in the Bowen Basin are reservoired within late Permian to early Jurassic sandstones in the Taroom trough and Permian sandstones in the Denison Trough (Boreham,1995). Coalseam gas and bitumen are sourced and reservoired within the late Permian coalseams throughout the Bowen Basin (Golding *et al*.,1996).

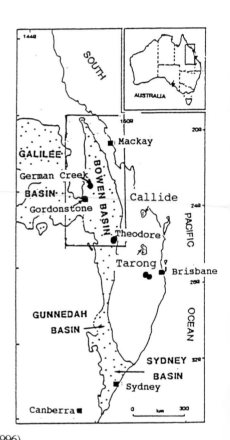

Figure 1. Location map

METHODS OF STUDY

Samples
Coal samples were collected from drill cores and high walls of coal exploration and mining areas in the north, central and southern Bowen Basin, as well as from the Triassic coals of Callide and Tarong sub- Basins (Fig.1), and overlying Jurassic coals of the Clarence-Moreton Basin. Samples with >96% vitrinite from bright bands were used for experiments (Table 1).

TABLE 1. Details of samples used for pyrolysis

Sample no.	Location/ age	Initial Ro	Maceral composition %				
			Vit.	Inert.	Lipt.	Bit.	Min.
CAL	Callide sub-Basin Triassic	0.5	59.3	28.2	2.0	5.9	4.6
TH4	Theodore Southern Bowen Basin, Permian	0.5	78.0	8.0	4.0	7.0	3.0
TAR	Tarong Basin Triassic	0.6	79.6	10.6	2.0	4.0	3.8
GCA (GOR)	Gordonstone Central Bowen Basin, Permian	0.8	78.0	8.0	<0.1	14.0	<0.1
247	"	1.4	52.0	22.0	<0.1	12.0	<0.1
OC-95	Oaky Creek, Central Bowen Basin,Permian	0.9	62.0	14.6	<0.1	9.4	14.0
EMCC	Ebenezer mine,Jurassic	0.5	49.1	3.0	13.1	3.6	30.8
JPCC	Jeebropilly mine, Jurassic	0.5	56.7	2.6	7.7	6.9	25.8

Experimental
(i) *Py-gc-ms* under open, dry conditions was carried out on coals using the technique described by Boreham *et al* (1998). Briefly, 15 mg of powdered sample was heated at 10 deg. C in a helium atmosphere, in quartz tubes within a mini-furnace (GC Chromatography, UK) over the temperature range of 300-550 deg.C. Pre-absorbed gases and liquids were thermally extracted from the coal sample by rapid heating to 300 deg.C. Commencing from this temperature stepwise pyrolysis with 30 deg.C intervals was applied. Between each interval the temperature of the mini-furnace was dropped to 50 deg.C below the final pyrolysis temperature after which the pyrolysis products were analysed by gc-ms.
(ii) *Confined py-gc-ms* was carried out in gold, sealed capsules following experimental procedures and techniques of Monthioux & Landais (1987), Freund *et al.* (1993), and Hill

et al. (1994). Pressures applied were 2-3 bar, 250 bar, and 500 bar at temperatures of 300,350,400,450,500, 550 degrees C at heating rates of 10 degrees C /min.

(iii) *Flash pyrolysis - ms* was based on a variant of the flash pyrolytic method where static isothermal pyrolysis occurs in an inert atmosphere. The sample size is adjusted depending upon sample type and the pyrolysis temperature. In general, sample size ranges from 20 to 200 mg. for coal samples and 10 to 30 mg for bitumen samples. Samples were placed in Pyrex tubes with glass wool plugs. Each sample is connected to the inlet system of an online quadrupole mass spectrometer, and the sample and inlet purged with ultra-high purity argon prior to pyrolysis. However, the samples are not evacuated prior to argon purge, allowing the presence of small amounts of atmospheric gases in the tube. The evolved head space gas is analyzed by injection into the on-line quadrupole mass spectrometer. Identification of gas species is by monitoring selected m/z peak intensities. After applying correction factors for minor peaks from other compounds, the corrected peak intensities are converted into partial pressure data. The partial pressure data is normalized to sample weight and to an internal standard that is the partial pressure of the Ar purge gas employed. The resulting data is converted to a dry air and CO2 free basis (DACF) by subtracting the partial pressure of N2, O2, Ar, H2O and CO2, and calculating the remaining detected gases on a percentage partial pressure basis.

Petrology and Electron Microscopy

Macerals in bright bands used for pyrolysis were quantified using standard technique as outlined in Stach *et al.*(1975).

Residues of samples heated to 400, 450, 500, 550, 600 ,700 and 800 C by TGA and py-gc-ms were collected for microanalysis, petrology, TEM and stable isotope analysis. (Residues of >600 C. heating were obtained from vitrinite, using Thermo-Gravimetric techniques in J.D.Saxby's laboratory, N.Ryde, NSW)

All residues from experimental coal maturation were observed and recorded in reflected white light and fluorescence mode, at wave length of 546 nm, using an MPV-2 microscope. Reflectance (Ro=random) was measured on all samples prior to heating, and on resulting residues .

Transmission electron microscopy (TEM)

 TEM was applied to ultra thin sections of coal prior to heating as well as residues from artificially matured coals, following the method outlined by Glikson & Taylor (1986). Coal macerals as well as residues were also observed in TEM, as dispersed particles on carbon film not involving sectioning. Representative samples from original coals of different rank were also observed in TEM . All changes were recorded visually and represented by photomicrographs. Observations were carried out on a Hitachi-H-800 , Jeol 1010 and Jeol 1210 electron microscopes, Jeol 6400 field emission SEM equipped with Energy Dispersive Spectrometer (EDS) were used in back-scatter mode (directly comparable to reflected light) for elemental analysis of mineral inclusions. The same coal blocks were used for maceral analysis and reflectance determinations. Extracted bitumen processed for TEM was used as reference material for mixed samples.

Elemental analysis was carried out routinely on separate lithotypes of initial coals and subsequent residues using a Carlo Erba Analyser Model 1106.

Bitumen extraction for pyrolysis from complimentary coal samples, representative of rank from sub-bituminous to medium volatile bituminous was carried out using chloroform in a Soxtech System HT2. Bitumen content was quantified by gravimetric methods for the solvent extractable fraction, and by petrological methods using fluorescence mode for all bitumen before and after extraction.

Results

GENERATION OF HYDROCARBON SPECIES FROM VITRINITE

open system pyrolysis

Fig.2 (Table 2) shows the yields of hydrocarbon species (methane/C1, wet gas/C2-C5, light oil/C6-C14, heavy oil/C15+) generated from vitrinite in three coals of different initial rank: A sub-bituminous coal (TH4) of 0.5% Ro, at the beginning of the oil window, produces highest light oil yields, followed by heavy oil (bitumen), and only minor quantities of methane. A coal (GCA) at peak oil generation (0.8%Ro) shows equal amounts of light and heavy oils peaking at a common pyrolysis temperature, 450 deg.C. In comparison, the lower rank coal shows peak heavy oil at this same temperature while peak light oil yields occur at lower temperature, namely 420 deg.C.

TABLE 2.   Mass loss during open pyrolysis from coal
of different rank

Sample	Initial Ro%	Initial mass (mg.)	Final mass (mg.)	Total mass loss (mg.)	mass loss (%)
CAL	0.5	30.85	17.95	12.91	43
TH4	0.5	20.60	13.16	7.44	36
TAR	0.6	26.03	17.04	8.99	35
GCA	0.8	11.35	7.73	3.62	32
247	1.4	18.17	15.88	3.29	18

Both light and heavy oil generation continues to higher temperatures from vitrinite of the high volatile bituminous rank coal compared to that from the sub-bituminous coal. Li *et al.* (1993) also obtained significantly higher bitumen yields from a high volatile bituminous vitrinite at 500 C, than at 400 C. In the present study it appears that the high volatile bituminous coal had exceptionally high yield of heavy oil, as is also evident from observations of bitumen micro-veins and cleat infill, as well as significant quantities of extractable bitumen. It is noteworthy that the maturation path of the high volatile bituminous coal (GCA-GOR) on a Van-Krevelen diagram (Fig.3) closely resembles the trend of the maturation path of type I kerogen, possibly due to the high input from emplaced heavy oils. The same coal also involves the lowest O/C ratio, and a strong correlation between initial rank and decreasing O/C ratio, most likely due to decarboxylation reactions over the range 0.5n to 0.8%Ro. In a higher rank coal (1.4%Ro) a significant drop in the yield of all hydrocarbon species is evident (Fig.2c), with both heavy and light oils peaking at close to 490 deg.C. A bitumen to light oil ratio of 1 :1.2 is evident in the sub-bituminous and high volatile bituminous coals. The higher rank medium volatile coal (Ro=1.4%) displays a ratio of bitumen to light oil of 1:2. These ratios are typical in parts of the world where hydrothermal processes were responsible for organic thermal maturation compared to burial induced maturation (Altbaumer *et al.*, 1981).

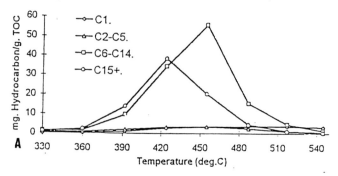

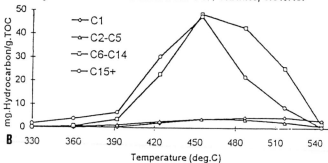

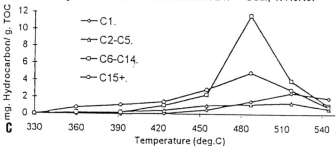

Figures 2A-C. Quantitaties of hydrocarbon species generated from vitrinite of different rank, in open system pyrolysis

Methane generation is generally characterised by low yields in sub-bituminous, as well as high volatile bituminous coals, peaking at significantly higher temperature than oil, at 520 deg.C, and remains at a more or less constant level to the limiting temperature of the experiment, 540 deg.C.

Generally the ratio of extractable to non-extractable bitumen decreases with rank, as evident from higher quantities of extractable bitumen obtained from the immature Jurassic coals (JPCC, EMCC) compared to Late Triassic (CAL) and Permian (TH4) coals of the same rank, and further emphasised in Permian coals with increasing rank (Fig.2).

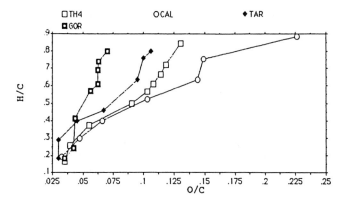

Figure 3A. Atomic H/C versus O/C of vitrinite residues from coals of different rank

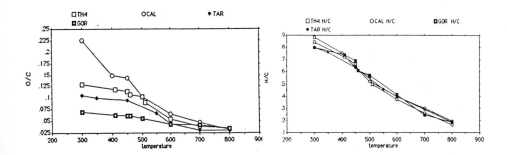

Figure 3B. Atomic O/C & H/C evolution with temperature of vitrinites from coals of different rank

Figs.4-7 show the partial pressure percentages of the gases generated by flash pyrolysis of the bitumen and the parent coal of Permian and Jurassic samples. It is evident that bitumen generates higher yields of hydrocarbon gases compared to the parent coals, with methane generally comprising over a third of the C1-C7 yields. However, the ratio of methane generated from bitumen to that generated from whole coal was higher in the Permian than in the Jurassic coals, regardless of coal rank (Figs. 5-7). In samples where higher methane generation from char (inertinite) occurred rather than commonly from vitrinite (Table 3) supports petrological data that show prevelance of inclusions of bitumen within char cavities (Plate fig.1). The bitumen inclusions in this instance are the source of methane. Generally, however whole coal samples would represent direct generation of methane from vitrinite.

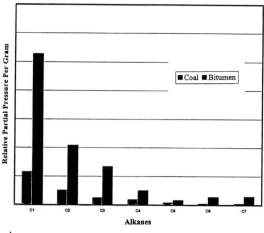

Figure 4. Comparative pyrolysis of coal and bitumen from JPCC

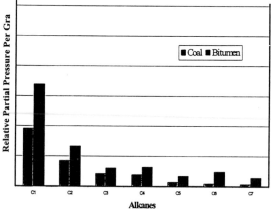

Figure 5. Comparative pyrolysis of GOR-GCA vitrinite and bitumen

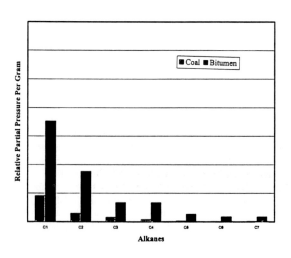

Figure 6. Comparative pyrolysis of coal and bitumen from 247

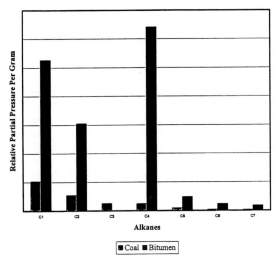

Figure 7. Comparative pyrolysis of vitrinite
and bitumen from TH4

TABLE 3. Dry air and CO_2 free relative partial pressure percentage of
hydrocarbon gases generated at 350 degrees C. Sample OC95

Alkanes	Vitrinite	Inertinite	Bitumen
CH_4	13.37	2.38	8.94
C_2H_6	6.85	3.63	9.25
C_3H_8	4.21	1.19	n.d.
C_4H_{10}	3.14	0.56	55.04
C_5H_{12}	1.10	n.d.	2.14
C_6H_{14}	0.47	n.d.	0.35
C_7H_{16}	0.83	0.15	0.20

Pyrolysis in anhydrous confined system at varying pressure
Methane generation from sub-bituminous coal TH4 shows a strong pressure dependance with decreasing methane yields with increasing pressure (Table 4, Fig.8). Quantitavely significantly higher gas yields were obtained from heating TH4 coal at 420 C and 2-3 bar pressure than from open pyrolysis of the same coal. Maximum methane yield from TH4 coal in open system pyrolysis was about 4 mg/gTOC compared with close to 12 mg/gTOC in open system pyrolysis was about 4 mg/gTOC compared with close to 12 mg/gTOC methane at 2-3 bar, at approximately the same temperature. At higher pressure, 250 and 500

bar, methane generation dropped less dramatically, to slightly lower levels than in open pyrolysis at the same temperature (500 C). Results from confined pyrolysis of GCA vitrinite show quantities of methane generated at 500 bar pressure equal to those from open system pyrolysis at ambient pressure at the same temperature, but increased at higher temperature as a result of cracking.

Different results seem to dominate heavy oil (bitumen) generation. Peak bitumen generation occurs at 420 deg.C in the open system pyrolysis of TH4 coal (Fig.2a), as well as in confined pyrolysis at 2-3 bar pressure, with 40 mg/gTOC in the former compared to only 25mg/gTOC in the latter. On the other hand confined pyrolysis at 250 bar pressure produced higher yields of bitumen at lower temperature than at atmospheric pressure (Fig.8). Higher pressure (500 bar) lowered bitumen yields, and increased peak temperature. In other words, accelerated maturation occurred at 250 bar whereas retardation of maturation resulted when 500 bar pressure was applied in a confined system.

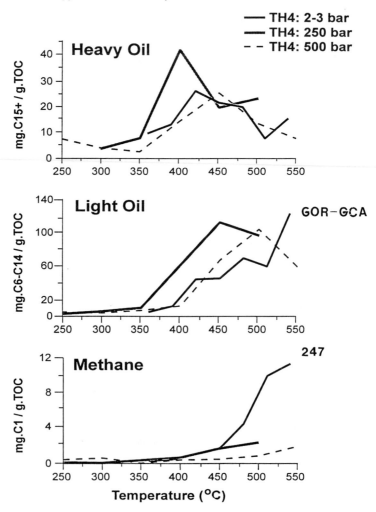

Figure 8. Pyrolysis of vitrinite in 3 coals in confined and pressurised system

Generation of bitumen therefore occurs at a slightly lower temperature in TH4 when 250 bar pressure has been applied compared to open system pyrolysis, with similar quantities of bitumen generated. At 500 bar a significant drop in yield is observed, as well as a shift of peak generation to higher temperature (Fig.8a). From these results it appears that decreasing methane and light oil production as pressure is increased, is related to retardation of cracking of bitumen retained within the confined system. The overall drop in bitumen production corresponds to increase in methane generation (Fig.8c).

Peak bitumen generation from GCA-GOR vitrinite (0.8%Ro) at 500 bar pressure occurs at a lower temperature than in TH4 , with lower quantities generated. This coal being of higher rank than TH4 has already generated copious amounts of bitumen prior to artifical maturation.

TABLE 4. Anhydrous pyrolysis of vitrinite

Sample	Pressure bar	Temp. deg.C	Ro %	Methane	Light-oil mg/g TOC	Bitumen
TH4	open system	420	0.5	2.0	35.0	40.0
	" "	450		3.0	55.0	22.0
	" "	510		4.0	3.0	0.1
	2-3	420		1.0	42.0	25.0
	2-3	550		11.8	122.0	16.0
	250	400		60.0	43.0	0.7
	250	450		1.5	112.0	20.0
	250	500		2.1	96.1	23.5
	500	450		0.2	66.5	25.6
	500	500		0.6	110.0	14.0
	500	550		1.5	60.0	10.0
GCA-GOR open	system	450	0.8	3.0	48.0	48.0
	"	510		5.0	25.0	10.0
	500	400		0.1	15.0	22.0
	500	450		2.5	20.0	7.0
	500	500		5.5	120.0	10.0

Generation of light oil from TH4 vitrinite occurs in confined pyrolysis system with application of 2-3 bar pressure at about 540 deg.C with about twice as high yields as in open system. Light oil in the open system pyrolysis peaks at temperature almost 100 degrees lower than in the confined system with 2-3- bar pressure. These differences can be explained as being due to the absence of secondary cracking in open system. Within these systems temperature and pressure have pronounced effect on species and quantities of hydrocarbons generated, with increase in pressure lowering temperature (e.g. during heavy oil generation).

PETROLOGY AND TEM

The primary macerals vitrinite and inertinite (char) were the basis of the present study, as well as the secondary maceral bitumen (Robert,1988; Jacob,1989, cf.bituminite of Taylor *et al.*,1991). The coals of the Bowen Basin display a diversity in inertinite to vitrinite ratios throughout the basin, as well as in the samples used for this study (Table 1). All Bowen Basin coals have an input from bitumen, varying from <5 % to 28 %, with usually a 1:4 ratio of solvent extractable to non-extractable. Highest bitumen concentrations occur at peak oil generation, of 0.8 %Ro (e.g.GOR-GCA coal, Figs.9). This compares favourably to solvent-extractable yields from petroleum exploration wells in the Bowen and Surat basins, where maximum yields occur between 0.8 to 0.9% Ro (Fig.9A).

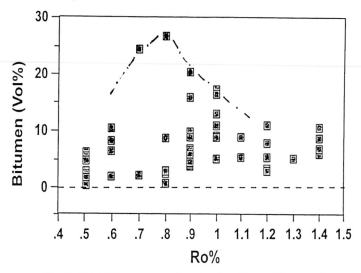

Fig.9: Total bitumen content and rank in Bowen Basin coals

Perhydrous vitrinite is the predominant vitrinite in most Australian coals, with varying degrees of oil proneness (Boreham,1995). Vitrinite in Late Triassic CAL coal (0.5% Ro) has generated early hydrocarbons as indicated by the presence of exsudatinite and bitumen in the coal prior to heating (Plate fig.2), and as also reported in an earlier study (Glikson &

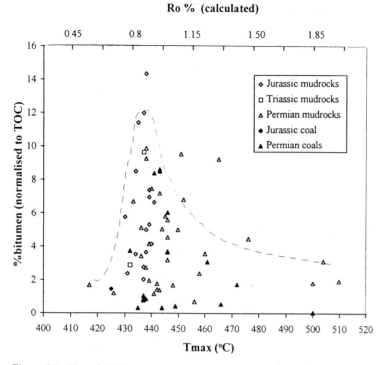

Ro % (calculated)

Figure 9A. Plot of % bitumen content (normalised to TOC) versus Rock-Eval Tmax (deg.C). Ro (%) is calculated from linear regression analysis (Ro=0.01718 Tmax - 6.749; Ro=0.97). Data from Boreham (1994).

Fielding,1991). Residue from CAL coal to 450 deg.C pyrolysis showed little distinguishable change when observed in reflected light with the exception of elevated reflectance values (Table 5). TEM, however, showed faint 'charring of vitrinite, bitumen displaying 'flow structures', and mobilized mineral matter often aligned along flow lines (Plate fig.3). These features suggest oil generation and the retention of heavy-end products within the coal. The observation of bitumen within inertinite would demonstrate primary migration by diffusion, and explain consistent presence of bitumen in inertinite cavities. Only minor charring / coking has occured in CAL coal at 450 deg.C. The same coal after 600 deg.C observed in TEM showed fairly well defined char formation (Plate fig.4). The char appeared to be organised into 'molecular orientation domaines' towards a mosaic texture such as has been observed and documented in earlier TEM studies (Ayache & Oberlin 1990; Fortin & Rouzaud 1994). Minerals have dispersed at this stage, and may only be detected in TEM as sub-micron size particles, residing within the mosaic texture. At 700 - 800 deg.C the charring and mosaic formation are completed, and an ordered 'molecular orientation' is clearly visible (Plate fig.5) closely resembling naturally matured coals (Golding et al. 1996). TEM observation of the residues enables the reconstruction of a three dimensional model of a typical 'mosaic' texture (Fig. 10). The ultra-thin ridges 'U' (lighter, electron transparent) forming cell-like organisation in a channel-like structure (C) bordered by thickenings (T) along the 'valleys' bordering the 'ridges'. This type of texture produces large surface areas for oil and gas diffusion, explaining the presence of bitumen within micro and macro cavities in char (Plate fig.6), the latter being a non- oil generating component of coal. The same microporosity may provide storage for methane. A recent study (Lawrie *et al*, 1997) obtained identical texture by applying atomic force microscopy to inertinites.

TABLE 5. Rank by Ro (random) % of vitrinite residue
from different heating temperatures

Temperature deg.C	CAL	TH4	Ro% TAR	GCA	247
initial	0.5	0.5	0.6	0.8	1.4
350	0.6	0.6	0.8	0.8	1.4
400	0.7	0.8	0.9	0.8	1.4
450	1.0	1.1	1.2	1.5	1.9
500	1.4	1.4	n.d.	2.0	2.4
550	n.d.	2.0	2.1	n.d.	n.d.
600	2.5	2.7	2.6	2.8	4.0
700	3.5	3.9	n.d.	4.5	5.1
800	5.0	6.5	4.6	5.5	6.5

Coking of vitrinite and mosaic formation in CAL coal, are recognised in residue from 600 deg.C but no earlier. On the other hand, coal TH4 of the same initial rank displayed some coking and newly generated bitumen at 450 deg.C , and complete coking by 500 deg.C. At 600 deg.C TH4 coal is seen in light microscopy to have developed coarse mosaic texture, and newly generated bitumen. Although small amounts of bitumen (C15+ hydrocarbons) are expected to be generated at these high temperatures, the potential exists for significantly more bitumen to be retained in the quasi-closed mosaic texture. Both coals are of the same initial rank (0.5% Ro), but CAL coal has high inherent mineral matter content, often at a sub-micron scale size. On the other hand TH4 coal rendered mineral-free vitrinite. Retardation of fluidity / plasticity by the presence of clay minerals has been demonstrated in a study by Saxby *et al.* (1992). This may be the reason for poorly developed coking of CAL compared to TH4 coal of the same rank. TEM of TH4 residue from vitrinite after 600 deg.C shows a well developed mosaic texture.

For TAR coal of initial 0.6% Ro, coking occurred at 500 deg.C. TEM of the residue showed partial charring and conversion of some vitrinite to inertinite, and prevalent bitumen with 'flow-aligned' minerals (Plate fig.7). At 600 deg.C, charring is complete, and is identical in mosaic texture from CAL vitrinite at 700-800 deg.C, equivalent to mosaic textutres documented in earlier studies of coal evolution (Brooks & Taylor 1966; Shibaoka & Ueda 1978).

GOR coal differs from the other coals by a significantly higher bitumen presence, as infill of micro cleats (Plate fig.8), char cavities, and frequently forming veins (Golding *et al.*1996) . The coal is already at peak oil generation at 0.8 % Ro, and yielded up to 10% extractable bitumen. Petrological analysis showed up to 40% bitumen content (commonly 15-20%), in pure form or adsorbed on clay indicating that only a fraction of the total bitumen is accessible to solvent extractraction. Bitumen infill of micro fractures has been used in an earlier study (Hvoslef *et al.*,1987) as indication of hydrocarbon migration in coals. The same study noted that a larger proportion of the bitumen remains entrapped in the coal matrix after solvent extraction. Clearly, the immature bitumen lends itself to solvent-extraction as varified by fluorescence and supported by FTIR of initial composition of bitumen in the Bowen Basin coals, as well as its gradual aromatisation during maturation (Mastalerz *et al.* 1996).

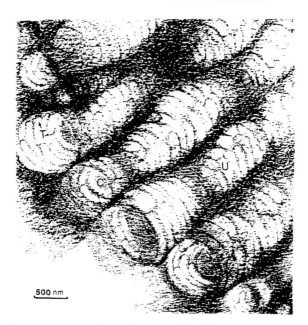

Figure 10. 3 - D reconstruction of mosaic texture in inertinite
T= Thickenings /'valleys'
U= Ultra-thin regions
C= Pseudo-channels

GOR coal generated significant quantities of oil and additional bitumen at 400-450 deg.C as demonstrated by py-gc-ms results (Fig. 2b), and observed in light microscopy of the residue. Char formation and coking became pronounced already at 500 deg.C, and displayed complete mosaic texture at 600 deg.C. Numerous small and large pores characterise this coal even in residue after 800 deg.C heating (Plate fig.9) due to intense and continuous generation and expulsion of volatile hydrocarbons. This is also confirmed from pyrolysis results of the present study (Fig.2), as well as earlier studies of regions subjected to rapid high thermal processes, which have shown long chain n-alkanes to have generated up to 1.7% Ro (Altbaumer *et al.*. 1981).

In TEM observations GOR vitrinite prior to heating is characterised by breakdown of fine sub-micron lamellae which characterise perhydrous vitrinite (Taylor 1966; Glikson & Fielding, 1991). In the GOR coal now at peak oil generation, the start of formation of mosaic texture can be observed (Plate fig.10a-d). In this 'naturally' matured coal, several stages are seen to co-exist at the same rank. This may be an indication of short-lived rapid heating at relatively high temperature, as also evident from studies of associated minerals (Golding *et al.*. 1996). Artificially matured samples do not show these features since heating is uniform throughout the sample.

TEM of GOR vitrinite residue after 600 deg.C displays some of the characteristic mosaic observed in the residues of the other coals. However it differs in having bitumen trapped within the mosaic texture (Plate fig. 11a, b), in agreement with Hvoslef *et al.*'s study (1987) who demonstrated entrapped 'saturates' within the coal matrix during rapid heating. The presence of newly formed mesophase within the mosaic texture after 800 deg.C (Plate fig.12) suggests that previously trapped bitumen has been fluidised at temperatures >600 deg.C, and converted into pyrobitumen. It has been suggested (Lewan, 1992) that bitumen is transformed mainly into an insoluble pyrobitumen in anhydrous pyrolysis, whereas in hydrous pyrolysis it would decompose into oil which may be either expelled or further cracked into methane in a confined system.

Coal 247 with initial highest rank (1.4% Ro) is susceptible to coking at the relatively low temperature of 400 deg. C. By 600 deg.C large pores still persist within a very fine lamellar mosaic matrix (Plate fig.13) characteristic of coking of high rank coal (Benedict & Thompson, 1980). Fine mosaic and pyrolitic carbon predominate in residue after 800 deg.C, closely resembling coals observed by the authors in proximity to intrusions in numerous locations in the Bowen Basin.

Residues from confined pyrolysis at varying pressure
TEM of the residue from TH4 vitrinite developed mosaic texture already at 400 deg.C when and 250 bar confined pressure, compared to a similar texture at 500 deg.C in open system pyrolysis at atmospheric pressure. GOR-GCA vitrinite developed uniform breakdown of lipidic laminae at 250 deg.C and 500 bar (Plate fig.14). Charring towards a mosaic texture was well advanced at 400 deg.C and 500 bar (Plate fig.15). It appears that physico-chemical changes in vitrinite occur earlier when heated in confined and pressurised system. The most obvious difference between the residue obtained from vitrinite in confined pyrolysis and the open pyrolysis residue of the bitumen-rich GOR-GCA coal is the absence of trapped bitumen, within the mosaic. These observations are a testimony to breakdown/cracking of bitumen within the confined system. A pronounced coarsening of mosaic textures is also observed.

MATURATION PATH DETERMINED BY VITRINITE REFLECTANCE & H/C ATOMIC RATIOS

The two sub-bituminous coals CAL and TH4 (Table 1) which had 0.5% Ro, initially followed a similar maturation path up to the temperature of 600 deg.C (Fig.11, Table 5). At 600 deg.C CAL coal begins to diverge from the common path, and at 700 deg.C and above the two coals follow distincly different paths with CAL coal having consistently lower reflectance. This retardation of maturation in CAL coal may be attributed to the consistent presence of mineral inclusions in CAL vitrinite to no less than 4%, predominantly clays and silicate minerals. Maturation retardation has been attributed to mineral presence (Saxby *et al.* 1992). On the VanKrevelen diagram (Fig.3A, Table 6), the paths of the two coals are very distinct initially where the higher oxygen content of the CAL coal is evident up to 500 deg.C. At higher temperatures, the maturation paths of the H/C and O/C atomic ratios for the two coals show similar elemental compositions (Figs.3A,3B).

The maturation path (Fig.11) of TAR coal (0.6 %Ro) closely follows that of CAL and TH4 coals up to 600 C . At 700 C, TAR coal path overlaps that of TH4 coal, but diverges from CAL coal, to follow more closely that of TH4, and at 800 C is slightly below TH4. The elemental composition of vitrinite of TAR coal parallels that of TH4, but with an overall higher H/C, and lower O/C up to 700 C. By 800 C, the elemental composition of TH4 coal merges with that of the other coals at that temperature.

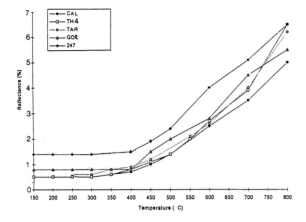

Figure 11. Maturation path of vitrinite obtained from
residues after open pyrolysis

TABLE 6. Atomic H/C and O/C of initial vitrinites and
their residues from different heating temperatures

| Temperature | CAL | | TH4 | | TAR | | GCA (GOR) | |
deg.C	H/C	O/C	H/C	O/C	H/C	O/C	H/C	O/C
initial	0.88	0.23	0.84	0.13	0.80	0.11	0.80	0.07
350	n.d.	n.d.	n.d.	n.d.	0.76	0.10	n.d.	n.d.
400	0.76	0.15	0.72	0.12	n.d.	n.d.	0.74	0.06
450	0.64	0.14	0.66	0.11	0.63	,0.09	0.70	0.06
500	0.52	0.10	0.57	0.10	n.d.	n.d.	0.57	0.05
550	n.d.	n.d.	n.d.	n.d.	0.49	0.06	n.d.	n.d.
600	0.40	0.06	0.37	0.05	0.39	0.04	0.41	0.04
700	0.30	0.04	0.37	0.05	0.39	0.04	0.41	0.04
800	0.19	0.03	0.16	0.03	0.18	0.03	0.18	0.03

GCA-GOR coal follows a similar maturation trend (Figs.3,11) as TAR coal, except always at
a higher vitrinite reflectance up to 800 deg.C. The reflectance offset is due to the initial
higher rank (Ro=0.8%) of the unheated GOR coal. This is further illustrated in 247 coal
where the higher initial reflectance of Ro=1.4% preserves an offset, and follows a
consistantly higher maturation level up to 800 C where it merges with TAR and GOR coals.
The elemental composition of GOR coal when plotted on the VanKrevelen diagram has a
similar H/C atomic ratio compared with the lower rank coals but is characterised by a low
O/C ratio (Fig.3B). The higher H/C atomic ratio of GOR coal is due to both high bitumen
content and the perhydrous nature of the vitrinite. The excellent hydrocarbon generation
potential of GOR is further illustrated from py-gc-ms analysis. The GOR coal generates
higher hydrocarbon yields compared to the lower rank TH4 coal.

Comparison of maturation paths by reflectance and elemental analysis highlights subtle differences between the various vitrinites, being graphically illustrated on the VanKrevelen diagram (Fig.3). The differences between maturation paths of CAL and TH4 coals of same rank, reflect not only differences in vitrinite composition, but also the presence or lack of minerals and functional groups, emphasised by O/C atomic ratios. Most pronounced loss of oxygen functional groups is observed between 300 - 450 deg. C. The low O/C atomic ratios obtained for the GOR coal are reflected in low yields of CO2 on pyrolysis (Fig.12).

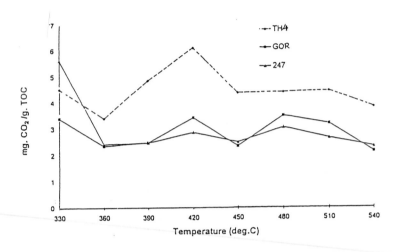

Figure 12. CO2 yields from open pyrolysis of vitrinites

The differences in maturation paths of the coals shown by our study confirm an earlier study by Khavari-Khorasani & Michelsen, (1994). There, maturation, as expressed by reflectance evolution, is not a sole function of chemical changes, but is more likely a function of the structural heterogeneity of vitrinite. The heterogeneity of vitrinite in the Bowen basin coals has been varified by our TEM studies.

Maturation profiles and iso-reflectance contours drawn for some coal seams of the Bowen Basin do not support uniform thermal maturation through burial and step-wise subsidence, but rather a rapid and variable thermal effect at high temperature.

The irregularity in vitrinite reflectance profiles (Figs.13,14), and presence of localised thermal aureoles (Fig.15) point to a non uniform heat transfer by hydrothermal processes (depending on numerous variables, not solely rock conductivity) associated with intrusives and volcanic activity. This is further emphasised by observing changes occurring to organic matter, particularly coking structures. The predominant cokes/chars throughout the Bowen Basin display a characteristic mixture of coarse mesophase and pitch (pyro-bitumen). A mixture of the two entities appears characteristically as a bubble-like texture typical of perturbance during the plastic phase of the vitrinite.

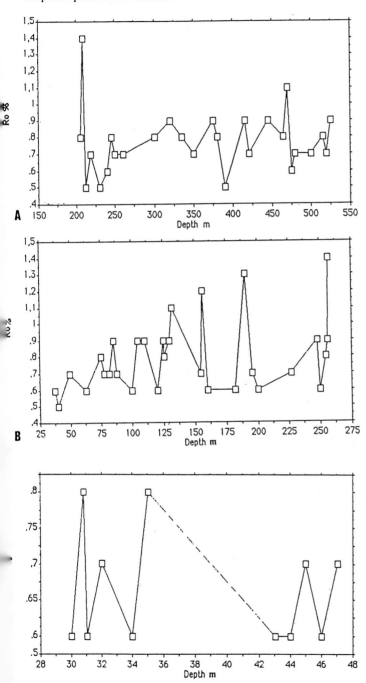

Figures 13A,B. reflectance profiles in Permian coal seams in drillholes from central Bowen Basin

Figure 14. Reflectance profile of a single seam, northern Bowen Basin

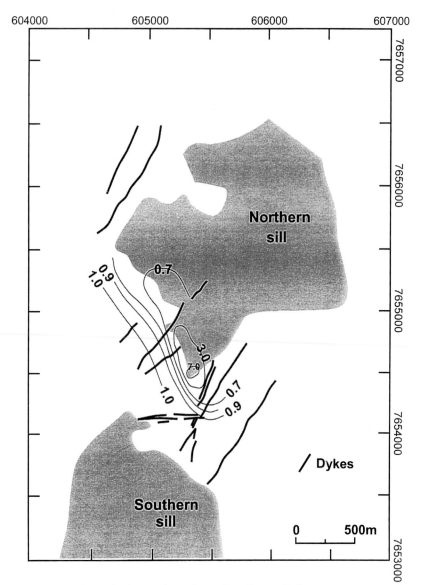

Figure 15. Thermal aureoles in a coalseam in northern Bowen Basin

Discussion

The present study of maturation simulation of vitrinites from coals of different rank suggests that open pyrolysis can be used to model rapid heating. A similarity of the coals obtained from pyrolysis to those naturally matured in the central and northern Bowen Basin coals in terms of structural and textural relationships, and possibly the characteristic high bitumen input supports their rapid heating. Heating mode was most likely through convective heat transfer by hydrothermal processes also supported by cleat mineralisation (Golding et al., 1996), rather than heating by step-wise subsidence and sediment loading over long geological time. Hydrous pyrolysis has been argued to simulate more closely thermal maturation, and shown to lead to high bitumen yield (Winters et al.,1983; Wanzl,1988; Kaiser & Wanzl, 1988; Lewan, 1992). Water has been shown to be an important factor in organic matter conversion in open systems up to 350 deg. C, by promoting hydrogen transfer reactions (Cypres & Furfari,1985). At very high temperature this may not be of significance (Mansut & Landais,1995). On the other hand, Lewan (1992) has shown that addition of water in confined pyrolysis unlike open pyrolysis does not effect yields of hydrocarbons. When comparing yields in our study of open non-hydrous pyrolysis to those from hydrous pyrolysis studies of others there appears to be no significant difference in methane production.

It has been argued that lithostatic pressure lowers the temperature of processes in natural systems (Henley,1985). Therefore a lower temperature is expected for hydrocarbon generation when high pressure is applied. However, our experiments show that this general rule may apply to a particular hydrocarbon species under specific conditions, but does not always apply to all hydrocarbons generated. For example, the difference in temperature between peak bitumen generation from TH4 at 500 bar and GCA-GOR at the same pressure is believed to be due to the higher rank of the GCA-GOR coal, having already reached peak bitumen generation prior to pyrolysis. Both retardation and enhancement of maturation processes were documented in different confined pyrolysis studies with varying pressure. Our studies demonstrate that both are valid at different pressures for different hydrocarbon species. At 2-3 bar and 250 bar peak bitumen generation is at a lower or same temperature as in open system pyrolysis. It has been shown by van Heek and Hodek (1994) that decrease in bitumen and light oil is inhibited by pressure due to inability of products to diffuse through the particles. Therefore, recombination of the primary products results in increased char and gas yields at the expense of oil. As these type of experiments are conducted in confined systems under static water pressure conditions they are not representative of dynamic hydrothermal conditions which by definition are open to semi-open systems.

Methane, a predominant gas obtained from cracking of bitumen in deeper parts of basins world-wide has been reported extensively in the literature (Hunt, 1979 ; Tissot & Welte, 1982, and many others). In the eastern Australian coals studied, yields of very immature bitumen were significant in Jurassic coals with lower methane yields than from pyrolysis of Triassic and Permian bitumens. The presence of mature bitumen in significant concentrations within Permo-Triassic coal matrix supports earlier (than Jurassic) thermal events responsible for hydrocarbon generation. As shown by the pyrolysis experiments much higher tempertures are needed to generate methane. The late Triassic hydrothermal event which has been shown to be of high temperature, up to 300 deg.C (Golding et al, this volume) may have been the major driving mechanism for bitumen cracking and methane generation. Later hydrothermal events in the Cretaceous and Teriary may have contributed to the generation of the immature bitumen found in the coals today.

Reactions controlling hydrocarbon generation in hydrothermal systems are shown to be primarily temperature dependant, as demonstrated from studies of active systems (Seewald 1994; Simoneit,1985; Simoneit,1994, Simoneit et al..1992). Hydrothermal hydrocarbon generation occurs at rapid heating at varying temperatures, often at high temperature, depending on the active system. Open pyrolysis, particularly hydropyrolysis therefore best simulates maturation of organic matter and hydrocarbon generation in hydrothermal systems. On the other hand artificial maturation under confined pyrolysis conditions suggests a closer resemblance to burial-induced maturation due to subsidence during basin evolution (Behar & Hatcher 1995). Hydrothermal systems almost by definition are open to semi-open systems; open to fluid circulation. Both, active hydrothermal systems as well as pyrolysis, where rapid maturation at high temperature occurs are characterised by high bitumen yields, as well as the potentially (depending on temperature) several hydrocarbon species being generated simultaneously from immature organic matter. Albaumer et al. (1981) reported higher bitumen to oil+gas ratios in geological hydrothermal systems (e.g.Bramsche Massif) compared to steadily subsiding basins subjected to conduction - driven thermal processes (e.g.Douala Basin).

Our study demonstrates significant oil generation from perhydrous vitrinite supporting an earlier study with similar results (Saxby & Shibaoka, 1986; Glikson & Fielding, 1991). One of the first studies to compare the geochemical and petrographic characteristics of coals (Bertrand, 1984) noted a discrepancy between geochemical indicators of petroleum potential from exinite. The same study (Bertrand 1984) noted oil generation from a perhydrous vitrinite. Collinson et al. (1994) also demonstarted oil generation from oil-prone vitrinites. Hydropyrolysis of vitrinite and exinite (Cagniant et al., 1990) produced equal yields of heavy oil from both macerals. Bowen Basin coals are dominated by perhydrous vitrinite (Glikson & Fielding 1991b) shown here to have H/C atomic ratio of 0.8 - 1.0. Vitrinite has been shown to be the source of the bitumen in these coals (Golding et al. 1996, Mastalerz et al. 1996), and its initial pronounced aliphatic character with gradual aromatisation with rank demonstrated by Mastalerz et al. (1996). A thermobalance pyrolysis study (Cagniant et al.. 1990) of a high volatile bituminous coal obtained 60% bitumen and light oil from some vitrinites, further supporting this maceral to be a major source of oil. Bitumen retained within the coal matrix may be subjected to further cracking with increase in temperature to produce coalbed methane as also suggested in an earlier study by Miyazaki (1995), and demonstrated by flash pyrolysis and microscopic observations of residues in the present study.

The close resemblance of microscopical observations of artificially matured vitrinites to naturally matured Bowen Basin vitrinites supports rapid convective heating at relatively high temperature (150-200 deg.C generally, and up to 300 deg.C in localised areas) as the means of geological maturation. Permian and Triassic coals from the Bowen Basin have a high contribution of bitumen closely resembling other Gondwana coals (Saxena et al. 1990). Maturation simulation pyrolysis in our study also generated significant amount of bitumen supporting petrological observations of all the coals . A close modern-day analogue with high fluid to rock ratios are active hydrothermal systems studied and documented by Simoneit (1985,1994a,1994b) and Peter et al. (1991). These studies showed that during hydrothermal generation of oils, the heavy-end products are deposited directly as bitumens. We attribute the significant presence of bitumen in the Bowen Basin coals and in other Gondwana coals (Saxena et al. 1991) to be the result of hydrothermal maturation rather than solely burial - driven maturation, as also supported by stable and radiogenic isotope geochemistry of mineralisation associated with the bitumen (Golding et al. 1996).

Summary and conclusions

Perhydrous vitrinite has initial relatively high H/C ratio in all Bowen Basin coals, and has been shown in laboratory simulation of hydrocarbon generation to yield significant amounts of gas and liquid hydrocarbons up to 600 deg.C. At higher temperatures, only methane and CO_2 were measurable.

Flash pyrolysis of coals and extractable bitumen support the latter as a major source of methane.

Maturation path of vitrinite from coals of different rank displayed a similar trend, however variations occurred due to differences in elemental composition, structural heterogeneity in vitrinites, and possibly presence of inherent minerals.

Vitrinite has been shown to be the major source of liquid and gaseous hydrocarbons in Bowen Basin coals. As maturation of vitrinite proceeds, charring occurs, resulting in formation of mosaic textures. The latter creates very high potential volume for gas storage. The ultra-thin areas in the textural make-up of the mosaic explain diffusion of light and heavy oils through char (inertinite).

The present study supports hydrocarbon generation from coals in hydrothermal systems. In the Bowen basin such systems were probably active during the Late Triassic, Cretaceous and Tertiary. The coals have been shown to be anomalously high in bitumen compared to San Juan basin, U.S (Levine, 1993). Bitumen accommodated within the changing coal microtexture may crack into methane if subjected to further thermal events.

Acknowledgments

This study was made possible thanks to grants by QTSC (Queensland Transmission & Supply Corporation) and ERDC (Energy & Research Development Corporation). We thank J.D. Saxby and P.Chatfield (CSIRO, North Ryde,NSW) for making available facilities and assisting L.S.Szabo with thermogravimetric techniques. Likewise we wish to thank R.M.Bustin for making available his laboratory facilities for L.S.Szabo to carry out confined heating experiments. We thank Lynne Milne for assistance with sample preparation and George Blazac for elemental analysis .3-D reconstruction drawing of mosaic texture by Micke is gratefully acknowledged . C.J.Boreham publishes with the permission of the Executive Director, AGSO .

References

Altbaumer F.J., Leythauser D. & Schaefer R.G. 1981: Effect of geologically rapid heating on maturation and hydrocarbon generation in Lower Jurassic shales from NW-Germany: *Advances in Org.Geochem.*; 80-86.
Ayache J. & Oberlin A. 1990: Thermal simulations of the evolution of carbonaceous oil-source rocks. *J.Analyt.&Applied Pyrolysis* 17:329-356.

Barker C.E. 1991: Implications for organic maturation studies of evidence for geologically rapid increase and stabilization of vitrinite reflectance at peak temperature: Cerro Pieto geothermal system. *AAPG Bull.*75(12): 1852-1863.

Barker M.R. 1995: An Interdisciplinary Study of Coal Measures from the Reids Dome beds, Lexington Dome, Minerva Hills, Central Queensland. Honours Thesis, University of Queensland: 134pp.

Behar F. & Hatcher P.G.1995: Artificial coalification of a fossil wood from brown coal by confined system pyrolysis. *Energy & Fuels* 9 (6): 984-994.

Benedict L.G. & Thompson R.R. 1980: Coke/carbon reactions in the study of factors affecting coke quality *Int. J. Coal Geol..* 1: 19-34.

Bertrand P. 1984: Geochemical and petrographic characterization of humic coals in relation to their petrographic nature. *Organic Geochemistry* 6: 481-488.

Boreham C.J. 1994: Origin of petroleum in the Bowen and Surat Basins: implications for source, maturation and migration. *AGSO Record* 1994/42. 106pp. unpublished.

Boreham C.J. 1995: Origin of petroleum in the Bowen Basin and Surat Basins: Geochemistry revisited. *APEA J.* 1995; 579-612.

Boreham C.J., Korsch R & Carmichael D. 1996: The significance of mid-Cretaceous burial and uplift on maturation and petroleum generation in the Bowen Basins, eastern Australia. In: *Mesozoic Geology of the Eastern Australian Plate conference*, Brisbane, australia 23-26 sept. Geol.Soc.Austral.Inc., Extended Abstracts 43; 104-113.

Boreham C.J., Golding S.D. & Glikson M.1997: Factors controlling gas generation and retention in Bowen Basin coals. *Abstract 18th European Association of Organic Geochemists' Conference*, Maastricht.

Boreham C.J., Horsfield B. & Schenk H.J.1998: Predicting the quantities of oil and gas generated from Permian coals, Bowen basin, Australia using pyrolytic methods. *Marine and Petroleum Geology* (in press).

Brooks J.D. & Taylor G.H. 1966: development of order in the formation of coke. In : *Chemistry and Physics of Coal*. Walker (ed.) Marcel Dekker Inc.N.Y.:243-286.

Cagniant D., Wilhelm, K. T., Van Heek K.H. & Wanz W. 1990: Thermobalance hydropyrolysis of a bituminous coal sample and its macerals vitrinite and exinite; structural analysis of relevant tars. *Fuel* 69: 1496 - 1501.

Collinson M.E.,Van Bergen P.F. Scott A.C. & Leeuw J.W. 1994: The oil generating potential of plants from coal and coal-bearing strata through time: A review with new evidence from Carboniferous plants. In: Coal and Coal-bearing Strata as Oil-Prone Source Rocks, Scott A.C. & Fleet A.J. (eds.) *Geological Soc. Special Publ.* 77; 31-70.

Cypres R. & Furfari S. 1985: Direct post-cracking of volatiles from coal hydropyrolysis. *Fuel* 64; 33-44.

Crossley J.M.1994: Geology of the Newlands Eastern creek South Area, Central Queensland. Honours Thesis University of Queensland: 82pp.

Fortin F.& Rouzaud J-N 1994: Different mechanisms of coke microtexture formation during coking coal carbonization. *Fuel* 73(6):795-809.

Freund H., Clouse J.A. & Otten G.A. 1993: The effect of pressure on the kinetics of kerogen. *Energy & Fuels* 7; 1088-1094.

Glikson M. & Taylor G.H. 1986: Cyanobacterial mats; major contributors to the organic matter in the Toolebuc Formation oil shales. *Geological Society of Australia* Special publication No. 12, 273 - 286.

Glikson M. & Fielding C. 1991a: The Late Triassic Callide coal measures, Queensland, Australia: Coal petrology and depositional environments. *Int. J. Coal Geol.* 18:313-332.

Glikson M. & Fielding 1991b: Re-Assessment of rank and maceral composition of Queensland coals: implications for hydrocarbon generation and other unexplained phenomena. Queensland Coal Symposium, Brisbane, August, 1991. *The Australasian Inst.of Mining & Metallurgy*, 5/91; 23-30.

Glikson M., Golding S.D.,Lawrie G., Szabo L.S., Fong C., Baublys K.A., Saxby J.D. & Chatfield P. 1995: Hydrocarbon generation in Permian coals of Queensland, Australia: Source of coalseam gas. In: *Proceedings Bowen Basin Symposium*, Follington,Beeston & Hamilton (eds.), Geological Soc. Australia Publication: 205-216.

Golding S.D.,Glikson M.,Collerson K.D.,Zhao J.X.,Baublys K. & Crossley J.M.1996: Nature and source of carbonate mineralisation in Bowen Basin coals; implications for origin of coalseam gases. In: *Mesozoic Geology of the Eastern Australian Plate*. Geological Soc.Australia Publ. : 205-212.

Heek van K.H. & Hodek W. 1994: Structure and pyrolysis behaviour of different coals and relevant model substances. Fuel 73(6);887-896.

Henley R.W. 1985: The geothermal framework for epithermal deposits. In: Berger & Bethke (eds.), *Geology and Geochemistry of Epithermal Systems. Soc.Econ,Geology;* 1-24.

Hill R.P., Jenden P.D., Tang Y.C., Teerman S.C. & Kaplan I.R. 1994: Influence of pressure on pyrolysis of coal. In: Mukhypadhyay & Dow (eds.), *Vitrinite Reflectance as a Maturity Parameter. Am.Chem.Soc.Symp.*Series 570; 161-193.

Hunt J.M. 1979: *Petroleum Geochemistry and Geology.* W.H.Freeman & Co. 617pp.

Hvoslef S., Larter S.R. & Leythauser D. 1987: Aspects of generation and migration of hydrocarbons from coal-bearing strata of the Hitra formation, Haltenbanken area, offshore Norway. *Organic Geochemistry* 13(1-3): 525-536.

Jacob H., 1989: Classification, structure, genesis and practical importance of natural of natural solid bitumen ("migrabitumen"). *Int.J.Coal Geol.* 11; 65-79.

Kaiser M. & Wanzl W. 1988: Characterisation and comparison of liquid products from coal pyrolysis in laboratory and process development units. *Fuel Proc.Tech.* 20; 23-32.

Khavari-Khorasani G.& Michelsen J. 1994: The effects of overpressure, lithology, chemistry and heating rate on vitrinite reflectance evolution, and its relationship with oil generation. *APEA J.* 1994: 418-433.

Lawrie G.A., Gentle I.R., Fong C. & Glikson M. 1997: Atomic force microscopy studies of Bowen Basin coal macerals. *Fuel* 76 (14/15); 1519-1526.

Lewan M.D. 1993: Laboratory simulation of petroleum formation; *hydrous pyrolysis:* In: *Organic Geochemistry*, Engel & Macko (eds.): 419-441.

Li C-Z, Bartle K.D. & Kandiyoti R. 1993: Characterisation of tars from variable heating rate pyrolysis of maceral concentrates. *Fuel* 72; 3-11.

Levine J.R.1993: Coalification: the evolution of coal as source rock and reservoir rock for oil and gas. In: Law & Rice (eds.), Hydrocarbons from Coal. *AAPG, studies in Geology*, 38;39-77.

Mallett C.W., Russell N. & McLennan T. 1990: Bowen Basin Geological History. In: Beeston (ed.), *Proc. Bowen Basin Symp.* AUSIMM; 5-20.

Mansuy L. & Landais P. 1995: Importance of the reacting medium in artificial maturation of a coal by confined pyrolysis.2.Water and polsr compounds. *Energy & Fuels* 9:809-821.

Mastalerz M., Glikson M. & Golding S.D. 1996: Source of coalbed methane in Bowen Basin, eastern Australia. *Proc.Geol.Soc.Am.* Denver, Colorado,September 1996.

Miyazaki S. 1995: Oil generation from coals and carbonaceous claystones in the Bass Basin. *Petrol. Expl.Soc.Austral.J.* 23:91-100.

Monthioux M. & Landais P. 1987: Evidence of free but trapped hydrocarbons in coals. *Fuel* 66; 1703-1708.

Peter J.M., Peltonen P., Scott B.D., Simoneit B.R.T. & Kawka U.E. 1991: 14C ages of hydrothermal petroleum and carbonate in Guaymas Basin, Gulf of California: Imlications for oil generation, expulsion and migration. *Geology* 19; 253-256.

Robert P. 1988: *Organic Metamorphism and Thermal History.* Elf-Aquitaine & D.Reidel Publ.; 96-101.

Saxby J.D. & Shibaoka M.1986: Coal and coal macerals as source rocks for oil and gas.*Applied Geochemistry* 1:25-36.

Saxby J.D., Bennett A.J.R., Corcoran J.F., Lambert Î.E. & Riley K.W. 1996: Petroleum generation: Simulation over six years of hydrocarbon formation from torbanite and brown coal in a subsiding basin. *Organic Geochemistry* 9(2) :69-81.

Saxby J.D., Chatfield P., Taylor G.H., Fitzgerald J.D., Kaplan I.R. & Lu S.I.1992: Effect of clay minerals on products from coal maturation. *Organic Geochemistry* 18(3):373-383.

Saxena R.,Navale G.K.B., Chandra D.&Prasad Y.V.S.1990: Spontaneous combustion of some Permian coal seams of India: An explanation vased on microscopic and physico-chemical properties. *Palaeobotanist* 38: 58-82.

Seewald J.S.1994: Evidence for metastable equilibrium between hydrocarbons under hydrothermal conditions. *Nature* 370: 285-287.

Shibaoka M. & Ueda S.1978: Formation and stability of mesophase during coal hydrogenation 1: Fornation of mesophase. *Fuel* 57:667-675.

Simoneit B.R.T.1985: Hydrothermal petroleum: Genesis, migration and deposition in Guaymas Basin, Gulf of California. *Canadian J.of Earth Sci.* 22: 1919-1929

Simoneit B.R.T. 1994A: Organic matter alteration and fluid migration in hydrothermal systems. . In: *Geofluids: Origin, Migration and Evolution of Fluids in Sedimentary Basins.*J.Parnell (ed.) Geological Soc. Publ. No. 78: 261 - 274.

Simoneit B.R.T. 1994B : Lipid/bitumen maturation by hydrothermal activity in sediments of Middle Valley, Leg 139. In: Mottl, Davis, Fisher & Slack (eds.) *Proceedings of the Ocean drilling Program,* Scientific Results, 139; 447-465.

Simoneit B.R.T.Kawka O.E. & Wang G.M. 1992: Biological maturation in contemporary hydrothermal systems, alteration of immature organic matter in zero geological time. In: *Biological Markers in Sediments and Petroleum.* Moldowan J., Philp R.P. & Albrecht P. (eds.) Prentice Hall, Euplewood Cliffs, NJ: 124 - 141.

Stach E.,Mackowsky M-Th, Teichmuller M., Taylor G.H., Chandra D. & Teichmuller R. 1975: Methods and tools of examination. In: *Coal Petrology*. Gebruder Borntraeger, Berlin: 239-309.

Taylor G.H. 1966: The electron microscopy of vitrinites. *Am. Chem.Soc.* Series 55; 274-283.

Taylor G.H., Liu S.Y. & Teichmuller M. 1991: Bituminite - a TEM study. *Int.J.Coal Geol.* 18; 71-85.

Tissot B.P.& Welte D.H. 1982: *Petroleum Formation and Occurrence.* Springer Verlag, Berlin; 699pp.

Veevers J.J. 1989: Middle/Late Triassic (230+5Ma) singularity in the stratigraphic and magmatic history of the Pangean heat anomaly. *Geology* 17(9): 784-787.

Wanzl W. 1988: Chemical reactions in thermal decomposition of coal. *Fuel processing Technology* 20; 317-336.

Winters J.C., Williams J.A. & Lewan M.D.1983: A laboratory study of petroleum generation by hydropyrolysis. *Adv.Org.Geochem.* 1981: 524-533.

Plate I

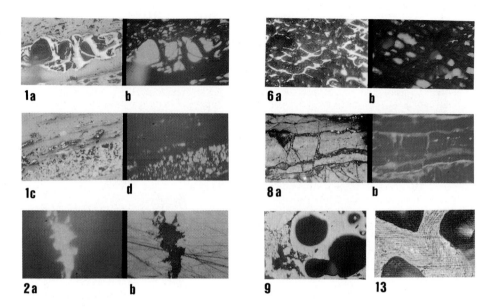

Fig.1: Bitumen infill of inertinite (char) cavities. Reflected white light (1a,c), and fluorescence mode (1b,d).

Fig.2: Bitumen infill of micro-fracture in vitrinite. Reflected white light (b), and fluorescence mode (a).

Fig.6: Newly generated bitumen within mosaic, formed from heated vitrinite after 600 degrees C. Reflected white light (a), and fluorescence mode (b).

Fig.8: Large and small pores in GOR-GCA char/inertinite formed from vitrinite after 800 degrees C. Reflected white light.

Fig.13: Lamellar mosaic with numerous pores characteristic of 247 coal, with initial higher reflectance (1.4%). Reflected white light.

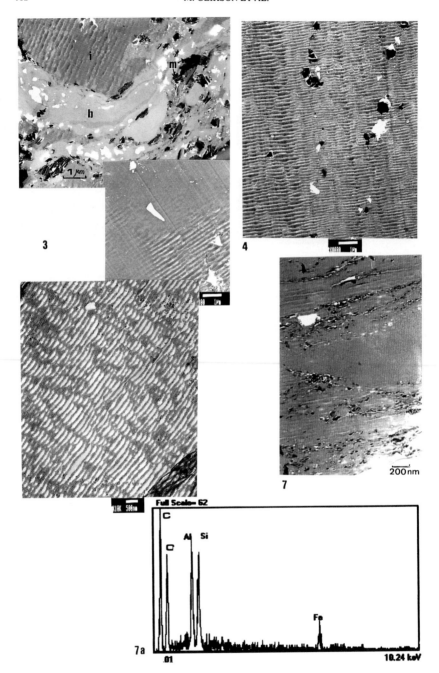

3

4

7

7a

Plate II

Transmission Electron Micrographs

Figs.1-7: TEM of residues from pyrolysed vitrinite.
Fig.3: Commencemnet of charring in vitrinite from a sub-bituminous coal (0.5% Ro) after 450 degrees C. bitumen (B) displaying flow structures, and mobilised minerals (M) aligned along 'flow lines'.
Fig.4: Vitrinite from sub-bituminous coal after pyrolysis to 600 degree C. showing well defined char (I) formation, and commencing organisation into 'molecular domaines' towards a mosaic texture.
Fig.5: The same cxoal as above after 800 degrees C. Charring completed, towards a complete mosaic texture representing ordered 'molecular domaines'.
Fig.7: Residue from TAR coal (initial Ro=0.6%) vitrinite after 500 degrees C. showing a greater proportion of vitrinite charring than in a lower rank coal (0.5% Ro) at similar temperature. Alignment of minerals along charring textures is evidence of their mobilisation.
Fig.7a: Spectrogram of EDS analysis of typical mineral composition in TAR coal.

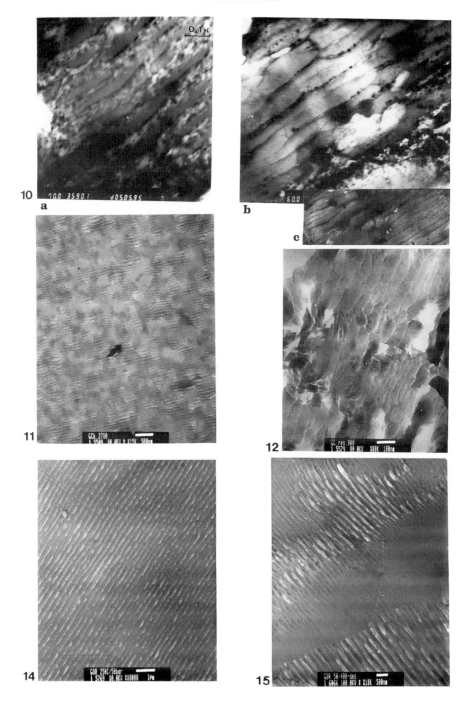

Plate III

Transmission Electron Micrographs

Figs.10a-c: GOR-GCA coal at peak oil generation before pyrolysis, showing naturally matured vitrinite. All stages towards mosaic formation can be observed in the same sample. **a;** break-down of 'lipidic' lamellae (Glikson & Fielding, 1991) during oil generation stage. **b;** condensation of non-oil generating lamellae towards char formation. **c;** complete charring and mosaic formation in the same sample.
Fig.11: GOR vitrinite at peak oil generation, after 600 degrees pyrolysis. Mosaic texture displays trapped bitumen (light, less electron-dense areas) within mosaic.
Fig.12: GOR vitrinite After 800 degrees C with pyrobitumen from previously fluidised bitumen 'adsorbed' on mosaic/molecular domaines of charred vitrinite.
Fig.14: Uniform breakdown of lipidic lamellae in GOR vitrinite, in confined pyrolysis at 250 degrees and 500 bar pressure.
Fig.15: GOR-GCA vitrinite after confined pyrolysis at 400 degrees C. and 500 bar pressure, characterised by well advanced coarse mosaic formation (MC).

THE RELATION BETWEEN GAS IN COAL SEAMS AND ARTIFICIAL COALIFICATION GAS UNDER HYDROTHERMAL PRESSURE SYSTEMS

TOYOHIKO YAMASAKI

Waseda University Advance Science & Engineering Research Center,

3-4-1, Ohkubo, Shinjuku-Ku, Tokyo 169-8555, Japan

AND

SUSAN A. ROCES

De La Salle University,

2401-Taft Ave, Manila, Philippines

1. INTRODUCTION

Extensive studies have been carried out on the genetic chemical reaction of coals by many famous researchers. Some of the previous studies are as follows: H. Potonei[1] initiated the study of coalification. F. Fischer, et. al.[2] supported the origin of lignin theory. F. Bergins[3], Erasmus[4] and R. V. Wheeler, et. al.[5] supported the cellulose theory. Tropsch[6], Berl and Schmit[7], Horn and Sustsmann[8] and Fuch and Horn[9] developed the systematic investigation on artificial coalification. Berl and Schmidt[10] experiments on coalification were checked and confirmed by Schuhmacher and Van Krevelen[11]. Two researches on this topic were studied in our country. W. Funasaka and C. Yokokawa[12] discussed the relationship between the genetic origin of coal and its chemical properties at atmospheric conditions by artificial coalification experiments. T. Yamazaki and R. Abe[13] reported the experimental results of the gas evolutions from raw coal during the artificial coalification process following the procedure presented by C. Yokokawa.

The authors discussed in this study the produced gas and the changed in the chemical and physical properties of coal during artificial coalification. The effects of hydrothermal reaction on the process will also be discussed. This study was divided into two stages. In the first stage experiment, the authors discussed the effect of temperature on the heating conditions using the same heating calorific value and the effect using different reaction rate of coalification expressed by tangencial lines at starting point. The heating path was kept constant in every runs in the second stage experiment so that the experimental results could be easily analyzed using the thermochemical treatment. It was found out from the results of the first stage experiment that the amount of evolved gases from coal were dependent on the reaction rate of coalification when the coalification grade was advanced under various reaction rate. From second experiments, evoloed gas is related to the heating temperature and time history in making the high grade bituminous coal. The different evolved gases produced using the autoclave and that of the natural free gas in coal was studied using the Taiheiyo coal by means of mass spectrometry.

2. EXPERIMENT AND PROCEDURES

Sample was prepared by mixing coal and water (paste coal) in which the coal and water ratio was 5 : 2. These paste coals were kept in a constant humidity container. Sample used in this study weighed 15 grams. A rotary steel autoclave that can stand high pressure and temperature was used in the experiment. The inside volume of the autoclave was 500 ml with the usual pressure at 300 kg/cm^2, temperature at 400 °C.

The paste sample of 15 g and 350 ml. distilled water was put to the autoclave. This was heated under the constant heating condition at 400 °C which was controlled by the transformer.

(1). First Stage Experimental Procedure[13]

The heater current was controlled by a transformer under constant voltage of the circuit at the start of the experiment. The temperatures were raised as the results of the program until it became stable as shown in Figure 1. In this procedure, it was planned to be an equal area as shown enclosed area by the curves (1), (2), (3), (4), and (5) as shown in Figure 1. These circuits areas were constant and defined as coalification index. The reaction rate index is the tangent of curve at the initial point of each curve as shown as Figure 1. Assuming the first order reaction to apply, the chemical reaction product x was propotional to the reaction time t shown as follows:

$$x = kt \tag{1}$$

where: k is the reaction constant and this is a first order reaction. However:

$$k = 2^{\frac{\theta}{10}} \cdot K \tag{2}$$

where: θ is the temperature and K is a constant
Therefore:

$$x = 2^{\frac{\theta}{10}} \cdot K \cdot t \qquad (3)$$

Temperatures were raised depending on curve(1)(2)(3)(4)and(5) as shown in Figure 1. The reaction product x was measured every $10\,°C$ and was shown as x_i. Wherein, i is from 1 to n as shown in equation (4).

$$\sum_{i=1}^{n} x_i = X = K \cdot \sum_{i=1}^{n} 2^{\frac{\theta_i}{10}} \cdot t_i \qquad (4)$$

Therefore:

$$\frac{X}{K} = \sum_{i=1}^{n} 2^{\frac{\theta_i}{10}} \cdot t_i \qquad (5)$$

The ratio, X/K, was called the productivity index of each product and these were shown in Tables 1, 2, 3 and 4.

(2). Second Stage Procedure[14]

The heating programs were changed in the second stage procedure as shown by the curves in Figure 6. The sample was rapidly heated in the autoclave. When the autoclave temperature was stable, autoclave kept each constant coalification temperature at $220\,°C$, $260\,°C$, $280\,°C$, $300\,°C$, $320\,°C$ and $340\,°C$. The evolved gases during the exposed time in each heating program were accumulated and analyzed by chemical analysis and mass-spectrometry.

3. RESULTS OF EXPERIMENTS AND DISCUSSION

3.1. RESULTS OF THE FIRST STAGE EXPERIMENT

Nakago, Yumoto and two kinds of Taiheiyo coals was used as the sample in the first stage experiments. The objective of this experiment was to determine the relationship between the coalification speed index which is tangent of each heating curves at initial point and the evolved gases such as CO_2, CH_4, H_2, C_nH_m and total gas.

The experimental results are shown in Tables 1, 2, 3 and 4, and these are also shown in Figures 2, 3, 4, and 5. The total evolved gas from coal in each experiments was propotional to the ratio, X/K, in each experiments as shown in Figures 2 to 5. Black lines show the relationship between the reaction rate and evolved gas while the dotted lines show the relationship between the reaction productivity index and total evolved gas. It was confirmed that the reaction rate equation was realized in coalification properties by means of Bergius methods.

3.2. RESULTS OF THE SECOND STAGE EXPERIMENT

The experimental results of the second stage experiment were shown in Table 7 to 10 and Figueres 7, 8. It was assumed that the reaction rate is constant in the first stage experiment as shown by equations (1) and (2). In the second stage experiment, however, it was proven that equation (1) is not correct. From these experiment it was confirmed that the following equation for the reaction product is applicable.

$$x = K_2 \cdot t^b \tag{6}$$

Wherein, K_2 and b were constants. The relation between temperature and evolved gas are shown in Figure 7. The black lines shows the different incli-nation starting at 300 °C. In this equation, the heating time, K_2, and b are constant depending on the temperature of the composition gas. In the case of methane, the constant b is equal as shown in Figure 8 in every temper-ature, however, total gas was varied in each temperature. Coal proximate analysis and plastic properties were confirmed in this experiment. These results were shown in Table 5.

4. COMPOSITION OF COAL SEAM GAS AND ARTIFICIAL COALIFICATION GAS

4.1. DIFFERENCE IN CHEMICAL COMPOSITION OF COAL SEAM GASES

Two kinds of coal seam gases that were produced from Taiheiyo and Ak-abera coal mines were compared with those of the chemical composition gases by mean of mass-spectrometry. The experimental results are shown in Table 6. The hydrocarbon gas evolved from Taiheiyo coal seam were almost methane. However, gases evolved from Akabera coal seam contain more higher hydrocarbon gases such as C_2H_6, C_2H_4, C_3H_8. Gases such sa N_2 and O_2 evolved were dependent on the air at the time of sampling.

4.2. COMPARISION BETWEEN COAL SEAM GAS AND ARTIFICIAL COALIFICATION GAS

Table 11 shows the compare the natural coal seam gas with artificial coal-ification gas at reaction temperatures of 270 °C, 306 °C and 345 °C. From these results, natural gas contains more methane than other hydrocarbon gases. However, the methane gas evolved from the artificial coalification gas was relatively lower ratio as compared with other hydrocarbon gases. Unsaturated hydrocarbon gases were evolved from the artificial gases while natural gas evolved only saturated hydrocarbon gases. The carbon dioxide is not discussed because it will be produced through the coal oxidation process by the air in autoclave when enclose the sample.

5. CONCLUSION

The following arrangements were applied in the first stage experiment:

1. The heated calorific values on every coal samples were equalized by keeping the electrical voltage constant, hence, the calorific values signify the coalification index.
2. The coalification index were considered constant. These constant values were the areas of the heating curve of Temperature versus Time plot.
3. The initial tangential lines of the heating curves were defined as the reaction rate index of coalification.

And then the following result was confirmed. The evolved gase from coal was not proportional to the reaction rate, but instead, it was proportional to the productivity index which was calculated using equation no. (5). The following conclusions were gathered from the results of the second stage experiment:

1. It was confirmed that the total evolved gas from coal was proportional to the exponential temperature as shown by the following equation.

$$x = K_3 \cdot e^{a\theta} \tag{7}$$

2. In the both logarithmic scal, the evolved gases were proportional to the time.
3. Majority of the chemical composition of natural coal seam gas were almost methane and others were traces of higher saturated hydrocarbon gases such as C_2H_6 or C_3H_8.
4. Gases from the artificial gasification evolved CO_2, CO, H_2 and many kinds of hydrocarbon gases in higher percentage as compared with the natural coal seam gas content.

6. ACKNOWLEDGEMENT

The authors would like to appreciate deceased Dr. Ryonosuke Abe who had directed us in our hydrothermal experimental studies. We would like to present this paper to Dr. Abe. We would also like to thank Mr. M.Furukawa and Mr.K.Kaneko who are contributed to the experimental works of this study.

References

1. Potanie, H.,(1910) *Die Entstehung der Steinkohle überhaupt,* **5** Aufl., (Berlin)
2. Fisher,H. & Schrader,H.(1920) *Alte und neue Ansichten über die Urspnungsstoffe der Kohle,* Aph. Kennt. Kohle,**5**,543,553,559
3. Bergius,F.(1920) *Die Anvendung von Hohen Drücken bei Chemisch-Technischen Vorgangen,* Z.Electrochem. **18**, 660-662
4. Erasmus, P.(1938) *Über die Bildung u. der Chemishen Bau der Kohlen,* (Stauttgart)
5. Wheeler, R.V. & Clark,A.H.(1913) *The volatile Constituent of Coal, Part 3,* J. Chem. Soc., **103**, 1704-1715
6. Tropsh, H. & A. von Philippovien, *Über die Kunstliche Inkohlung von Cellulose und Lignin in Gegenwart von Wasser,* Ges. Abh. Kennt. Kohle, **7**,84-102
7. Berls, E., Schmidt, A., A & Koch, H.,(1930) *Über die Entstehung der Kohlen,* Z.Angew.Chem.,**43**,1018-1019
8. Horn, O. & Stustman, H.,(1931) *Über die Kunstlichen Steinkohlen,* Brenn Chem., **12**,409-412
9. Fuchs, W. & Horn, O.,(1931) *Zur Frage der Enstehung der Steinkohlen,* Z. Angew Chem. **44**,180-184
10. Berl, E. & Schmidt, H.,(1928) *Über das Verhalten der Cellulose bei der Drucherhitzung mit Wasser,* Ann, **461**, 192-220
11. Schuhmacher, J. P. & von Krevelen, D.W., *Coal Science* (Elsevier), 104
12. Funasaka, W. & Yokokawa, C.(1950) *Fac. Kyoto Univ.,* **12**, 128
13. Yamasaki, T. & Abe, R.(1956) *On the Relation Between the Gas in Coal Seams and the Coalfication Speed Under the High Pressure,* J.Min. I. Jap.,**72**,189-193
14. Yamasaki, T. & Abe, R.(1957) *On the Relation Between the Gas in Coal Seams and the Coalification Speed Under the High Pressure*(2nd Part), J.Min.I.Jap, **73**, 285-287

TABLE 1. Coalification Gas Analysis of Upper Nakago Coal in 1st Stage Experiment

Experiment No.		21	20	19	18	17
Temp ℃		270	300	348	348	347
Fuel Ratio		1.12	1.30	1.39	1.47	1.51
Coalification Index*		1.54	1.50	1.52	1.52	1.48
Reaction Rate Index**		2.10	2.50	2.80	3.00	4.00
Productivit Index***		21×230	587×230	2960×230	5170×230	6113×230
Evolved	total	280	350	430	480	520
Gas	correct+	73.2	124.0	224.0	287.0	310.3
CO_2	%	24.8	26.4	35.5	39.4	39.5
	cc	69.0	93.0	152.0	189.1	206.4
C_nH_m	%	0	0.9	1.6	1.7	1.1
	cc	0	3.0	7.0	8.2	5.7
CO	%	0	0.2	0	0.4	0.1
	cc	0	1.0	0	1.9	0.5
CH_4	%	1.5	6.8	12.7	14.6	16.4
	cc	4.2	24.0	55.0	70.1	85.2
H_2	%	0	0.9	2.4	3.0	2.4
	cc	0	3.0	10.0	14.4	12.5
O_2	%	1.0	1.3	1.4	0.5	3.4
	cc	2.8	4.0	6.0	2.4	17.7
N_2	%	72.7	63.5	46.4	40.4	37.1
	cc	204.0	222.0	200.0	193.9	192.0

(by Chemical Analysis)

* Circuit area of curve(1)(2)(3)(4)and(5)in Fig6
** Tangent of the curve at initial point
*** X/K
+ Subtracted O_2 and N_2 from total evolved gas

TABLE 2. Coalification Gas Analysis of Yumoto Coal in 1st Stage Experiment

Experimen No.		27	28	26	23	24
Temp ℃		265	306	345	345	345
Fuel Ratio		0.89	1.01	1.01	1.05	1.05
Coalification Index*		1.54	1.50	1.52	1.51	1.48
Reaction Rate Index**		2.00	2.50	2.80	3.00	4.00
Productivit Index***		21×230	587×230	2960×230	5170×230	6113×230
Evolved	total	230	300·	326	360	392
Gas	correct +	18.2	86.4	125.1	162.3	188.2
CO_2	%	6.8	19.0	25.6	24.8	25.4
	cc	15.6	57.0	73.5	89.3	99.6
C_nH_m	%	0.4	1.2	2.0	1.4	2.2
	cc	0.9	3.6	6.5	5.0	8.6
CO	%	0.2	0.2	0.2	0.6	0.2
	cc	0.5	0.6	0.7	2.2	0.8
CH_4	%	0.5	6.7	16.2	17.2	19.0
	cc	1.2	20.1	42.8	61.9	74.5
H_2	%	0	1.7	0.5	1.1	1.2
	cc	0	5.1	1.6	3.9	4.7
O_2	%	1.0	0.6	1.6	0.8	1.0
	cc	2.3	1.8	5.2	2.9	3.9
N_2	%	91.1	70.6	53.9	54.1	51.0
	cc	209.5	211.8	195.7	194.8	199.9

(by Chemical Analysis)

* Circuit area of curve(1)(2)(3)(4)and(5)in Fig6
** Tangent of the curve at initial point
*** X/K
+ Subtracted O_2 and N_2 from total evolved gas

TABLE 3. Coalification Gas Analysis of Main Taiheiyo Coal in 1st Stage Experiment

Experiment No.		42	43	41	40	39
Temp ℃		270	310	344	342	343
Fuel Ratio		0.86	0.92	0.88	1.05	1.03
Coalification Index*		1.54	1.52	1.52	1.52	1.48
Reaction Rate Index**		2.00	2.50	2.80	3.00	4.00
Productivit Index***		21×230	587×230	2960×230	5170×230	6113×230
Evolved	total	300	350	430	458	455
Gas	correct +	90.0	113.5	211.5	247.3	275.7
CO_2	%	24.8	27.8	33.0	33.8	37.2
	cc	74.4	86.2	141.9	154.8	169.3
C_nH_m	%	0.6	0.8	1.6	1.7	1.8
	cc	1.8	2.8	6.9	7.8	8.2
CO	%	0.1	0.2	0.2	0.1	0.3
	cc	0.3	0.7	0.9	0.5	1.4
CH_4	%	3.7	7.3	14.4	18.4	20.1
$+ C_2H_6$	cc	11.1	22.6	65.5	83.7	91.5
H_2	%	0	0	0	0	1.2
	cc	0	0	0	0	5.5
O_2	%	0.8	0.4	0.6	0.7	1.0
	cc	2.4	1.2	2.6	3.2	4.6
N_2	%	70.0	63.5	50.2	45.3	38.4
	cc	210.0	236.5	215.9	207.5	174.7

(by Chemical Analysis)

* Circuit area of curve(1)(2)(3)(4)and(5)in Fig6
** Tangent of the curve at initial point
*** X/K
+ Subtracted O_2 and N_2 from total evolved gas

TABLE 4. Coalification Gas Analysis of upper Taiheiyo Coal in 1st Stage Experiment

Experimen No.		42	43	41	40	39
Temp °C		270	311	.344	343	343
Fuel Ratio		1.03	1.01	1.07	1.15	1.15
Coalification Index*		1.54	1.50	1.52	1.52	1.48
Reaction Rate Index**		2.00	2.50	2.80	3.00	4.00
Productivit Index***		21×230	587×230	2960×230	5170×230	6113×230
Evolved	total	270	350	400	445	450
Gas	correct +	64.4	139.8	194.3	237.6	274.2
CO_2	%	20.6	30.4	38.5	39.8	40.6
	cc	55.6	106.4	154.3	177.1	182.7
C_nH_m	%	0.2	1.0	1.4	1.5	1.4
	cc	0.5	3.5	5.6	6.7	6.3
CO	%	0.6	0.2	0.3	0.2	0.1
	cc	1.6	0.7	1.2	0.9	0.5
CH_4	%	1.9	4.2	7.4	7.6	13.3
$+ C_2H_6$	cc	5.1	14.7	29.6	33.8	59.9
H_2	%	0	0.8	1.0	4.0	5.2
	cc	0	2.8	4.0	17.8	23.4
O_2	%	0.6	0.6	0.5	0.3	0.3
	cc	1.6	2.1	2.0	1.3	1.4
N_2	%	76.1	62.8	50.9	46.6	39.1
	cc	205.6	210.2	203.7	207.4	175.8

(by Chemical Analysis)

* Circuit area of curve(1)(2)(3)(4)and(5)in Fig6
** Tangent of the curve at initial point
*** X/K
+ Subtracted O_2 and N_2 from total evolved gas

TABLE 5. Proximate Analysis of Product Coal

No	Temp ℃	Time min	moisture %	Valile %	Fixcarbon %	Ash %	Fuel Ratio	Plasticity
1	340	100	4.01	47.08	43.46	5.45	0.92	agglomerate
2	340	160	3.18	47.17	44.42	5.23	0.94	
3	340	220	3.68	45.85	44.57	5.90	0.98	
4	340	340	3.22	44.09	46.67	6.02	1.06	
5	340	510	3.52	44.40	46.29	5.79	1.04	
6	320	100	4.66	46.15	44.40	4.79	0.96	
7	320	540	3.17	44.35	46.05	6.43	1.04	
8	300	100	5.09	46.70	42.38	5.83	0.91	a little agglomerate
9	300	570	4.74	44.83	45.21	5.22	1.01	
10	280	100	5.29	46.44	42.60	5.67	0.91	non plastic
11	280	600	3.27	47.94	42.85	5.94	0.89	a little agglomerate
12	260	650	4.91	47.89	42.53	4.67	0.88	
13	220	60	5.88	47.39	42.13	4.60	0.88	non plastic
14	220	180	5.88	46.55	41.16	6.51	0.88	
15	220	360	5.66	46.69	41.20	6.45	0.88	
16	220	540	5.61	46.37	40.06	7.96	0.87	
17	220	760	5.04	47.63	41.71	5.62	0.88	

TABLE 6. Gas Analysis of Coal Seam Gas by Mass-spectrometry

	Main Coal Seam of Harutory Pit of Taiheiyo Coal Mine	No.11th Coal Seam of 1st Pit of Akabera Coal Mine
H_2	0 %	0 %
CH_4	90.66	94.69
C_2H_4		0.20
C_2H_6		2.69
C_3H_6		tr
C_3H_8		0.16
S conpounds	0	0
N2	6.30	1.37
Ar	0.09	0.07
O_2	0.95	0.21
CO_2	0.31	0.32
CO	0	0
others *	1.69	0.29
total	100.00	100.00

* depend on the moisture effects

TABLE 7. Coalification Gas Analysis of Taiheiyo Coal in 2nd Stage Experiment

Experimental No.		1	2	3	4	5
Max. Temp. ℃		340	340	340	340	340
Max Press. Kg/cm^2		140	147	148	149	152
Reaction Time min		100	160	220	340	510
Fuel Ratio		0.92	0.94	0.98	1.15	1.20
Evolved	total	290	325	350	380	390
Gas	corrected	63.83	115.47	140.10	184.94	182.32
CO_2	%	18.0	24.8	25.0	30.0	28.2
	cc	52.20	80.60	87.50	114.00	109.98
C_nH_m	%	0.2	0.8	1.0	1.2	1.8
	cc	0.58	2.60	3.50	4.59	7.02
CO	%	0.2	0.4	0.3	tr	0.9
	cc	0.58	1.30	1.05		3.51
CH_4	%	3.61	8.40	11.18	15.29	13.68
	cc	10.47	27.30	39.13	58.10	53.35
H_2	%	0	1.13	2.55	2.18	2.17
	cc		3.67	8.92	8.28	8.46
O_2	%	0.8	0.4	0.3	0.2	0.7
	cc	2.32	1.30	1.05	0.76	2.73
N_2	%	77.19	64.07	59.67	51.13	52.55
	cc	223.85	208.23	208.85	194.30	204.95

(by Chemical Analysis)

TABLE 8. Coalification Gas Analysis of Taiheiyo Coal in 2nd Stage Experiment

Experimental No.		6	7	8	9
Max. Temp. ℃		320	320	300	300
Max Press. Kg/cm^2		118	114	86	90
Reaction Time min		100	540	100	570
Fuel Ratio		0.96	1.04	0.91	1.04
Evolved	total	285	350	255	282
Gas	corrected	62.47	134.54	31.11	63.39
CO_2	%	19.2	26.0	12.0	19.2
	cc	54.72	91.00	30.60	54.14
C_nH_m	%	0.2	1.0	0	0.6
	cc	0.57	3.50		1.96
CO	%	0.2	0.3	0.2	0.2
	cc	0.57	1.05	0.51	0.56
CH_4	%	2.32	8.91	0	1.43
	cc	6.61	31.19		4.03
H_2	%	0	2.23	0	1.03
	cc		7.80		2.90
O_2	%	0.6	0.2	1.0	0.8
	cc	1.71	0.70	2.55	2.26
N_2	%	77.48	61.36	86.80	76.74
	cc	220.82	214.76	221.34	216.41

(by Chemical Analysis)

TABLE 9. Coalification Gas Analysis of Taiheiyo Coal in 2nd Stage Experiment

Experimental No.		10	11	12
Max. Temp. ℃		280	280	260
Max Press. Kg/cm^2		67	67	51
Reaction Time min		100	600	650
Fuel Ratio		0.91	0.89	0.88
Evolved	total	230	275	255
Gas	corrected	23.46	41.56	37.56
CO_2	%	10.0	14.6	10.6
	cc	23.00	40.15	27.03
C_nH_m	%	0	0.3	0
	cc	0	0.83	0
CO	%	0.2	0.2	0.2
	cc	0.46	0.55	0.51
CH_4	%	0	0	3.93
	cc			10.02
H_2	%	0	0	0
	cc			
O_2	%	0.8	0.9	0.6
	cc	1.84	2.47	1.53
	%	89.0	84.0	84.67
N_2	cc	204.7	231.0	215.91

(by Chemical Analysis)

TABLE 10. Coalification Gas Analysis of Taiheiyo Coal in 2nd Stage Experiment

Experimental No.		13	14	15	16	17
Max. Temp. ℃		220	220	220	220	220
Max Press. Kg/cm^2		22	24	23	25	25
Reaction Time min		60	180	360	540	760
Fuel Ratio		0.88	0.88	0.88	0.88	0.88
Evolved	total	215	225	225	240	240
Gas	corrected	9.03	10.35	12.15	17.04	16.80
CO_2	%	4.2	4.6	5.4	7.0	7.0
	cc	9.03	10.35	12.15	16.80	16.80
C_nH_m	%	0	0	0	0	0
	cc					
CO	%	0	0	0	0.1	0
	cc				0.24	
H_2	%	0	0	0	0	0
	cc					
CH_4	%	0	0	0	0	0
	cc					
O_2	%	1.2	1.2	0.6	0.6	0.8
	cc	2.58	2.70	1.35	1.44	1.92
N_2	%	94.6	94.2	94.0	92.3	92.2
	cc	203.39	211.95	211.50	221.52	221.28

(by Chemical Analysis)

TABLE 11. Gas Analisis of Coal Seam Gas and Artificial Coalification Gas by Mass-spectrometry

Compornents	Coal Seam Gas	Artificial Coalification Gas		
		270 ℃	306 ℃	345 ℃
Methane	84.42 %	1.63%	6.13%	11.37%
Ethane	-	0.18	1.03	3.11
Propane	-	0.94	1.74	3.59
Propylene	0.08	0.34	0.93	1.56
Butane	-	-	0.41	1.28
Butene	-	-	0.37	0.46
Hydrogen	-	-	-	0.96
Carbon dioxide	0.08	22.39	35.98	41.97
Nitrogen	14.54	74.52	53.41	37.50
Oxygen	0.88	-	-	-

Fig. 1   The Heating Curves of the Reaction Rate
of Coalification during the 1st Experiment

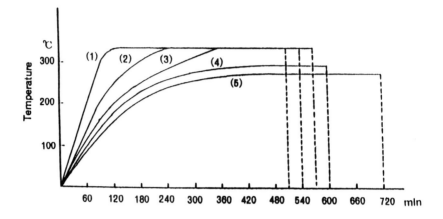

Fig. 2   Relationship between Evolved Gas and Reaction Rate Index
of Coalification or Reaction Product Index on Nakago Coal

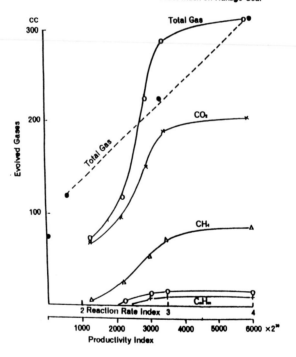

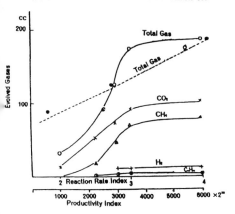

Fig. 3 Relationship between Evolved Gas and Reaction Rate Index
of Coalification or Reaction Product Index on Yumoto Coal

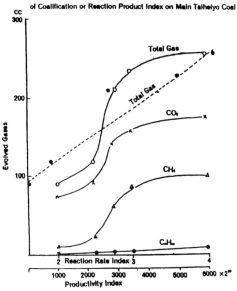

Fig. 4 Relationship between Evolved Gas and Reaction Rate Index
of Coalification or Reaction Product Index on Main Taiheiyo Coal

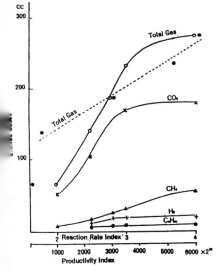

Fig. 5 Relationship between Evolved Gas and Reaction Rate Index
of Coalification or Reaction Product Index on Upper Taiheiyo Coal

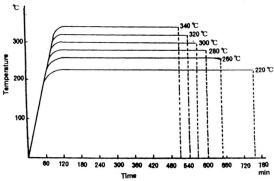

Fig. 6 The Heating Curve under Constant Temperature
during the 2nd Experiment

Fig. 7   The Relationship between Temperature and Total Evolved Gas

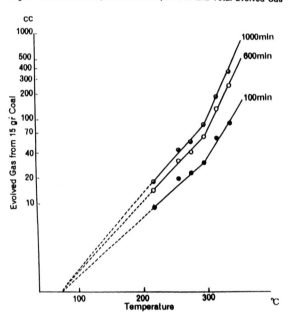

Fig. 8   The Relationship between Evolved Gas and Time

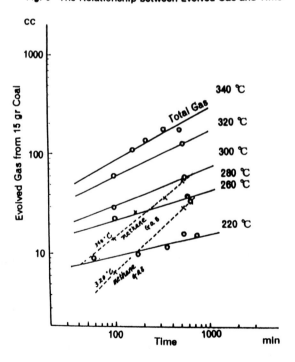

COAL BED GAS CONTENT AND GAS UNDERSATURATION

The significance of Self-generated Thermogenic Gas, Sorption Equilibria, Biogenic Gas and Diffusive Gas Loss

G. KHAVARI-KHORASANI[1] and J.K.MICHELSEN[2]

[1] *PETEC, Prof. Olav Hanssens vei 15, P.O.Box 2503, 4004 Stavanger, Norway;* [2]*STATOIL, 4035 Stavanger, Norway*

1.0 Abstract

Undersaturation of coal with respect to gas is a major economic risk in the coal bed gas exploration. This communication addresses the following questions: (1) How much thermogenic gas is formed from coals at different levels of maturity? (2) What is the minimum rank at which a given coal has generated sufficient thermogenic gas to be saturated at reservoir pressure and temperature conditions? (3) What is the effect of uplift on sorption equilibria? (4) What is the role of biogenic gas for coal bed gas saturation? and (5) what is the magnitude of the diffusive gas loss? Results and conclusions of this work do not include coals which are within a conventional gas reservoir. Kinetic compositional modeling results, calibrated with coal-specific data, show that the vitrinite reflectance (R_o) level at which a coal has generated sufficient thermogenic gas to be saturated at reservoir pressure-temperature (PT) conditions varies from less than 0.60% to more than 1.4%. The coals rich in linear aliphatic polymers (LAP) become saturated by self-generated thermogenic gas at reflectance levels below 0.75%. The self-generated gas from coals which are dominated by "humic" aromatic polymers (HAP) does not reach the saturation sorption level, below a vitrinite reflectance of around 0.9-1.2%. The vitrinite-rich coals of Jurassic, Cretaceous and Tertiary age are more likely to have an appreciable contribution from LAP material, than are the coals of Palaeozoic age. Coals which are dominated by semi-inertinites may not become saturated by self-generated gas even at very high ranks ($R_o>1.4\%$). The importance of biogenic gas for saturating a severely undersaturated coal is questionable. Very strong biogenic signature is observed in strongly undersaturated coals, which have lost significant amounts of thermogenic gas. Evaluation of sorption (ad- & ab-sorption) and desorption experiments and modeling results indicates that coal can sorb larger quantities of gas in deep basin positions, and those which have generated enough gas to do so, will desorb gas during uplift. A critical factor which can lead to undersaturation, of an initially saturated coal bed, is diffusive gas loss. The effect of diffusive losses can be significant and should be accounted for in the risk assessment.

2.0 Introduction

One of the critical factors required for the economic coal bed gas production is a high hydrocarbon gas content of the coal reservoir. The coals become saturated with self-generated thermogenic gas, when the amount of gas formed exceeds the retention capacity. The gas adsorption capacity increases with increasing coal rank, however, there are no consistent trends regarding the role of coal maceral composition on the adsorption capacity (e.g., van Krevelen, 1961; Kim, 1977; Yee *et al.*, 1993; Lamberson and Bustin, 1993; Rice, 1993). The gas content of the coal is not only dependent on thermogenic gas generation, and hence prediction of gas content of coal beds is not straight forward. In this paper we address the following:

(1) The role of self-generated thermogenic gas, compared to the gas requirements for coals to be saturated at reservoir pressure-temperature (PT) conditions.

(2) the role of uplift and sorption equilibria in enhancing/ inhibiting undersaturation conditions,

(3) the role of biogenic gas in saturating a severely undersaturated coal, and

(4) the magnitude of the diffusive gas loss.

The following terms are used in this paper.

Transformation ratio and coal maturity–We define the coal maturity as the transformation ratio obtained from the kinetic parameters for petroleum generation. When transformation ratio, TR≈0, the coal is immature. When TR≈1, the coal generating capacity from primary cracking of coal kerogen is exhausted. Vitrinite reflectance kinetics has a poor relationship with petroleum generation kinetics (Khavari-Khorasani and Michelsen, 1994). Nevertheless, since R_0 data are commonly used and widely available, we present kinetic compositional modeling results for gas generation against the R_o, calculated from the vitrinite reflectance kinetics of Burham and Sweeney (1989).

Petroleum–The term petroleum, used in this paper, refers to all organic molecules, or mixtures of such molecules, which are the main constituents of oil and gas, regardless of phase states (e.g., gas, liquid, ad- or ab-sorbed states). Petroleum generated from coal is a mixture of varying quantities of C_1 (methane) to $C_{30}+$ hydrocarbons, as well as asphaltenes (heavier fraction of non-hydrocarbons) and "resins" (lighter fraction of non-hydrocarbons). C_1-C_4 molecules will be mainly in a gas phase at very shallow basin positions or at the surface.

The samples used in this study are examples from Tertiary coals from SE Asia, USA, Columbia, and Svalbard; Cretaceous of USA; Carboniferous of Poland, and Permian of Australia and South Africa. The relevant features of the coals are presented later, together with their gas generating properties.

The results, shown in this paper, were obtained by the *PetroTrak PCP* and *PetroTrak CBM* software packages. The background numerical and calibration methods have been described elsewhere (Khavari-Khorasani *et al.*, 1998). The coals used for obtaining compositional calibrations are immature to very low maturity samples.

3.0 Results and Discussion

3.1 THE ROLE OF THERMOGENIC GAS GENERATION

The amount of petroleum generated at different levels of maturity from a coal is dependent on the polymeric composition of the coal, which controls (a) the thermal stability of the coal with respect to petroleum generation, (b) the composition of generated petroleum, and (c) the petroleum yield of the coal. Therefore, we first discuss the relationship between the coal petrographic and polymeric compositions.

3.1.1 *Coal Macromolecules, Polymeric Composition, and Dominant Macerals*

It is well known that coal is an aggregate of different polymers. Numerous works have addressed the petroleum generating capacity of kerogens, including coals, and the relationship with the nature of their pyrolysis products (e.g., Larter and Sentfle, 1985; Boreham *et al.*, 1988; Horsfield, 1989; Boreham and Powell, 1991; Khavari-Khorasani and Michelsen, 1991). Based on the macromolecular structure of coals, as reflected by pyrolysis gas chromatography, the coal polymers can be best divided into two major classes.

(1) Truly lignin-derived or lignin-like structures and their oxidation products. We refer to these polymers, collectively, as "humic" aromatic polymers (HAP). The pyrolysates of these polymers are dominated by light hydrocarbons, and have high concentration of phenols and alkyl phenols, and/or aromatics. Such products are typically released from coals rich in telocollinitic vitrinite and their precursor macerals (Figure 1[a1-a2]), and coals enriched in inertinites (Figure 1[b1-b2]).

(2) The "Linear" aliphatic polymers (LAP). The pyrolysates have a high relative concentration of C_{19}+ hydrocarbons, and display a "waxy" nature. Such products are released during pyrolysis of liptinite-rich coals (Figure 1[c1-c2]). However, many coals which are rich in desmocollinitic vitrinite can also be enriched in LAP material (e.g., Figure 1[d5-d9]. This is particularly the case for some Mesozoic and Tertiary coals.

The polymeric composition of desmocollinitic vitrinites varies significantly, between the two end-member polymer classes in the coal, as illustrated by flash pyrogram examples in Figure 1[d1-d9], for coals having 80% or more desmocollinite. This is related to the fact that desmocollinite precursors contain detrital fragments of mixtures of LAP and HAP, which can be detected microscopically at the brown coal stage of coalification, but which become "masked" at higher ranks, as the coal undergoes a process known as "geochemical gelification". This process results in the transformation of the vitrinite-precursor organic matter into a physical entity known as maceral vitrinite (e.g., Stach *et al.*, 1982). Hence, desmocollinitic vitrinites can display large variation in their chemical make up, depending on the relative contribution from LAP and HAP material.

3.1.2 *Relative Thermal Stability of Coal Macerals with Respect to Petroleum Generation*

There are large variations in the thermal stability of coal macerals (Figure 2). The peak petroleum generation (transformation ratio$\approx$0.5) from different macerals vary from $R_o \approx 0.6$ to more than 1.3%. In general, the more enriched a coal is in telocollinitic vitrinite, the higher the thermal stability for generating petroleum. With increasing content of LAP-enriched desmocollinite and/or liptinites (with exception of alginite) the average thermal stability of the coal will decrease and generation will be shifted to lower R_o levels.

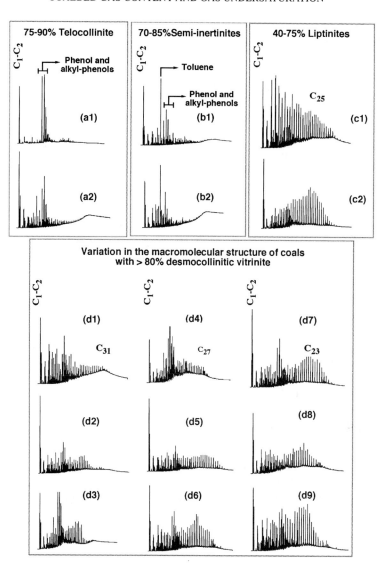

FIGURE 1. Examples of flash pyrograms of immature to low maturity coal kerogens–[a1]:A typical fossil wood, the precursor of telocollinitic vitrinite from Victorian brown coal; [a2]: a Carboniferous coal rich in telocollinitic vitrinite from Poland; [b1]: a South African Permian coal rich in semi-inertinite; [b2]: an Australian Permian coal rich in semi-inertinite; [c1]: a European Carboniferous cannel-algal coal; [c2]: a Tertiary coal rich in liptodetrinite as well as desmocollinite from SE Asia. [d1-d9]: Coals rich in desmocollinitic vitrinite, [d1] & [d4]: Carboniferous of Poland; [d2]: Permian of Australia; [d3]: Permian of South Africa; [d5]: Tertiary of Columbia; [d6]: Cretaceous of USA; [d7]: Tertiary of NW Svalbard; [d8]: Tertiary of USA, and [d9]: Tertiary of SE Asia.

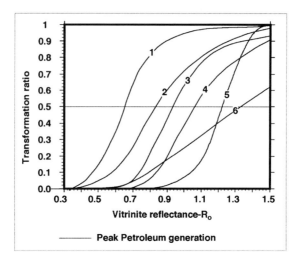

FIGURE 2. Transformation ratio (different stages of petroleum generation) of coal macerals versus vitrinite reflectance (R_o). 1: Suberinite; 2: sporinite/cutinite; 3: "perhydrous" desmocollinite; 4: "orthohydrous" desmocollinite; 5: alginite; and 6: telocollinitic vitrinite.

The observed variations in Figure 2 has important implications for the amounts of gas generated at different reflectance levels from different coals (see later data and discussion)

3.1.3 Quantitative Gas Generation from Coals: Kinetic Compositional Modeling

Figure 3 shows calculated gas generation curves versus the transformation ratio for the coals listed in Table 1.

The gas generation curves in Figure 3 are obtained from kinetic compositional modeling, and the results are only related to primary cracking of coal kerogen, and do not include secondary cracking of the retained petroleum in the coals. Furthermore, the differences in relative concentration of the LAP material in the starting coals are shown by low temperature flash pyrograms (525°C) of the coals. The starting coals are all of very low maturity. Hence, the systematic differences in the concentration of LAP material in the coals, detected from the low temperature flash pyrolysate products (see pyrograms in Figure 3), are not related to any maturity differences between the samples.

TABLE 1. Sample code, dominant maceral, and relative concentration of LAP products in flash pyrolysates of the coals, which were used to obtain the compositional calibration data for the kinetic compositional modeling (Figure 3).

Sample code	Dominant maceral (>80% by volume)	Relative concentration of LAP products in flash pyrolysates*
(a)	Suberinite + Desmocollinite	High
(b)	Desmocollinite	High
(c)	Liptodetrinite + Desmocollinite	High
(d)	Desmocollinite	Intermediate
(e)	Desmocollinite	Intermediate
(f)	Telocollinite	Low

*The corresponding flash pyrograms are shown in Figure 3

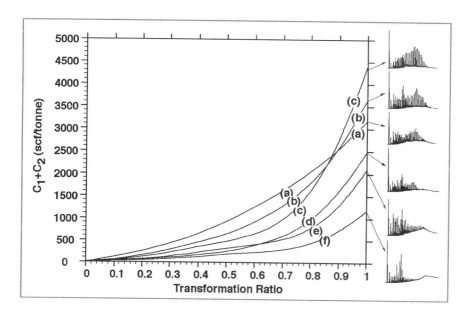

FIGURE 3. The cumulative self-generated thermogenic gas with increasing transformation ratio for the coals listed in Table 1.

The gas generation results (Figure 3) are shown only for C_1+C_2 hydrocarbon gases. This is because coal bed hydrocarbon gases are mainly methane and to a much lesser extent ethane, and higher hydrocarbon gases are commonly negligible. Figure 3 shows the following:

(1) The cumulative amounts of gas generated at any level of maturity, from coals with a high concentration of LAP material is significantly higher (Figure 3[a-c]), compared to a coal which is mainly composed of HAP (Figure 3[f]). The coals with intermediate polymeric features have gas generating properties intermediate between the former and latter coals. (Figures 3[d-e]).

(2) The gas generation from some coals shows a uniform evolution (e.g., Figure 3[a]), while for most other coals it shows a significantly more rapid increase above a transformation ratio of TR≈0.7 (Figure 3[b-f]). The contrast between the thermal evolution pattern of the former (a) and the latter coals (b-f) is related to the relative thermal stability of C_1-C_2 precursors versus the total product yield.

3.1.4 Saturation Sorption Level at Viable Coal Bed Gas Reservoir Conditions

The relationship between the adsorption capacity, R_o and maceral composition for a number of coals from both the northern and southern hemispheres, varying in age from Carboniferous to Tertiary, is shown in Figure 4. The indicated saturation sorption levels are the required gas content for the coals to be saturated at the coal bed gas reservoir PT conditions.

The saturation sorption level shows a general increase, with increasing R_o. For a given R_o level, the variation in the sorption capacity is significant in coals below R_o≈1.3%, and the differences are strongly reduced with increasing R_o above 1.4%. This is consistent with the known chemical and physical properties of coals and how they change with rank (e.g., van Krevelen, 1961, 1993). The large differences which are observed in the polymeric composition and in the chemical and physical properties of lower rank coals, will be strongly reduced above R_o≈1.4%, as the polymeric make up of the high rank coals is transformed into increasingly more condensed, well ordered aromatic structures, (Blayden et al., 1944; Franklin, 1951; Diamond, 1957; van Krevelen, 1961, 1993).

In Figure 5 the gray shaded area is based on data from Figure 4, and shows the common range of saturation sorption levels (at coal bed gas reservoir PT conditions) for global coals in R_o range 0.5-1.5%. Below we evaluate what is the minimum R_o at which a coal has generated enough thermogenic gas to reach the saturation level.

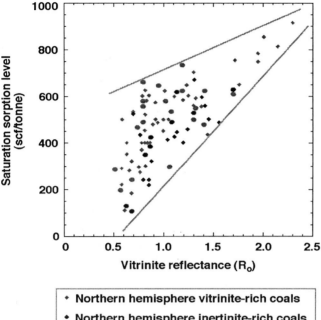

FIGURE 4. Saturation sorption levels, at viable coal bed gas reservoir PT conditions, as a function of R_o, for vitrinite- and inertinite-rich coals from the northern and southern hemispheres.

3.1.5 Self-generated Thermogenic Gas Generation as a Function of R_o, Compared With Gas Requirements for Saturation Conditions

In Figure 5 the common range of saturation sorption levels (at reservoir PT conditions) for global coals is compared with the kinetic compositional modeling results for generation of C_1-C_2 hydrocarbon gases from the coals listed in Table 2. Figure 5 shows the following:

(1) Disregarding diffusive gas losses, the minimum coal rank at which a coal has generated sufficient thermogenic gas to be saturated at reservoir PT conditions vary from less than $R_o \approx 0.6$ to more than 1.4%.

TABLE 2. Sample code, sample age, and dominant maceral(s) of the coals, which were used
to obtain the compositional calibration data for the kinetic compositional modeling (Figure 5)

Coal sample code	Coal sample age	Dominant maceral(s)
(1) SE Asia	Tertiary	Desmocollinite + Liptodetrinite
(2) USA	Tertiary	Desmocollinite + Suberinite
(3) USA	Cretaceous	Desmocollinite
(4) Norway (offshore)	Jurassic	Desmocollinite
(5) Australia	Permian	Desmocollinite
(6) Poland	Carboniferous	Desmocollinite
(7) Poland	Carboniferous	Telocollinite
(8) Australia	Permian	Semi-inertinite

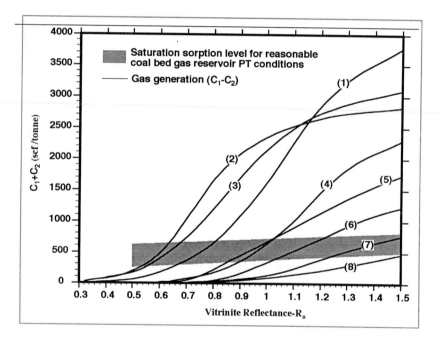

FIGURE 5. The common range of saturation sorption levels (at reservoir PT conditions), as a
function of R_o (the gray shaded area), for global coals (based on data in Figure 4), compared
with the quantities of self-generated thermogenic gas from the coals listed in Table 2.

(2) The desmocollinite-rich coals of Cretaceous and Tertiary age can be saturated by self-generated gas at reflectance levels below $R_o \approx 0.75\%$. For desmocollinite-rich coals of Palaeozoic age to be saturated by self-generated gas, they must attain a higher rank ($R_o \approx 0.9$ or higher). The inertinite-rich coals are unlikely to be saturated by self-generated gas even at high ranks.

(3) The coal dominated by telocollinitic vitrinite does not become saturated by self generated gas before a higher rank is reached ($R_o \approx 1$ or higher).

3.2 THE ROLE OF UPLIFT IN SORPTION EQUILIBRIA AND COAL BED GAS SATURATION AND UNDERSATURATION

Some authors (Scott et al., 1994) have concluded that the equilibrium sorption level increases as a coal is uplifted from the depths where thermogenic gas is formed to PT levels where CBM production is possible. This implies that when a coal saturated with light hydrocarbons is uplifted towards viable coal bed gas production depths, additional gas is required to saturate the coal. The conclusion of Scott et al. (1994) is based on extrapolation of the Langmuir model (pure monolayer adsorption) to high PT conditions. Below we evaluate this hypothesis.

Langmuir isotherms are measured at constant and normally low temperatures (25-75°C). Theoretically, equilibrium sorption level at constant pressure will decrease as a function of increasing temperature. The key question to evaluate for coal bed gas is, therefore, the equilibrium sorption level along typical pressure-temperature gradients in sedimentary basins.

Adsorption behavior for coal bed gas is normally discussed assuming single layer adsorption of a pure gas (methane) at a constant temperature. This treatment is probably adequate for methane-dominated fluids of uplifted coals, at coal bed gas reservoir PT conditions. However, the fluids generated from coal are commonly a complex mixture of C_1 (methane) to $C_{30}+$ compounds (e.g., Figure 1[d1-d9]), with saturation pressure of the fluids close to the actual pressure. Under these conditions the monolayer model is inadequate and cannot be used to extrapolate from low PT conditions of (uplifted) coal reservoirs to PT conditions of generation. For a single component fluid, the Brunauer, Emmett and Teller (BET) equation predicts that total adsorption rapidly increases above the monolayer capacity as the pressure increases, and that even at very high temperatures the total adsorption exceeds monolayer capacity, as the pressure approaches the saturation pressure (Gregg and Sing, 1982).

In early sorption studies of coal, the coal mass was assumed to be a homogeneous, highly cross linked aromatic crystalline-like polymer (Moffat & Weale, 1955). The coal was assumed to have a porosity corresponding to the helium volume which could be injected into the coal. Because of the short apparent equilibration times for gas sorption, the sorption was assumed to be exclusively adsorption. However, Moffat and Weale (1955) observed significant coal volume changes during gas sorption under pressure conditions varying from 0-700atm. Such volume changes are difficult to explain as related to adsorption, but can be explained by absorption. In adsorption gas will reside on surfaces of empty micropores. However, in absorption material is added to the solid solution, and the solution must expand according to the volume the new mass is occupying in the solution.

In the experiments and calculations of Moffat & Weale (1955) only synthetic semi-graphite (carbon black) displayed behavior compatible with a pure adsorption process. But all coals in the rank range of interest to coal bed gas displayed an adsorption incompatible behavior, at pressures above 200atm. Moffat & Weale (1955) attributed the adsorption incompatible behavior of the coals to penetration of methane at high pressures, into porosity not available to helium at lower pressures. However, Moffat & Weale (1955) and most authors at the time did not acknowledge the fact that coal can contain considerable amount of linear aliphatic polymers, and therefore, overlooked the importance of absorption. Extrapolation of the temperature-dependency of measured adsorption from low pressures (up to 150-200atm.) to higher pressures (>200 atm.,) would have been applicable (with appropriate PT corrections), if coal macromolecular structure was similar to that of true lignin-derived or lignin-like polymers (e.g., Figures 1[a1-a2]), or to that of carbon black.

Later authors, e.g., Reucroft and Patel (1983,1986), Milewska-Duda & Duda (1994) realized, from a combination of chemical analysis and sorption behavior, that coal represents a heterogeneous mixture of everything from strongly cross-linked aromatic polymers to linear aliphatic polymers. This fact has also been long recognized by numerous pyrolysis experiments of coals (see examples in Figures 1[a1-a2], 1[b1-b2], 1[c1-c2], 1[d1-d9]).

Milewska-Duda and Duda (1994) developed a viable dual ad/ab-sorption model, and from a combination of sorption/swelling measurements and modeling, showed that for coals varying in rank from high-volatile bituminous to low-volatile bituminous, already at pressures around ≈45 atm., <u>ab</u>sorption accounted for 10-20% of the total methane sorption. The authors further showed that the <u>ab</u>sorption capacity increased almost linearly with increasing pressure, in the pressure range used (0 to ≈45 atm.) (Figure 6).

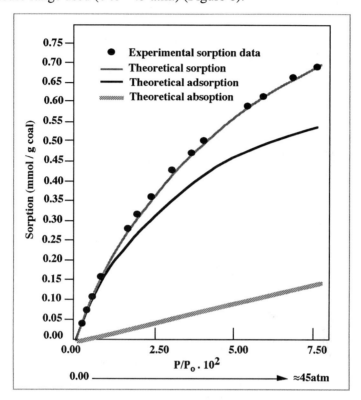

FIGURE 6. Sorption isotherm of a coal, including experimental sorption data and theoretical sorption, compared with theoretical ad- and ab-sorption, at pressures below ≈50atm. (modified from Milewska-Duda and Duda, 1994).

At low pressures up to 150-200 atm., sorption of methane can be approximated by the Langmuir isotherm. Hence, for CBM resource assessment and enhanced recovery, the sorption can adequately be described as an adsorption process. If stress-perm-shrinkage evaluations are important, it should be noted that the volume change of the coal matrix is strongly affected by <u>ab</u>sorption (Reucroft & Patel, 1983; Milewska-Duda & Duda, 1994).

It is implicit that in the potential coal bed gas sites, the coal seams of bituminous rank have been extensively uplifted. The adsorption capacity of coal and carbonaceous material is typically reached at around 200 atm. Adsorption-incompatible behavior of coals above 200atm., shown by Moffat & Weale

(1955), as well as the experiments and modeling results of Milewska-Duda & Duda (1994) show that the Langmuir model cannot be applied to coals at fluid pressures above 200atm ($\approx$2880psi). Similarly, extrapolation of the temperature-dependency of measured adsorption from low pressures (up to 150-200atm.) to high pressures (similar to those prevailed at deeper parts of the sedimentary basins) will be misleading, because at pressures above $\approx$200atm. absorption become important. Furthermore, at high PT of generation, the total adsorption will exceed the monolayer capacity, as the fluids are approaching their saturation pressure (Gregg and Sing, 1982).

No studies are known to experimentally determine absorption equilibria of light hydrocarbons at high pressure and temperature. The main difference between the organic polymers which (*under high pressures*) display, adsorption-compatible behavior (e.g., carbon black) and those which display a strong adsorption-incompatible behavior (e.g., coals with potential for coal bed gas) is the concentration of linear aliphatic polymers. The former is void of, while the latter contains varying amount of such polymers. The aliphatic kerogen in coals (i.e., LAP material in desmocollinitic vitrinites and in liptinites), as well as retained free oil, will hold varying volumes of methane in solution.

Heavy oils are materials which in many respects resemble aliphatic polymers, even though a heavy oil will represent a more dilute polymer solution than the aliphatic kerogen. PVT properties of heavy oils are well studied and correlations are available to predict methane solubility in such material and the PT-dependency of the methane solubility (Standing, 1950). While we cannot predict levels of methane absorption in kerogen from such correlations, we can possibly get an idea of the PT dependency of the methane absorption in coal kerogen. For heavy oils, the bubble point pressure line in a TP (T=x, P=y) projection has always a positive slope. The slope is low, but for temperatures up to 250°C the slope is significant, indicating that along any PT gradient in a sedimentary basin, the absorption capacity of methane will increase with increasing depth. Furthermore, the higher the amount of methane absorbed, the higher the shrinkage of the solution when it is depressurized.

Hence, it is indicated that coals can sorb larger quantities of methane at deeper basin positions, and those which have generated sufficient methane to do so, will desorb methane (and shrink due to desorption from an absorbed state) during uplift.

3.3 THE MAGNITUDE OF GAS LOSS

After the generation is exhausted, free gas can coexist with the coal matrix if for example, the coal beds are forming or are part of a capillary trap (i.e, when coal beds are either a seal for a conventional reservoir or are located within a traditional petroleum reservoir). This paper does not address coal reservoirs which are within conventional reservoirs.

As long as there is no free petroleum fluid phase in a coal bed, the gas loss from coal is mainly by molecular diffusion. We do not imply that diffusion does not occur before the generation is ceased. However, as long as generation continues, the rate of generation is typically larger than the rate of diffusion, and hence diffusive loses are compensated. The medium for diffusion can be water, grain boundaries and organic matter.

In Figures 7a-b we show examples of some undersaturated coals, from the Silesian and Sydney basins. Here the saturation sorption level of the coals, are compared with their present day gas content (from desorption experiments). The kinetic compositional modeling results, calibrated with data from the low maturity counterparts of the coals, indicate that (1) the quantities of gas generated from the concerned coals exceeds their saturation sorption level, (2) the initially saturated coal seams have became undersaturated with time, due to diffusive gas loss. The results in Figures 7a-b cannot be generalized, however, the implication is clear:

The magnitude of diffusive gas loss from coals can be large and should be included in the risk assessment.

In some sections it is observed that the content of gas in the deepest, more mature coal seams is lower than the overlying, or relatively shallower coal seams. An example of this is shown in Figure 8, from a given coal section (Figure 8[i]). Here the two coal seams which were evaluated for the coal bed gas potential are those marked "B" and "C". The most mature coal seam "C" has a considerably lower gas content than the seam in the overlying sediments ("B") (Figure 8[ii]. The interval between the coal seam "B" and "C" are rich in thin coal layers uniformly distributed, and a number of coal seams or organic rich carbonaceous shales are also present in the shallowest part of the section, overlying the coal seam "B". The self-generated gas from the concerned coal seams "B" and "C" is more than that required to saturate them at reservoir PT conditions (Figure 8[ii]), but they are both undersaturated. The minimum gas loss is equivalent to the difference between the saturation sorption level and the actual gas content of the coals.

Of interest is the pattern of undersaturation, where the deepest coal seam "C" is much more strongly undersaturated than the overlying coal seam "B" (Figure 8[ii]), despite that the former has generated around 300scf/tonne more gas than the latter. Simulation of the diffusive gas loss, shown in Figure 9, predicts that such a pattern can easily develop from diffusive losses.

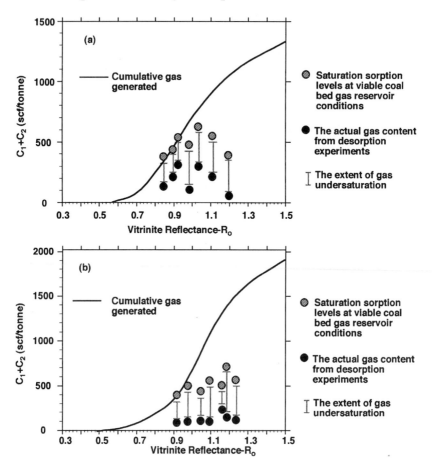

FIGURE 7. The saturation sorption levels (at viable reservoir PT conditions), and the actual gas content of the coals (from desorption experiments) for some undersaturated coals from the Silesian Basin, Poland (a), and the Sydney Basin, Australia (b). The corresponding gas generation curves show that the quantities of the generated thermogenic gas exceed the saturation capacity of the coals. The coals have lost large amounts of gas with time, after the generation had ceased. The gas generation curves are based on compositional modeling with calibration data from the low maturity counterparts of the concerned coals, and represent a good average for the coals.

It should be emphasized that in simulating the diffusive gas loss, it is not only the coal seams of concern which must be accounted for, but also the entire gas budget in the system. That is, all coal layers and all organic containing sediments which could have contributed to gas generation, should be accounted for in the mass balance.

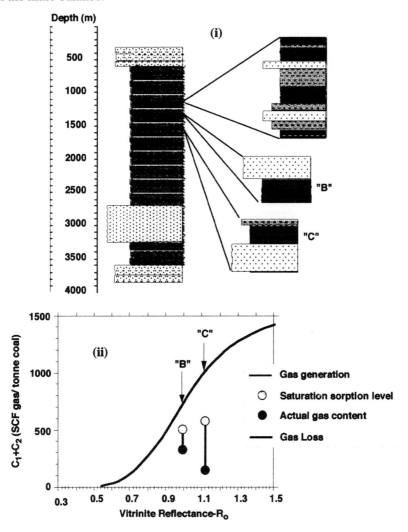

FIGURE 8. (i) A coal-bearing section, from which the coal bed gas potential of two thick coal seams "B" and "C" were evaluated and found to be undersaturated. (ii) A comparison between the saturation sorption levels and the actual gas content of the two coals. Note that the deeper, higher maturity coal seam "C", is more strongly undersaturated than the lower maturity coal seam "B". The gas generation curve is based on compositional modeling with calibration data from the low maturity counterparts of the concerned coals, and represents a good average for the coals. Since in both coals the quantities of generated thermogenic gas exceed the sorption capacity, the diffusive gas loss is directly translated to the difference between the saturation sorption level and the actual gas content of the coals.

The diffusive gas loss from the section in Figure 8[i] was estimated[1], based on the available geological information regarding the uplift of the section. That is, the coals had uplifted from their maximum burial depth (≈4500m) to around 2000m in the time period of 300-270Ma. This period of uplift had been followed by a stable period (270-60Ma), and then followed by a younger uplift in the period of 60-0Ma, which moved the coals to their present day depth of burial. The simulation results are shown in Figure 9.

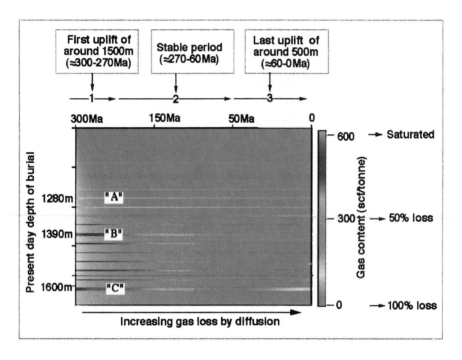

FIGURE 9. Simulation of the diffusive gas loss from the coal seams "B", and "C", in the section shown in Figure 8[i]. At the beginning of simulation both coal seams were gas saturated (saturation sorption level≈600scf/tonne). At the end of simulation (present day) 75% of the initial gas is lost from the coal seam "C", compared to 50% from the coals seam "B". The undersaturation pattern observed in Figure 8[ii], can be developed by diffusive gas loss, and is mainly related to the distribution of gas in the system and the position of the coal seams. Note that the shallowest coal seams (e.g., "A") have also become strongly undersaturated.

---

1. The diffusive gas loss was estimated taking into account the gas budget in the entire coal-bearing sequence, porosity and tortuosity of the sediments, burial and thermal history. The diffusion coefficients were corrected for temperature and viscosity. At the beginning of simulation the concerned coal seams was gas saturated, with a gas content of 600 scf/tonne, and the water in the surrounding sediments were also gas saturated. Methane iso-fugasity was always enforced between the coal and water. No free gas occurred in the simulation.

The deepest, highest rank coal seam "C" became already undersaturated during the first uplift (300-270Ma), at the same time that the shallower coal seam "B" stayed close to saturation. During the stable period ($\approx$270-60Ma) the coal seam "B" also became undersaturated with time, but to a lesser extent, compared to the coal seam "C". The last uplift (60-0Ma) which moved the coals to their present depth of burial, resulted in severe undersaturation of the deeper coal seam "C" which has lost more than 75% of its initial gas content. The coal seam "B" has lost around 50% of its initial gas, having today a higher gas content than the underlying coal seam "C". It should be noted that the gas loss is the net loss, i.e., total loss minus gained. Despite that all coal seams will be subjected to diffusive gas loss, the coal seams located at the center of a gas generating sequence (in terms of generation capacity), retain a higher quantity of the gas, compared to the deepest and the shallowest coals (Figure 9). For the given section, this pattern follows from the uniform distribution of coal seams and their effect on the total gas budget of the system. However, the observed pattern should not be generalized, since diffusion is controlled by many additional geological factors.

A comprehensive discussion of the main geological factors which control the extent of diffusive gas loss will be presented elsewhere.

3.4   THE ROLE OF BIOGENIC METHANE FOR COAL BED GAS

Since potential coal bed gas reservoirs have been uplifted, and are located at relatively shallow depths, it is possible that the presence of permeable sediments surrounding the coals can result in the introduction of bacteria into the coals. Below we evaluate the role of biogenic gas in coal bed gas sites.

An outline for why biogenic coal bed methane, has mainly been produced at late stages during uplift or at present coal bed positions, was given by Scott *et al.*, (1994). In addition to the factors discussed by the above authors, there is another factor which should be added to the arguments. During subsidence and generation, the temperatures are high. Hence the small quantities of biogenic methane which subside with the coal will not survive, because the residence time of a methane molecule in the coal is smallest at the highest temperatures.

Most of the biogenic methane molecules in shallow coal deposits or at coal bed gas sites must have been introduced or formed *in situ* at late stages, i.e, during and after the uplift of the coal to the shallower parts of the sedimentary basin. The question is if the biogenic methane generation in coal beds is sufficient to saturate an undersaturated coal.

3.4.1   *Observations from San Juan Basin Coals*

The isotope data of the north-central San Juan Basin coals display an isotope signature which is consistent with the presence of some biogenic gas in the coal bed gas mixtures (Scott *et al.,* 1994). However, the biogenic isotope signature of the San Juan coal bed gases is weak and mostly overlapped with the thermogenic signature. The important question to be addressed is how much thermogenic gas has been generated, how much biogenic gas is present in the mixture, and when and what fraction of the earlier thermogenic gases has been lost through diffusive losses. Scott *et al.* (1994), based on the data from artificial coalification experiments of north Dakota lignite (Tang *et al.,* 1991) and isotope data from the San Juan Basin coal bed gases, deduced that 15-30% of gases in the high productivity fairway of the basin is of late biogenic origin. However, the lignite from north Dakota is significantly different from the San Juan Basin coals. That is, much more thermogenic gas has been formed from the San Juan coals than that suggested from the data of the north Dakota lignite. In Figure 10 we show an example from the kinetic compositional modeling of gas generation, with calibration data from a low maturity San Juan Basin coal.

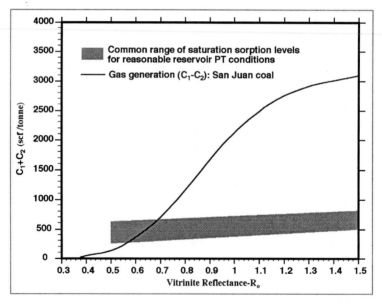

FIGURE 10. The quantities of self-generated thermogenic gas with increasing R_o, from compositional calibration data from a San Juan Basin coal, compared with the common range of saturation sorption levels (at viable reservoir PT conditions) for global coals. San Juan coals with $R_o \approx 0.7$-1.5% have generated far more thermogenic gas than that needed for the coals to be saturated at the coal bed gas reservoir PT conditions.

A comparison between the north-central Fruitland R_0 map (Rice, 1993; Scott *et al.*, 1994) and the compositional kinetic modeling results (Figure 10) shows the following:

(1) The San Juan coals, in R_0 range of 0.7 to 1.5%, have generated around 700-3000 scf /tonne thermogenic gas, quantities which are far above the saturation sorption level of the coals (Figure 10). These results are consistent (a) with the interpretation of Scott *et al.* (1994) that these most prolific coal bed gas deposits known today, are dominantly of thermogenic origin, and (b) with the good correspondence between the gas wetness and R_0 variations in the Basin (Scott *et al*, 1991; Rice, 1993).

(2) The San Juan coals with $R_0 \approx 0.7\%$ and higher have been initially fully saturated by self-generated thermogenic gas. Hence we conclude that the minor quantities of late biogenic gas in the north-central San Juan coals must be directly proportional to the diffusive loss of the self-generated thermogenic gas.

3.4.2 Observations from Sydney Basin Coals

Data published by Smith and Pallasser (1996) for Sydney Basin coals show a strong biogenic methane signature in the coals. However, the authors did not present any data on the quantities of gas which has been recovered, and it must be noted that their study was focussed on gases released during the mining. The R_0 range of coals used in their study was between 0.8-1.2%. The kinetic compositional modeling results, calibrated with the coal specific compositional data, show that the Sydney Basin average coals (with respect to generation capacity), in R_0 range of 0.8-1.2%, have generated at least $\approx$200-1400 scf/tonne *in-situ* thermogenic gas. A strong biogenic signature in the Sydney Basin coals is associated with strongly undersaturated coals, such as those shown in Figure 7b. These coals have lost significant amounts of self-generated thermogenic gas (Figure 7b), which have not been compensated by the later biogenic gas formation. The factors which result in introduction of bacteria to the coal (e.g., the presence of permeable sediments in the coal-bearing strata) are the same factors which lead to large diffusive gas loss (Khavari-Khorasani, 1997).

Hence, a very strong biogenic isotope signature of coal bed gas could be indicative of a "leaky" coal bed reservoir, and could present an enhanced risk of undersaturation.

4.0 Conclusions

The vitrinite reflectance level at which a coal has generated sufficient *in situ* thermogenic gas to be saturated at reservoir pressure-temperature (PT) conditions, varies largely from less than 0.60 to more than 1.4%. This rank is shifted to $R_0 < 0.75\%$ in coals rich in liptinites or in desmocollinitic vitrinites with appreciable quantities of linear aliphatic polymers (LAP). In contrast, coals rich in telocollinite, or in other vitrinites which are mainly originated from "humic" aromatic polymers (HAP), cannot be saturated by self-generated gas until they reach higher ranks ($R_0 \approx 0.9\text{-}1.2\%$). Coals which are dominated by semi-inertinite may not become saturated by self-generated gas even at very high ranks ($R_0 > 1.4\%$).

Many undersaturated coals have been once saturated, but have lost significant amounts of gas. In coal bed gas reservoirs (which are not part of a traditional gas reservoir), the gas loss is mainly by diffusion, and the magnitude of diffusive gas loss can be large and should be evaluated in the exploration risk assessment.

In some sections it is observed that the content of gas in the deepest, more mature coal seams is lower than that in the overlying coal seams. Simulation results predict that this pattern can be developed by the diffusive gas loss. The coal seams located at the center of a gas generating sequence (with respect to generation capacity), will display a smaller (net) gas loss, compared to the deepest and the shallowest coals.

It has been reported that the San Juan prolific coal bed gas deposits, which are dominantly of thermogenic origin, to have a small contribution from biogenic gas. However, the importance of biogenic gas for saturating a severely undersaturated coal, is questionable. Very strong biogenic signature, such as those reported for some of the Sydney Basin coals, is observed in strongly undersaturated coals which have lost significant amounts of self-generated thermogenic gas. The same factors which introduce bacteria into the coal (high permeability sediments) are amongst the factors which strongly enhance the loss of the thermogenic gas, via diffusion. Hence, a very strong biogenic isotope signature could present an enhanced risk of undersaturation.

It has been earlier suggested that as temperature and pressure decrease during uplift, the sorption isotherm of the coal changes, allowing the coal to sorb more gas, resulting in undersaturation. This suggestion is from an extrapolation of the temperature-dependency of measured adsorption from low pressures to high pressures. We have evaluated this hypothesis, and conclude that the monolayer adsorption (e.g., Langmuir model) cannot be used to extrapolate from low PT conditions of (uplifted) coal reservoirs to PT conditions of generation. At pressures higher than $\approx$200atm. ($\approx$2800psi), the change in sorption level will be to a large extent controlled by absorption equilibria, and in addition, the total adsorption will exceed the monolayer capacity. It is indicated that coals can sorb larger quantities of gas at deeper basin positions. During uplift the coal will desorb gas, and will shrink due to desorption from an absorbed state.

Acknowledgments–This paper is continuation of an ongoing research, which has been largely benefited from review and suggestions of Bill Hanson and John Seidle. We wish to acknowledge the PETEC management for continuation of the work and for support of the publication.

5.0 References

Blayden, H.E., Gibson, J. and Riley, H.L. (1944) An x-ray study of the structure of coals, cokes and chars. *Proceedings Conference on Ultra-fine structures of coals and cokes*. British Coal Utilization Research Association, London, 176-231.

Boreham. C.J., Powell, T.G., and Hutton, A.C. (1988) Chemical and petrographic characterization of the Australian Tertiary Duaringa oil shale deposit. *Fuel*, **67**, 1369-1377.

Boreham. C.J., Powell, T.G. (1991) Variation in pyrolysate of sediments from the Jurassic Walloon coal measures, eastern Australia as a function of thermal maturation. *Organic Geochemistry*, **17**, 723-733.

Diamond, R. (1957) X-ray diffraction data for large aromatic molecules. *Acta Cryst.*, **10**, 359.

Gregg, S.J., and Sing, K.S.W. (1982) *Adsorption, Surface Area and Porosity*, Academic Press, 303p.

Franklin, R.E. (1951) Crystallite growth in graphitizing and non-graphitizing carbons. *Proceedings of the Royal Society*, Series A, 209, 196-218.

Horsfield, B. (1989) Practical criteria for classifying kerogen: Some observations from pyrolysis gas chromatography. *Geochimica et Cosmochimica Acta*, 53, 891-901.

Kim, A. G. (1977) Estimating methane content of bituminous coalbeds from adsorption data. *US Department of the Interior, Bureau of Mines Research Investigation* 8245, 1-22.

Khavari-Khorasani, G. (1997) Coal bed methane: gas saturation and cleating properties. Abstracts:1997 *AAPG Annual Convention*, April 97, Dallas.

Khavari-Khorasani, G. and Michelsen, J.K. (1994) The effect of overpressure, lithology, chemistry and heating rate on vitrinite reflectance evolution, and its relationship with oil generation. *Australian Petroleum Exploration Association Journal*, 418-434.

Khavari-Khorasani, G., Dolson, C., and Michelsen J.K. (1998). The factors controlling the abundance and migration of heavy versus light oils, as constrained by data from the Gulf of Suez– Part I The effect of expelled petroleum composition, PVT and the petroleum system geometry. *Organic Geochemistry*, 23 p. In press.

Khavari-Khorasani and Michelsen, J.K. (1991) Geological and laboratory evidence for early generation of large amounts of liquid hydrocarbons from suberinite and subereous components. *Organic Geochemistry*, **17**, 849-863.

van Krevelen, D.W. (1961) *Coal*. Amsterdam: Elsevier, 514 p.

van Krevelen, D.W. (1993) *Coal*. Amsterdam: Elsevier, 979 p.

Lamberson, M.N. and Bustin, R.M. (1993) Coalbed methane characteristics of gates formation coals, Northeastern British Columbia: Effect of maceral composition. *American Association of Petroleum Geologists Bulletin*, **12**, 2062-2076.

Larter, S. R. and Sentfle, J.T. (1985) Improved kerogen typing for petroleum source rock analysis. *Nature*, **318**, 277-280.

Milewska-Duda, J, Ceglarska-Stefanska, G., and Duda, J. (1994) A comparison of theoretical and empirical expansion of coals in the high pressure sorption of methane. *Fuel*, **73**, 975-979.

Moffat, D.H. and Weale, K.E. (1955) Sorption by coal of methane at high pressures. *Fuel*, **34**, 449-462.

Reucroft P.J.and Patel H.(1983) Surface area and swellability of coal. *Fuel*, **62**, 279-284.

Reucroft P.J.and Patel H.(1986) Gas induced swelling in coal. *Fuel*, **65**, 816-820.

Rice, D.D (1993) Composition and origin of coalbed gas, in B.E. Law and D.D Rice (eds.) *Hydrocarbons from coal*. AAPG Studies in Geology, **38**, 159-184.

Scott, A.D., Kaiser, W.R., and Ayers, W.B. (1991) Composition, distribution and origin of Fruitland Formation and Pictured Cliffs Sandstone gases, San Juan basin, Colorado and New mexico, in S.D. Schwochow, D.K. Murray, and M.F. Fahy (eds.) *Coalbed methane of western North America*, Rocky Mountain Association of Geologists, 93-108.

Scott, A.D., Kaiser, W.R., and Ayers, W.B. (1994) Thermogenic and Biogenic gases, San Juan basin, Colorado and New Mexico - Implications for coalbed producibility. *American Association of Petroleum Geologists Bulletin*, **78**, 8. 1186-1209.

Smith, J.W. and Pallaser, R.J. (1996) Microbial origin of Australian coalbed methane. *American Association of Petroleum Geologists Bulletin*, **80**. 891-897.

Stach, E., Mackowsky, M.Th., Teichmüller, M. taylor, G.H., Chandra, D., and Teichmüller, R. (1982) *Stach's Textbook of Coal Petrology*. Gebrüder Borntraeger, Berlin, 535p.

Standing, M.B. (1950) *Volumetric and phase behavior of oil field hydrocarbon systems.* Reinhold Publishing Corporation, 123p.

Tang, Y., Jenden, P.D., and Teerman, S.C. (1991) Thermogenic methane formation in low-rank coals–Published models and results from laboratory pyrolysis of lignite. In Manning D. (ed.) *Organic Geochemistry Advances and Applications in Energy and the natural Environment,* 15th Meeting of the European Association of Organic Geochemists. Abstracts, 329-331.

Yee, D., Seidle J.P. and Hanson, W.B. (1993) Gas sorption on coal and measurement of gas content, in B.E. Law and D.D Rice (eds.) *Hydrocarbons from coal.* AAPG Studies in Geology, **38**, 203-218.

HIGHER HYDROCARBON GASES IN SOUTHERN SYDNEY BASIN COALS

M.M. FAIZ[1], A. SAGHAFI[2], N. R. SHERWOOD[1]

[1]*CSIRO Petroleum*
PO Box 136
North Ryde, NSW 1670
[2]*CSIRO Division of Energy Technology*
PO Box 136
North Ryde, NSW 1670

Abstract

Coal seams of the Sydney Basin contain large volumes of gas, mainly methane (CH_4) and carbon dioxide (CO_2) with subordinate amounts of heavier hydrocarbons (C_{2+}). The desorbable gas content of the Sydney Basin coals ranges up to about 20 m^3/t, and its abundance is mainly related to depth and geological structure.

Gas isotope data indicate that most of the hydrocarbons presently occurring within the coal measures sequence was generated during coalification between the Permian and Late Cretaceous or as a result of post-Cretaceous bacterial activity, or both. Most of the CO_2 was introduced into the sequence in association with intermittent igneous activity between the Triassic and the Tertiary.

In the Sydney Basin coals, the CH_4 content of the gas ranges up to 100% whereas the C_{2+} ranges up to 12% (by volume). A systematic variation in the various hydrocarbon components with depth is evident in most parts of the basin. Hydrocarbons, at depths shallower than about 600 m and close to the basin margins are essentially dry (depleted C_{2+}). At greater depth, however, the quantity of C_{2+} in the gas increases regularly with depth up to 12% at about 1200 m. Carbon isotope data for the southern part of the basin show that CH_4 becomes isotopically light towards shallower parts and close to the margins of the basin.

Two different hypotheses are proposed to explain these variations in gas composition. One hypothesis emphasises the role of gas fractionation during migration due to variations in molecular diffusivity and solubility of the components. In this context, lighter and more rapidly diffusing components such as CH_4 can migrate readily to shallower depths and as a result the CH_4/C_{2+} ratio is expected to increase towards the upper part of the sequence.

However, in the Sydney Basin, such preferential migration cannot totally account for the abrupt scarcity of C_{2+} at depths less than about 600 m. The lack of C_{2+} gases at < 500 m indicates that they may have been eliminated by a process which is confined to shallow depths. At these depths CH_4 is isotopically light ($\delta^{13}C$ values range generally between -75‰ and -50‰ PDB) signifying a strong biogenic influence. Worldwide experience shows that biogenically formed gases are generally dry. Therefore, both the molecular and carbon isotope compositions indicate that in the Sydney Basin the lack of C_{2+} at shallow depths is probably due to bacterial alteration of the gases.

233

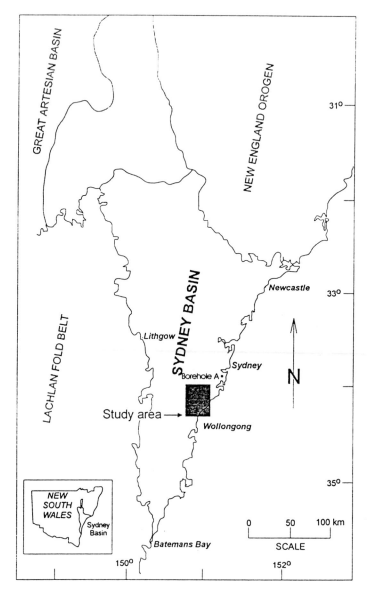

Figure 1 Location map of the Sydney Basin showing the main study area and Borehole A.

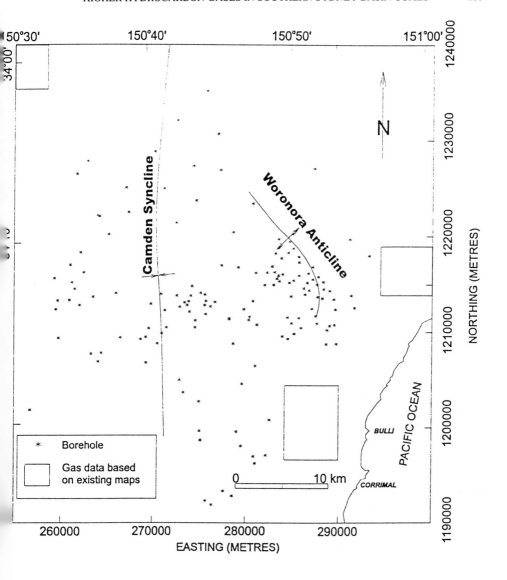

Figure 2 Borehole and other data locations in the main study area.

1. Introduction

The Permian coals of the Sydney Basin contain large amounts of gas and its distribution is mainly controlled by geological structure, depth and coal rank (Faiz and Hutton, 1995 and 1997). The origin of gas in these coals has been a subject of much discussion over the past 4 decades with several hypotheses being presented by researchers such as Hargraves (1963), Smith et al. (1984; 1992), Faiz and Hutton (1995). It is widely accepted that most of the CO_2 in the coal seams of the Sydney Basin were derived from magmatic sources (e.g. Hargraves, 1963; Smith et al. 1984; Faiz and Hutton, 1995). On the basis of carbon and hydrogen isotopes for CH_4 and CO_2, Smith et al (1991; 1996) suggested that, in contrary to conventional wisdom, the majority of CH_4 in Sydney and Bowen Basin coals is derived from bacterial reduction of CO_2. However, detailed work on the distribution of higher hydrocarbons (C_{2+}) and the lateral distribution of isotopic composition of these gases was not addressed. The purpose of the present paper is to investigate the distribution of C_{2+} and gas isotopic compositions in a view to better understand the origin of hydrocarbon gases in Sydney Basin coals.

The present study is mainly based on data from about 100 boreholes and about 12 coal mines from the southern Sydney Basin, complemented by data from 3 boreholes from the northern part of the Sydney Basin. The Sydney Basin and borehole locations in the main study area are shown in Figures 1 and 2, respectively.

2. Geology and Structure

The Sydney Basin represents a retro-arc foreland basin which covers an area of about 37000 km². The sediments in the basin range from Early Permian to Middle Triassic which unconformably overlie Lower to Middle Paleozoic magmatic and metasedimentary rock of the Lachlan Fold Belt. The Illawarra Coal Measures which is the main coal bearing sequence in the southern Sydney Basin, was deposited in fluvio-deltaic environments during the Late Permian (Figure 3). The coals are composed of mainly vitrinite and inertinite with minor amounts of liptinite (< 0.3% by volume), and the rank varies from high volatile bituminous to medium-low volatile bituminous.

The broad N-S trending Camden Syncline marks the main structural element of the basin (Figure 4). The eastern limb of the Camden Syncline is superimposed with a series of NW-SE, trending gentle anticlines and synclines. The western limb consists of N-S and NNW-SSE trending monoclines and NE-SW trending synclines. The folding in the region is gentle with most structures plunging northwest with regional dips of commonly less than 5°.

The Illawarra Coal Measures outcrop along the western and southern margins of the basin as well as along the escarpments of the eastern coast. Depth to the coal measures generally increases towards the north central parts of the basin where it reaches up to about 800 m along the northern parts of the Camden syncline.

Age	Group	Subgroup	Formation
TRIASSIC	Wianamatta Group		Bringelly Shale Minchinbury Sandstone Ashfield Shale
			Mittagong Formation
			Hawkesbury Sandstone
	Narrabeen Group	Gosford Subgroup	Newport Formation Garie Formation
		Clifton Subgroup	Bald Hill Claystone Bulgo Sandstone Stanwell Park Claystone Scarborough Sandstone Wombarra Claystone Coalcliff Sandstone
PERMIAN	Illawarra Coal Measures	Sydney Subgroup	Bulli Coal Eckersley Formation Wongawilli Coal Kembla Sandstone Allans Creek Formation Appin Formation Tongarra Coal Wilton Formation
		Cumberland Subgroup	Erins Vale Formation Pheasants Nest Formation
	Shoalhaven Group		Broughton Formation Berry Siltstone Nowra Sandstone Wandrawandian Siltstone
		Conjola Subgroup	Snapper Point Formation Pebbly Beach Formation Wasp Head Formation
	Clyde and Yarrunga Coal Measures		
	Talaterang Group		Yadboro and Tallong Conglomerates Pigeon House Creek Siltstone

Figure 3 Stratigraphy of the southern Sydney Basin (compiled from the Standing Committee on Coalfield Geology of New South Wales, 1971; Herbert, 1980; Carr, 1983).

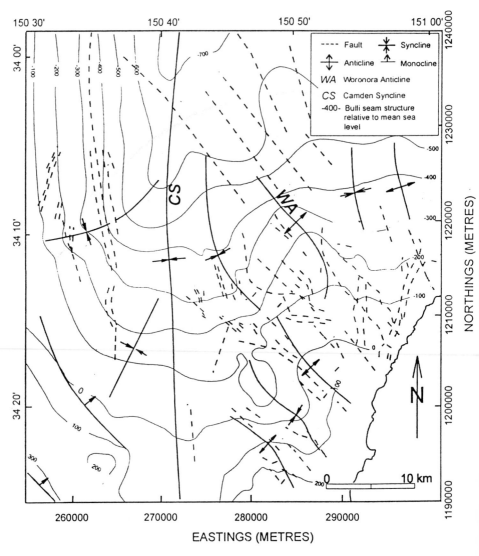

Figure 4 Bulli seam structure contours and the major structures in the study area (compiled from Bunny, 1972; Wilson, 1975; Shepherd, 1990 and Clark, 1992).

3.   Gas in the Southern Sydney Basin Coals

3.1 GAS GENERATION HISTORY

The coals of the southern Sydney Basin sequence attained ranks up to medium volatile bituminous coal as a result of high paleoheat fluxes (up to 105 mWm^{-2}) and deep burial (up to 3 km) during the Cretaceous (Faiz and Hutton, 1993 and 1997). Between the Early Triassic and Middle Jurassic, large volumes of CO_2 and H_2O and subsidiary amounts of hydrocarbons were generated from coals and organic matter dispersed in other sediments. The major phase of CH_4 and other hydrocarbon generation occurred during the Middle Jurassic and Late Cretaceous at burial depths greater than about 2 km (Faiz and Hutton, 1997). Towards the end of Cretaceous the basin sequence was rapidly uplifted and as a result, about 1.5 to 2.5 km of post-Permian sediments probably were eroded (Crawford et al., 1980; Faiz and Hutton, 1993).

3.2 COAL SEAM GAS CONTENT

The southern Sydney Basin coals contain up to about 20 m^3/t of gas (Figure 5); distribution in the Bulli seam is shown in Figure 6. This gas mainly consists of CH_4 and CO_2 with subsidiary amounts of C_{2+} (ethane, propane and butane) and N_2. At depths shallower than about 200 m the gas content of the coals is very small (< 0.1 m^3/t) due to severe degassing of.

The CH_4 and CO_2 in the coals generally account for > 90% of the in situ gas. Carbon isotope data indicate that in areas where the CO_2 content is > 10%, the gas has been primarily derived from magmatic sources (Smith et al., 1984). The distribution of CO_2 and CH_4 in the coal seams of the southern Sydney Basin are related to geological structure, depth and proximity to igneous intrusions (Hargraves and Lunarzewski, 1985; Faiz and Hutton, 1995). In structural highs and near some faults, coals generally contain high amounts of CO_2 whereas in the structural lows CH_4 is dominant (Figure 7). These variations are related to solubility and migration properties of these two gases; a more detailed discussion on the distribution of CO_2 in coal seams in the southern Sydney Basin is given in Faiz and Hutton (1995 and 1997).

3.3 HYDROCARBON GASES

Methane is the main hydrocarbon gas desorbed from the coals of the southern Sydney Basin at ambient pressure and temperature (NTP) conditions. The ethane (C_2H_6) content ranges up to about 10% (up to 1.5 m^3/t) of the total gas. Other higher hydrocarbons such as propane and butane, which in total range in volume up to about 1.5% of the gas, also occur in the coals. Although higher amounts of C_{2+} is likely to be sorbed in the coals, these hydrocarbons are not likely to be desorbed at NPT which the measurements were made, because they are strongly bonded to micropore surfaces.

As shown in Figure 8, C_2H_6 content of the gas in the southern areas and proximal to the basin margins of the southern Sydney Basin is very small (< 1%) but increases towards the north where the depth to the coal seams is greater. Figures 9a and 9b show that the content of C_{2+} at depths shallower than about 600 m is very low but increases rapidly with increasing depth. Limited data from the northern part of the Sydney Basin show a similar trend (Figures 10a and 10b). In Borehole A, which is located proximal to the coastal margin of the Sydney Basin, the coals lack C_{2+} gases down to depths of about 1100 m (Figure 10a). No strong correlation is apparent between the C_{2+} content with coal composition, rank or geological structure.

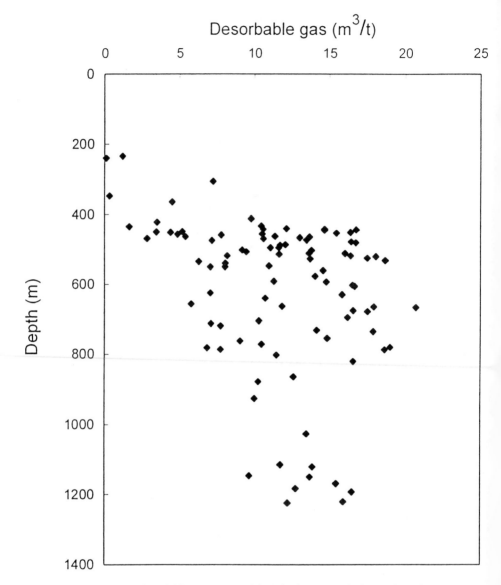

Figure 5 Desorbable gas content and depth for the southern Sydney Basin coals.

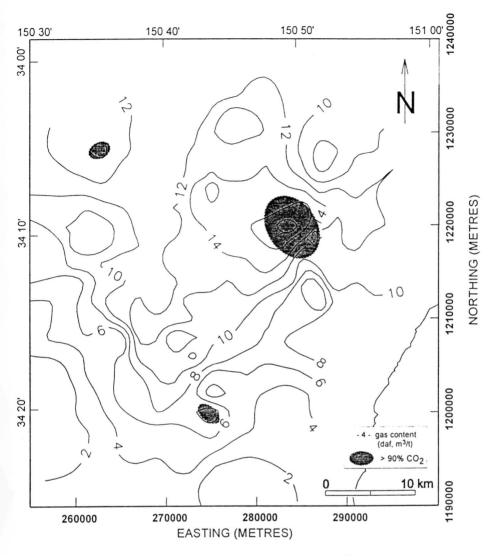

Figure 6 Isopleth map showing the desorbable gas content of the Bulli seam.

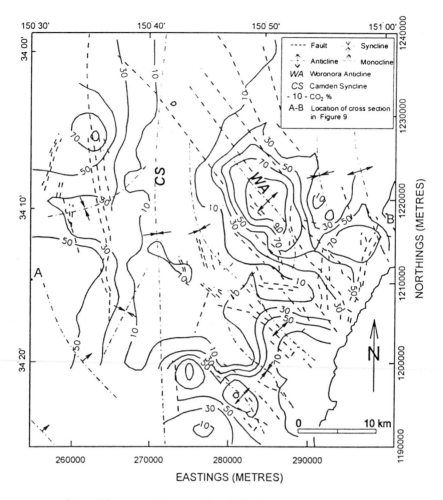

Figure 7 Structure and the proportion of CO_2 in gas desorbed from the Bulli seam.

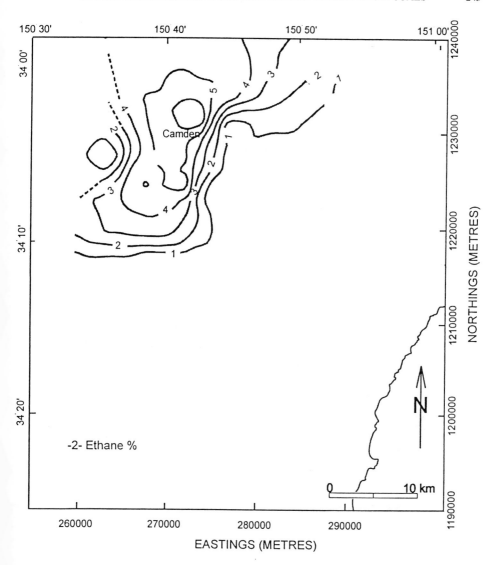

Figure 8 The proportion of ethane in gas desorbed from the Bulli seam.

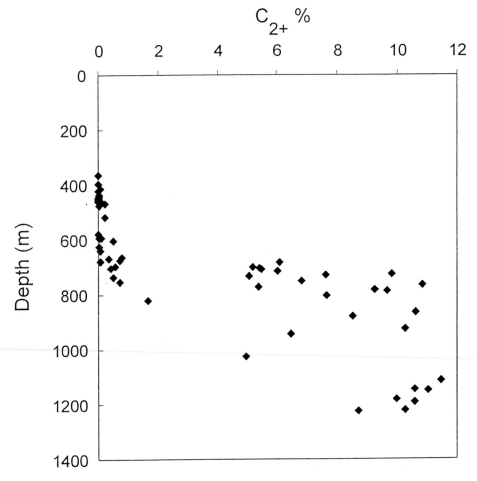

Figure 9a The proportion of ethane and higher hydrocarbons desorbed from southern Sydney Basin coals vs depth.

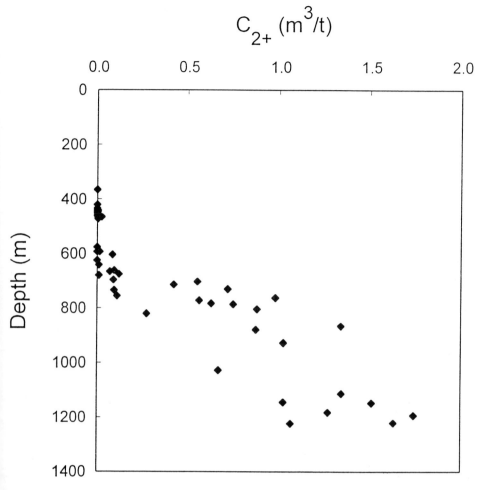

Figure 9b The ethane and higher hydrocarbons content (m³/t) desorbed from southern Sydney Basin coals vs depth.

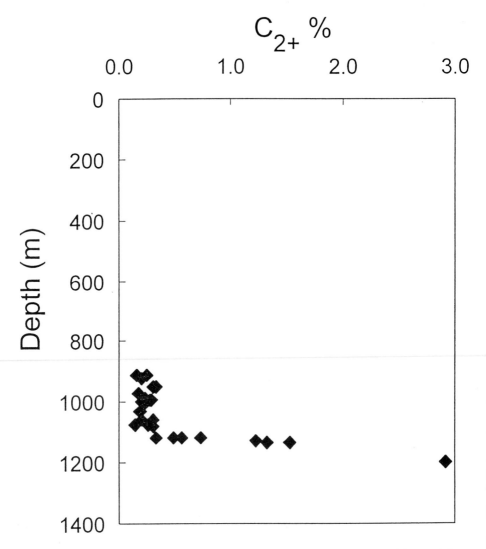

Figure 10a The proportion of ethane and higher hydrocarbons desorbed from coals in Borehole A (located north of the main study area) vs depth.

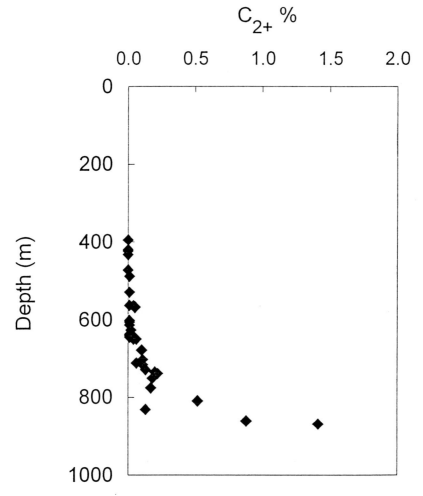

Figure 10b The proportion of ethane and higher hydrocarbons desorbed from 2 boreholes in the Newcastle area (in the northern part of the Sydney Basin) vs depth.

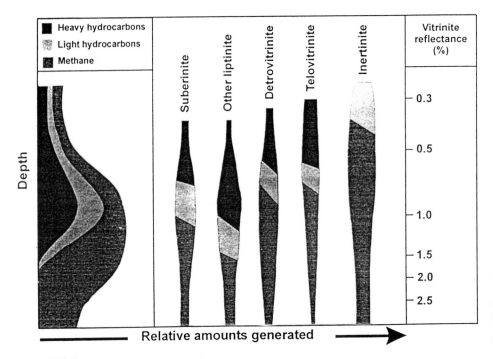

Figure 11 A generalised zones of hydrocarbon generation from coal maceral types as a function of maturity (modified from Murchison, 1987).

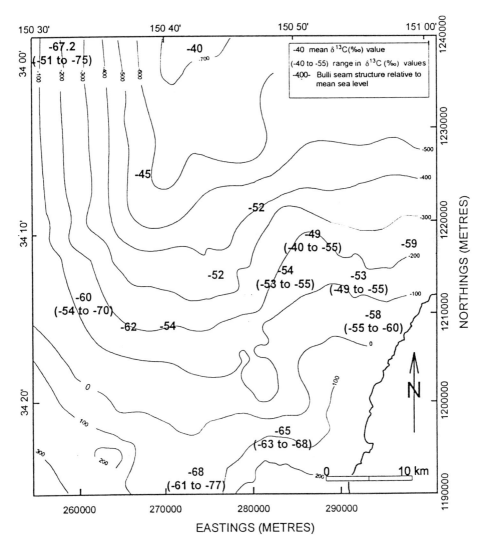

Figure 12 The $\delta^{13}C$ (‰, PDB) values for CH_4 desorbed from coals and Bulli seam structure contours.

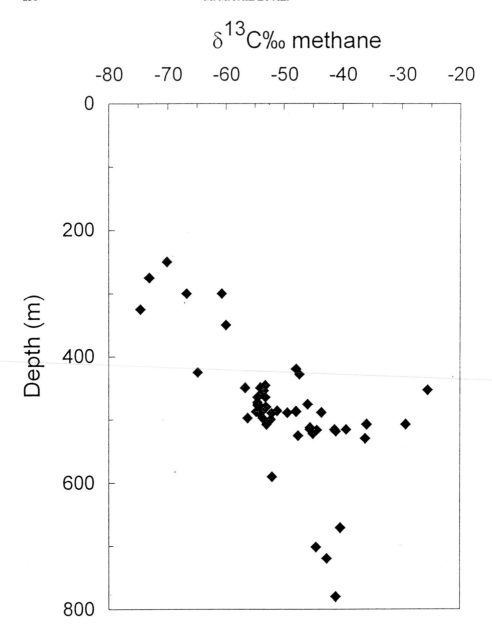

Figure 13 The δ¹³C (‰, PDB) values for CH₄ desorbed from coals vs depth for southern Sydney Basin.

4. Discussion

The volumes of hydrocarbons generated during coalification of sediments is mainly related to rank and maceral composition (Figure 11). The Permian coals of the study area mainly consist of vitrinite and inertinite with < 0.3% liptinite (Faiz, 1993). In general, during maturation of these coals the main hydrocarbon gas formed from these coals is CH_4. In general terms, most of the higher hydrocarbon generation normally occurs during sub-bituminous to high volatile bituminous coal stages, between vitrinite reflectances of about 0.5% and 0.8% (Figure 11). Beyond these ranks, the amount of C_{2+} generated decreases and the amount of CH_4 increases. However, in the southern Sydney Basin no systematic relationship exists between the C_{2+} content and vitrinite reflectance; this indicates that the present day distribution of these gases is not controlled by maturity. Although originally the gas composition within the basin may have been primarily related to the thermal maturity, subsequent processes such as migration and secondary alteration are likely to have caused redistribution of gases. The marked lack of C_{2+} gases in coals at shallow depths and close to basin margins indicates that these gases may have been eliminated from the basin since the post-Cretaceous uplift and cooling.

In German coals Colombo et al. (1970) and Knuper and Hukel (1971) observed a similar lack of C_{2+} at shallow depths and attributed this to gas fractionation during migration and strong degassing of the coals at shallow depths. Rice (1993) and Scott et al. (1994) stated that secondary bacterial alteration of wet gases causes the depletion of these at shallow depths in many coal bearing basins. The $\delta^{13}C$ values of CH_4 from the coal seam gases in the southern Sydney Basin range from -75‰ to -30‰ PDB (Smith et al, 1984). It is generally believed that $\delta^{13}C$ values of less than -55‰ is indicative of biogenic gas and isotopically heavier gas as thermogenic (Rice and Claypool, 1981). Based on C and H isotope data for CH_4 and CO_2, Smith et al. (1992) and Smith and Pallasser (1996) postulated that CH_4 present in the Sydney Basin is biogenically formed by microbial reduction of CO_2 and implied that most of the hydrocarbons that were generated during coalification, have been expelled from the sequence.

In the southern Sydney Basin, most of the CH_4 with $\delta^{13}C$ values < -55‰, were recorded from depths < 500 m. The isotopically lightest CH_4 has $\delta^{13}C$ values in the range of -75‰ to -60‰ and were mainly recorded from coals close to the basin margins or close to outcrops of the Illawarra Coal Measures (Figure 12). Methane in coals at > 500 m depth is generally isotopically heavier, indicative of a strong thermogenic influence (Figure 13). The regional variations indicate that the increase in the $\delta^{13}C$ values of CH_4 is accompanied by an increase in the C_2H_6 (Figures 8 and 12).

The variations in molecular and isotopic compositions of the Sydney Basin coal seam gases may be explained by way of two different hypotheses: one involves gas fractionation during migration and the other involves bacterial alteration of gases at shallow depths.

As a result of the post Cretaceous uplift and erosion of post Permian sediments on the order of 2 km, a large amount of gas formed during coalification was likely to have been expelled from the coals and other strata due to the decrease in pressure and increased permeability. At these shallow depths, almost all of the original gas, which includes CH_4 and C_{2+} gases, was probably expelled. The general lack of gas in coals at depths shallower than 200 m is clear evidence for such degassing. As a result of the expulsion of these shallow gases, those from deeper parts of the basin may would have migrated to the shallow depths and replaced the original gas. During upward migration of gas, both molecular and isotopic fractionation occur due to variations in molecular weight and diffusion rates. In this context, light hydrocarbons that diffuse faster would migrate farther than heavier and slower diffusing molecules. Therefore, CH_4 will migrate faster and concentrate at shallower depths than the C_2H_6 or higher hydrocarbons would. This process could therefore give rise to a systematic increase in the C_{2+} content with increase in depth as occurs in the Sydney Basin.

Similarly, if strong isotope fractionation occurred during migration, a concentration of $\delta^{12}C$ at shallow depths relative to $\delta^{13}C$ can be envisaged. However, the extent of isotopic fractionation during gas migration in geological systems and especially coal seams is not clearly understood. Research by Feux (1980) has shown that isotope fractionation during gas migration is minor. Experimental work conducted by Smith et al. (1985) and Gould et al. (1987) show that during desorption of gas from coal, isotope fractionation occurs but the rate changes with sample type. These experiments indicate that strong isotope fractionation occurs when CH_4 desorbs from run of mine samples enclosed in a canister. For example, within about 2 months of desorption, the $\delta^{13}C$ values changed from about -73‰ to -63‰. For the borehole and pillar samples studied however the rate of change in the isotopic composition of the desorbed gas with time is less than half of that for the run of mine samples. The data presented by these experiments are not sufficiently conclusive as to whether strong isotope fractionation occurs during gas migration within coal seams over long geological periods. It is also uncertain whether isotope fractionation in the order of -20‰ within about 250 m of migration, as present in the Sydney Basin, can be strictly attributed to preferential migration only. Furthermore, the abrupt changes in C_{2+}, at about 500 m, is not likely to have been caused only by preferential migration.

At shallow depths, where C_{2+} gases are lacking, the $\delta^{13}C$ values (-55‰ to -75‰) indicate a biogenic affinity for CH_4. The elimination of C_{2+} at shallow depths therefore may have been caused by alteration of the gases to CH_4 by bacterial activity. In the southern Sydney Basin such bacterial alteration of gas in shallow coals was possible since the Late Cretaceous uplift. During the Tertiary and Recent most of the coal seams would have been subjected to meteoric water flow in the shallow parts and close to the margins of the basin. Microbes that were carried in the meteoric water may have interacted with the organic compounds in the coal and the sorbed C_{2+} gases, to produce isotopically light CH_4. As the extent of biodegradation decreases with increasing depth, the C_{2+} gas in the coals is expected to increase, as is the case in the southern Sydney Basin. In this context, the molecular and isotopic compositions of the hydrocarbon gases indicate widespread microbial alteration between the surface and 600 m, and especially adjacent to margins of the southern Sydney Basin. The CH_4 in gases of deeper coals is wetter and isotopically heavier; it typically represents original gas formed during coalification prior to the Late Cretaceous. The abrupt lack of C_{2+} gases at depths shallower than 600 m implies that there could be a sharp boundary below which the microbes may not have been prevalent. A similar process may have been active in the northern part of the Sydney Basin where there is a lack of wet gas at shallow depths.

One of the questions that can be posed about the shallower CH_4 being bacterially derived relates to the feasibility of bacterial penetration of micropores < 2 nm in diameter, where the majority of gas in coal is stored. The most common size of bacteria is about 50 nm to 5 µm in diameter and therefore the direct alteration of gases sorbed on the micropore surfaces of the coals by these microbes would not be possible. However, these bacteria are capable of metabolising gas in the macro-pores (> 50 nm in diameter) and other organic compounds in the coal. The data of Scott et al. (1994) suggest that in the San Juan Basin, USA, coal seams may have acted as permeable aquifers where bacteria have been transported in groundwater to depths in the order of 3 km and up to 56 km basinward from the recharge zone. Similarly, such conditions may have prevailed in the Sydney Basin during periods of increased permeability, particularly soon after the

Late Cretaceous uplifts At these times meteoric water carrying microbes capable of metabolising C_{2+} gases and other organic compounds may have entered the coals from the margins of the basin.

5. Summary

The distribution of hydrocarbon gases in coals of the Sydney Basin is related to depth and proximity to outcrops. In the southern Sydney Basin, at depths shallower than 600 m, the gas desorbed from coals is primarily CH_4 with little or no C_{2+} gases. The C_{2+} components increase with increasing depth and away from outcrops of the coal seams.

Variations in the molecular and isotopic compositions of the coal seam gas of the southern Sydney Basin indicate that, at depths shallower than about 600 m most of the original gas from coalification probably has been expelled from the coals since the Late Cretaceous uplift. Subsequently these gases have been replaced by gases derived from microbial alteration of organic compounds in the coal and C_{2+} gases.

The precise timing and mechanism of the introduction of microbes into the coals, and the processes involved in the alteration of the sorbed gas and other organic compounds in medium volatile bituminous coals are not well understood. A better understanding of the fundamentals of isotopic fractionation of gas during diffusion and of the groundwater flow history in the basin is necessary to clearly explain variations in hydrocarbon gas distribution in the coal seams.

6. Acknowledgments

The authors thank the Australian Gas and Light Company, BHP Collieries Division (Illawarra), Dept of Mineral Resources (NSW), Kembla Coal and Coke Pty Ltd. and Metropolitan and South Bulli Collieries for providing most of the data used in this study. Special thanks are due to Bulent Agrali, John Anderson, Malcolm Bocking, Marcelle Burton, Barry Clark, Alan Cook, John Hanes, Adrian Hutton, Ripu Lama, Peter Lamb, Paul Maddocks, Ian Stone, David Titheridge and Jeff Wood for their assistance and co-operation. We also thank Carol Buckingham for drafting figures. The staff of the School of Geosciences, University of Wollongong is greatly acknowledged for their assistance.

7. References

Bunny, M.R. (1972) *Geology and coal resources of the Southern Catchment Coal Reserve, southern Sydney Basin, New South Wales*. N.S.W. Geological Survey, Bulletin 22, 146.

Carr, P.F. (1983) A reappraisal of stratigraphy of the upper Shoalhaven Group and the lower Illawarra Coal Measures, southern Sydney Basin, New South Wales, *106th Proceedings of the Linean Society of NSW*, 257-297.

Clark, B. (1992) Depositional environment of the Bulli Coal and its effect on quality characteristics. MSc thesis, University of Wollongong, unpublished.

Colombo U., Gazzarrinin, F, Gofiantini, R., Kneuper, G, Teichmuller, M and Teichmuller, R. (1970) Carbon isotope study on methane from German coal deposits. In: Hobson, G.D. and Speers, G.C. (eds), *Advances In Organic Geochemistry*. Oxford, Pergamon Press, 1-26.

Crawford, E.A., Herbert, C., Taylor, G., Helby, R., Morgan R. and Ferguson J. (1980) Diatremes of the Sydney Basin, in Herbert, C. and Helby, R.J. (eds). *A Guide to the Sydney Basin. New South Wales Geological Survey Bulletin* 26, Sydney. 294-323.

Faiz, M.M. (1993) *Geological controls on the distribution of coal seam gases in the southern Sydney Basin.* PhD thesis, University of Wollongong, 326, unpublished.

Faiz, M.M and Hutton, A.C. (1997) Coal seam gas in the southern Sydney Basin. *APPEA Journal* 37(1). 415-428.

Faiz, M.M. and Hutton. A.C. (1995) Geological controls on the distribution of CH_4 and CO_2 in coal seam gas of the Southern Coalfield. NSW. Australia. in Lama R.D. (ed.). *Proceedings of the International Symposium-cum-workshop on management and control of high gas emissions and outbursts in underground coal mines.* Wollongong, NSW, Australia, 375-383.

Faiz, M.M. and Hutton. A.C., (1993) Two kilometres of post-Permian sediment. Did it exist? *Proceeding of the 27th Newcastle Symposium. 'Advances in the study of the Sydney Basin'.* The University of Newcastle, 221-228.

Faiz, M.M., Aziz, N.I., Hutton, A.C. and Jones. B. (1992) Porosity and gas sorption capacity of some eastern Australian coals. *Symposium on coalbed methane research and development in Australia 4.* James Cook University of North Queensland, Townsville, Queensland, Australia, 9-20.

Faiz, M.M., and Cook. A.C., (1991) Influence of coal type, rank and depth on the gas sorption capacity of coal in the Southern Coalfield. in Bamberry, W.J. and Depers, A.M. (eds), *Gas in Australian Coals.* Geological Society of Australia Symposium Proceedings 2, 19-29.

Feux, A. N. (1980) Experimental evidence against an appreciable isotopic fractionation of methane during migration, in Douglas. A.G. and Maxwell. J.R. (eds), *Advances in Organic Geochemistry,* 1979: Oxford. Pergamon Press. 725-733.

Gould, K.W., Hart, G.N. and Smith, J.W. (1981) Technical note: Carbon dioxide in the Southern Coalfields N.S.W. - A factor in the evaluation of natural gas potential? *Bulletin of the Australian Institute of Mining and Metallurgy,* 279, 41-42.

Gould, K.W., Hargraves. A.J. And Smith, J.W. (1987) Variation in the composition of seam gases issuing from coal: *Bulletin of the Australia Institute of Mining and Metallurgy* 292, 69-73.

Hargraves, A. J. (1963) *Instantaneous outbursts of coal and gas.* PhD thesis. The University of Sydney, 323, unpublished.

Hargraves, A.J. And Lunarzewski, L.S. (1985) Seam gas drainage in Australia - review. *Bulletin of the Australian Institute of Mining and Metallurgy* 290(1), 55-70.

Herbert, C. (1980) 2. Depositional development of the Sydney Basin, in Herbert, C. and Helby, R.J. (eds), *A Guide to the Sydney Basin.* New South Wales Geological Survey Bulletin 26, Sydney, 10-52.

Jakeman, B.L. (1980) *Quantitative analysis of sedimentation and structure relating to the Permo-Triassic succession of the southern Sydney Basin.* PhD thesis, 499, unpublished.

Kenuper, G.K. and Hukel, B.A. (1972) Contribution to the geochemistry of mine gases in the Carboniferous coalfield of the Saar Region, Germany. *Advances in Organic Geochemistry,* 93-112.

Lohe, E.M. and Mclennan, T.P.T. (1991) An overview of the structural fabrics of the Sydney Basin, and comparison with the Bowen Basin. Proceedings of the *Twenty Fifth Symposium in the Study of the Sydney Basin,* Department of Geology, University of Newcastle, New South Wales, 12-21,

Rice, D.D. (1993) Composition and origins of coalbed gas. In; Law, B.E and Rice, D.D. (eds), *Hydrocarbons from Coal. AAPG Studies in Geology* **38,** 159-184.

Rice D.D. and Claypool, G.E. (1981) Generation, accumulation and resource potential of biogenic gas. *AAPG* **65(1),** 5-25.

Shepherd, J. (1990) Coal seam faults in the Southern Coalfield of New South Wales, in Hutton, A.C. and Depers, A.M. (eds), *Proceedings of the Workshop on the Southern and Western Coalfields of the Sydney Basin.* University of Wollongong, Wollongong, 93-102.

Scott, A.D., Kaiser, W.R. and Ayers, W.B.(1994) Thermogenic and secondary biogenic gases, San Juan Basin. Colorado and New Mexico- Implications for Coalbed Gas Producibility. *AAPG* **78(8),** 1186-1209.

Smith, J.W., Pallasser, R and Rigby, D. (1992) Mechanism of coalbed methane formation. *Symposium on coalbed methane research and development in Australia,* James Cook University of North Queensland, Townsville, Queensland, Australia, 63-73.

Smith, J.W. and Gould, K.W. (1980) An isotopic study of the role of carbon dioxide in outbursts in coal mines. *Geochemical Journal* **14,** 27-32.

Smith, J.W., Botz, R.W., Gould, K.W., Hart, G.H., Hunt, J.W. and Rigby, D. (1984) *Outburst and gas drainage investigations,* NERDDP Final Report, 321, unpublished.

Smith, J.W. and Pallasser, J. (1996) Microbial origin of Australian coalbed methane. *AAPG* **80(6),** 891-897.

SMITH, J.W., Gould, K.W., Hart, G.H., and Rigby, D. (1985) Isotopic studies of Australian natural and coal seam gas: *Bulletin of Australasian Institute of Mining and Metallurgy* **290,** 43-51.

Standing Committee On Coalfield Geology Of New South Wales (1971) *Report of combined subcommittees for Southern and South-western Coalfields.* New South Wales Geological Survey. Records 13. 443-447.

Waller, S.F. (1994) A comparative study between coalbed methane basins, with a focus on PEL 2, Sydney Basin, Australia. *Proceedings of the 28th Newcastle Symposium, 'Advances in the study of the Sydney Basin',* The University of Newcastle, 51-59.

Wilson. R.G. (1969) Illawarra Coal Measures. Southern Coalfield, in Packham. G.H. (ed.). *Geology of New South Wales,* Geological Society of Australia, Journal **16(1),** 370-379.

Wilson, R.G. (1975) Southern Coalfield, in Travis, D.M. and King, K. (eds), *Australasian Institute of Mining and Metallurgy, Monograph 6, Coal,* 206-218.

SOURCE AND TIMING OF COAL SEAM GAS GENERATION IN BOWEN BASIN COALS

S.D. GOLDING, K.A. BAUBLYS, M. GLIKSON, I.T. UYSAL
Department of Earth Sciences, The University of Queensland, QLD 4072
C.J. BOREHAM
Petroleum and Marine Division, AGSO, GPO Box 378, Canberra, ACT 2601

Abstract

Coal seam gases collected from Bowen Basin cores have moderately negative methane carbon isotope compositions (-51 ± 9 per mil) which overlap the published range for Australian coal seam methane of -60 ± 11 per mil. No systematic relationship between coal rank and methane $\delta^{13}C$ value is apparent. A thermogenic origin for methane has been assigned when its carbon isotope composition is heavier than -60 per mil, although biogenic methane generated in closed systems may have similar $\delta^{13}C$ values from -60 to -40 per mil depending on the methanogenic pathway and the carbon isotope composition of the source. Subordinate inputs from biogenic methane could account for some of the isotopic variability of the desorbed methane; however, a good correlation between desorbed methane volumes and bitumen/pyrobitumen content suggests that much of the methane sorbed in the coal was produced by secondary cracking of bitumen.

Bowen Basin methane $\delta^{13}C$ values are typically some 20 to 30 per mil lighter than vitrinite and inertinite $\delta^{13}C$ values. Carbon isotope compositions of vitrinites and inertinites in sub-bituminous coals become less negative with increasing rank as a result of the preferential loss of the lighter isotope of carbon during maturation. Vitrinite reflectance and maceral carbon isotope compositions often display an anomolous trend for coals in the high to medium volatile bituminous rank as a result of the presence of adsorbed isotopically light methane. Thus, a sharp distinction in isotope systematics distinguishes coals that are within peak oil generation from those that are at or below the threshold of oil generation. Bitumen may show equal or higher concentrations within inertinite cell cavities as in vitrinite cleats and affect carbon isotope compositions of vitrinites and inertinites.

Compositional data for desorbed coal seam gases from the Bowen Basin show that ethane and the other wet gases are a minor component. On the other hand, high concentrations of wet gases are produced during pyrolysis of coals. This discrepancy between the proportion of wet-gas components produced during pyrolysis and that observed in many naturally matured coals may be the result of preferential migration of wet gas components. Alternatively, the wet gas components may be thermally cracked to methane at higher maturation levels or diluted by additional methane produced by secondary cracking of bitumen.

Previous vitrinite reflectance and clay mineral diagenesis studies indicate that thermal maturation of the Late Permian coals in the central and northern Bowen Basin occurred largely as a result of a short-lived hydrothermal event in the Late Triassic rather than during maximum burial in the Middle Triassic as previously thought. Textural relationships at a variety of scales and the observation that coals in proximity to Cretaceous intrusions are highly mineralised with carbonates and sulfides suggest several periods of hydrothermal activity. It is concluded, therefore, that thermal maturation of coal in the Bowen Basin to form oil and gas was caused predominantly by transient thermal and fluid flow events in the Mesozoic. High temperatures associated with transient hydrothermal events and the potential for secondary cracking of bitumen may provide an explanation for anomolous gas compositions and isotope systematics.

1. Introduction

Methane is generated from organic matter including coals mainly as a result of biological decomposition or thermal maturation (cf. Schoell, 1980). During early diagenesis, the most important mechanisms of methane generation are microbially-mediated acetate fermentation and the reduction of carbon dioxide (Rice and Claypool, 1981; Whiticar et al., 1986; Jenden and Kaplan, 1986). With increasing temperature and pressure, thermal degradation and cracking reactions produce carbon dioxide and methane (e.g., Schoell, 1980; Rice and Claypool, 1981). Methane is the main constituent of thermogenic gas from overmature source rocks (Hunt, 1996). Original thermogenic gases in coals may be modified at shallow levels by secondary processes such as oxidation of the wet gas component or mixing with biogenic gases (e.g., Rice, 1993; Scott et al., 1994). In addition, magmatic carbon dioxide or other natural gases may be adsorbed by coal seams resulting in complex gas mixtures (e.g., Smith et al., 1982, 1985).

Stable carbon isotope compositions have been routinely used to establish the origin and follow the evolution of natural gases and petroleum and coal seam methane. Artificial maturation studies on a variety of organic components show that there can be up to a -30 per mil difference between the parent organic matter and the lightest methane; however, the carbon isotope composition of methane varies significantly depending on the mechanism of formation and the heterogeneity of the methane-generating moieties (Sackett, 1978; Chung and Sackett, 1980; Rooney et al., 1995). A microbial origin for methane is generally associated with carbon isotope values less than -60 per mil, attributed to the large kinetic isotope effect during methanogenesis (Hunt, 1996). A thermogenic origin for methane has been assigned when its carbon isotope composition is heavier than -60 per mil, although biogenic methane generated in closed systems may have similar $\delta^{13}C$ values from -60 to -40 per mil depending on the methanogenic pathway and the carbon isotope composition of the source.

The source of coal seam gas in Australian Permo-Triassic coal basins is a subject of controversy. Most studies support thermal cracking of coal macerals through burial effect as the trigger and source of coal seam gases (e.g., Mallett et al., 1990). Recently, bacterial reduction of carbon dioxide by methanogens was put forward as a major process responsible for coal seam gas generation in Sydney and Bowen Basin coals (Smith *et al.*, 1992; Smith and Palasser, 1996). These authors have worked extensively on isotopic characterisation of Australian coals and coal seam gases and were the first to recognize that there were two major sources of carbon dioxide in these gases, one of seam gas origin and the other of magmatic origin (e.g., Smith *et al.*, 1982, 1985). Smith and coworkers have generated a large isotopic data base for coal seam gas from the Sydney Basin and to a lesser extent the Bowen Basin; however, sample selection was driven largely by interest in outbursting in underground coal mines. Their conclusion that much of the methane in shallow coal seams was of microbial origin was based on anomolously large carbon isotope fractionations between methane and ethane as well as light carbon isotope compositions less than -60 per mil for some methane samples. Because several sources and processes may be involved in gas generation, the current study used a range of techniques to establish the thermal history of the coals as well as maceral composition, gas chemistry and isotope composition. Aspects of this work have been presented elsewhere and in this volume (Glikson *et al.*, 1995, 1998; Golding *et al.*, 1998; Boreham *et al.*, 1998).

2. Methods of Study

2.1 SAMPLING

The samples for this study were collected from the Rangal, Fort Cooper and Moranbah Coal Measures in the central and northern Bowen Basin and from the Baralaba Coal Measures in the southernmost part of the basin (Theodore area) (Fig. 1). Desorption of gases was carried out on core samples which had been immediately stored after drilling in stainless steel cylinders. Cores were desorbed over a time of 600 hours.

2.2 ORGANIC PETROLOGY

Organic petrology and specialised characterisation of coals was undertaken using facilities in the Department of Earth Sciences and The Centre for Microscopy and Microanalysis at The University of Queensland. Organic petrology was carried out on polished coal blocks in reflected white light, and fluorescence mode, using an MPV-2 photomicroscope at wave length of 546nm. The same coal blocks were used for maceral analysis and reflectance determinations.

2.3 ELEMENTAL ANALYSIS

Elemental analysis was carried out routinely on separate coal lithotypes using a Carlo Erba Analyser Model 1106 in the Department of Chemistry, The University of Queensland.

2.4 BITUMEN EXTRACTION

Bitumen extraction from coal cores for stable isotope analysis was carried out in the Department of Earth Sciences, The University of Queensland using chloroform in a

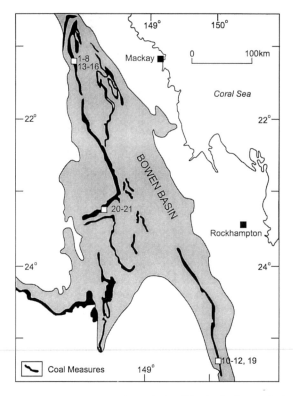

Figure 1. The distribution of coal measures and sample localities in the Bowen Basin (open squares with numbers correspond to Q series samples in Table 1).

Soxtech System HT2. Bitumen content was quantified by gravimetric methods for extractable fraction, and by petrological methods using fluorescence mode for all bitumen before and after extraction.

2.5 STABLE ISOTOPE ANALYSIS

Stable isotope analyses were undertaken in the Stable Isotope Geochemistry Laboratory, The University of Queensland and in the Australian Geological Survey Organisation Isotope and Organic Geochemistry Unit. Separate vitrinite and inertinite fractions as well as separated bitumens were combusted with cuprous oxide in sealed vicor tubes at 1040°C to produce carbon dioxide for carbon isotope analysis (Grady et $al.$, 1984). Carbon isotopic ratios were measured on a Micromass 602E dual inlet isotope ratio mass spectrometer. Methane gas was quantitatively converted to CO_2 after passing through a Finnigan combustion interface at 1000°C. The CO_2 was introduced via an open split into a Finnigan 252 continuous flow isotope ratio mass spectrometer and isotopically analysed using the m/z 44 response and m/z 45/44 ratio trace. Carbon isotope analyses are reported in per mil relative to PDB with analytical uncertainties of better than ±0.2 (2 SD) per mil.

3. Results and Discussion

3.1 NATIVE GAS DESORPTION WITH COAL RANK AND CHARACTER

Gas distribution patterns in the Bowen and Sydney Basin are known to be complex; however, gas contents generally correlate with coal rank at the regional scale (e.g., Bocking and Weber, 1993). Our desorption studies confirm the importance of rank but also suggest that maceral composition is an important indicator of probable gas content. Although there is a general increase in gas content with rank, the highest and lowest gas contents are for the highest rank coals (Fig. 2). This anomaly suggests that other factors such as coal composition are more important than previously believed. Petrological studies of Bowen Basin coals indicate two major groups of maceral populations dominated respectively by vitrinite and inertinite (Glikson et al., 1995). In addition, heavy oils generated from Bowen Basin coals during maturation are retained in microcleats and inertinite cell cavities as the secondary maceral bitumen. There is a good correlation between native gas desorption and bitumen content although the rank parameter is also highlighted (Fig. 3).

3.2 CARBON ISOTOPES

The carbon isotope compositions of coal seam gases desorbed from Bowen Basin coal cores as well as vitrinite, inertinite and bitumen separated from these coals are given in Table 1. The $\delta^{13}C$ values of methane typically vary systematically through the desorption, with the earliest desorbed gases several per mil lighter than those subsequently desorbed from the same core (e.g., 9407Q005V02 C2/CH$_4$ and C3/CH$_4$; -45 and -42 respectively). The $\delta^{13}C$ values of the coal seam methane range from -67 to -36 per mil, with a mean of -51±9 per mil which overlaps the published range for Australian coal seam methane of -60±11 per mil (Smith et al., 1992). The methane $\delta^{13}C$ values are typically some 20 to 30 per mil lighter than the vitrinite and inertinite $\delta^{13}C$ values; however, no systematic relationship between coal rank and methane $\delta^{13}C$ value is apparent (Table 1).

Increasing rank (and decreasing H/C) is followed generally by isotopically 'heavier' (i.e., less negative) vitrinites in sub-bituminous coals before oil generation (Fig. 4). The simplest explanation for the trend of heavier carbon isotope compositions of the major maceral groups with maturity is that the lighter isotope of carbon was preferentially lost during maturation as a result of Rayleigh fractionation (cf. Clayton, 1991). Studies of kerogen in both contact and regionally metamorphosed sediments show similar changes in carbon isotope composition with maturity as measured by vitrinite reflectance (Simoneit et al., 1981; Clayton and Bostick, 1985).

In samples where bitumen concentrations are relatively low, inertinites have less negative carbon isotope compositions than those of the corresponding vitrinites (Fig. 5). Differentiation between inertinite and vitrinite carbon isotope compositions is often masked, however, by higher bitumen concentrations in the former. Carbon isotopic compositions of extracted bitumens are similar to or lighter than those of the associated vitrinites (Table 1), explaining why a high bitumen concentration in inertinite reduces the

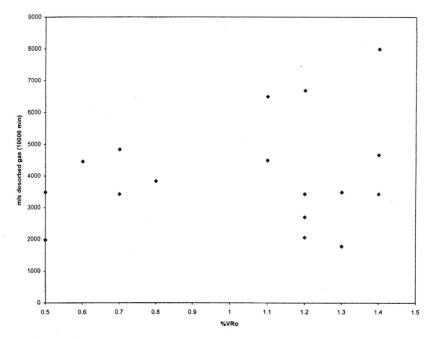

Figure 2. Native gas desorption versus % VRo for Bowen Basin coal cores.

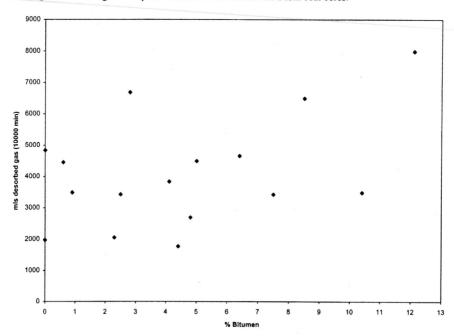

Figure 3. Native gas desorption versus % bitumen for Bowen Basin coal cores.

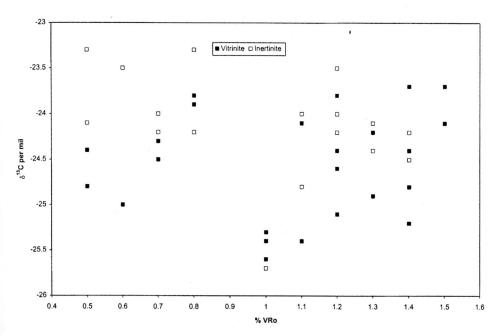

Figure 4. Carbon isotope compositions of macerals versus % VRo for Bowen Basin coal cores.

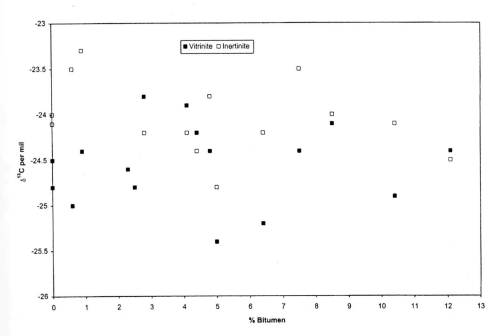

Figure 5. Carbon isotope compositions of macerals versus % bitumen for Bowen Basin coal cores.

Table 1. Carbon isotope compositions of desorbed coal seam gases and coal constituents separated from Bowen Basin core and mine samples. C2/CH$_4$ is the second gas sample collected during a desorption.

Sample Number	Formation	Description	δ^{13}C ‰ PDB	VRo %
9406Q001V08	Rangal	Vitrinite Inertinite	-23.9 -24.2	0.8
9406Q002V09	Fort Cooper	Vitrinite Inertinite	-24.3 -24.2	0.7
9406Q003V10	Fort Cooper	Vitrinite Inertinite	-24.1 -24.0	1.1
9407Q004V01	Moranbah	Vitrinite Inertinite	-25.4 -24.8	1.1
9407Q005V02	Moranbah	Vitrinite Dull/Shale C2/CH$_4$ C3/CH$_4$	-24.4 -24.5 -45 -41	1.4
9407Q006V03	Moranbah	C2/CH$_4$ C3/CH$_4$	-45 -42	
9407Q007V04	Moranbah	Vitrinite Inertinite	-25.0, -24.8 -23.9, -24.3	1.3
9407Q008V05	Moranbah	Vitrinite Inertinite C2/CH$_4$	-25.2 -24.2 -67	1.4
9408Q010V09	Rangal	Vitrinite Inertinite C2/CH$_4$	-25.0 -23.5 -57	0.6
9409Q011V08	Rangal	Vitrinite Inertinite	-24.5 -24.0	0.7
9409Q012V10	Rangal	Vitrinite Inertinite C2/CH$_4$ C3/CH$_4$	-24.4 -23.3 -57 -55	0.5
9410Q013V01	Moranbah	Vitrinite C2/CH$_4$	-24.8 -56, -55	1.4
9410Q014V02	Moranbah	Vitrinite Inertinite	-24.2 -24.4	1.3
9410Q015V03	Moranbah	Vitrinite Inertinite	-24.4 -23.5	1.2
9410Q016V04	Moranbah	Vitrinite	-24.6	1.2
9410Q019V08	Rangal	Vitrinite Inertinite C2/CH$_4$	-24.8 -24.1 -36	0.5

Table 1. Continued.

Sample Number	Formation	Description	δ¹³C ‰ PDB	VRo %
9510Q020V07	German Creek	Vitrinite Inertinite	-24.4 -23.8	1.2
9510Q021V09	German Creek	Vitrinite Inertinite	-23.8 -24.2	1.2
Gordonstone	German Creek	Vitrinite Inertinite Bitumen	-23.8 -23.3 -24.6	0.8
Oakey	German Creek	Bitumen Residue	-25.3 -24.6	1.0
G Ck 117 72.3	German Creek	Vitrinite Bitumen	-23.7 -24.3	1.4
G Ck 247-1	German Creek	Bitumen Residue	-25.6 -24.8	1.4
G Ck 251 114.9	German Creek	Vitrinite Inertinite	-25.1 -24.0	1.2
G Ck 200/120	German Creek	Vitrinite Inertinite	-24.1 -23.7	1.5
G Ck 305/220	German Creek	Vitrinite Inertinite	-23.7 -23.7	1.5
Cook	Rangal	Vitrinite Inertinite	-25.6, -25.4 -25.7, -25.6	1.0

The $\delta^{13}C$ values above are reported as δ¹³C ‰ PDB.

difference between inertinite and vitrinite carbon isotope compositions. Our petrological studies and those of Masterlerz *et al.* (1993) indicate that inertinite in higher rank coals is largely the product of intense thermal degradation of the associated vitrinite. Rayleigh fractionation of carbon isotopes as a consequence of thermal degradation would enrich the inertinite product in ^{12}C relative to the vitrinite precursor.

Vitrinite reflectance and carbon isotope compositions often display an anomolous trend for coals in the high to medium volatile bituminous rank, with lighter $\delta^{13}C$ values for both vitrinite and inertinite from 1.0 to 1.4 % reflectance (Fig. 4). This anomaly can be explained only as a result of the presence of adsorbed isotopically light methane. In sub-bituminous coals, a more 'normal' positive correlation is obtained between carbon isotope composition and vitrinite reflectance. Thus, a sharp distinction in isotope systematics versus vitrinite reflectance marks coals that are within peak oil generation from those coals that are at the threshold or below the oil window.

The majority of eastern Australian coal seam gases analysed by Smith *et al.* (1982, 1985) and Smith and Palasser (1995, 1996), and those we have collected from the Bowen Basin have moderately negative methane carbon isotope compositions consistent with a thermogenic origin. Pyrolysis studies indicate that thermogenic processes can generate isotopically light methane with $\delta^{13}C$ values about -50 per mil from coals only at low levels of overturn of the solid organic components (Boreham et al., 1998). Carbon isotope depletions in ^{13}C are further enhanced, however, by trapping of gases over selected rank levels. Subordinate inputs from biogenic methane could also account for some of the isotopic variability of the desorbed methane (i.e., -51 $\pm$ 9 per mil, 1σ); however, the good correlation between desorbed methane volumes and bitumen/pyrobitumen content (Fig. 3) suggests that much of the methane sorbed in the coal was produced by secondary cracking of bitumen.

Apparent equilibrium carbon isotope fractionations between methane and carbon dioxide when the latter comprises 10% or less of the seam gas have been reported in the Australian coal seam gas literature (Smith et al., 1982, 1992, 1998; Smith and Palasser, 1995). These isotope systematics appear to require a gas-generating process able to produce methane and carbon dioxide in isotopic equilibrium. Evidence for ^{13}C exchange between carbon dioxide and methane comes primarily from studies of geothermal systems (Giggenbach, 1982), whereas laboratory studies are more equivocal (Chung and Sackett, 1979). Smith et al. (1998) have recently shown that the thermal decomposition of acetic acid at temperatures about 300°C does produce methane and carbon dioxide in isotopic equilibrium. This new information is consistent with evidence for coal maturation and oil and gas generation in the Bowen Basin largely as a result of transient high temperature hydrothermal events (Golding et al., 1998). Other coal seam gases with large carbon isotope fractionations between methane and carbon dioxide are possibly of mixed thermogenic and biogenic origin; however, the good correlation observed between rank and coal seam gas contents at the regional scale in the Bowen and Sydney Basin coals (cf. Bocking and Weber, 1993) suggests these isotope systematics are the result of late stage, shallow level modification of original thermogenic/hydrothermal seam gases rather than bacterial reduction of carbon dioxide and other near surface oxidation products of coal.

3.3 HYDROCARBON GAS COMPOSITIONS

Open system pyrolysis of coal demonstrates that there are two stages of thermogenic methane generation from Bowen Basin coals (Boreham et al., 1998). The first and major stage shows a steady increase in methane generation maximising at 570°C, corresponding to a VR of 2-2.5%. Heavier (C_{2+}) hydrocarbons are generated up to 570°C after which only the C_1 gases are produced. The main phase of heavy hydrocarbon generation occurs between 420 and 510°C. Over this temperature range, methane accounts for only a minor component, whereas the wet gases (C_2-C_5) are equally or more abundant than the liquid hydrocarbons. Compositional data for desorbed coal seam gases from the Bowen Basin show that ethane and the other wet gases are a very minor component. Apart from two samples which have high nitrogen contents, the methane content of the gases desorbed from the cores ranges from 87.00% to 99.90 %. The gases are essentially dry as carbon dioxide and nitrogen are the predominant other gases present. This discrepancy between the proportion of wet-gas components produced during pyrolysis and that observed in naturally matured coals may be the result of preferential migration of wet gas components at the time of generation or subsequently. Alternatively, the wet gas components may be thermally cracked to methane at higher maturation levels or diluted by additional methane produced by secondary cracking of bitumen. Bacterial alteration of the heavier hydrocarbons at shallow depths could also account for the methane-dominated gas compositions (Faiz and Hutton, 1997).

3.4 TIMING OF COAL SEAM GAS GENERATION

Vitrinite reflectance and clay mineral diagenesis studies indicate that thermal maturation of the Late Permian coals in the Bowen Basin occurred largely as a result of a short-lived hydrothermal event in the Late Triassic rather than during maximum burial in the Middle Triassic as previously thought (Uysal et al., 1997; Golding et al., 1998). This is why reflectance values are higher along the present eastern margin of the central and northern Bowen Basin rather than in the depocentre of Triassic sedimentation in the south-east of the basin. Coalification gradients are also abnormally high, generally more than 50°C/km throughout the basin (e.g., Baker, 1989; Faraj et al., 1996). K-Ar ages of illite/smectite mixed-layer clays from shallower parts of the Late Permian coal measures in the Theodore area in the south-east of the basin record a second short-lived thermal event from about 145-140 Ma which accompanied renewed subsidence in the Surat Basin (Uysal et al., 1997). In addition, thermal aureoles are developed above Cretaceous intrusions which were emplaced episodically from 135-125 and 110-100 Ma in the central and northern Bowen Basin (Golding et al., 1998).

Clay mineralisation in the Late Permian coal measures is Middle to Late Triassic in age, whereas the absolute timing of carbonate mineralisation in these coals is not established because of a lack of suitable minerals for dating. Calcite mineralisation on butt cleats clearly crosscuts clay mineralisation on face cleats, whereas various ferroan carbonates predate clay mineralisation (Golding et al., 1998). This suggests several episodes of carbonate mineralisation of the coals as does the complex thermal history of the region and the observation that coals in proximity to Cretaceous intrusions are highly mineralised with carbonates and sulfides. The intimate association between bitumen and calcite mineralisation confirms that hydrocarbon generation at least locally accompanied carbonate emplacement (Golding et al., 1998). This conclusion is supported by the carbon and oxygen isotope systematics of the calcite mineralisation. Calcites exhibit an extremely wide range of $\delta^{13}C$ values which are positively correlated with a narrower range of $\delta^{18}O$ values (Golding et al., 1998). The slope of the correlation trend is too steep to be explained by temperature variation alone; however, carbonate deposition from a fluid containing both oxidized and reduced carbon species provides an explanation for the correlation trend and the very positive $\delta^{13}C$ values of many of the calcites (Golding et al., 1998). It is concluded, therefore, that thermal maturation of coal in the Bowen Basin to form oil and gas was caused predominantly by transient thermal and fluid flow events in the Mesozoic. High temperatures associated with transient hydrothermal events and the potential for secondary cracking of methane may provide an explanation for anomolous gas compositions and isotope systematics.

4. Summary

The majority of eastern Australian coal seam gases including those we have collected from the Bowen Basin have moderately negative methane carbon isotope compositions consistent with a thermogenic origin. Subordinate inputs from biogenic methane could account for some of the isotopic variability of the desorbed methane as well as methane-dominated gas compositions; however, the good correlation between desorbed methane volumes and bitumen/pyrobitumen content for the Bowen Basin cores suggests that much of the methane sorbed in the coal was produced by secondary cracking of bitumen. Bitumen, heavy oil generated from Bowen Basin coals during thermal/hydrothermal maturation may show equal or higher concentrations within inertinite cell cavities as in vitrinite cleats and effect carbon isotope compositions of vitrinites and inertinites.

5. Acknowledgments

This study was made possible thanks to grants by QTSC (Queensland Transmission and Supply Corporation) and ERDC (Energy and Research Development Corporation). Shell and CapCoal, M.G.C. Resources, Newlands, Oaky Creek, Gordonstone and Yarrabee Coal are acknowledged for considerable in kind contribution through provision of coal core, access to mines and data sets and on site accomodation.

6. References

Baker J.C. 1989. Petrology, diagenesis and reservoir quality of the Aldebaran Sandstone, Denison Trough, east-central Queensland (unpublished PhD thesis), The University of Queensland, 255pp.

Bocking M.A. & Weber C.R., 1993. Coalbed methane in the Sydney basin, Australia. *International Coalbed Methane Symposium,* Birmingham, Alabama, Volume I:15-23.

Boreham C.J., Golding S.D. & Glikson M., 1998. Factors controlling gas generation and retention in Bowen Basin coals. *Advances in Organic Geochemistry,* in press.

Chung H.M. & Sackett W.M., 1979. Use of stable carbon isotope compositions of pyrolytically derived methane as maturity indicies for carbonaceous materials. *Geochimica et Cosmochimica Acta,* 43:1979-1988.

Chung H.M. & Sackett W.M., 1980. Carbon isotope effects during the pyrolytic formation of early methane from carbonaceous materials. *in:* Douglas A.G. and Maxwell J.R. (eds) *Advances in Organic Geochemistry,* 1979, Physics and Chemistry of the Earth, 12:705-710.

Clayton J.L. & Bostick N.H., 1985. Temperature effects on kerogen and on molecular and isotopic composition of organic matter in Pierre Shale near an igneous dike. *Advances in Organic Geochemistry,* 10:131-143.

Clayton C., 1991. Carbon isotope fractionation during natural gas generation from kerogen. *Marine and Petroleum Geology,* 8:232-240.

Faiz M.M. & Hutton A.C., 1997. Coal seam gas in the southern Sydney Basin, New South Wales. *Australian Petroleum Production and Exploration Association Journal,* 415-428p.

Faraj B.S., Fielding C.R. & Mackinnon I.D.R., 1996. Cleat mineralisation of Upper Permian Baralaba/Rangal Coal Measures, Bowen Basin, Australia. *in:* Gayer R. and Harris I. (eds) *Coalbed Methane and Coal Geology,* Geological Society Special Publication, 109:151-164.

Giggenbach W.F., 1982. Carbon-13 exchange between CO_2 and CH_4 under geothermal conditions. *Geochimica et Cosmochimica Acta,* 46:159-165.

Glikson M., Golding S.D., Lawrie G., Szabo L.S., Fong C., Baublys K.A., Saxby J.D. & Chatfield P., 1995. Hydrocarbon generation in Permian coals of Queensland, Australia: source of coalseam gas. *in:* Follington I.L., Beeston J.W. and Hamilton L.H. (eds) *Proceedings Bowen Basin Symposium,* Geological Society Australia Inc. Coal Geology Group, Brisbane, 205-216p.

Glikson M., Golding S.D., Saxby J.D., Boreham C.J. & Szabo L.S., 1998. Artificial maturation of vitrinites; oil and gas generation by thermal techniques and residue characterisation by transmission electron microscopy (TEM), carbon isotope geochemistry and elemental composition. *Organic Geochemistry,* in press.

Golding, S.D., Collerson, K.D., Uysal, U.T., Glikson, M., Baublys, K. & Zhao, J.X., 1998. Nature and source of carbonate mineralization in Bowen Basin coals: implications for the origin of coal seam methane and other hydrocarbons sourced from coal. *in:* Glikson, M. and Mastalerz, M. (eds), *Organic Matter and Mineralisation: Thermal Alteration, Hydrocarbon Generation and Role in Metallogenesis.* Chapman and Hall, London, in press.

Grady M.M., Swart, P.K. & Pillinger C.T., 1982. The variable carbon isotopic composition of Type 3 ordinary chondrites. *Journal Geophysical Research,* Volume 87 Supplement:A289-A296.

Hunt J.M., 1996. *Petroleum Geochemistry and Geology,* 2nd edition. W.H. Freeman, New York, 743pp.

Jenden P.D. & Kaplan I.R., 1986. Comparison of microbial gases from the Middle America Trench and Scripps Submarine Canyon: implications for the origin of natural gas. *Applied Geochemistry,* 1:631-646

Mallett C.W., Russell N. & McLennan T., 1990. Bowen Basin geological history. *in:* Beeston J.W. (ed.) *Bowen Basin Symposium,* Geological Society Australia, Queensland Division, 5-20p.

Mastalerz M., Wilkes K.R., Bustin R.M. & Ross J.V., 1993. The effect of temperature, pressure and strain on carbonization in high-volatile bituminous and anthracitic coals. *Organic Geochemistry,* 20(2):315-325.

Rice D.E. & Claypool G.E., 1981. Generation, accumulation and resource potential of biogenic gas. *American Association Petroleum Geologists Bulletin,* 65:5-25.

Rice D.D., 1993. Composition and origins of coalbed methane. *in:* Law B.E. and Rice D.D. (eds) *Hydrocarbons from Coal,* American Association Petroleum Geologists Studies in Geology 38, 159-184.

Rooney M.A., Claypool G.E. & Chung H.M., 1995. Modeling thermogenic gas generation using carbon isotope ratios of natural gas hydrocarbons. *Chemical Geology,* 126:219-232.

Sackett W.M., 1978. Carbon and hydrogen isotope effects during the thermocatalytic production of hydrocarbons in laboratory simulation experiments. *Geochimica et Cosmochimica Acta,* 42:571-580.

Scott A.R., Kaiser W.R. & Ayers W.B., 1994. Thermogenic and secondary biogenic gases, San Juan Basin, Colorado and New Mexico: Implications for coalbed gas produceability, *American Association Petroleum Geologists Bulletin,* 78:1186-1209.

Schoell M., 1980. The hydrogen and carbon isotope composition of methane from natural gases of various origins. *Geochemica et Cosmochimica Acta,* 44:649-661.

Simoneit B.R.T., Brenner S., Peters K.E. & Kaplan I.R., 1981. Thermal alteration of Cretaceous black shale by diabase intrusions in the Eastern Atlantic: II effects on bitumen and kerogen. *Geochemica et Cosmochimica Acta,* 45:1581-1602.

Smith J., Gould K.W. & Rigby D., 1982. The isotope geochemistry of Australian coals. *Organic Geochemistry,* 3:111-131.

Smith J., Gould K.W., Hart, G. & Rigby D., 1985. Isotopic studies of Australian natural and coal seam gases. *Proceedings of the Australasian Institute of Mining and Metallurgy,* 290:43-51.

Smith J.W., Palasser R.J. & Rigby D., 1992. Mechanisms for coalbed methane formation. *Coalbed Methane Symposium,* Townsville, 63-73p.

Smith J.W. & Palasser R.J., 1995. CO$_2$- A key factor in gas characterisation. *in:* Pajeres J.A. and Tascon J.M.D. (eds) *Coal Science,* Amsterdam, Elsevier Science, Volume 1:91-94.

Smith J.W. & Palasser R.J., 1996. Microbiological origin of Australian coalbed methane. *American Association Petroleum Geologists Bulletin,* 80:891-897.

Smith J.W., Palasser R.J. & Pang, L.S.K., 1998. Thermal reactions of acetic acid $^{13}C/^{12}C$ partitioning between CO$_2$ and CH$_4$. *Organic Geochemistry,* in press.

Uysal T., Golding S.D. & Glikson M., 1997. Carbonate and clay mineral diagenesis in Late Permian coal measures of the Bowen Basin, Queensland, Australia: implications for thermal and tectonic histories. *Abstracts EUG 9,* European Union of Geosciences, Strassbourg, France, Terra Nova, 9: 572.

Whiticar M.J., Faber E. & Schoell M., 1986: Biogenic methane formation in marine and freshwater environments: CO2 reduction vs acetate fermentation-isotope evidence. *Geochemica et Cosmochimica Acta,* 50:693-709.

THE DEVELOPMENT OF AN UNDERSTANDING OF THE ORIGINS OF THE SYDNEY AND BOWEN BASIN GASES

J. W. SMITH

CSIRO Divisions of Petroleum Resources and Energy Technology
P.O. Box 136, North Ryde, NSW 1670, Australia

1. Abstract

It is remarkable that the origins of the coal-associated gases in the Sydney and Bowen Basins should have been so generally and consistently misunderstood. These gases occur at depths to some 600 m, vary widely in composition and most commonly occur as mixtures of very dry methane with carbon dioxide in all proportions. The origins of these gases remain in doubt, although a thermal maturation mechanism was initially agreed. Up-lifting and erosion have resulted in the loss of the bulk of this early-formed wet thermogenic gas. It has been almost completely displaced by very dry methane and now remains as traces in seam-enclosing sandstones and in the deepest coal measures. A biogenic origin, largely via carbon dioxide reduction, is often assigned to gases of this type. However in these Basins higher temperatures associated with hydrothermal activity, and anomalies in the behaviour of carbon dioxide, appear likely to curtail biogenic activity and suggest alternative gas sources.

2. Introduction

Looking back with the advantage of hindsight, virtually all the early chemical evidence (Hargraves, 1962, 1963) and the isotopic data of Smith et al. (1982) contra-indicated a traditional thermal origin for the great bulk of the coal seam gases. For example, the 2 or 3% of ethane and higher hydrocarbons which normally characterise seam gases from medium-rank bituminous coals is absent. Furthermore, the methane from Australian coals of this rank is markedly depleted in ^{13}C, relative to methane from other European coals of similar maturity. In these respects it should be noted that not all seam gas has this unusual composition. Hargraves (1963), and later (pers.comm 1980), commented on the very wet, petroleum-type gas found in traces in seam-enclosing sandstones. Smith et al. (1985) similarly reported traces of isotopically 'heavier' and 'wetter' gases from the deepest levels of coal measures e.g. the Blake and Bowen Seams at Collinsville, the Tongara and Wongawilli Seams in the South Sydney Basin and from the much deeper Moonshine and Bootleg bores at Wilton (Hart and Smith 1981, Gould et al.,1982). Faiz and Hutton (1997) comment similarly on these and other borehole data. All these deeper gases are considered to be remnants of an original wet thermogenic gas which once filled the basin but the bulk of which was lost to the surface after uplifting and erosion of some 1-2 km (Faiz and Hutton, 1997). Comparison of the isotopic compositions of the ethane in the wet remnant gas and of the dry methane in the gas currently occupying the overlying basin result in anomalously large $\Delta^{13}C(C_2H_6-CH_4)$ values. Such fractionations suggest the gas origins to be unrelated (Smith et al., 1985) and in this context imply the latter to be a more recent invader from an external source, as indicated in *Fig.1.*

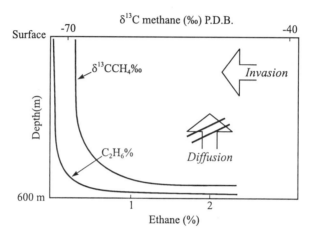

Figure 1. Variation of seam gas properties with depth of burial.

Such isotopic fractionations, and the huge volumes of methane relative to ethane now filling the basin, virtually demand invasion from a major secondary source of methane rather than an on-going slow diffusion of gases from possibly more deeply-buried sediments. With respect to that latter source, where coal seams been invaded by deep-seated, magmatic carbon dioxide and in mines where methane has been almost entirely displaced by such carbon dioxide e.g. Metropolitan Mine, Southern Coalfields N.S.W., higher ethane contents or higher ethane/methane ratios in mine gases have not been reported. If the invading carbon dioxide were scavenging hydrocarbon gases from petroliferous sediments, some increases in these values might have been expected.

3. Results and Discussion

The chemical recognition of two major sources of carbon dioxide, one a coal-seam product and the other magmatically sourced, was well established before carbon isotope measurement allowed a more precise identification of each to be made. As shown in *Fig.2*, when the carbon dioxide content of the seam gas was 10% or less, δCO_2 values commonly ranged from -8 to +20 per.mil. At all higher carbon dioxide contents, δCO_2 values averaged -7 ± 2 per.mil. However very recently it has become evident that it is not only in ^{13}C content but also in reactivity that the two forms appear to differ. In *Fig.3a and 3b*, a most general presentation, the isotopic compositions of methane and carbon dioxide in some 300 seam gas samples from the two basins have been subdivided and similarly compared on the basis of their air-free carbon dioxide contents. On that basis, as shown in *Fig.4*, the gases are seen to congregrate strongly at either end of the compositional spectrum.

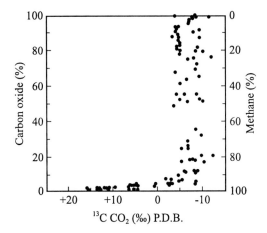

Figure 2. Relationship between concentration and $\delta^{13}C$ value of CO_2, Sydney Basin gases (after Gould et al., 1981).

Gas sample selection was certainly influenced by interest in outbursting situations, however this distribution may largely reflect normality with the carbon dioxide content of gases being 5% or less, other than where invasions of magmatic gas occur. With that understanding, *Fig.3a* shows that when carbon dioxide comprises 10% or less of the seam gas (see *Fig.2*), it is always in a state approaching isotopic equilibrium with methane. This is demonstrated by the line AB in *Fig.3a*. This, the mean calculated isotopic fractionation between the two gases, $\Delta^{13}C(CO_2\text{-}CH_4)$, is 54 ± 2 per.mil, irrespective of the ^{13}C contents of the gases. The correlation coefft. of 0.39 illustrates a considerable degree of confidence. By way of contrast, as illustrated in *Fig.3b*, in gases with more, often very much more, than 10% of largely magmatic carbon dioxide (again see *Fig.2*) the $\delta^{13}C(CO_2)$ is relatively constant at -7.5 per.mil. In this case the carbon dioxide is substantially unrelated to, and unaffected by the isotopic composition of the methane. That is, the magmatic carbon dioxide is essentially inert. Laboratory studies of the interaction of methane and carbon dioxide under a wide range of conditions at temperatures to 500° C also provide no evidence of chemical or isotopic reaction between these gases (Sackett and Chung, 1979).

On the premise that the methane is not a maturation product but is generated at shallow depths after up-lifting, mechanisms for methane formation via the biogenic reduction of carbon dioxide, have been widely proposed. In these, the carbon dioxide and possibly other low molecular weight oxidation products of coal are first generated by reaction of the coal with air episodically transported underground by shallow aquifers. Then later, when strongly reducing conditions have been re-established, methanogen activity reduces carbon dioxide to methane (Rice 1993, Smith and Pallasser 1993, Scott et al., 1994, Kotarba and Rice, 1995).

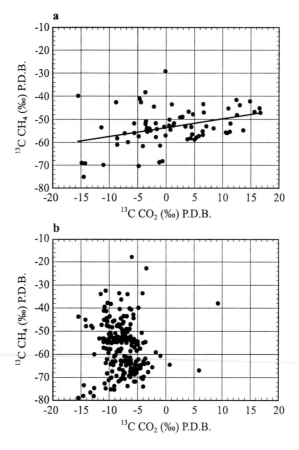

Figure 3. Relationship between δ¹³C values of CO₂ and CH₄ for Sydney/Bowen in gases
(after Smith and Pallasser 1995) a) less than 10 % CO₂ b) more than 10 % CO₂.

One difficulty with this model is the resistance of the readily available deep-seated magmatic carbon dioxide to reduction to methane. Explanations for this might include;

   a) only the oxidation products of coal contain nutrients essential for methanogen activity and carbon dioxide reduction,

   b) magmatic gases lack nutrients and/or contain toxic components,

   c) this is a question of timing, with magmatic invasion post-dating the environmentally favourable reductive stage,

   d) biogenic reduction of carbon dioxide is not the major mechanism for methane generation and thus,

   e) the generation of methane by the continuing thermal, non-biogenic/biogenic re-working of other simple coal oxidation products, rather than carbon dioxide, becomes an alternative mechanism.

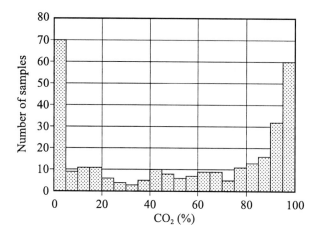

Figure 4. Histogram - distribution of carbon dioxide contents of Sydney/Bowen Basin gases.

Although it has been previously suggested that product coal seam methane and carbon dioxide might have been *formed* in isotopic equilibrium (Smith et al., 1982) no evidence for such a reaction had been described. However more recently the thermal decomposition of acetic acid at temperatures of 300° C and above has been shown (*Fig.5a and 5b*) to result in product methane and carbon dioxide in isotopic equilibrium (Kharaka et al., 1983; Smith et al., 1998). The present limiting temperature near 300° C, remains a serious impediment to natural gas generation, other than at extended reaction times. In this respect, the recent higher maturation temperatures associated with episodic rapid heating rates (Glikson et al., 1998) together with an understanding of the timing of clay and carbonate mineralisations, indicate that transient hydrothermal activity associated with magmatic intrusions have significantly increased the maturity of the Bowen basin coal (Golding et al., 1998). Presumably thermal gas yields might have correspondingly been increased.

This new information on the likely elevated temperatures in the Bowen basin requires serious consideration to be given to a chemical, in addition to, a biogenic origin for these gases. The higher temperatures accompanying hydrothermal conditions appear likely to favour chemical over biogenic reactions and to add another dimension to possible gas-forming mechanisms. Such a mechanism although still dependent on temperature, water flow, oxygen levels and coal geochemistry to generate a suitable mix of interacting simple coal oxidation products has an advantage over the biogenic model. It does not rely on a critical reversal of oxidising/reducing conditions and the proliferation of methanogens to generate methane. In the same vein, in view of the well-documented resistance of coals to direct biogenic attack (Couch, 1987), a preceeding or accompanying chemical oxidation of the coal surface appears likely to favour increased product gas yields.

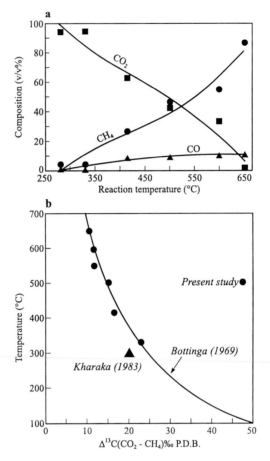

Figure 5. Thermal decomposition of acetic acid (after Smith et al in Press)
        a) Product gas yields, b) Variation of isotopic equilibria with reaction
        temperature.

4. Conclusions

Wet thermogenic gases which once filled the Bowen and Sydney basins have, following
uplifting and erosion, been almost totally displaced by dry methane. Only traces of the
original gas remain in enclosing sandstones and the most deeply buried coals.
Mechanisms for production of the dry displacing methane include;

1) in accord with the most widely accepted origin and δCH_4 values - the biogenic
   reduction of carbon dioxide,
2) in view of the anomolous reactivities of 'carbon dioxide' and higher basin
   temperatures related to hydrothermal activity - the chemical/biogenic reworking of
   simple coal oxidation products e.g. acetic acid, and
3) again with respect to hydrothermal activity and accompanying higher temperatures -
   the direct generation of highly-cracked thermogenic gas.

On the basis of the present analytical data some differences appear to exist in the nature and compositions of the Sydney and Bowen basin gases, for example, recent evidence suggests that higher temperatures and correspondingly thermally altered products and residues may occur in the latter. More analyses are required to confirm not only whether such conclusions are real or result from sampling techniques, but also the mechanisms controlling gas production.

5. Acknowledgements

Thanks are due to Drs S. C. George and J. R. Smith for most useful comments on the content and structure of this work.

6. References

Couch, G.R. (1987) Biotechnology and coal, *I.E.A.Coal Research,* 7-56.

Faiz, M.M. and Hutton, A.C. (1997) Coal seam gas in the Southern Sydney Basin, New South Wales, *Australian Petroleum Production and Exploration Association Journal,* 415-428.

Glikson, M., Boreham, C.J., and Theide, D. (1998) Coal composition, temperature and heating rates: a determining factor in hydrocarbon species generated and prediction of primary/secondary migration, *Abstracts of Papers,* International Conference on Coal Seam Gas and Oil, Brisbane, Australia, March 23-25.

Golding, S.D., Boreham, C.J., Uysal, I.T., Baublys, K.A. and Glikson, M. (1998) Source and timing of coal seam gas generation in Bowen Basin coals, *Abstracts of Papers,* International Conference on Coal Seam Gas and Oil, Brisbane, Australia, March 23-25.

Gould, K.W., Hart, G.N. and Smith, J.W. (1981) Technical Note: Carbon Dioxide in the Southern Coalfields N.S.W., *Proceedings of the Australasian Institute Of Mining and Metallurgy* **279**, 41-42.

Gould, K.W., Hart, G.N. and Smith, J.W. (1982) Gas Prospecting, *Investigation Report 1336R* CSIRO Division of Coal Research, 8p.

Hargraves, A.J. (1962) Gas in face coal, *Proceedings of the Australasian Institute of Mining and Metallurgy* **203**, 7-43.

Hargraves, A.J. (1963) Some variations in the Bulli Seam, *Proceedings of the Australasian Institute of Mining and Metallurgy* **208**, 251-283.

Hart, G. and Smith, J.W. (1981) Gas Prospecting, *Investigation Report 1256R* CSIRO Division of Coal Research, 5p.

Kharaka, Y., Carothers, W. and Rosenbauer, R.J.(1983) Thermal decarboxylation of acetic acid: implications for origin of natural gas, *Geochimica et Cosmochimica Acta* **47**, 379-402.

Kotarba, M.J. and Rice, D.D. (1995) Microbial methane and endogenic carbon dioxide in Lower Silesian Coal Basin, SW Poland; in Lama, R.D.(ed) *High Gas Emissions and Outbursts in Underground Mines.* Australian Coal Association, Wollongong, NSW, Australia.

Rice, D.D.(1993) Compositions and origins of coalbed gas; in Law, B.E. and Rice, D.D.(eds) *Hydrocarbons from Coal,* AAPG Studies in Geology **38**,159-184.

Sackett, W.M. and Chung, H.M. (1979) Experimental confirmation of the lack of carbon isotope exchange between methane and carbon oxides at high temperatures, *Geochimica et Cosmochimica Acta* **43**, 273-276.

Scott, A.R., Kaiser, W.R. and Ayers, W.B. (1994) Thermogenic and secondary biogenic gases, San Juan Basin, Colorado and New Mexico Implications for Coalbed Gas Producibility, *American Association of Petroleum Geologists Bulletin* **78**,1186-1209.

Smith, J.W., Gould, K.W. and Rigby, D. (1982) The stable isotope geochemistry of Australian coals, *Organic Geochemistry* **3**, 111-131.

Smith, J.W., Gould, K.W., Hart, G. and Rigby, D. (1985) Isotopic studies of Australian natural and coal seam gases, *Proceedings of the Australasian Institue of Mining and Metallurgy* **290**, 43-51.

Smith, J.W. and Pallasser, R.J. (1993) Microbial origin of coal seam methane, *Abstracts of Papers* 206th ACS National Meeting, Chicago, Part 1, GEOC 79.

Smith, J.W. and Pallasser, R.J. (1995) CO_2 - A key factor in gas characterization, in J.A. Pajares and J.M.D. Tascon, *Coal Science,* Amsterdam, Elsevier Science.Vol **1**, 91-94.

Smith, J.W., Pallasser., R.J. and Pang, L.S.K. (1998) Thermal reactions of acetic acid $^{13}C/^{12}C$ partitioning between CO_2 and CH_4, *Organic Geochemistry* (in press).

MINERAL-CATALYZED FORMATION OF NATURAL GAS DURING COAL MATURATION

Calvin H. Bartholomew[1], Steven J. Butala, Juan Carlos Medina,
Milton L. Lee, Terrence Q. Taylor, and Dallan B. Andrus
Departments of Chemical Engineering and Chemistry & Biochemistry
Brigham Young University, Provo, UT 84602

[1]To whom correspondence should be addressed

1. Abstract

Kinetic data from the literature were used to predict formation rates and product yields of oil and gas at typical low-temperature conditions of coal maturation. The data indicate that gas formation rates from hydrocarbon thermolysis reactions are several orders of magnitude too low to have generated known reserves of coalbed gas. By contrast, acid-mineral-catalyzed cracking, transition-metal-catalyzed hydrogenolysis of liquid hydrocarbons, and transition-metal-catalyzed CO_2 hydrogenation form gas at sufficiently high rates in geologic time and at geologic conditions to account for formation of enormous reserves of coal-bed gas. Rates of gas production in these reactions are 5 to 10 orders of magnitude higher than those predicted from thermolysis. The gaseous product compositions for metal-catalyzed hydrogenolysis of hydrocarbon liquids and for CO_2 hydrogenation are nearly the same as those of natural gases, while those from thermal and catalytic cracking are vastly different. From chemical analysis of a pair of gas-producing and non-gas-producing coals it was found that significant, comparable amounts of iron are present. The available data are consistent with a model involving thermal and catalytic cracking of kerogen to oil followed by iron-metal-catalyzed hydrogenolysis of oil to natural gas; in CO_2-containing coal gases, natural gas may also be formed by iron-catalyzed CO_2 hydrogenation.

2. Introduction

Coal seam reservoirs are important potential sources of natural gas. In fact, world-wide resources are estimated to be about 3,000-12,000 trillion cubic feet (TCF), while U.S. reserves are estimated to be 800-1,200 TCF [Davidson et al., 1995; Gayer and Harris, 1996; Tyler and Scott, 1998]. While enormous, relatively little of this resource is being produced, in part due to the absence of reliable exploration methods. Accordingly, the potential for expanded production is considerable, if more reliable geologic indicators could be discovered.

It is commonly assumed that oil and hydrocarbon gases were formed in coal seams by thermolysis (cracking) of coal organic matter [Philippi; 1965; James, 1983; Tissot & Welte, 1984; Kissin, 1987; Takach et al., 1987; Barker, 1990; Ungerer, 1990; and Hunt, 1991]. This model is supported by experiments showing high rates of gas and oil production during pyrolysis of coal kerogen at 200-250°C, e.g., those reported by Harwood et al. [1977] and Tannenbaum and Kaplan [1985]. Recently, however, the reliability of the thermogenic model for oil-gas and coal-gas generation has been questioned [Mango, 1992; Mango et al., 1994; Shock, 1997; Nelson et al., 1998] because (1) thermolysis of model organic compounds is too slow to account for the present reserves even over hundreds of millions of years [Jackson et al., 1995; Butala et al., 1997 and 1998], (2) the hydrocarbon product distribution obtained during thermolysis of model organic compounds is much different than

279

natural gas [Evans and Felbeck, 1983; Espitalie *et al.*, 1988; Horsfield *et al.*, 1991; Mango *et al.*, 1994; Butala *et al.*, 1997] and (3) results of artificial maturation experiments indicate
that raw (mineral-containing) coal generates hydrocarbon gas at substantially higher rates than demineralized coal [Lu and Kaplan, 1990; Tang *et al.*, 1996]. This suggests that mineral catalysis may play a crucial role in hydrocarbon gas formation during coal maturation.

The objective of this work was to evaluate potential roles of transition metal minerals in catalyzing coal-bed methane formation. In the first phase of this study, rate and product selectivity data for hydrocarbon thermolysis and mineral-catalyzed cracking or synthesis reactions were compiled in a comprehensive review of technical literature sources [Butala *et al.*, 1997]. In the second phase of this work, two potential methane-forming reactions, CO_2 hydrogenation and liquid olefin hydrogenolysis, both catalyzed by metallic iron were studied under simulated geologic conditions. Concentrations and chemical compositions of iron minerals in a gas-producing coal and non-producing coal were studied by XRF.

It should be emphasized that this work addresses thermogenic methane formation, *i.e.* methane formed within coal seams at $T > 100°C$ from organic matter in the coal. It does not address methane produced by indigenous bacteria, decomposition of acetic acid and/or, or other alternative means.

3. Experimental

3.1 CATALYST AND REACTANT MATERIALS

A 10% Fe/silica catalyst was prepared by multiple impregnation of silica (Grade M-5 fumed silica, Cabosil) with an aqueous solution of $Fe(NO_3)_3 \cdot 9H_2O$ to incipient wetness followed by drying at 90°C for 1 h. A 0.8 g sample of the dried catalyst was reduced *in situ* (in the batch reactor) in 35 mL min^{-1} of flowing hydrogen for 150 h at 200°C and rereduced for 41 h at 200°C in between reaction tests. H_2 and CO_2 gases (obtained from Air Liquide and Scott Specialty Gases, respectively) were further purified by passing through molecular sieve purifiers to remove moisture and hydrocarbons.   1-Dodecene (Aldrich Chemical, 99.9%) was not further purified.

3.2   BATCH REACTOR SYSTEM

Reactions were carried out in a 500 mL Pyrex batch reactor equipped with a magnetic-stirring heating mantel and tubular openings at the top for a pressure transducer, PEEK tube for sampling, thermocouple well, and stopcock for either evacuating or charging hydrogen gas to the vessel. The PEEK tube was connected to a 6-position sampling valve with an electronic actuator. Gas samples were routed by the valve to a 10 µL external loop for sampling to a gas chromatograph (HP 5890) equipped with a Carboxen 1006 PLOT column and thermal conductivity and flame ionization detectors. A peristaltic pump was used to provide a continuous circulation of gases in the reactor through the sampling loop. Reactor temperature, pressure, and gas concentration were monitored, while temperature was controlled using a Pentium 120 PC with an analog-to-digital converter and Visual Basic software. Further details of this system are provided elsewhere [Medina *et al.*, 1998].

Two reactions, 1-dodecene hydrogenolysis and CO_2 hydrogenation, were carried out for 40 to 100 h periods in the batch reactor at a total pressure of 1.2 atm using 0.8 g of pre-reduced 10% Fe/silica catalyst. The olefin hydrogenolysis reaction was initiated by either: (a) charging 30 mL of 1-dodecene to the evacuated reactor containing the pre-reduced catalyst at 25°C, following which the glass vessel was pressurized to 1.2 atm of hydrogen and then heated to 180°C or (b) adding hydrogen followed by charging 1-dodecane dropwise at 180°C so that only hydrocarbon vapor and hydrogen contacted the sample at a H_2/C molar ratio exceeding 5. Gaseous products were collected and analyzed every hour in the gas chromatograph. CO_2 hydrogenation was similarly carried out at 180 or 192°C and a total pressure of 1.2 atm with a reactant gas having a molar H_2/CO_2 ratio of 4.0.

4. Results

4.1 RESULTS OF LITERATURE SURVEY OF KINETIC DATA FOR GAS AND OIL FORMATION IN RELEVANT THERMAL AND CATALYTIC REACTIONS

Our survey of the literature uncovered relevant rate data for methane formation from (1) thermal cracking of liquid hydrocarbons, paraffinic oil and aromatic oil, (2) low-temperature pyrolysis of various coal and petroleum kerogens, (3) acid-mineral cracking of liquid hydrocarbons, (4) hydrogenolysis of 1-octadecene on a sedimentary rock containing transition metal compounds, (5) hydrogenolysis of C_3-C_6 hydrocarbons on supported iron and nickel metal catalysts, (6) CO_2 hydrogenation on iron and nickel metal catalysts, and (7) steam reforming of hydrocarbons on nickel catalysts. From these data, rates of methane production for thermal and catalytic reactions and activation energies for geologic conditions applicable to coal beds were calculated to the extent allowed by the published information.

Representative methane formation rates at 180°C, estimated times for reaching 10% conversion at this temperature, and activation energies for these relevant thermal and catalytic reactions are summarized in Table 1. Representative times for 10% conversion to methane for thermolysis of *n*-hexane and for three catalytic reactions are shown as a function of temperature in Figure 1. Data for steam reforming are not included since methane formation rates were estimated to be negligible at temperatures below 300°C.

Three important assumptions were made in these calculations: (1) about 10% of the carbon in the coal is converted to methane during maturation, (2) coal has a typical porosity of 20% and (3) one-third of the pore volume is occupied by liquid or vapor hydrocarbons and the remainder by water vapor and gases.

The first assumption of 10% conversion of hydrocarbons to methane is based on an estimate of gas generation potential during maturation of 150 cm^3(STP)/g coal for a U.S. carboniferous coal [Rice, 1986] which corresponds to 0.082 g-C/g-coal or 0.11 mole-CH_4/mole-C in the coal. The second assumption is confirmed by measured porosities for several U.S. coals of low to medium rank of 15-20% mainly in the macropore range

TABLE 1. Summary of Calculated Methane Formation Rates from Literature Data for Thermal and Catalytic Reactions Under Geologic Conditions (180°C).[a]

Reaction	Rate[a], 180°C $(g_{meth}/g_{coal}\,y)^a$	t [b] (y)	E_{act} (kcal/m)	Ref.
Thermal				
Thermal cracking, $n\text{-}C_6$	1.8×10^{-12}	9.2×10^7	69	Domine, 91
Thermal cracking, $n\text{-}C_{16}H_{34}$	1.1×10^{-12}	2.0×10^7	60	Goldstein, 83
Thermal cracking, aromatic	1.3×10^{-10}	8.1×10^5	65	Ungerer, 87
Thermal cracking, paraffinic	7.9×10^{-13}	1.3×10^8	69	Ungerer, 87
Catalytic				
Pyrolysis of ND Lignite[c]	1.2×10^{-5}	490	48	Tang et al.
Pyrolysis of kerogen (Type	1.2×10^{-5}	1,400	--	Tannenbaum,
Acid mineral cat. cracking,	2.1×10^{-6}	10	35	Goldstein, 83
Sedim. rock hydrogenol.,	3.7×10^{-5}	146 d [b]	31	Mango, 94
Hydrogenolysis $n\text{-}C_5$, Ni	5.6×10^{-4}	91 d	31	Somorjai, 94
Hydrogenolysis $C_{18}{}^=$, Ni	1.8×10^{-4}	29 d	50	Mango, 96
Hydrogenolysis $n\text{-}C_5$, Fe	1.8×10^{-4}	285 d	25	Somorjai, 94
Hydrogenolysis $C_{18}{}^=$, Fe	1.8×10^{-5}	299 d	--	Mango, 96
Hydrogenolysis $C_{12}{}^{=d}$, Fe[e]	1.7×10^{-5}	9.6 y	--	This work
Hydrogenolysis $C_{12}{}^=$, Fe[f]	2.9×10^{-5}	5.6 y	--	This work
CO_2 hydrogenation, Ni	5.4×10^{-1}	2 h [b]	19	Weatherbee,
CO_2 hydrogenation, Fe	1.1×10^{-1}	8 h	15	Weatherbee,
CO_2 hydrogenation, Fe	1.3×10^{-2}	36 h	17	This work

[a] Units of g-methane per g-coal or source rock per year; assumes 20% porosity and that pores are one-third filled with HC vapor (or liquid) or 1% CO_2 and 4% H_2 (remainder is water); coal is assumed to contain 100 ppm of surface Ni or Fe. Reaction is assumed to be first order.
[b] y = years, d = days, h = hours.
[c] North Dakota Lignite; mineral-containing.
[d] $C_{18}{}^=$ 1-octadecene; $C_{12}{}^=$ = 1-dodecene.
[e] Excess liquid 1-dodecene
[f] H_2/1-dodecene = 5.

[White et al., 1991]. The third assumption is reasonable since medium to low rank coals contains 3-10 wt% moisture and 3-20 wt% occluded hydrocarbons [Rice, 1994] and oil and gas generated at low ranks are readily accommodated in the macropores [Cook, 98].

It is evident that thermal cracking of mineral-free aromatic and paraffinic oils to methane occurs at very slow rates relative to those for acid catalytic cracking or catalytic hydrogenolysis. Indeed, times for 10% conversion of aromatic or paraffinic oil by thermolysis at 180°C are on the order of millions to hundred-millions of years. Conversely, acid-mineral-catalyzed cracking of C_{16}, Fe- or Ni-catalyzed hydrogenolysis of C_3-C_{18} hydrocarbons, and Fe- or Ni-catalyzed CO_2 hydrogenation occur at rates that are 5-12 orders of magnitude higher than thermal cracking of hydrocarbons to methane. For example, assuming only 100 ppm of Fe metal surface atoms to be present in the coal, 10% conversion of n-pentane to methane by Fe-

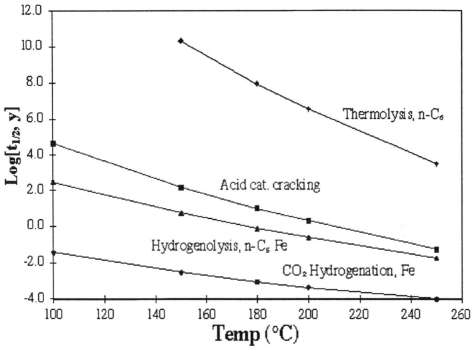

Figure 1. Times for 50% conversion of hydrocarbons or CO_2 to methane as a function of temperature for thermolysis of n-C_5, acid-catalyzed cracking of n-C_{16}, hydrogenolysis of n-C_3-C_6, and CO_2 hydrogenation.

TABLE 2. Comparison of Oil-to-gas and Kerogen-to-oil Rates for Thermolysis Versus Catalysis Under Geologic Conditions (180°C)

Reaction	Rate[a], 180°C	t (10%conv.) (y)[b]	E_{act} (kcal/mol)	Ref.
Thermal cracking, n-C_6 to C_1	1.8×10^{-12}	9.2×10^7	69	Domine, 91
Thermal cracking, n-C_{16} to C_1	1.1×10^{-12}	2.0×10^7	60	Goldstein, 83
Thermal cracking, kerogen (Type III) to C_{12+} oil[c]	3.8×10^{-7}	42,000	--	Lu and Kaplan, 90
Acid mineral cat. cracking, C_{16} to C_1	2.1×10^{-6}	10	35	Goldstein, 83
Hydrogenolysis C_{18}* to C_1, Fe	1.1×10^{-5}	0.8	50	Mango, 96

[a] Rate has units of grams methane or grams oil per gram coal or source rock per year; assume 20% porosity and that pores are one-third-filled with HC vapor and 50% H_2O; 700 ppm H_2; 100 ppm surface Fe. Reaction assumed to be first order.

[b] y = years.

[c] Rate extrapolated from 300 to 180°C using E_{act} = 48 kcal/mol [Tang *et al.*, 96].

catalyzed hydrogenolysis occurs in about 91 d while 10% conversion of CO_2 to methane by Fe-catalyzed hydrogenation occurs in only 8-36 h at 180°C under simulated geologic conditions. In other words, production of methane in coal beds by Fe- (or Ni-) catalyzed hydrogenolysis of liquid hydrocarbons or CO_2 methanation is a likely scenario, while thermolysis of hydrocarbons is unlikely to contribute measurably to coal-bed methane, even over long periods of geologic time at relatively high coal-bed temperatures, *e.g.* 150-200°C [Hunt, 1979]. Nevertheless, pyrolysis of lignite or kerogen to methane occurs at geologically significant rates, *i.e.*, 10% conversion in 490-1,400 y. However, the relatively high rate of methane formation during kerogen pyrolysis may be explained by metal catalysis in view of the difficulty of completely removing metal impurities, *e.g.*, organometallic complexes or suspended, colloidal Fe/SiO_2 or Fe/Al_2O_3. That pyrolysis of mineral-containing lignite and acid-treated kerogen occur at very similar rates (Table 1) supports this conclusion.

While literature data rule out any practical contribution of hydrocarbon thermolysis to formation of methane at coal bed depths, they nevertheless establish the possibility of thermal decomposition of kerogen to liquids as shown by data in Table 2. For example, thermal cracking of coal kerogen to C_{12+} liquids is found to occur at geologically measurable rates, *i.e.*, 10% conversion to C_{12+} liquids in 42,000 years at 180°C. Accordingly, these data indicate that thermolytic conversion of kerogen to oil may occur in parallel with acid-mineral-catalyzed production of oil, followed by metal-catalyzed conversion of oil to gas.

Distinguished by their extremely low rates at 160-200°C, thermal cracking reactions of liquid hydrocarbons are also characterized by relatively high activation energies of 60-69 kcal/mol (see Table 1); this observation is consistent with what one might predict for thermal scission of C-C bonds. On the other hand, activation energies for catalytic cracking and synthesis reactions are generally substantially lower (15-50 kcal/mol), as one might expect for catalytic C-C or C-O bond breaking.

While this section has focused thus far on rates of gas and oil formation in coal seams, it is equally important to consider if the product distributions observed for thermal

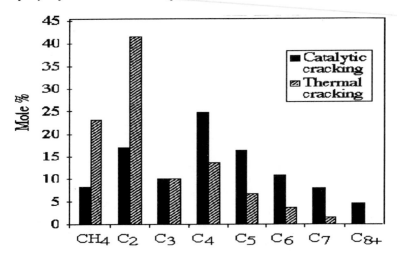

Figure 2. Product distributions for thermal cracking and catalytic cracking (on Houdry M-46) of 1-octadecene at 500 and 190°C, respectively [Mango *et al.*, 1994].

or catalytic routes are consistent with those measured in coal seams. Methane concentrations in U.S. coal mines are typically 85-95%. In contrast, thermal cracking of 1-octadecene at 500°C produces only 23% methane along with significant fractions of C_2- C_7 hydrocarbons (see Fig. 2 and Table 3). While no rate data are available at lower temperatures expected in coal beds (100-200°C) for liquid hydrocarbon thermolysis (rates are too low to permit measurement within a humanly reasonable time frame), the fraction of methane is expected to be significantly lower and the fraction of liquid hydrocarbons significantly higher under these milder conditions [Ungerer et al., 1987; Horsfield et al., 1991; Jackson et al., 1995].

TABLE 3. Gaseous Product Distributions for Thermal and Catalytic Cracking of Hydrocarbons Compared with the Compositions of Coal-bed Gas.

Reaction	Mole % of Gaseous Product						
	CH	C_2	C_3	C_4	C_5	C_6	C_7
Thermolysis, C_{18}[a]	23	41	10	14	7	4	1
Acid mineral catalytic	8	17	10	25	16	11	13
Pyrolysis of Type III	48	15	3	5	22	2	5
Pyrolysis of Rocky Mt.	38	17	17	12	10	6	
Mordenite-catalyzed	43	8	10	39			
Catalytic	90	4	1.3	0.7			
C_{12}[=], 180°C on Fe[f]	86	14					
Catalytic hydrogenation	94						
Coal-bed gas, Piceance[g]	90	3					
Coal-bed gas, Mary	96	0.0	0.7				

[a] Mango et al. [1994]; heated in glass for 1 h.
[b] Mango et al. [1994]; Houdry catalytic cracking catalyst, M-46; reaction for 24 h.
[c] Harwood [1977]; "mineral-free, pre-extracted kerogen" was pyrolyzed in evacuated glass tube for 1 week; analysis of impurities in kerogen was not provided.
[d] Lu and Kaplan [1990].
[e] Mango et al. [1994]; catalysis by carbonaceous rock (Monterey Formation) at 190°C and 1 atm of pure H_2.
[f] This study.
[g] Rogers [1994]; D Coal Seam, Piceance Basin; also contains 6.35 mol% CO_2.
[h] Rogers [1994]; Mary Lee Seam, Warrior Basin; also contains 0.1 mol% CO_2 and 0.01% H_2.

Products of catalytic cracking at 190°C are characterized by gas and liquid hydrocarbons having a maximum in the distribution around C_4 to C_5 with only 8% methane (see Fig. 2 and Table 3). Products of low-temperature pyrolysis of kerogen contain 40-60% methane, a product possibly characteristic of combined acid-catalytic cracking and metal-catalyzed hydrogenolysis (see Table 3). On the other hand, the product compositions for Fe- or Ni-catalyzed olefin hydrogenolysis or CO_2 methanation measured by Mango et al. [1994] and in this study are methane rich (90-94 mol%) and closely match those for U.S. coal gases (See Fig. 3 and Table 3).

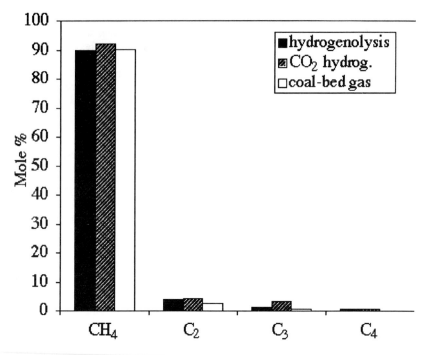

Figure 3.   Product distributions for catalytic hydrogenolysis of 1-octadecene [Mango *et al.*, 1994] and CO$_2$ hydrogenation [this work] on iron catalysts at 190 and 180°C, respectively, compared with the composition of a typical coal gas [Rogers, 1994].

4.2 EXPERIMENTAL MEASUREMENTS OF RATES OF ALKENE HYDROGENOLYSIS AND CO$_2$ METHANATION UNDER SIMULATED GEOLOGIC CONDITIONS

4.2.1   *Hydrogenolysis of 1-dodecene.* For reaction in the liquid slurry, the reaction rate during the first five hours was 0.15 mol$_{CH_4}$/ mol$_{surf.Fe}$–d, where mol$_{surf.Fe}$ refers to the moles of surface iron determined by H$_2$ chemisorption uptake; this rate decreased steadily (probably due to catalyst deactivation) for the next 15 hours at which point no further methane was formed and the amount of methane remained constant. During the first few minutes of reaction, only methane (86 mol%) and ethane (14 mol%) were observed; however, after 10 hours of reaction, significant quantities of ethylene and propylene were formed (18 and 23 mol%, respectively). At 43 hours methane, ethylene, and ethane were present (61, 18 and 21 mol%, respectively); since the quantity of methane did not change significantly after about 10 h, it is apparent that the quantities of ethylene and ethane were continuing to increase with time after 10 hours. The rate reported in Table 1 for 1-dodecene hydrogenolysis on Fe was extrapolated to geologic conditions by assuming 100 ppm of surface iron to be present in the coal, 20% coal-particle porosity, and pores filled with a liquid or vapor containing 33% hydrocarbons (remainder water and gases). No products were observed over 100 h of reaction at 180°C if no catalyst or the unreduced Fe$_2$O$_3$/SiO$_2$ was present.

4.2.2   *Methanation of CO$_2$.*   CO$_2$ conversion, methane selectivity, and catalyst activity data obtained in a batch reactor in this study for CO$_2$ hydrogenation on 10% Fe/silica at 180 and 192°C are listed in Table 4; methane production rates are plotted with time at these two temperatures in Figure 4. Methane production rate (the slope of the methane production versus time - see curves in Figure 4), after an 8-10 h induction period, is constant over 55 h at 192°C; however, methane production rate at 180°C, after a similar induction period (4-5 h), is constant for less than 30 h. The observed decrease in methane production rate at 180°C and t > 35 h is probably due to deactivation of the catalyst by carbon. Methanation activities, *i.e.*, rates of methane production per mole of surface iron per day were obtained from the linear fits of the data in Figure 4, based on an iron metal dispersion (fraction of metal atoms exposed to the surface and therefore available for catalysis) of 0.06. CO_2 conversions and methane production rates are very significant at these relatively low reaction temperatures. Again no products were observed in the absence of the reduced catalyst.

Methane selectivities for CO_2 hydrogenation listed in Table 4 are 9 and 52% at 180°C and 192°C, respectively. Other gas-phase reaction products observed in this work (not shown in Table 4) included principally CO with small fractions of ethane and propane. These observations are also consistent with the previous work [Weatherbee and Bartholomew, 1984] showing the product of CO_2 hydrogenation to consist of 40-53 mol% CO, 13-40% CH_4, and 4-6% C_2-C_5 hydrocarbons.

TABLE 4. Conversion, Selectivity, and Activity of 10% Fe/SiO$_2$ Hydrogenation (H_2/CO_2 = 3.5).

Temperatur	CO_2	H_2	CH_4	CH_4 yield	Catalyst
180	70[b]	68[b]	8.9[b]	6.2[b]	1.2[b]
192	25	34	52	13	2.0

[a] Moles of CH_4/(mole surface Fe-d).

[b] Average of 2 runs

Two aspects of the data in Table 4 are unexpected: (1) conversions of CO_2 and H_2 are about 2 and 10 times higher at 192 and 180°C, respectively, than required to produce the observed amounts of methane and (2) CO_2 and H_2 conversions are higher and methane selectivity is lower at 180°C relative to those at 192°C. A large fraction of the CO_2 was apparently converted to CO (since large CO peaks were observed but were not separated and quantified). Moreover, unexpectedly large uptakes of H_2 at the beginning of each run may be due to further reduction of iron oxides to iron metal in the catalyst. Indeed, we estimate that about 10% of the reactant H_2 was consumed in further reducing the iron catalyst within the first 2 hours of reaction and before any methane had been produced. Since it is known that iron catalysts form carbide in a H_2/CO mixture at 180-280°C and since CO_2 is known dissociate to CO and O when adsorbed [Weatherbee and Bartholomew, 1981], some of the CO_2 may have been decomposed to carbon on the surface which then further reacted. Thus, both carbiding and reduction of iron may explain the 5-10 h induction period for methane production (see Fig. 4). Moreover, at 180°C a significant fraction of the H_2 and CO_2 may also be converted to liquid hydrocarbons, since it is known that Fe/SiO$_2$ is an effective Fischer-Tropsch catalyst for producing hydrocarbon liquids from CO/H_2 syngas and since selectivity for methane is known to

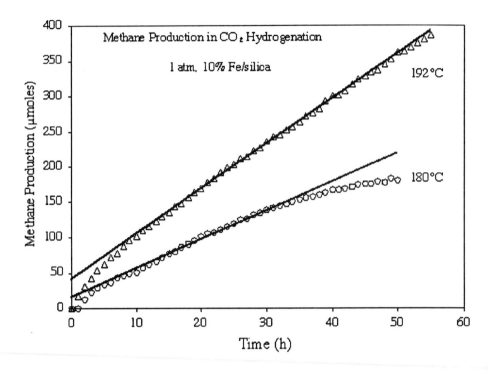

Figure 4.   Methane production during CO_2 hydrogenation on 10 wt.% Fe/silica at 180 and 192°C, 1
atm, and $H_2/CO_2 = 3.5$. Correlation coefficients for the linear fits of the data for 180 and
192°C are 0.999 and 0.994, respectively.

decrease, and that for liquids to increase, with decreasing temperature [Anderson, 1984]. In fact, at
180°C, the H_2/CO_2 ratio (after the initial drop due to catalyst reduction) increased steadily with time,
consistent with (1) CO formation, (2) carbiding and/or (3) formation of liquid hydrocarbons.
Nevertheless, it should be emphasized that based on our GC analysis, *90 mol% of the gaseous
hydrocarbon product (C_1-C_4) consists of methane at both reaction temperatures* (see Fig. 3).

4.3 XRF ANALYSIS OF MINERALS IN PRODUCING AND NONPRODUCING COALS

Ultimate analyses of Belle Ayr and Black Thunder coals on a moisture and ash-free basis are
summarized in Table 5; the data for the Black Thunder coal are an average of analyses of 90 samples
from a single bed, while that for Belle Ayr is from a single analysis. Carbon, hydrogen, nitrogen,
oxygen, sulfur, and ash contents are very similar for the two coals, suggesting that they are of similar
rank and reactivity.

TABLE 5.   Ultimate Analysis of Belle Ayr and Black Thunder Coals [Yin, 1997]

	Belle Ayr[a]	Black Thunder[b]
Carbon, maf[c]	74.6	74.8
Hydrogen, maf	5.5	5.3
Nitrogen, maf	1.2	1.0
Oxygen, maf	18.2	18.4
Sulfur, maf	0.5	0.5
Ash, mf[d]	6.3	6.6 (4.5 – 9.3)[e]

[a] Analysis, F.E Walker, September 4, 1974.
[b] Typical values reported by the Atlantic Richfield Co. for 90 samples from the Roland Bed.
[c] Moisture and ash-free basis.
[d] Moisture-free basis.
[e] Range of values.

Selected results from our XRF analysis of ash from the two coals are summarized in Table 6. The ash contents of Belle Ayr and Black Thunder coals of 5.4 and 4.7% measured in this study are comparable to those listed in Table 5. The iron content of the Belle Ayr coal of 0.39% measured in this work is almost a factor of two higher than the value of 0.20% reported by Yin [1977], while that for the Black Thunder of 0.29% is close to the value of 0.26% reported by Slater [1998]. The difference in iron content of Belle Ayr coal between this and the earlier work is attributed to generally large variabilities of coal compositions. Fe contents of the same samples of Belle Ayr and Black Thunder coals (e.g. Table 6) were found by PIXE analysis to be 0.30 and 0.24%, respectively [Medina et al., 1998]. Accordingly, when compared with the PIXE data for the same samples, the XRF values from this study may be as much as 10-20% high; nevertheless, both XRF and PIXE data are consistent with a trend of 25-30% higher iron content for the Belle Ayr coal.

TABLE 6.   Results of XRF Analysis ([a] dry basis) of Belle Ayr and Black Thunder Coals.

	Belle Ayr	Black Thunder
Ash Content, wt. % (gravimetric)[a]	5.4	4.7
Fe, wt % (XRF)[a]	0.39	0.29

5. Discussion

5.1 PLAUSIBLE REACTIONS FOR PRODUCING NATURAL GAS IN COAL BEDS

Plausible reactions for producing natural gas in coal beds must meet the following criteria: (1) produce methane at significant rates under geologic conditions within geologically significant times for coal maturation (i.e., less than 1-10 million years) and (2) produce gases rich in methane.

According to Hunt [1984], methane formation from catagenesis is significant in the range of 100-220°C with a maximum at 150°C—corresponding to a depth of 4500 m [Tissot and Welte, 1984;

Ungerer *et al.*, 1987]. Maximum generation of methane apparently occurs in medium-volatile bituminous and low-volatile bituminous coals [Hunt, 1979] and is high for coals having a reflectance (R_o) higher than about 2% [Tissot and Welte, 1978]. These conclusions are supported by the recently developed model of Tang *et al.* [1996] indicating that T > 120°C and R_o ≥ 0.9% are required for a minimum threshold methane production of 300 ft^3/ton (9 cm^3/g). Accordingly, the most important methane-producing reactions should occur at temperatures in the range of 120-180°C. While the prevailing wisdom holds that coal-bed gas is produced by thermal decomposition of light oils [*e.g.*, Ungerer, 1990; Hunt, 1991], a careful analysis of rate data from the previous literature (see Tables 1 and 2 and Figure 1) provides strong evidence that reaction rates for thermal cracking of light to medium hydrocarbons are orders of magnitude too slow to have produced significant quantities of methane within coal maturation times. On the other hand, hydrocarbon decomposition or methanation reactions catalyzed by acidic or transition-metal minerals could have produced known reserves of coal-bed gas within days to thousands of years, *i.e.* well within maturation times (see Tables 1 and 2, Fig. 1). The experiments in this study confirm that rates of olefin hydrogenolysis and CO_2 methanation are very significant in the presence of reduced, dispersed iron under conditions simulating coals beds, *i.e.* 20-50% conversion of CO_2 or 1-dodecene occurs on 10% Fe/SiO_2 in 40-60 hours at 180°C; moreover, when extrapolated to reactant and catalyst concentrations at the lower end of what might be expected for a coal bed (*e.g.*, only 100 ppm surface iron), 10% conversion times of 1-10 months for olefin hydrogenolysis and 2-36 hours for CO_2 hydrogenation are predicted.

The rate of methane production for 1-dodecene hydrogenolysis on reduced 10% Fe/silica at 180°C is in excellent agreement with that reported by Mango [1996] for 1-octadecene hydrogenolysis on reduced $Fe(AcAc)_3$ (see Table 1). Rates of CO_2 hydrogenation on 10% Fe/silica from this study obtained in a batch system are in fair agreement with those extrapolated from the higher temperature data of Weatherbee and Bartholomew obtained in a flow reactor for 15% Fe/silica [1984]. However, the data reported in this study are, to our best knowledge, the first to be obtained at low temperatures in a closed system. Moreover, the concept of CO_2 hydrogenation as a reaction for production of methane in coal beds is proposed in this paper for the first time. While Mango and Hightower [1997] have suggested that hydrogenolysis of oil catalyzed by transition metal minerals may be a unique route to natural gas of high methane content, our results provide the possibility of an additional catalytic route to "dry" natural gas, *i.e.* CO_2 methanation on Ni and Fe minerals, in basins in which both CO_2 and H_2 concentrations are significant.

The second criterion for identifying plausible reactions that produce natural gas in coal beds is a high selectivity to methane. While catalytic cracking on acidic minerals, *e.g.*, silica-aluminas or montmorillinites, occurs at rates comparable to those of alkane/alkene hydrogenolysis (see Tables 1 and 2, and Figure 1), products contain substantially less methane than does natural gas (see Table 2 and Fig. 2). Thus, while catalytic cracking may contribute to formation of methane, light hydrocarbons, and oils in coal beds, subsequent hydrogenolysis of these light hydrocarbons and oils by transition metal catalysis is necessary to realize methane contents of typical natural gases. Similarly, while formation of hydrocarbon liquids by thermolysis occurs at geologically relevant rates (see Table 2), these hydrocarbons would have to be further processed by metal-catalyzed hydrogenolysis to form typical coal-bed gas.

Accordingly, we conclude that thermal and catalytic cracking of hydrocarbons may contribute to the formation of light hydrocarbons and hydrocarbon liquids; however, the most plausible reactions for forming natural gas of high methane content in coal beds are alkane/alkene hydrogenolysis and CO_2 methanation.

Metal-catalyzed hydrocarbon hydrogenolysis or CO_2 hydrogenation requires that gas-phase hydrogen be present. Formation of significant amounts of H_2 gas during mild pyrolysis of kerogens and coal is well documented [Harwood, 1977; Tannenbaum and Kaplan, 1985; Lu and Kaplan, 1990]. While mechanisms of hydrogen formation in coal seam reservoirs are uncertain, coal-gas hydrogen contents are nevertheless generally significant. A survey of analytical data for natural gases compiled by the US Bureau of Mines from 1917 to 1992 reveals that 253 of 1067 gas samples from the Western US were found to contain hydrogen gas in the range of 0.1 to 0.3%; a few samples contained more than 1% and as high as 4%, while a large number of these samples were reported to contain only "trace amounts of H_2" [Yin, 1998]. However, even "trace amounts" could be significant, since only 500 ppm (0.05%) of hydrogen corresponds to a partial pressure of 0.7 atm in a geologic formation at 1400 atm, a concentration adequate for driving catalytic hydrogenation or CO_2 methanation at the rates estimated earlier (Table 1). Moreover, hydrogenolysis of liquid hydrocarbons can proceed in the absence of gas-phase hydrogen, possibly by hydrogen transfer from larger hydrocarbons [Mango and Hightower, 1997].

There are a number of potential routes for the formation of gas-phase hydrogen during coal maturation including: (1) free-radical dehydrogenation of aromatic clusters during coal maturation [Butala et al., 1998], (2) the water-gas-shift reaction (assuming a source of CO), and (3) the steam-carbon reaction (which might provide CO at extreme temperatures and depths). However, there are too few data to enable drawing definitive conclusions regarding the viability of any of these routes. Rate calculations based on available literature data for nickel-catalyzed steam-reforming of hydrocarbons lead us to conclude that this reaction is not a viable route to formation of hydrogen at typical coal bed temperatures.

The comparisons in this work of rates and selectivities for methane production (Tables 1-3 and Figures 1-3) include data obtained in both batch (closed) and flow (open) systems. There are no fundamental problems in comparing rates for a single stoichiometric reaction obtained in these two different kinds of reactors (systems) as long as the comparisons are made at the same temperatures, reactant partial pressures, and conversions, as well as at similar reaction/residence times [Levenspiel, 1972]. If these same conditions are met, product selectivities will be significantly different. This is also borne out by the reasonably good agreement between rates and selectivities obtained for olefin hydrogenolysis in batch systems and CO_2 hydrogenation [Mango, 1996; this study] with those obtained in a flow (open) system [Weatherbee and Bartholomew, 1984; Somorjai, 1994] (see Tables 1 and 3). Accordingly, the comparisons in this study involving single reactions (i.e. model compounds to oil, kerogen to oil, or oil to gas) are fundamentally sound.

However, as pointed out by Tang et al., rate and selectivity data obtained for a sequence of reactions, as in pyrolysis of coal (kerogen $\rightarrow$ oil $\rightarrow$ gas) are different when measured in a low-residence-time open (flow) system relative to those obtained in a closed (batch) system. Thermogenic oil and gas formation in coal seams, probably involving sequential reactions, is apparently best modeled as a closed system, especially since oil and gas formation and storage occur mainly within the pores of the coal [Nelson et al., 1998; Cook, 1998] (see section below on proposed model).

5.2 COAL-BED MINERALS AS CATALYSTS FOR GAS AND OIL FORMATION

The concept of catalytic generation of natural gas in coal mines presupposes that suitable catalysts are available in the coal to (1) convert coal kerogen to hydrocarbon liquids and (2) convert hydrocarbon liquids to natural gas. From the previous discussion the identification of acidic and transition-metal minerals in coals or in adjacent strata would provide this link.

There are apparently two kinds of acidic minerals available for catalytic cracking of hydrocarbons in coal beds: (1) those present in the coal matrix, e.g. silica-aluminas or aluminosilicates (e.g. see Table 7) and (2) those present in the nearby strata, e.g. natural clay minerals such as bentonite and montmorillonite [Saxby *et al.,* 1992]. Generation of hydrocarbon gases and condensates during catalytic cracking by these and closely related minerals under geologically-relevant conditions has been demonstrated in a number of studies [*e.g.,* Goldstein, 1983; Tannenbaum and Kaplan, 1985; Saxby *et al.,* 1992].

The previous work implicates reduced oxides of nickel and iron (see Tables 1 and 2) as important minerals for methane production. Most coals contain significant amounts of iron minerals as pyrites, oxides, carbonates, clays/micas (*e.g.,* illite), and/or organometallic compounds (porphyrins, ferrous acetate, and ferrous iron associated with carboxylic groups) [Lefelhocz *et al.,* 1967; Schafer, 1977; Huffman and Huggins, 1978; Montano, 1981; Taneja and Jones, 1984; Herod, *et al.,* 1996].

Our calculations (Table 1) show that only 100 ppm of reduced, surface iron can catalyze high rates of methane formation by hydrocarbon hydrogenolysis or CO_2 methanation at geologic temperatures. Assuming that iron at low concentrations would be highly dispersed, *i.e.* have a significant fraction of iron atoms exposed to the surface (*e.g.* 5-15%), it is reasonable to conclude that as much as 100 ppm of reduced, surface iron could be available in selected coals for catalyzing methane-forming reactions.

Although nickel occurs in coals at only ppm levels, even 1-10 ppm of reduced, surface nickel could be significant in decomposing oil to methane at geologic conditions within coal maturation times. For example, Mango and Hightower [1997] estimate a half-life for light crude oil in contact with 1 ppm of active nickel to be approximately 350,000 years at 175°C and 45,000 years at 200°C. Accordingly, nickel catalysis could play a role in methane formation during coal maturation. However, since iron-catalyzed methane formation by hydrogenolysis or CO_2 hydrogenation occurs at rates roughly comparable to the corresponding nickel-catalyzed reactions at the same catalyst concentration (see Table 1) and since iron is present at 2-3 orders of magnitude higher levels in coal, we expect that catalysis of methane-forming reactions by reduced iron minerals is a more likely scenario.

Catalysis of gas and oil formation by minerals in the sedimentary strata in close vicinity to coals seams is also a possibility. That sedimentary rocks contain iron and nickel minerals is reasonably well documented; for example, a sedimentary source rock from the Monterey Formation (California) was found to contain 350 ppm nickel [Mango *et al.,* 1994] and was found to effectively catalyze hydrogenolysis of 1-octadecene at 190°C. Catalysis of gas and oil formation in sedimentary rock would require either a significant dispersed organic matter (dom) content in the rock or migration of liquid hydrocarbons and light gases from the coal to the nearby rock. However, there is lack of agreement regarding the importance of dom in associated clastic rock relative to dom in coal beds [Cook, 1998]. Migration of hydrocarbons from coals of lower rank (vitrinite reflectance of 0.4 to 0.65%) is considered to be relatively easy [Cook, 1998]; however, according to Cook [1998] in higher ranks of coal, *i.e.,* those most likely to produce natural gas, oil-like compounds tend be trapped within the coals and gradually cracked to gases. Cook's observation is supported by the rate calculations of this study showing that gas formation in coals occurs at least as rapidly or more rapidly than oil formation. Accordingly, we conclude that gas and oil formation in surrounding strata is less likely than in coal seams and that oil formed in coals is likely to be converted to gas as quickly as it is formed.

In any catalytic process it is important to know the chemical state of the active catalytic phase. It is especially important in coal-bed methane formation for purposes of (1) modeling the process and (2) using catalytic minerals as fingerprints for natural-gas exploration. Mango and coworkers [1994, 1996, 1997] have reported that transition metals oxides, *e.g.,* NiO and Fe_3O_4, are the active catalytic phases for hydrogenolysis of liquid hydrocarbons; however, in every instance their catalyst precursors, *e.g.,* NiO/silica and Fe(AcAc)$_3$ were reduced at 200-400°C for 24 hours in flowing H_2, conditions which in our experience would effect nearly complete reduction of the oxides or salts to the metal. Moreover, we observed no methane formation in reaction tests involving unreduced iron oxides.

Furthermore, there is considerable experimental evidence that Group VIII metals, rather than the oxides, are the active catalytic phases for hydrocarbon hydrogenolysis and CO_2 hydrogenation [Weatherbee and Bartholomew, 1981, 1984; Somorjai 1994; Ribeiro et al., 1997]. While Mango and Hightower show that water partially inhibits the hydrogenolysis of petroleum on reduced Ni/silica, i.e., reduces the reaction rate about 30-40%, they found the effect to be reversed when water was replaced with H_2, indicating a weak inhibition rather than oxidation of the metal surface.

In this study, we observed high rates of hydrogenolysis and CO_2 hydrogenation on a 10% Fe/silica catalyst reduced at only 200°C (for 150 h). Moreover, we observed a sizable H_2 chemisorption uptake on this catalyst (41 μmol/g), indicating that a significant fraction of the iron had been reduced to the metal; this was also confirmed in an ongoing companion study in this laboratory using Mössbauer spectroscopy. On the other hand, we observed no production of gas for hydrogenolysis or CO_2 hydrogenation tests on the dried catalyst precursor consisting of $Fe(NO_3)_3 \bullet xH_2O$ and Fe_2O_3/silica. Thus, we conclude on the basis of previous experience and the results of this study that the active catalytic phases for methane production in coal beds include well-dispersed Fe and Ni metals. We propose that oxides or carbonates, e.g., Fe_2O_3 or $FeCO_3$, dispersed in the pores of high-surface-area aluminosilicates, clay silicates or silica in the coal, are partially reduced to dispersed Fe during coal maturation in the presence of hydrogen and may be subsequently reoxidized to the oxides or carbonates after mining and exposure to air. Accordingly, we propose that the presence of dispersed iron oxides or carbonates in coals may be a key to identifying catalyzed natural gas formation; thus, analysis for these minerals could be used as a gas exploration guide.

5.3 PROPOSED CHEMICAL MODEL FOR COAL GAS/OIL FORMATION

We propose a new chemical model for coal-bed gas and oil formation (see Figure 5) in which coal kerogen is cracked thermally or catalytically to form oil and hydrogen gas; oil is then hydrocracked to methane on Fe and Ni metals present in reduced, dispersed coal minerals. Alternatively, CO_2 formed during catagenesis reacts catalytically with hydrogen present in coal gas in the presence of metal minerals to form methane.

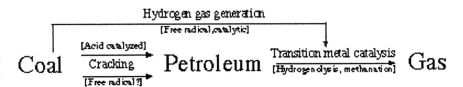

Figure 5.   Proposed chemical model for catalytic production of natural gas in coal beds.

Our model is consistent with and in several aspects similar to the recently proposed model of Mango et al. [1997] for natural gas generation in petroleum source rock. In their model, Ni metal catalyzes hydrogenolysis of petroleum to olefins and olefins to principally methane at low temperature geologic conditions. Our model is supported by calculations reported in this paper, based on our comprehensive survey of the literature, of gas and liquid formation rates. These calculations demonstrate that catalytic hydrogenolysis and CO_2 hydrogenation are the most plausible reactions to have generated natural gas in coal formations. It is also supported by our experimental work reported in this paper showing that Fe metal catalyzes methane formation at high rates under typical coal-bed conditions via hydrogenolysis of 1-dodecene or CO_2 methanation.

The kinetic data reported in this study indicate that gas formation by catalytic hydrogenolysis is rapid relative to formation of oil by catalytic cracking and thermogenic means (especially the latter), *i.e.*, oil formation is probably the rate-determining step. Accordingly, if suitable minerals are present, gas may be formed as quickly as oil is formed by cracking reactions; if, however, acidic minerals are present in adequate concentrations but reduced iron or H_2 are not, oil may build up and ultimately undergo expulsion [Cook, 1998). Gas formation by CO_2 hydrogenation is extremely rapid but relies on significant concentrations of hydrogen and CO_2 being present simultaneously. Since CO_2 is generated by decarboxylation in relatively large quantities during maturation, generation of adequate H_2 for reducing iron oxides to iron metal and reacting with CO_2 will probably be the rate-determining process.

It is well documented that coal has adequate macroporosity for storing significant amounts of oil and microporosity for storing significant quantities of H_2, CO_2 and CH_4 gases [Beamesh *et al.*, 1998; Cook, 1998, White *et al.*, 1991]. However, models of oil and gas formation predict that the quantities of oil and gas produced during maturation will generally exceed actual adsorbed or producible amounts by an order of magnitude or more [Rightmeyer, 1984; Rice, 1993; Tang, 1996]. The kinetic data base assembled in this study is consistent with these model predictions, since (1) only 10% of the carbon in coal need be converted to methane to exceed its adsorption capacity by roughly a factor of ten and (2) times for 10% conversion of coal kerogen to oil and oil to gas by catalytic routes are orders of magnitude smaller than times projected for gas formation during coal maturation. If gas and oil can be trapped or adsorbed in tight surrounding formations, it may be possible to produce more gas and oil than those quantities adsorbed in the pores. Thus, the sometimes observations of coal-bed methane contents exceeding adsorption capacity are explained.

Previously reported models of coal-bed methane formation have not included data for catalytic reactions. It is hoped that the data set provided by this work will be useful in refining such models.

6. Conclusions

a. Rates based on kinetic data assembled from the literature predict that at typical coal-bed temperatures, gas formation rates from hydrocarbon thermolysis reactions are several orders of magnitude too low to account for the formation of known coal-seam natural gas reserves during coal maturation.

b. Acid-mineral-catalyzed cracking, iron- or nickel-metal-catalyzed hydrogenolysis of liquid hydrocarbons, and iron- or nickel-metal-catalyzed CO_2 hydrogenation occur at sufficiently high rates at 180°C to account for formation of enormous reserves of coal-bed gas. Rates of gas production in these reactions are 5-10 orders of magnitude more rapid than those predicted from thermolysis.

c. Thermolysis and acid-mineral catalysis have low selectivities for methane, and hence the composition of hydrocarbon products from these reactions are substantially different than for natural gas. On the other hand, the compositions of gaseous products from metal-catalyzed hydrogenolysis and CO_2 hydrogenation are almost identical to typical compositions of coal-bed gas.

d. Much of the iron in many coals is tied-up as pyrites, *i.e.*, iron sulfides, having no measurable activity for hydrogenolysis or CO_2 hydrogenation; nevertheless, significant fractions of the iron are available in a number of coals as reducible oxides, clays or carbonates.

e. XRF analysis of a pair of gas-producing and non-producing coals reveals that significant and approximately comparable amounts of iron are present in both coals;

f. From the results of this study and previous studies, it is concluded that the active catalytic phases for methane production in coal beds include well-dispersed Fe and Ni metal crystallites, although there is as yet no direct evidence in coal beds.

g. Calculated rates from literature data and our experimental data support the concept that natural gas is formed in coal seams by either CO_2 methanation or liquid hydrocarbon hydrogenolysis on reduced iron minerals present in the coal. An important implication of our analysis is that iron-mineral catalysis rather than homogeneous thermolysis leads to natural gas formation during coal maturation. This, in turn, suggests the possibility of using analyses of coal minerals, *e.g.*, iron

h. oxide, carbonate, and metal contents, rather than currently-used coal thermal maturity parameters for gas resource assessment and exploration.

7. Acknowledgements

The authors gratefully acknowledge the technical assistance of Dr. P. Yin and Professor R. Surdam of the Institute for Energy Research, University of Wyoming, in obtaining coal samples, coal data, and coal-gas compositions and Dr. Charles R. Nelson of the Gas Research Institute in providing useful guidance and perspectives. The authors also gratefully acknowledge financial support from the Gas Research Institute under Contract No 5093-260-2764.

8. References

Anderson, R. B., *The Fischer-Tropsch Synthesis,* Academic Press, 1984.

Barker, C. *Calculated volume and pressure changes during the thermal cracking of oil to gas in reservoirs,* AAPG Bulletin 74 (1990) 1254-1261.

Beamish, B.B., Crosdale, P.J. and Moore, T.A. *Fundamentals of methane sorption by New Zeland coals,* Lecture at the International Conference on Coal Seam Gas and Oil, Brisbane Australia, March 23-25, 1998.

Butala, S.J., Medina, J.C., Bowerbank, C.R., Lee, M.L., Felt, S.A., Taylor, T.Q., Andrus, D.B., Bartholomew, C.H., Yin, P., and Surdam, R.C. *Catalytic effects of mineral matter on natural gas formation during coal maturation,* Gas Research Institute, 1997.

Butala, S.J., Medina, J.C., Lee, M.L., Taylor, T.Q., Andrus, D., Bartholomew, C.H. Yin, P. and Surdam, R. *Chemical indicators for mineral-catalyzed coal seam gas producibility sweet spots,* Gas Research Institute, Annual Report for 1997, March 5,1998.

Butala, S.J., Taylor, T.Q., Medina, J.C., Bartholomew, C.H. and Lee, M.L. *Mechanisms and kinetics of reactions leading to natural gas formation during coal maturation,* manuscript in preparation, 1998.

Cook, A.C., *Oil occurrence, source rocks and generation history of some coal-bearing tertiary basins,* Plenary Lecture at the International Conference on Coal Seam Gas and Oil, Brisbane Australia, March 23-25, 1998.

Davidson, R.M., Sloss, L.L., and Clarke, L.B. *Coalbed methane extraction,* IEA Coal Research, London, 1995.

Dominé, F. *High pressure pyrolysis of n-hexane, 2,4-dimethylpentane and 1-phenylbutane. Is pressure an important geochemical parameter?,* Org. Geochem. 17 (1991) 619-634.

Espitalié, J., Ungerer, P., Irwin, I., and Marquis, F. *Primary cracking of kerogens. Experimenting and modeling C_1, C_2-C_5, C_6- and C_{15+} classes of hydrocarbons formed,* F. Org. Geochem. 13 (1988) 893-899.

Evans R.J. and Felbeck, G.T. Jr. *High temperature simulation of petroleum formation-I. The pyrolysis of Green River Shale,* Org. Geochem. 4 (1983) 135-144.

Gayer, R. and Harris, I. *Coalbed methane and coal geology,* The Geological Society, London, Special Publication No. 109, 1996.

Goldstein, T.P., *Geocatalytic reactions in formation and maturation of petroleum,* Am. Assoc. Petr. Geol. Bul. 67 (1983) 152-159.

Harwood, M. H., Burkholder, J.B. and Hunter, M., *Absorption cross sections and self reaction kinetics of the IO radical,* Journal of Physical Chemistry 101 (1997) 853-863.

Harwood, R.J., *Oil and gas generation by laboratory pyrolysis of kerogen,* Am. Assoc. Pet. Geol. Bull., 61 (1977) 2082-2102.

Herod, A.J., Gibb, T.C., Herod, A.A., Xu, B., Zhang, S. and Kandiyoti, R., *Iron complexes by Mössbauer spectroscopy in extracts from Point of Ayr coal,* Fuel 75 (1996) 437-442.

Horsfield, B., Schenk, H.J., Mills, N. and Weite, D.H. *An investigation of the in-reservoir conversion of oil to gas: compositional and kinetic findings from closed-system programmed-temperature pyrolysis,.* Org. Geochem. 19 (1991) 191-204.

Hunt, J.M. *Petroleum geochemistry and geology,* W.H. Freeman, 1979.

Hunt, J.M. *Generation of gas and oil from coal and other terrestrial organic matter,* Org. Geochem. 17 (1991) 673-680.

Jackson, K.J., Burnham, A.K., Braun, R.L. and Knauss, K.G., *Temperature and pressure dependence of N-hexadecane cracking,* Org. Geochem. 23 (1995) 941-953.

James, A.T., Bull. Am. Ass. Petrol. Geol. 67 (1983) 1176-1191.

Kissin, Y.V. *Catagenesis and composition of Petroleum: Origin of n-alkanes and isoalkanes in petroleum crudes,* Geochim. Cosmochim. Acta 51 (1987) 2445-2457.

Lefelhocz, J.F., Friedel, R.A. and Kohman, T.P. *Mössbauer spectroscopy of iron in coal*, Geochim. Cosmochim. Acta **31** (1967) 2261-2273.

Levine, J.R., *Coalification: the evolution of coal as a source rock and reservoir rock for oil and gas*, eds. B.E. Law and D.D. Rice *Hydrocarbons from Coal AAPG studies in geology #38*, Chapter 3.

Levenspiel, O., *Chemical Reaction Engineering*, John Wiley & Sons, INC., New York, Second Edition (1972).

Lu, S.T. and Kaplan, I.R. *Hydrocarbon-generating potential of humic coals from dry pyrolysis*, Am. Assoc. Petrol., Geol. Bulletin **74** (1990) 163-173.

Mango, F. D., Hightower, J.W., *The catalytic decomposition of petroleum into natural gas*, Geochimica et Cosmochimica Acta, **61** (1997) 5347-5350.

Mango, F.D., Hightower, J.W., James, A.T. *Role of minerals in the thermal alteration of organic matter-I: Generation of Gases and Condensates Under Dry Conditions*, Nature **368** (1994) 536-538.

Mango, F.D., *Transition metal catalysis in generation of natural gas*, Org. Geochem. **24** (1996) 977

Mango, F.D., *Transition metal catalysis in the generation of petroleum and natural gas.*, Geochim. Cosmochim. Acta. **53** (1992) 553-555.

McNab, J.G., Smith, P.V. and Betts, R.L., *Evolution of petroleum* , Ind. Eng. Chem. **44**, (1952) 2556-2563.

Medina, J.C., Andrus, D.B., Taylor, T.Q., Butala, S.J., Bartholomew, C.H., and Lee, M.L. Manuscript submitted (1998).

Montano, P.A., *Application of Mössbauer spectroscopy to coal characterization and utilization*, in *Mössbauer Spectroscopy and its chemical applications*, eds. J. G. Stevens, and G. K. Shenoy, Adv. Chem. Series, **194** (1981) 135-175.

Nelson, C.R., Li, W., Lazar, I.M., Larson, K.H., Malik, A., and Lee, M.L. *Geochemical significance of n-alkane compositional-trait variations in coal*, Energy & Fuels **12** (1998) 277-283.

Philippi, G.T., *On the depth, time and mechanism of petroleum generation*, Geochim. Cosmochim. Acta. **29** (1965) 1021-1049.

Ribeiro, F. H., Schach von Wittenau, A.E., Bartholomew, C. H., and Somorjai, G.A., *Reproducibility of turnover rates in heterogeneous metal catalysis: compilation of data and guidelines for data analysis*, Catal.Rev.—Sci.Eng. **39** (1997) 49-76.

Rice, D.D, *Composition and origins of coalbed gas in hydrocarbons from coal*, eds. B.E. Law and D.D. Rice, *Hydrocarbons from Coal AAPG studies in geology #38*, The American Association of Petroleum Geologists, Tulsa, OK, Chapter 7 (1994).

Rightmeyer, C.T., *In coalbed methane resources of the United States*, eds. C.T. Rightmire, G.E. Eddy, and J.N. Kirr, American Association of Petroleum Geologists, (1984).

Rogers, R.E., *Coalbed Methane: principles and practice*, PTR Prentice Hall, Englewood Cliffs, NJ, (1994).

Saxby, J. D., Chatfield, P. Taylor, G. H., FitzGerald, J. D. Kaplan, I. R., and Lu, S.-T., *Effect of clay minerals on products from coal maturation*, Organic Chem. **18** (1992) 373-383.

Saxby, J.D. and Riley, K.W. *Petroleum generation by laboratory-scale pyrolysis over six years simulating conditions in a*

Schafer, H.N.S. *Organically bound iron in brown coals*, Fuel **56** (1977) 45-46.

Shock, E.L. *Catalysing methane production*, Nature **368** (1994) 499-500.

Slater, P.N., *Pilot-Scale investigation of coal ash transformations under staged combustion*. Ph.D. Dissertation, Brigham Young University, August 1998.

Somorjai, G.A. *Introduction to surface chemistry and catalysis*, Wiley, New York (1994) 526-555.

Takach, N.E., Barker, C., and Kemp, M.K. *Stability of natural gas in deep subsurface: Thermodynamic calculation of equilibrium compositions*, Bull. Am. Ass. Petrol. Geol. **71** (1987) 322-333.

Taneja, S.P. and Jones, C.H.W. *Mössbauer studies of iron-bearing minerals in coal and coal ash*, Fuel **63** (1984) 695-701.

Tang, Y., Jenden, P.D., Nigrini, A., and Terman, S.C. *Modeling Early Methane Generation in Coal*, Energy and Fuel **10** (1996) 659-671.

Tannenbaum, E. and Kaplan, I.R. *Low-Mr hydrocarbons generated during hydrous and dry pyrolysis of kerogen*, Nature **317** (1985) 708-709.

Tannenbaum, E. and Kaplan, I.R. *Role of minerals in the thermal alteration of organic matter-I: generation of gases and condensates under dry conditions*, Geochim. Cosmochim. Acta **49** (1985) 2589-2604.

Tissot, B.P. and Welte, D.H. *Petroleum Formation and Occurrence* , Springer, New York (1984).

Tyler, R. and Scott, A. *Developing coalbed methane exploration fairways along the north slope of rural Alaska*, AAPG Annual Convention Abstract (1988).

Ungerer, P., Behar, F., Billalaba, M., Heum, O.R. and Audiber, A., *Kinetic modeling of oil cracking*, Org. Geochem. **13** (1987) 857-868.

Ungerer, P., *State of the art of research in kinetic modeling of oil formation and expulsion*, Org. Geochem. **16** (1990) 1-25.

Weatherbee, G.D. and Bartholomew, C.H., *Hydrogenation of CO_2 on group VIII metals: IV. specific activities and selectivities of silica-supported Co, Fe, and Ru*, Catalysis **87** (1984) 352-362.

Weatherbee, G.D. and Bartholomew, C.H., *Hydrogenation of CO_2 on group VIII metals: I. specific activity of Ni/SiO_2*, Catalysis **68** (1981) 67-76.

White, W.E., Bartholomew, C.H., Hecker, W.C. and Smith, D.M. *Changes in surface area, pore structure and density during formation of high-temperature chars from representative U.S. coals*, Adsorption Science and Technology **7** (1991)180-209.

Yin, P., *Data extracted from the US Bureau of Mines data*, February 19, (1998).

Yin, P.,*Data extracted from Wyoming Geological Survey*, August 11, (1997).

THE ROLE OF IN-SITU STRESS IN COALBED METHANE EXPLORATION

J. Enever* (1), D. Casey (2), M. Bocking (3)

(1) CSIRO Petroleum, Melbourne
(2) D.A. Casey and Associates, Sydney
(3) Pacific Power, Sydney

Five years and more of experience gained during coalbed methane exploration in Australia has pointed to the important role that the in-situ stress state has on coal seam permeability, and hence gas/water producability. Similar understanding is also emerging from other parts of the world. The aim of this contribution is to elaborate the context in which stress appears to influence coal seam permeability in Australia, and to investigate some of the mechanisms controlling the in-situ stress state in Australian coal basins. The paper, while generic, draws particularly on experience gained during exploration of the Glouscester Basin by Pacific Power.

Figure 1 summarises much of the available field data for Australian coals, representing the relationship between coal seam permeability and the corresponding minimum effective stress in the coals (minimum total stress - reservoir pressure). The relatively severe adverse impact of increasing effective stress on permeability is evident in Figure 1, leading to the obvious conclusion that successful coalbed methane exploration in Australia must revolve largely around locating regions of low effective stress, given that in the majority of cases, reasonable gas content can be assumed. Experience suggests that reservoir pressures in Australian coal seams are more often than not around "normal". Under these circumstances, low effective stress translates to low total stress. The challenge, therefore, becomes to identify geological environments producing low minimum total stress in coal.

Figure 2 summarises the results of a number of measurements of the minimum total stress component magnitude in Australian coal basins. Most of the data points lie between approximately 50% and 85% of the corresponding vertical overburden pressure based on the present depth of cover and a nominal overburden S.G. of 2.5. Below about 600 to 800 metres depth of cover, Figure 2 suggests that the minimum total stress in coal seams may tend toward the upper of these bounds, or even the equivalent overburden pressure. Minimum total stress magnitudes less than the equivalent overburden pressure suggests that the minimum stress component is oriented horizontally or sub-horizontally. The apparent tendency for the minimum total stress to increase relative to overburden pressure with depth may be due to an increase in horizontal stress magnitude related to higher coal seam stiffness and/or confinement at depth. Ultimately, the vertical stress component (overburden pressure) could become the minimum total stress component in deeper coal seams.

The range in minimum total stress magnitude above 600 to 800 metres in Figure 2 is very significant with regard to coalbed methane exploration, in that the range of total stress suggested, when converted to an equivalent range of effective stress by subtraction of the corresponding reservoir pressures (assuming generally normally pressured seams) represents a potentially very wide range in permeability (Figure 1). The reason for the apparent range in total stress magnitude above 600 to 800 metres is open to debate. The authors suggest three possible explanations:

- the impact of varying coal stiffness, compared to the stiffness of the bounding rock units, on the distribution of horizontal stress between coal and rock;
- the impact of seam thickness on the relative confinement offered to the coal by the bounding rocks, and hence the proportion of stress conducted through the coal compared to the rock;
- the impact of the regional and/or local geological environment on the magnitude of tectonic stress existing in a particular region, embracing both coal seam and rock units.

In a very general sense, the implications of the first two issues with respect to coalbed methane exploration are that soft coals and thick seams should be targeted, as these are more likely to be consistent with lower stress magnitudes, and hence higher permeabilities. The latter issue requires further consideration.

A large number of direct rock stress measurements have now been made in Australia's major coal basins (Sydney and Bowen Basins) both from the surface by hydraulic fracturing (allowing study of trends with depth) and from underground by overcoring (allowing study of the spatial distribution at the level of coal measures). Compilation of this information is revealing regional trends of tectonic stress that can be used as a starting point to highlight areas where relatively lower rock stress may exist. At another level of detail, it is possible to discern from the stress measurement data base the impact of specific geological structures on the local stress field.

The net stress environment existing at any particular location will be a product of the interaction of the underlying regional stress field and local perturbations deriving from specific geological features. In the context of coalbed methane exploration, the relative importance of regional stress patterns and the impact that specific geological structures have on the local stress field is still to be resolved.

Figure 3 illustrates an example of the impact of flexure associated with an anticline on the horizontal stress magnitude in rock and the minimum stress magnitude (presumably horizontal or sub-horizontal) in coal. The style of trend with depth of the stress data is symptomatic of flexural stresses superimposing themselves on a base level horizontal compression, with slippage occurring at various points in the sequence on sub-horizontal planes. Figure 3, along with a number of similar examples from the author's experience, suggests parallelism between the stress occurring in rock and coal with respect to the impact of geological processes on the stress field.

In general terms, examples such as Figure 3 imply a degree of continuity between the response of rock units and associated coal seams to the imposition of tectonic stresses and to the impact of specific geological features causing changes to the stress field.

The Glouscester Basin, situated inland from Taree on the NSW north coast, contains a number of coal seams, some relatively thick, in a tightly folded graben structure. The coals are tentatively considered to be contemporaneous with formation of the Hunter Valley coal measures, although currently completely isolated from them. Pacific Power have a current exploration program in the area to investigate coalbed methane prospectivity. This has involved the drilling of five fully cored holes to date, on a relatively close spacing, generally to depths around 500 metres (one hole has been drilled to approximately 900 metres). Each of the holes has been tested for coal seam permeability as well as the horizontal rock stress field (orientation and magnitude) and minimum coal stress magnitude.

Figure 4 summarises the information relating to the orientation of the horizontal stress field measured in rock. The generally good agreement between the orientation of the major horizontal secondary principal stress and the predominant structural trend revealed from seismic work is a reflection of the stress relief normal to the latter direction associated with the relatively severe flexure impacting the sequence (Figure 5). The resultant trend of rock stress magnitude with depth derived for the Glouscester Basin (Figure 6) compared to the trend exhibited through a nominally comparable sequence in the Hunter Valley (Figure 7) indicates a markedly different situation, with the influence of stress relief obvious in the case of the Glouscester Basin. In both areas, the measurements of rock stress were obtained over approximately the same depth window, above the apparently significant depth limit of 600-800 metres discussed above.

Figure 8 summarises the average minimum horizontal principal rock stress component magnitude and minimum coal stress magnitude, normalisal for depth, on a hole by hole basis for the Glouscester Basin data. Figure 8 suggests reasonably consistent behaviour between holes for the rock units. The overall average rock stress magnitude (approximately 80% of nominal overburden pressure) suggests an essentially relaxed basin environment, with the signature of any tectonic stress component that may have existed historically largely obliterated by the impact of flexure.

In the case of the coal stresses, Figure 8 indicates consistent average behaviour from hole to hole, except for PGSD2 which exhibits a markedly higher average stress magnitude. The reason for this is open to conjecture, however, Figure 9 does reveal the distribution of coal seam thickness in PGSD2 to be markedly different to the other holes (relatively greater incidence of thinner seams). Apart from PGSD2, the overall average coal stress magnitude (approximately 55% of nominal overburden pressure) is toward the lower bound of the range discussed above. Not surprisingly, the coal seam permeabilities measured in these holes were generally reasonably high. Figure 10 shows the results of well tests in the Glouscester holes in relation to the global Australian data base.

The major findings illustrated by the material presented here are:

• that the effective stress (total stress minus reservoir pressure) existing in a coal seam profoundly influences the net permeability of the coal seam;
• that the minimum total stress in Australian coal seams generally varies between approximately 50 and 85 percent of overburden pressure to depths of around 600-800 metres, after which it tends toward the upper of these bounds;
• that the range from 50 to 85 percent of overburden can be equivalent to several orders of magnitude difference in permeability, when the total stress range is translated to an equivalent effective stress range for normal reservoir pressure conditions;
• that the cause for the variation between 50 and 85 percent of overburden appears to be ·related to prevailing geological conditions, coal seam thickness and coal stiffness;
• that the total stress condition in a coal seam is often a reflection of the stress condition existing in the bounding rocks, which in turn can be related to the prevailing geological environment.

At the most general level, the above finings lead to the following key conclusions of importance when framing coalbed methane exploration strategies:

• exploration should generally focus on coal seams above about 600 to 800 metres depth, subject to adequate gas content and gas desorption characteristics;
• exploration should focus on relatively softer coal seams of relatively greater thickness;
• exploration should focus in areas where the geological environment has led to relief of any tectonic stress component, subject to gas having not consequently migrated out from the target zone.

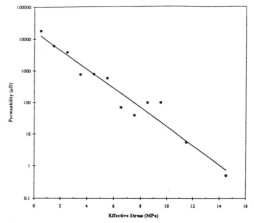

Figure 1:   Relationship between permeability and effective stress
             for Australian Coals (determined from field testing).

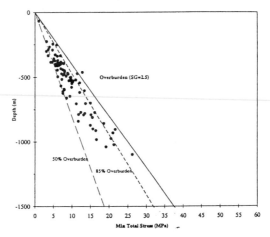

Figure 2:   Minimum total stress in a range of Australian coals
             versus depth.

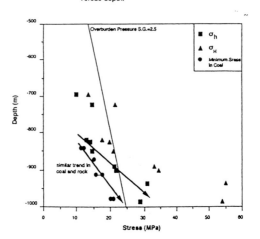

Figure 3:   Illustration of the impact of strata flexure on the
             horizontal stress field in a rock/coal sequence.

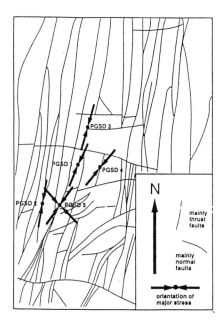

Figure 4:   Summary of the hole by hole average orientation of the
            major horizontal secondary principal stress in the
            Glouscester C.B.M. exploration area (the two vectors
            shown for PGSD5 reflect a bimodal orientation
            distribution not evident in other holes).

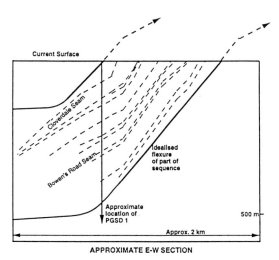

APPROXIMATE E-W SECTION

Figure 5:   Cross section showing relatively severe flexure of
            seams in exploration area.

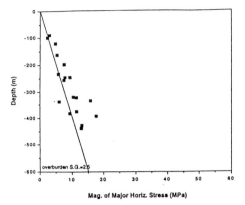

Figure 6:   Trend of major horizontal rock stress magnitude with
            depth from measurements in Glouscester Basin.

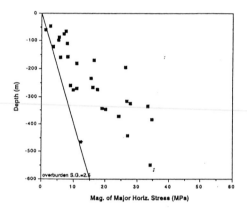

Figure 7:   Trend of major horizontal rock stress magnitude with
            depth through comparable window in Hunter Valley.

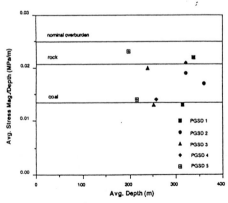

Figure 8:   Summary of average minimum horizontal rock stress
            and minimum coal stress on a hole by hole basis for
            exploration area.

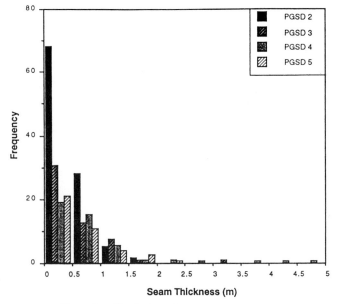

Figure 9: Distribution of coal seam thicknesses in the Glouscester exploration holes (PGSD2-5).

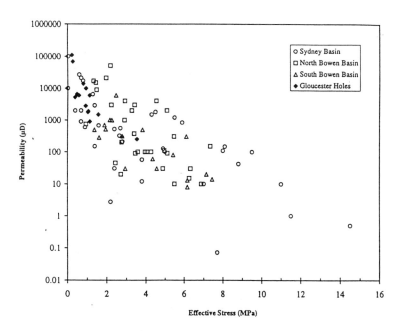

Figure 10: Results of well tests in Glouscester holes compared to other comparable Australian data.

MECHANICAL AND THERMAL CONTROL OF CLEATING AND SHEARING IN COAL: EXAMPLES FROM THE ALABAMA COALBED METHANE FIELDS, USA

JACK C. PASHIN AND RICHARD E. CARROLL
Geological Survey of Alabama
P. O. Box O, Tuscaloosa, AL 35486-9780, USA

JOSEPH R. HATCH AND MARTIN B. GOLDHABER
U.S. Geological Survey
Denver Federal Center, Denver CO 80225

1. Introduction

Natural fractures provide most of the interconnected macroporosity in coal. Therefore, understanding the characteristics of these fractures and the associated mechanisms of formation is essential for effective coalbed methane exploration and field management. Natural fractures in coal can be divided into two general types: cleat and shear structures. Cleat has been studied for more than a century, yet the mechanisms of cleat formation remain poorly understood (see reviews by Close, 1993; Laubach *et al.*, 1998). An important aspect of cleating is that systematic fracturing of coal is takes place in concert with devolatization and concomitant shrinkage of the coal matrix during thermal maturation (Ammosov and Eremin, 1960). Coal, furthermore, is a mechanically weak rock type that is subject to bedding-plane shear between more competent beds like shale, sandstone, and limestone. Yet, the significance of shear structures in coal has only begun to attract scientific interest (Hathaway and Gayer, 1996; Pashin, 1998).

Multidisciplinary studies of sedimentary basins can help increase our understanding of the mechanisms of cleating and shearing in coal by linking the morphology, distribution, and directionality of natural fractures with other geologic variables, including stratigraphy, structure, coal rank, and mineralization. This paper applies this type of approach to coalbed methane reservoirs in Alabama, where intensive coalbed methane development has been underway since 1980 (fig. 1). In this area, coalbed methane is produced from thin (<3 m; 10 ft) coal beds in the Lower Pennsylvanian (Westphalian A) Pottsville Formation of the Black Warrior foreland basin and the Cahaba synclinorium of the southern Appalachian thrust belt. The long production history and complex tectonic framework of this region provides a rich setting for comparing fracture patterns with proven production trends.

305

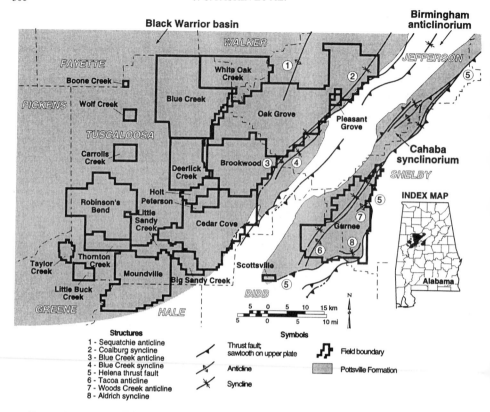

Figure 1. Location of the Alabama coalbed methane fields in the Black Warrior basin and the Cahaba synclinorium.

This paper synthesizes the results of more than a decade of research on coalbed methane reservoirs in Alabama (e.g., Ward *et al.*, 1984; Pashin and Hinkle, 1997) and is intended to highlight the numerous mechanical and thermal variables that contributed to the development of cleat and shear structures in these strata. The objective of this synthesis is not to provide a template that can be readily applied to other areas with significant coalbed methane potential. Rather, this synthesis is designed to use the Black Warrior basin and Cahaba synclinorium as an example of how the stratigraphy, structure, and thermal evolution of sedimentary basins can result in unique fracture patterns that control the distribution and producibility of coalbed methane.

2. Methods

Characterization of the morphology, distribution, and mineralization of natural fractures in Alabama coal are based mainly on field and core observations, and measurements of directionality are from Ward *et al.* (1984), who measured more than

1,600 joint and cleat orientations in the Black Warrior basin. Few cleat data were collected in the Cahaba synclinorium because of limited outcrop and shearing. Measurements of cleat spacing are based solely on observations of primary cleats as recognized in outcrops and cores throughout the study area. Vector-mean azimuths of cleat orientation and spacing information were plotted on maps, and cleat systems were defined. Similar methods were used to analyze shear fractures, and polished blocks of coal were especially useful for identifying and classifying the fractures. Locations of key outcrops mentioned in the text are given in table 1.

Maps of coal rank in the Black Warrior basin and the Cahaba synclinorium are based on mean-maximum vitrinite reflectance (R_o) and volatile-matter content (dmmf); the maps have been updated from Winston (1990) and Pashin *et al.* (1995) to incorporate new information. The rank classification and correlation of rank parameters used in this report is that of Stach *et al.* (1982). General cross sections were constructed and combined stratigraphic data from geophysical well logs with all information on coal rank available near the lines of cross section to determine the timing of coalification relative to structural deformation. Final interpretations of the mechanical and thermal causes of fracturing in Alabama coal are based on comparison of fracture patterns with stratigraphic, structural, and thermal maturation patterns.

TABLE 1. Register of localities mentioned in this report.

Locality	Name	7.5-minute quadrangle	Location
1	Flat Top Mine	Dora	SW1/4SW1/4, sec. 19, T. 16 S., R. 4 W.
2	Jim Walter Resources No. 5 Mine	Brookwood	sec. 20, T. 20 S., R. 7 W.
3	Chestnut Ridge	Greenwood	SW1/4SW1/4, sec. 18, T. 20 S., R. 3 W.
4	Jagger Mine	McCalla	SW1/4, sec. 28, T. 19 S., R. 5 W.
5	Area 6 Mine	Abernant	NW1/4, sec. 10, T. 20 S., R. 6 W.
6	Kellerman Mine	Brookwood	NW1/4, sec. 28, T. 20 S., R. 7 W.
7	Hoover	Helena	NW1/4NE1/4, sec. 24, T. 19 S., R. 3 W.
8	Warrior	Warrior	NE1/4NE1/4NW1/4, sec. 23, T. 14 S. R. 3 W.

3. Regional Setting

Twenty one coalbed methane fields have been established in Alabama; 19 are in the eastern part of the Black Warrior basin, and 2 are in the Cahaba synclinorium (fig. 1). Cumulative production of coalbed methane now exceeds 860 Bcf, and more than 2,800 wells are currently producing. Vertical coalbed methane wells have been drilled in all 20 fields, and mine-related degasification operations that include horizontal and gob wells are in Oak Grove and Brookwood fields. Coalbed methane resources have been estimated between 10 and 20 Tcf (Hewitt, 1984; McFall *et al.*, 1986; Telle and Thompson, 1987), and reserves have been estimated at slightly more than 2.5 Tcf (Rice, 1995).

3.1. STRUCTURE

The Black Warrior basin is a triangular foreland basin that formed adjacent to the juncture of the Appalachian and Ouachita orogenic belts during the late Paleozoic Alleghanian orogeny (Thomas, 1985a, 1988). The basin can be characterized as a southwest dipping homocline that is broken by numerous northwest-striking normal faults (fig. 2). Normal faults in the eastern part of the basin generally have surface traces shorter than 10 km and have vertical separations less than 100 m; they have been interpreted as thin-skinned structures that are detached near the base of the Pottsville Formation (Wang et al., 1993; Pashin et al., 1995). Farther west, however, major faults have surface traces longer than 40 km, have vertical separations greater than 300 m, and offset crystalline basement (Thomas, 1988). Alleghanian folds, including the Sequatchie and Blue Creek anticlines (fig. 1), are superimposed on the southeast margin of the homocline and are detached no deeper than the Cambrian shale overlying basement (Rodgers, 1950; Thomas, 1985b).

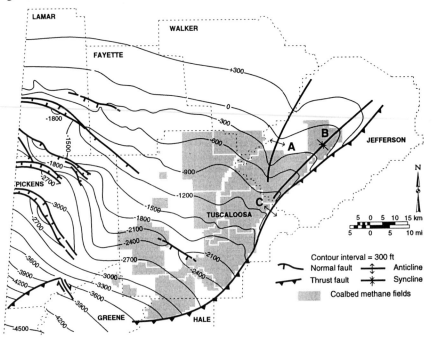

Figure 2. Generalized structural contour map of the top of the Mary Lee coal zone in the Black Warrior basin (after Pashin *et al.*, 1991).

The Cahaba synclinorium is the southernmost structural coal basin of the Appalachian thrust belt and lies approximately 10 km southeast of the Black Warrior basin (fig. 1). Strata in the Cahaba synclinorium generally strike N. 40° E. and generally dip less than 20° SE, away from a major frontal thrust ramp within the Birmingham anticlinorium (Thomas, 1991). Structure where coalbed methane is produced in the southwestern part of the synclinorium is dominated by broad, open folds, such as the Tacoa and Woods Creek anticlines, that have wavelengths greater than 3 km and have vertical relief less than 0.75 km. These folds were apparently developed above shallow detachments within the Mississippian-Pennsylvanian synorogenic stratigraphy (Pashin et al., 1995; Pashin and Groshong, in press).

3.2. STRATIGRAPHY

The Pottsville Formation in the Black Warrior basin is composed principally of shale, sandstone, and coal and is locally thicker than 2,000 m. Economic coal and coalbed methane resources are mainly in the upper part of the Pottsville Formation, which contains numerous coarsening-upward, flooding-surface-bound cycles dominated by fluvial-deltaic deposits (Pashin, 1994a, b, 1998). Coal beds in the Black Warrior basin are bright-banded, typically thinner than 3 m, and have an average thickness of less than 0.3 m; many wells penetrate more than 40 coal beds (Pashin et al., 1991). The Pottsville Formation in the Cahaba synclinorium is locally thicker than 2,500 m and generally resembles that in the Black Warrior basin. However, the frequency of marine strata decreases upward in the Cahaba section (Pashin et al., 1995). The Pottsville Formation is overlain with pronounced angular unconformity by unconsolidated sand and gravel of the Upper Cretaceous Tuscaloosa Group in the western two thirds of the Black Warrior basin and at the southwest end of the Cahaba synclinorium.

3.3. RANK PATTERNS

Coal in the Black Warrior basin and Cahaba synclinorium ranges in rank from high-volatile C to low-volatile bituminous (figs. 3, 4), and woody material in unconsolidated Cretaceous sediment is lignitic (e.g., Semmes, 1929; Carroll et al., 1995). High-volatile C bituminous coal in the Mary Lee coal zone is restricted to the north-central part of the Black Warrior basin in northern Walker County and in an elongate area extending southward from western Fayette County into northeastern Pickens County (fig. 3). High-volatile B bituminous coal is present between the 0.7- and 0.8-percent vitrinite-reflectance contours, and most coal in the Black Warrior basin of Alabama is of high-volatile A bituminous rank. Some medium-volatile bituminous coal is in the extreme southwestern part of the map area. A larger, elliptical area of medium- and low-volatile bituminous coal extends along the southeastern margin of the basin in eastern Tuscaloosa and western Jefferson counties. This area centered in Oak Grove and Brookwood fields and is a major source of metallurgical coal. A generalized cross section shows, however, that coal only reaches a maximum rank of medium-volatile bituminous at the surface (fig. 5).

Comparison with the structural contour map reveals little relationship between the regional structure and the rank pattern (figs. 2, 3). In all but the southwestern part of the basin in Alabama, rank contours cut squarely across structure contours. This relationship is especially apparent in cross sections through the elliptical area of medium- and low-volatile bituminous coal (fig. 5). Here, lines of equal rank cut across bedding in all areas, including the forelimb of the Blue Creek anticline.

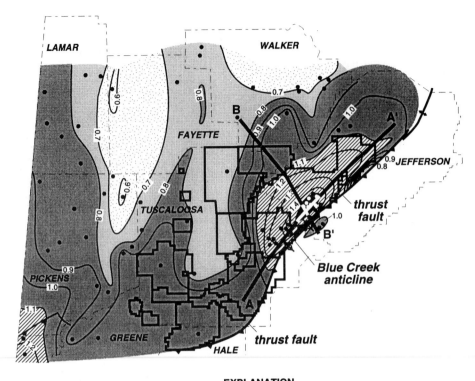

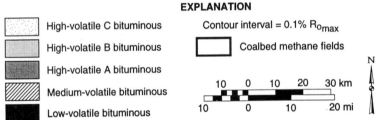

Figure 3. Coal rank of the Mary Lee coal zone in the Black Warrior basin based on vitrinite reflectance (after Winston, 1990).

A map of coal rank at the surface of the Cahaba synclinorium is based on a combination of volatile-matter and vitrinite-reflectance data (Pashin *et al.*, 1995) (fig. 4). Coal at the surface ranges in rank from high-volatile C bituminous in the western part of the coal field to medium-volatile bituminous in a small area in the center of the map. Within individual beds, however, rank in the Cahaba field generally increases toward the southeast (Levine and Telle, 1991). As in most of the Black Warrior basin, coal rank in the Cahaba coal field increases with depth and, in the structurally deepest part of the coal field, the Gould coal zone reaches low-volatile bituminous rank (Levine and Telle, 1991). In general, the rank pattern at the surface of the Cahaba synclinorium

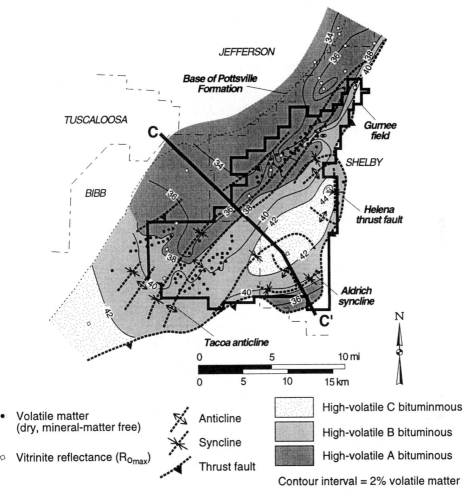

Figure 4. Coal rank at the surface of the Cahaba synclinorium based on volatile matter content (dmmf) and vitrinite reflectance (after Pashin *et al.*, 1995).

is a reflection of the regional structural pattern. A cross section through the synclinorium confirms that lines of equal rank are folded but not as strongly as bedding in the Aldrich syncline (fig. 5).

4. Cleat Systems

Cleat is simply a miner's term for closely spaced joints in coal. Joints are among the most abundant structures in the earth's crust (Pollard and Aydin, 1988) and can be

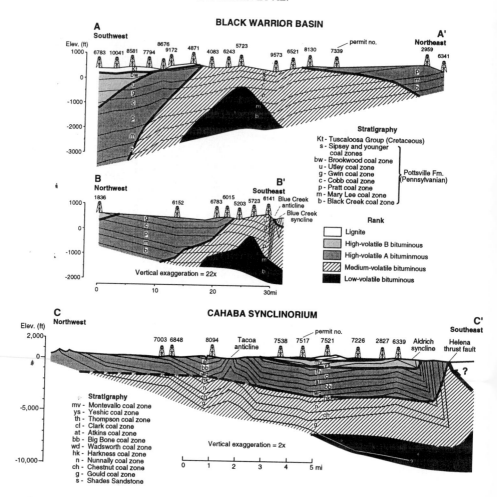

Figure 5. Generalized cross sections showing the relationship of rank to structure in the Black Warrior basin (A-A', B-B') and Cahaba synclinorium (C-C'). Note that isorank lines are oblique to bedding.

classified as nonsystematic and systematic (Hodgson, 1961; Groshong, 1988). Nonsystematic joints have no preferred orientation, whereas systematic joints form subparallel sets. Systematic joints form regionally extensive fracture systems in sedimentary basins. Cleat has all the characteristics of orthogonal joint systems. Orthogonal joint systems characteristically form perpendicular to bedding and contain two types of fractures: systematic joints and cross joints. In coal, systematic joints are called face cleat, and cross joints are called butt cleat. Systematic joints can have planar and rough surfaces and have greater continuity than cross joints. Cross joints more commonly have rough surfaces and tend to be discontinuous; cross joints commonly terminate at systematic joints. Cross joints commonly have a curvilinear plan, especially where they originate between systematic joints. However, curving cross joints typically intersect systematic joints at nearly right angles.

Coal in the Black Warrior basin typically contains orthogonal cleat systems, but the characteristics of the cleat systems vary markedly. In most areas cleat systems are well-developed and can be characterized using a hierarchical system similar to that of Laubach and Tremain (1991). Primary face cleat surfaces extend laterally for more than 3 m, are spaced closer than 1 cm, and traverse the full thickness of each bed or bench of coal; they can even cut across non-coal partings. Secondary face cleats that penetrate only part of the coal are numerous, and tertiary cleats that are restricted to vitrain bands are abundant, as is typical of bright-banded coal (MacRae and Lawson, 1954). Face cleat surfaces are typically planar, but banding of the coal imparts some roughness to the surfaces. Cleat apertures widen slightly along vitrain bands, and faint vertical ribs and hackle plumes are common. In some samples, tiny plumose structures radiate from pinpoint irregularities on cleat surfaces. Butt cleat surfaces can have highly irregular surface morphologies and can similarly be divided into primary, secondary, and tertiary classes on the basis of vertical continuity.

Cleat systems are poorly developed in parts of northwestern Walker and eastern Fayette counties. In this area, primary cleats are spaced on the order of 10 cm, secondary and tertiary cleats are extremely few, and face and butt cleats are difficult to distinguish. Coal in this area also contains a high proportion of curving, nonsystematic joints that are morphologically similar to the primary cleats. Cleat spacing tends to decrease southeastward where the Pottsville Formation is exposed at the surface (McFall et al., 1986) (fig. 6). Cleats tend to be most closely spaced in eastern Tuscaloosa and western Jefferson counties; spacing less than 1 cm is common in this area.

4.1. ORIENTATION

Face cleat maintains uniform strike over large regions of many sedimentary basins (e.g., Nickelson and Hough, 1967; Laubach and Tremain, 1991; Kulander and Dean, 1993), and this is true where the Pottsville Formation is exposed in the eastern part of the Black Warrior basin (Ward et al., 1984). Ward et al. (1984) identified two cleat systems in the Black Warrior basin (fig. 6). A regional cleat system (System 1) with face cleats striking with a vector-mean azimuth of N. 62° E. extends throughout the Pottsville outcrop area. A local cleat system (System 2) with face cleats striking with a vector-mean azimuth of N. 36° W., by contrast, has been recognized only along the southeast basin margin.

Ward et al. (1984) also recognized regional and local joint systems that correspond partly to the cleat systems. Like the regional cleat system, a regional joint system (System 1) extends throughout the Pottsville outcrop area; however, the vector-mean azimuth of strike of the systematic joints is N. 47° E, or 15° NW of that of the joints. The localized joint system (System 2) is also developed along the southeast margin of

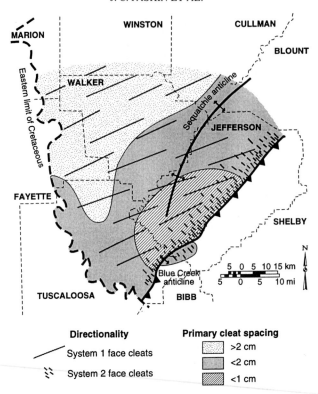

Directionality

—— System 1 face cleats

ʻʻ System 2 face cleats

Primary cleat spacing

[ ] >2 cm

[ ] <2 cm

[///] <1 cm

Figure 6. Cleat systems and primary cleat spacing in the eastern Black Warrior basin.

the Black Warrior basin but extends just northwest of the axial trace of the Sequatchie anticline. Systematic joints strike with a vector-mean azimuth of N. 64° W., or 28° SW of that of the localized cleat system.

Coal is cleated in the Cahaba synclinorium, but directional data are difficult to obtain because of limited or weathered exposure and predominance of sheared coal. Shear structures make primary cleats difficult to distinguish, although secondary and tertiary cleats are abundant.

4.2. MINERALIZATION

Cleat can be filled with a variety of authigenic minerals, including, calcite, pyrite, kaolinite, silica, and lead-zinc minerals (Hatch *et al.*, 1976; Spears, 1987; Faraj *et al.*, 1996). Pyrite, calcite, and kaolinite are the most important cleat filling minerals in the Black Warrior basin, and weathered exposures hamper identification of cleat-filling minerals in the Cahaba synclinorium. Pyrite characteristically occurs as partial cleat fills of coalesced framboids and as localized fine- to medium-crystalline euhedra; in few places does pyrite cover more than 5 percent of cleat surfaces. Coarsely crystalline nodules as large as pebbles were observed in coaly fault gouges. Framboidal pyrite is also abundant in fusain bands. A recent investigation of sulfide mineralization in the Black Warrior basin demonstrated that early authigenic pyrite in fusain bands and cleat apertures has been overgrown by arsenic-rich pyrite (Goldhaber *et al.*, 1997).

Calcite generally forms fine- to coarse-crystalline fracture fills that typically cover 25 to 75 percent of cleat surfaces. Most commonly, calcite is found in coal adjacent to faults. In a fresh highwall exposure of the Pratt coal zone at the Flat Top Mine (locality 1), for example, pervasive calcite cleat fill was observed in the footwall block of a normal fault. Calcite cleat fills become less pervasive with distance from the fault plane, and no cleat-filling minerals were observed more than 50 m from the fault, save for a few isolated patches of pyrite. Calcite cleat fills are especially common in cores from Moundville Field, which is the southernmost coalbed methane field and is near where Cambrian-Ordovician carbonate rocks are in thrust juxtaposition with the Pottsville Formation.

Kaolinite has been observed only locally as a cleat-filling mineral in the Black Warrior basin. In the Jim Walter No. 5 Mine in Brookwood Field (locality 2), where the Blue Creek bed of the Mary Lee coal zone is actively being mined and degassed, kaolinite was observed as a pervasive cleat-filling mineral. The mineralization extends for no more than 0.4 m vertically and 3 m laterally in the coal face, yet kaolinite fills cleat apertures that are in places wider than 2 mm.

5. Mechanical and Thermal Aspects of Cleating

Systematic joint and cleat systems have many mechanical properties in common. For example, both are opening-mode fractures that form perpendicular to the least principal horizontal stress direction and parallel to the greatest principal horizontal stress direction (Nickelsen and Hough, 1967; Engelder, 1985). Curving of butt cleats to intersect face cleats at right angles reflects interaction of growing cross joints with preexisting free surfaces, specifically systematic joints (e.g., Lachenbruch, 1962). Hackling of fracture surfaces confirms that the cleats are tension fractures, and ribbing of fracture surfaces indicates that cleat growth included frequent episodes of propagation and arrest. According to Lacazette and Engelder (1992), incremental fracture propagation under conditions of constant stress is favored if the fracturing fluid is a gas.

In the Black Warrior basin, consistent orientation of System 1 (fig. 6) cleats indicates that the least principal stress was oriented north-northwest during fracturing. However, the precise source of the regional stress field is unclear because the face cleats are oblique to all known structures and regional dip (figs. 5, 6). The different orientation of the regional joint and cleat systems indicates that the orientation of the least compressive stress in coal was rotated relative to that in the siliciclastic rocks. System 2 face cleats strike perpendicular to the axial traces of structures in the Appalachian thrust belt (fig. 6), suggesting a close relationship to folding and thrusting.

Tectonic stresses may help orient systematic fractures in coal, but internal forces related to shrinkage by devolatization during thermal maturation may be a more fundamental cause of cleat formation. Cleats are absent or poorly developed in lignite and appear to reach maximum abundance at an approximate rank of low-volatile bituminous (Ammosov and Eremin, 1960; Ting, 1977) (fig. 7). A lack of systematic fractures in some anthracite implies that cleats anneal by repolymerization at high rank (Levine, 1993), yet some extremely high rank coal contains systematic fractures (Law, 1993).

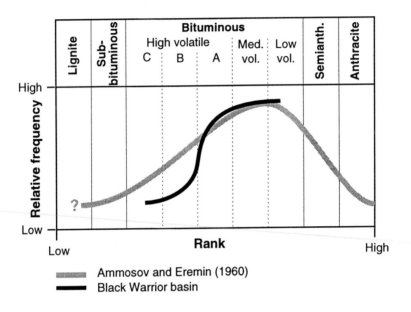

Figure 7. Relationship of cleat frequency to rank.

A significant relationship between rank and cleating is apparent in parts of the Black Warrior basin (figs. 6, 7). Most high-volatile B and C bituminous coal in the northern and western parts of the Pottsville outcrop area contains primary cleats spaced as widely as 10 cm, whereas all coal of high-volatile A bituminous and higher rank contains primary cleats spaced closer than 2 cm. Moerover, coal with primary cleats spaced closer than 1 cm is concentrated in the elliptical area where rank is medium volatile bituminous or higher. Thermogenic methane generation begins at a rank of high-volatile C bituminous and reaches a maximum rate at a rank of high volatile bituminous (Jüntgen and Karweil, 1966; Jüntgen and Klein, 1975). Therefore, one explanation of the jump in cleat frequency could be that a high rate of gas generation increased pore pressure to a critical level, thereby fracturing the coal.

Lopatin models based on wells in the Black Warrior basin indicate that the major maturation, and hence cleating, of Pottsville coal took place near maximum burial at the end of the Alleghanian orogeny, which was approximately 260 Ma (Hines, 1988; Carroll *et al.*, 1995). Regional variation of the geothermal gradient helps explain why isorank lines are oblique to structural contours and bedding in much of the basin (Levine and Telle, 1991) (figs. 2, 3, 5). Isorank lines cut sharply across the forelimb of the Blue Creek anticline along the southeast margin of the basin (fig. 5), indicating a strong post-kinematic component to coalification in the elliptical area of high-rank coal. Indeed, raising the rank above high-volatile A bituminous significantly enhanced the frequency of primary cleats. Mineralization of cleat systems in the Black Warrior basin was clearly a late event postdating regional fracturing of the coal, and the high arsenic content of some pyrite suggests a link with metamorphic fluids in the Appalachian orogen (Goldhaber et al., 1997).

6. Shear Structures

Shearing can give rise to complex and varied structural fabrics in coal that differ markedly from orthogonal cleat systems. In the Alabama coalbed methane fields, the most common structures in sheared coal are inclined fractures, faults, and folds (Pashin et al., 1995; Pashin, 1998) (fig. 8). These structures are commonly restricted to coal beds, although in some places other mechanically weak lithologies like underclay may be involved in the deformation. Inclined fractures and faults are most commonly developed in beds dipping 10° or steeper; they generally strike with bedding and dip approximately 45 to 80° relative to bedding; fracture spacing is similar to cleat spacing. In plan view, the fractures and faults generally have curvilinear surface traces and are convex updip. Polished slabs establish minor normal or reverse offset of coal banding along the fractures (fig. 9). Most fractures and faults dip in the direction of bedding, and fractures dipping opposite to bedding can be observed at some locations. In the Cahaba synclinorium at Chestnut Ridge (locality 3), for example, small-scale conjugate normal faults are developed in one coal bed.

Minor structures include shear cones and strike-slip faults, both of which were observed in the Blue Creek syncline. Shear cones resemble cone-in-cone structures and were first identified in coal by Price and Shaub (1963). In the Black Warrior basin, shear cones were recognized only at the Jagger Mine (locality 4), where they are in a bench of coal containing numerous small-scale thrust faults. Friability of the coal prevents determining if the shear cones are nested as in true cone-in-cone structures. Minor strike-slip faults with slickensides were observed at the Area 6 Mine (locality 5) near the southwest end of the Blue Creek syncline. These structures are vertical and form conjugate sets striking approximately 45° from bedding and the System 2 cleats. Importantly the faults cut across cleats of System 2.

6.1. RELATIONSHIP TO LARGE-SCALE STRUCTURES

Sheared coal is most common adjacent to normal faults and in the limbs of compressional folds. Inclined fractures and normal faults were observed in hanging-wall drag folds adjacent to some normal faults in the Black Warrior basin. These fractures generally dip steeper than 60° relative to bedding, are present only within 10 m of the

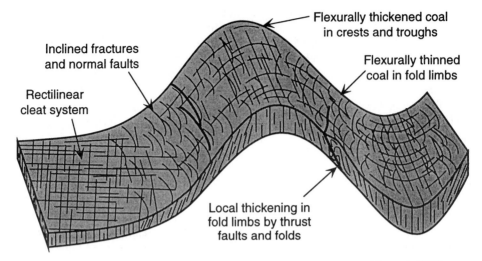

Figure 8. Flexural slip model of structures in folded coal beds (after Pashin *et al.*, 1995).

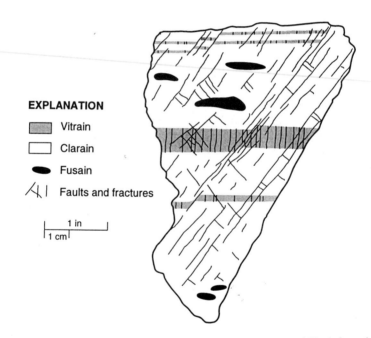

Figure 9. Sketch of small-scale normal faults and shear fractures in a coal block from the Jagger Mine (locality 4) (after Pashin *et al.*, 1991). Note preservation of tertiary cleats within vitrain bands.

major faults and appear to be little more than inclined cleats. A lateral transition from System 1 cleats to inclined fractures is especially well shown in a hanging-wall drag fold exposed at the Kellerman Mine (locality 6) in Brookwood Field (fig. 10).

Coal within the Alleghanian folds contains a diverse assemblage of tectonic structures that reflect the relative intensity of shearing. Inclined fractures and faults are abundant and can be of compressional or extensional origin. The inclined fractures and faults commonly cut across cleats and are thus of a different origin than the inclined cleats adjacent to normal faults. Small-scale folds and thrust faults are common in the coal and can be isolated structures or can be so closely spaced as to constitute the principal deformational fabric of the rock (fig. 11). Indeed, slickensided decollements and ramps are in parts of the Cahaba synclinorium spaced closer than 2 mm, forming an intricate lacework that makes the coal extremely friable. Thin decollement zones are present within the Blue Creek coal bed in the Jim Walter Resources No. 5 Mine (locality 2), which is located in southern Brookwood Field near the Alleghanian thrust front. Inclined fractures and faults of extensional and compressional origin are common within 20 cm of the decollements.

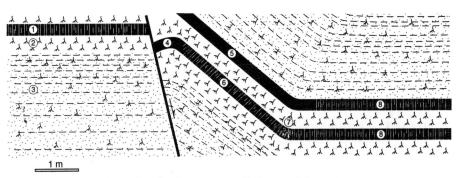

1 m

1 - Cleated coal, no mineralization	5 - Inaccessible
2 - Underclay	6 - Inclined cleats, no mineralization
3 - Interbedded shale and sandstone	7 - Pulverized coal, pyrite nodules
4 - Reverse drag structure, inaccessible	8 - Cleated coal, some pyrite cleat fills

Figure 10. Sketch of part of a highwall at the Kellerman Mine (locality 6) showing inclined cleats in a hanging-wall drag fold.

Compressional and extensional structures are irregularly distributed in the coal beds. An example of this is at the Jagger Mine where inclined fractures of extensional origin are present between thrust faults and folded zones. Within fold limbs, a single coal bed may be dominated by extensional structures at one location and by compressional structures at another. Multiple levels of deformation are also common within coal beds. Along U.S. Highway 31 at Hoover (locality 7), for example, an intensely folded interval separates benches of well-cleated coal (fig. 11).

Complex folds and thrust duplexes have also been observed within coal beds at several locations in the Appalachian thrust belt (figs. 11, 12). Although the coal is

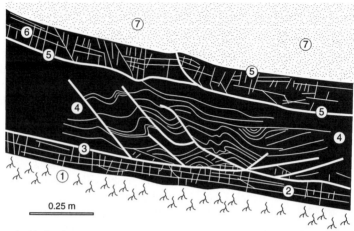

1 - Underclay   4 - Sheared coal with folds and thrust faults
2 - Well-cleated coal   5 - Roof thrust
3 - Floor thrust   6 - Cleated coal with minor thrusts
                             7 - Sandstone

Figure 11. Sketch of intensely folded and faulted coal between two thin benches of cleated coal at Hoover (locality 7).

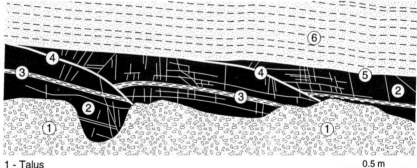

1 - Talus
2 - Coal with cleats, shear fractures, and bedding-parallel partings
3 - Shale parting
4 - Thrust faults
5 - Roof thrust
6 - Interbedded shale and sandstone

Figure 12. Sketch of passive-roof duplex in the Black Creek coal zone in the backlimb of the Sequatchie anticline at Warrior (locality 8).

folded at a macroscopic scale, careful examination reveals that the banding is not deformed and that deformation is accommodated entirely by small-scale faults and fractures. The best exposure of a thrust duplex is at Warrior (locality 8) in the backlimb of the Sequatchie anticline (fig. 12). Here, broad thrust horses with only 2 cm of displacement are exposed in well-cleated coal, and a roof thrust is present at the top of the coal bed. At other locations, substantial displacement within duplexes has resulted in significant thickening of coal. At the other end of the spectrum, cleavage duplexes were identified in some abandoned highwalls in the Blue Creek syncline. Cleavage duplexes were first described by Hathaway and Gayer (1996) for coal in which thrust-bound horses are present, but displacement is miniscule.

6.2. MINERALIZATION

Authigenic mineralization of shear structures was recognized at three mine exposures. At the Kellerman Mine (locality 6) (fig. 10), coal with inclined cleats is largely nonmineralized. However, pyrite fracture fills and nodules were observed in the hinge at the base of the hanging-wall drag fold, and pyrite cleat fills are common in more distal parts of the hanging wall. At the Area 6 Mine (locality 5), pyrite was recognized on the surfaces of the strike-slip faults. In the Jim Walter Resources No. 5 Mine, sulfate weathers preferentially out of the decollement zone, and kaolinite fills some cleats and inclined fractures above this zone. In addition to authigenic minerals, clay smear was seen along slickensided surfaces in some intensely deformed coal in the Cahaba synclinorium.

7. Mechanical and Thermal Aspects of Shearing in Coal

Shearing of coal between more competent beds reflects bedding-plane slip associated with regional folding and faulting. Among the most important effects of shearing are changes of coal thickness by extension and compression. In the case of flexural slip folding, incompetent beds like coal are redistributed by thinning in fold limbs and by thickening in crests and troughs (fig. 8). Flexural slip models apply well to thick, tightly folded coal beds in the Cordilleran thrust belt of western Canada (Bustin, 1982; Langenberg et al., 1992). Although flexural slip may operate at a large scale in the Alleghanian folds, thin coal, large fold wavelengths, and heterogeneously distributed extensional and compressional strain within individual fold limbs indicates that more local factors may be effective. One explanation of heterogeneous strain distributions is related to boundary conditions associated with changing bed geometry. Indeed, coal beds in the Pottsville Formation of Alabama pinch, swell, split, and undulate profusely (Ferm and Weisenfluh, 1989; Pashin, 1994c). Therefore, irregular contacts between coal and more competent beds may act as restraining and releasing fault bends (fig. 13).

Inclined cleats in drag folds adjacent to normal faults (fig. 10) demonstrate that cleating was affected by shear stresses associated with normal faulting in the Black Warrior basin. However, it is unclear whether the stress field responsible for the inclined cleats was contemporaneous with fault movement or was a residual field that remained after fault movement. Considering that cleats and shear structures coexist in more intensely deformed coal (figs. 9, 11), it is clear that normal and shear stresses are both effective fracturing mechanisms in folded regions. However, cross-cutting relationships between cleat and shear structures are indeterminate.

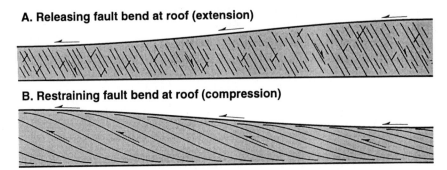

Figure 13. Possible influence of irregular contacts of coal with more competent rock types on the development of extensional and compressional structures by bedding-plane slip.

Development of shear cones at the Jagger Mine (locality 4) indicates that point stress was a factor during shearing and may have been the product of heterogeneous deformation of the coal at this locality. Strike-slip faults at the Area 6 Mine (locality 5) form a bisectrix between the strike of cleats and bedding and thus may reflect a shear component related to the normal stresses responsible for the System 2 cleats. Cross-cutting relationships, however, indicate that the faults postdate the cleats. Development of decollements in the Jim Walter Resources no. 5 Mine (locality 2) suggests that bedding-plane slip associated with Alleghanian folding was transmitted more than 5 km into unfolded parts of the Black Warrior basin.

Nearly all deformation that is observable in coal at the macroscopic scale is brittle, yet shrinkage of the coal matrix during devolatization is a significant source of ductile strain at the microscopic scale. Shear stress is another source of ductile strain that can give rise to significant reflectance anisotropy in vitrinite (e.g., Levine and Davis, 1989). Shrinkage probably influences formation of inclined cleats in the same way as in any other cleat system, but the influence of shrinkage on the development of other macroscopic shear structures is less clear. Partial folding of isorank lines with strata in the Cahaba synclinorium (fig. 5) indicates a major synkinematic component to coalification, thus significant shrinkage occurred during regional folding.

Mineralization appears to be restricted mainly to extensional shear structures and strike-slip faults, demonstrating that these structures were significant sources of permeability at the time of mineralization. The lack of authigenic mineralization in compressional folds and thrust faults, conversely, indicates that they remained closed at the time of authigenesis. Indeed, clay smear along slickensided surfaces in the Cahaba synclinorium indicates that only mechanical modes of mineralization were effective.

8. Implications for Reservoir Development

Cleat systems provide the major flow conduits within coalbed methane reservoirs and are thus a primary source of reservoir heterogeneity that impacts fluid flow. Reservoir tests from the Rock Creek completion test site in Oak Grove Field indicate that flow is favored in the face cleat direction in well-cleated coal of the Pratt coal zone (Koenig, 1989). In the same well, however, flow is favored along System 1 joints in siliciclastic rocks of the the Black Creek coal zone. Therefore, flow in fractured siliciclastic rocks can affect drainage of coalbed methane reservoirs.

Little evidence exists regarding the effect of shear structure on the reservoir properties of coal. Results from Rock Creek indicate essentially isotropic flow parallel to bedding in the Blue Creek bed, which has been sheared pervasively by bedding-plane slip in the backlimb of the Sequatchie anticline (Koenig, 1989). Considering the variety of shear structures that occur in coal, a large range of flow patterns may develop in sheared coalbed methane reservoirs. For example, closely spaced inclined fractures and faults may give rise to anisotropic flow patterns similar to those in well-cleated coal.

Economic production of coalbed methane has been achieved only from coal of high volatile A bituminous rank and higher in the Black Warrior basin (Pashin and Hinkle, 1997). A robust cleat system and the presence of a major thermogenic gas resource that is augmented by late biogenic gas (Rice, 1993: Scott, 1993) are key factors that contribute to economic production in Alabama. Conversely, limited permeability related to poor cleat development and insufficient thermal maturity of the coal may account for the lack of success in parts of the basin where coal is lower in rank than high volatile A bituminous.

The effect of mineralization on the reservoir properties of coal in the Black Warrior basin appears to be variable. Pyrite and kaolinite have such patchy distributions that they do not appear to be a major concern for reservoir development. Calcite, on the other hand, is the most pervasive cleat-filling mineral and may limit hydrologic communication across faults. Indeed, production mapping indicates that some reservoirs are segmented by normal faults (Pashin *et al.*, 1995; Pashin and Groshong, in press).

9. Summary and Conclusions

Cleat and shear structures are the major sources of macroporosity in coal of the Black Warrior basin and Cahaba synclinorium. Coal in most flat-lying beds contains a well-developed orthogonal cleat system. Primary cleats extend through entire coal beds or benches, secondary cleats penetrate only part of the coal, and tertiary cleats are restricted to vitrain bands. Coal of high-volatile C and B bituminous rank can contain widely spaced primary cleats and a high proportion of nonsystematic fractures, whereas coal of high-volatile A and through low-volatile bituminous rank is universally well cleated. This relationship suggests that cleating was driven by matrix shrinkage during devolatization and that the onset of major thermal gas generation at high-volatile A bituminous rank contributed to systematic fracturing. Along the frontal structures of the southern Appalachians, a localized cleat system that strikes northwest is superimposed on a regional system that strikes northeast. Secondary calcite, pyrite, and kaolinite are common in southern Appalachian coal and can plug cleat, and calcite is especially common adjacent to normal faults. Well testing indicates that gas drainage is influenced by joint and cleat systems and that cones of depression in some beds are elongate in the face-cleat direction.

Reservoir properties of cleat systems have been the focus of intensive investigation, yet folded coal beds contain abundant shear fractures. At the macroscopic scale, brittle deformation predominates in sheared coal. Shear fractures in Alabama coalbed methane reservoirs have spacing similar to cleat, generally strike with bedding, and are inclined 30 to 60 degrees relative to bedding; pyrite mineralization has been observed at only one locality. Reverse and normal dislocation of coal banding indicates that these structures can form in compressional and extensional stress regimes. Sheared coal beds can be internally complex. For example, a single bed may contain multiple structural levels with normal cleat, inclined fractures, and imbricate coal slices, all separated by decollement zones. Shear structures in coal are inferred to have formed by bedding-plane slip between more competent beds, and compressional and extensional strain are distributed heterogeneously in the limbs of folds. One explanation of this heterogeneity is that irregular contacts of coal with more competent beds function as restraining and releasing fault bends. Rank data indicate a major synkinematic component to coalification in the Cahaba synclinorium, thus matrix shrinkage may have influenced shearing in some coal. Limited well testing has been performed in sheared coal, but tests in one bed indicate that intense shearing can give rise to isotropic flow.

10. Acknowledgments

Parts of this research were funded by the U.S. Geological Survey as part of the National Coal Resources Data System under agreement 1434-HO-97-AG-01701. Critical reviews by R. H. Groshong, Jr. and M. Mastalerz substantially improved the paper.

11. References

Ammosov, I.I., and Eremin, I.V. (1960) *Fracturing in coal* (English vers., 1963), Israel Program in Science Translation, Tel Aviv, 111 p.

Bustin, R.M. (1982) The effect of shearing on the quality of some coals in the southeastern Canadian Cordillera, *Canada Institute of Mining and Metallurgy Bulletin* **841**, 76-83.

Carroll, R.E., Pashin, J.C., and Kugler, R.L. (1995) Burial history and source-rock characteristics of Upper Devonian through Pennsylvanian strata, Black Warrior basin, Alabama, *Alabama Geological Survey Circular* **187**, 29 p.

Close, J.C. (1993) Natural fractures in coal, *American Association of Petroleum Geologists Studies in Geology* **38**, 119-132.

Engelder, J.T. (1985) Loading paths to joint propagation during a tectonic cycle: An example from the Appalachian Plateau, USA. *J. Structural Geology* **7**, 45-476.

Faraj, S.M., Fielding, C.R., and MacKinnon, I.D.R. (1996) Cleat mineralization of Upper Permian Baralaba/Rangal coal measures, Bowen Basin, Australia, *Geological Society of London Special Publication* **109**, 151-164.

Ferm, J.C., and Weisenfluh, G.A. (1989) Evolution of some depositional models in Late Carboniferous rocks of the Appalachian coal fields, *Int. J. Coal Geology* **12**, 259-292.

Goldhaber, M.B., Hatch, J.R., Pashin, J.C., Offield, T.W., and Finkelman, R.B. (1997) Anomalous arsenic and fluorine concentrations in Carboniferous coal, Black Warrior basin, Alabama: Evidence for fluid expulsion during Alleghanian thrusting? *Geological Society of America Abstracts with Programs* **29**, A-51.

Groshong, R.H., Jr. (1988) Low-temperature deformation mechanisms and their interpretation, *Geological Society of America Bulletin* **100**, 1329-1360.

Hatch, J.R., Gluskoter, H.J., and Lindahl, P.C. (1976) Sphalerite in coals from the Illinois basin, *Economic Geology* **71**, 613-624.

Hathaway, T.M., and Gayer, R.A. (1996) Thrust-related permeability in the South Wales coalfield, *Geological Society of London Special Publication* **109**, 121-132.

Hewitt, J.L. (1984) Geologic overview, coal, and coalbed methane resources of the Warrior basin—Alabama and Mississippi, *American Association of Petroleum Geologists Studies in Geology* **17**, 73-104.

Hines, R.A., Jr. (1988) *Carboniferous Evolution of the Black Warrior Foreland Basin, Alabama and Mississippi* (dissertation): University of Alabama, Tuscaloosa, 231 p.

Hodgson, R.A. (1961) Classification of structures on joint surfaces, *American J. Science* **259**, 493-502.

Jüntgen, H., and Karweil, J. (1966) Gasbildung und gasspeicherung in steinkohlenflozen, Tiel I und II, *Erdöl und Köhle Petrochemistrie* **19**, 251-258.

Jüntgen, H., and Klein, J. (1975) Entstehung von erdgas gos kohligen sedimenten, *Erdöl und Köhle Petrochemistrie* **1**, 52-69.

Koenig, R.A. (1989) Hydrologic characterization of coal seams for optimal dewatering and methane drainage, *Quarterly Review of Methane from Coal Seams Technology* **7**, 30-31.

Kulander, B.R., and Dean, S.L. (1993) Coal-cleat domains and domain boundaries in the Allegheny Plateau of West Virginia, *American Association of Petroleum Geologists Bulletin* **77**, 1374-1388.

Lacazette, A., and Engelder, T. (1992) Fluid-driven cyclic propagation of a joint in the Ithaca siltstone, Appalachian basin, New York, *in* Evans B. and Wong, T.F., eds., *Fault Mechanics and Transport Properties of Rocks*, Academic Press, London, p. 297-324.

Lachenbruch, A.H. (1962) Mechanics of thermal contraction cracks in ice-wedge polygons in permafrost, *Geological Society of America Special Paper* **70**, 69 p.

Langenberg, W., MacDonald, D., and Kalkreuth, W. (1992) Sedimentologic and tectonic controls on coal quality of a thick coastal plain coal in the Foothills of Alberta, Canada, *Geological Society of America Special Paper* **267**, 101-116.

Laubach, S.E., Marrett, R.A., Olson, J. E., and Scott, A.R. (1998) Characteristics and orgins of coal cleat: A review, *Int. J. Coal Geology* **35**, 175-207.

Laubach, S.E., and Tremain, C.M. (1991) Regional coal fracture patterns and coalbed methane development, *Proc. 32nd U.S. Symposium on Rock Mechanics*, Balkema, Rotterdam, p. 851-859.

Law, B.E. (1993) The relationship between coal rank and cleat spacing: Implications for the prediction of permeability in coal, *1993 Int. Coalbed Methane Symposium Proc.*, University of Alabama, Tuscaloosa **2**, 435-441.

Levine, J. R. (1993) Coalification: The evolution of coal as a source rock and reservoir rock for oil and gas, *American Association of Petroleum Geologists Studies in Geology* **38**, 39-77.

Levine, J.R., and Davis, A. (1989) The relationship of coal optical fabrics to Alleghanian tectonic deformation in the central Appalachian fold-and-thrust belt, *Geological Society of America Bulletin* **101**, 1333-1347.

Levine, J.R., and Telle, W.R. (1991) Coal rank patterns in the Cahaba coal field and surrounding areas, and their significance, *Alabama Geological Society 28th Annual Field Trip Guidebook*, p. 99-117.

Macrae, J.C., and Lawson, W. (1954) The incidence of cleat fracture in some Yorkshire coal seams, *Trans. Leeds Geological Association* **6**, 224-227.

McFall, K.S., Wicks, D.E., and Kuuskraa, V.A. (1986) A geological assessment of natural gas from coal seams in the Warrior basin, Alabama - topical report (September 1985-September 1986), Lewin and Associates, Inc., Washington, D. C., Gas Research Institute contract no. 5084-214-1066, 80 p.

Nickelsen, R.P., and Hough, V.D. (1967) Jointing in the Appalachian Plateau of Pennsylvania, *Geological Society of America Bulletin* **78**, 609-630.

Pashin, J.C. (1994a) Flexurally influenced eustatic cycles in the Pottsville Formation (Lower Pennsylvanian), Black Warrior basin, Alabama, *Society of Economic Paleontologists and Mineralogists Concepts in Sedimentology and Paleontology* **4**, 89-105.

Pashin, J.C. (1994b) Cycles and stacking patterns in Carboniferous rocks of the Black Warrior foreland basin, *Gulf Coast Association of Geological Societies Trans.* **44**, 555-563.

Pashin, J.C. (1994c) Coal-body geometry and synsedimentary detachment folding in Oak Grove coalbed-methane field, Black Warrior basin, Alabama, *American Association of Petroleum Geologists Bulletin* **78**, 960-980.

Pashin, J.C. (1998) Stratigraphy and structure of coalbed methane reservoirs in the United States: An overview, *Int. J. Coal Geology* **35**, 207-238.

Pashin, J.C., Carroll, R.E., Barnett, R.L., and Beg, M.A. (1995) Geology and coal resources of the Cahaba coal field, *Alabama Geological Survey Bulletin* **163**, 49 p.

Pashin, J.C., and Groshong, R.H., Jr. (in press) Structural control of coalbed methane production in Alabama, *International Journal of Coal Geology*.

Pashin, J.C., Groshong, R.H., Jr., and Wang, S. (1995) Thin-skinned structures influence gas production in Alabama coalbed methane fields, *InterGas '95 Proceedings*, University of Alabama, Tuscaloosa, p. 39-52.

Pashin, J.C., and Hinkle, F. (1997) Coalbed methane in Alabama, *Alabama Geological Survey Circular* **192**, 71 p.

Pashin, J.C., Ward, W.E., II, Winston, R.B., Chandler, R.V., Bolin, D.E., Richter, K.E., Osborne, W.E., and Sarnecki, J.C. (1991) Regional analysis of the Black Creek-Cobb coalbed-methane target interval, Black Warrior basin, Alabama, *Alabama Geological Survey Bulletin* **145**, 127 p.

Pollard, D.D., and Aydin, A. (1988) Progress in understanding jointing over the last century, *Geological Society of America Bulletin* **100**, 1181-1204.

Price, P.H., and Shaub, B.M. (1963) Cone-in-cone in coal. *West Virginia Geologic and Economic Survey Report of Investigations* **22**, 9 p.

Rice, D.D. (1993) Composition and origins of coalbed gas, *American Association of Petroleum Geologists Studies in Geology* **38**, 159-184.

Rice, D.D. (1995) Geologic framework and description of coal-bed gas plays, *U.S. Geological Survey Digital Data Series* **DDS-30**, 103 p.

Rodgers, J. (1950) Mechanics of Appalachian folding as illustrated by the Sequatchie anticline, Tennessee and Alabama, *American Association of Petroleum Geologists Bulletin* **34**, 672-681.

Scott, A.R. (1993) Composition and origin of coalbed gases from selected basins in the United States, *1993 International Coalbed Methane Symposium Proceedings*, University of Alabama, Tuscaloosa, p. 207-222.

Semmes, D.R. (1929) Oil and gas in Alabama, *Alabama Geological Survey Special Report* **15**, 408 p.

Spears, D.A. (1987) Mineral matter in coals, with special reference to the Pennine Coalfields, *Geological Society of London Special Publication* **32**, 171-185.

Telle, W.R., and Thompson, D.A. (1987) Preliminary characterization of the coalbed methane potential of the Cahaba coal field, central Alabama, *1987 Coalbed Methane Symposium Proceedings*, University of Alabama, Tuscaloosa, p. 141-151.

Thomas, W.A. (1985a) The Appalachian-Ouachita connection: Paleozoic orogenic belt at the southern margin of North America, *Annual Review of Earth and Planetary Sciences* **13**, 175-199.

Thomas, W.A. (1985b) Northern Alabama sections, *University of Tennessee Department of Geological Sciences Studies in Geology* **12**, 54-60.

Thomas, W.A. (1988) The Black Warrior basin, *Geological Society of America, The Geology of North America* **D-2**, 471-492.

Thomas, W.A. (1991) Tectonic setting of the Cahaba synclinorium, *Alabama Geological Society 28th Annual Field Trip Guidebook*, p. 29-36.

Ting., F.T.C. (1977) Origin and spacing of cleats in coal beds, *J. Pressure Vessel Technology*, **November**, 624-626.

Wang, S., Groshong, R.H., Jr., and Pashin, J.C. (1993) Thin-skinned normal faults in Deerlick Creek coalbed-methane field, Black Warrior basin, Alabama, *Alabama Geological Society 30th Annual Field Trip Guidebook*, p. 69-78.

Ward, W.E., II, Drahovzal, J.A., and Evans, F.E., Jr. (1984) Fracture analyses in a selected area of the Warrior coal basin, Alabama, *Alabama Geological Survey Circular* **111**, 78 p.

Winston, R.B. (1990) Preliminary report on coal quality trends in upper Pottsville Formation coal groups and their relationships to coal resource development, coalbed methane occurrence, and geologic history in the Warrior coal basin, Alabama, *Alabama Geological Survey Circular* **152**, 53 p.

THE MICROSTRUCTURE OF PORE SPACE IN COALS OF DIFFERENT RANK

A small angle scattering and SEM study

A. P. RADLINSKI
Australian Geological Survey Organisation
GPO Box 378, Canberra Cit, ACT 2601, Australia

E. Z. RADLINSKA
Department of Applied Mathematics
Research School of Physical Sciences and Engineering
The Australian National University
GPO Box 4, Canberra, ACT 0200, Australia

Abstract

Scanning electron microscopy (SEM), small angle neutron scattering (SANS) and small angle X-ray scattering (SAXS) were used to analyse pore space for a series of Bowen Basin coals ranked in the range VR=0.7% toVR =3.1%. The linear pore size range assessed in the small angle scattering (SAS) experiments was 1.3 nm to 200nm. Using fractal analysis of the SAS data we showed that for the lower rank coals (VR less than 1%) the pore-coal interface is rough in the entire pore size range. For coals with VR larger than 1%, the interface becomes smooth on scales larger than 100 nm, but remains rough at smaller scales. The degree of roughness is decreasing with increasing coal rank and for anthracites the interface becomes smooth in the entire scale range.

These findings are quantified by the calculated values of the specific area of the internal pore-coal interface. Owing to the fractal character of the pore space for sizes above 10 nm, the specific surface area was obtained by an appropriate scaling procedure and depends on the size of measuring yardstick. For the particular case of coals used in this study, when measured with the 5 Å (0.5 nm, atomic size) yardstick the specific area decreases from about 200 m^2/g for low rank coals to 3 m^2/g for anthracites. The corresponding sorption capacity for methane at saturation is calculated to decrease from about 40 mg/g to 0.5 mg/g of coal. For a 60 Å (large molecule size) yardstick these limits are 20 m^2/g and 3 m^2/g, respectively.

Apart from the fractal micro-architecture at scales above 100 Å, an additional microstructural feature of the characteristic smallest dimension of 20Å was observed. This feature develops in coals as their maturity increases. As a consequence, for mature coals the specific internal surface area may be significantly larger than that calculated solely from fractal scaling.

SEM observations were performed on fragments of samples previously analysed by SAS. Samples were cleaved in the direction perpendicular to the bedding plane and internal surfaces of relatively large pores (of the order of 1 μm across) opening to the exposed surface were observed at magnifications from x500 to x500000. For low rank coals, structural features down to the size of 5 nm were observed. The wide size range of the observed structural detail is consistent with the notion of fractal microstructure of coals, as determined using small angle scattering techniques.

1. Introduction

Coal seam gases are an important natural resource, but also a mining safety hazard. At the formation pressure and temperature, over 90% of gas molecules are bound in the quasi-liquid state by physical adsorption forces at the coal/pore interface (Yee et al. 1993). During mining operations the pressure gets released and, consequently, the gas molecules desorb and migrate through the network of interconnected pores and cleats out of the coal seam. Consequently, there are two separate major issues in the coal seam gas problem: the structural one, related to the geometry and topology of the pore space on scale varying from molecular to macroscopic, and the geochemical one, related to the controls on the adsorption/desorption process of light hydrocarbons and other formation fluids.

In this work we concentrate on the microstructure of coals in the length scale range 1nm to 1mm, with emphasis on the lower end of this range. The internal surface area of the pore/coal interface (and the vast majority of sites for molecular adsorption) is concentrated in this small size region. The micro-architecture of the pore space may be unique for every coal formation. In general, it is very complex and not easily susceptible to observation. The main obstacle in experimental studies of the geometry of the pore space in coals (as well as other sedimentary rocks) has been the presence of the wide range of pore (or, more precisely, geometrical feature) sizes, varying from nearly-molecular (several nanometres) to macroscopic (of the order of millimetres).

Various methods have been used to study the microstructure of porous materials, including the molecular adsorption, mercury injection, scanning tunnelling microscopy (STM), transmission electron microscopy (TEM), scanning electron microscopy (SEM) and light microscope techniques, as well as the small angle scattering of X-rays (SAXS) and neutrons (SANS). Each of these methods has its limitations and poses specific interpretative difficulties. Therefore, in order to obtain reliable data it is best to employ at least two generically different techniques with at least partial spatial overlap. In this work we combined SEM with SAXS and SANS.

Scanning electron microscopy has been usually used for identification of major coal components. It was also applied to examine the variety of microstructures present in coals of various type, rank and orientation (Gamson and Beamish 1991). The major advantage of SEM is that it provides direct imaging of the structural features of interest. However, there is a practical limit to the number of 2D sections through the 3D pore space that one can analyse. Furthermore, the wealth of detail in rock microstructure is such that the spatial resolution of a single image is orders of magnitude lower than the size range of the structural features present in the coal sample. As a result, in order to gain insight into the rock microstructure it is necessary to examine images in a wide range of magnifications. Finally, the sample preparation process for SEM is invasive.

On the other hand, the small angle scattering methods are non-invasive, resolve a wide range of feature sizes in a single experimental run and provide data averaged over the entire sample volume (which is about 0.1 mm^3 for X-rays and 0.5 cm^3 for neutrons). The trade-off is that the structural information is indirect (in the form of the Fourier transform of the density correlation function), many details are lost, and the interpretation involves mathematical processing. Therefore, the combination of SEM and SAS has the potential to provide a comprehensive and complete picture of the microstructural characteristics of coal.

2. Samples and experimental facilities

Six coal samples originating from Bowen Basin, Queensland, Australia, were used in this study (table 1). These are Permian coals and the dominant maceral is vitrinite (Beeston et al. 1978). The vitrinite reflectance (VR) varied throughout

coal name	coal origin	RV (%)	ash (wt%)	SAXS sample thickness (mm)	SANS sample thickness (mm)
1C	German Creek	0.67	6.2	1.3	4.2
3C	Gregory (West Pit)	0.95	9.9	1.3	4.3
5C	Goonyella	1.12	12.7	1.2	3.7
7C	Peak Downs	1.4	15.3	1.1	3.7
9C	Norwich Park	1.6	10.9	1.2	3.8
11C	Nebo West Seam T50	3.05	4.5	1.3	4.0

Table 1. Coal samples used in this study.

the series of samples in the range 0.67 to 3.05 %.

Samples used for SAXS and SANS were solid square platelets (of the large surface area about $1 cm^2$) cut out parallel to the bedding plane. The direction of the bedding plane was assessed by visual inspection and was later checked against the 2D scattering pattern, which is expected to be isotropic for this particular orientation.   SAXS and SANS data in such geometry provide information about the in-bedding-plane micro-architecture of the coal matrix. A small fraction of data turned out to be anisotropic due to sample misalignment during the cutting process and was not used for further analysis.

Samples used for the scanning electron microscopy (SEM) work were cleaved from the SANS samples in the direction perpendicular to the bedding plane. Such orientation enables one to observe cross sections of the in-plane pores, which also give rise to the small angle scattering signal. SEM specimens were prepared in a standard way by vacuum coating with platinum.

The small angle scattering work was performed at the Center for Small-Angle Scattering Research, Solid State Division, Oak Ridge National Laboratory, USA. We used the W. C. Koehler 30 m SANS facility and the 10 m SAXS instrument.   Experimental procedures are described in detail elsewhere (Radlinski et al. 1996).   The SEM work was performed on a Hitachi 4500 machine at the Australian National University Electron Microscopy Unit.

3. Small Angle Scattering Technique

3.1 BACKGROUND

Small angle scattering techniques have been used for many years for microstructural research in materials science, polymer physics and surface chemistry. Coals were actually one of first materials studied by SAXS (Gray and Zinn 1930).   However, the small angle scattering theory was originally developed for particulate systems, like large molecules and molecular clusters (Guinier et al. 1955).

Wider application of these techniques in geosciences has been hindered by complex, non-particulate micro-architecture of sedimentary rocks (and coals in particular), limited Q-range of first SAXS and SANS machines and the lack of general theoretical approach to data interpretation. With the advent of fractal geometry, several groups used SEM and optical microscopy (Katz and Thompson 1985, Krohn and Thompson 1986, Krohn 1988a, Krohn 1988b), SAXS (Bale and Schmidt 1984, Bale et al. 1984) and later SANS (Mildner and Hall 1986, Wong et al. 1986, Gethner 1988) to demonstrate the underlying dilation symmetry in sedimentary rocks of various kinds: shales, carbonates and coals. This, as well as instrumental developments, paved the way for more specific applications of SANS and SAXS in source rock analysis (Radlinski et al. 1996), petroleum geochemistry (Radlinski et al. 1998a) and coal petrography.

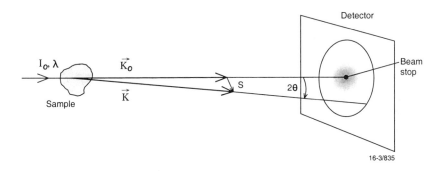

Figure 1. The principle of small angle scattering.

The principle of a SAS experiment is illustrated in figure 1. A flux of monochromatic neutrons or X-ray photons propagating in the direction of their wavevector $\mathbf{K_o}$ is elastically scattered inside the sample of volume V. In the scattering experiment one measures the intensity dI scattered in the direction $\mathbf{K}$, where by convention $\mathbf{K} - \mathbf{K_o} = \mathbf{S}$ and the quantity $\mathbf{Q} = 2\pi\mathbf{S}$ is called the scattering vector. The magnitude of the scattering vector is related to the radiation wavelength λ and the scattering angle Θ:

$$Q = (4\pi/\lambda)\sin\Theta \qquad (1)$$

The upper limit of Q in SAS experiments is about 5 degrees of arc, whereas the lower limit is determined by the divergence of the incident beam. According to Braggs law, the linear size of objects that contribute most to the scattering intensity at a given Q-value is:

$$2\pi/Q = \lambda/(2\sin\Theta) \qquad (2)$$

which shows that the choice of wavelength λ and the scattering angle Θ determine the suitability of experimental system for studying objects within a given size range. For example, the classical X-ray diffraction instruments (with λ of the order of 1 Å) use Θ-range 5 deg to 90 deg, which corresponds to the Q-range roughly 1 to 10 Å^{-1}. This Q-range corresponds to distances between the scattering planes in the range 0.5 to 6 Å. Small angle scattering machines are designed to access small values of Q, thus being capable of 'seeing' objects whose dimensions are large compared to interatomic distances: large molecules, molecular aggregates and small clusters. Since the ordering of many large-scale structures is non-crystalline, it is customary not to invoke the Braggs law (equation 2) and characterise the SAS machines simply by their Q-range. Pinhole geometry SANS and SAXS machines (as used in this work and shown schematically in figure 1) usually have Q_{min} of the order of 10^{-3} Å $^{-1}$. This enables one to study the geometry of objects in the size range 20 Å to several thousands of Å. The size range can be extended to about 20 μm using Bonse-Hart geometry experimental arrangements, which rely on the diffraction from perfect silicon crystals rather than the relatively simple geometry as in figure 1. (Lambard and Zemb 1991, Agamalian et al. 1997).

At the molecular level coals are complex mixtures of hundreds of different organic molecules and at least ten inorganic oxides present in significant amount. From the point of view of materials science these are 'dirty' systems. However, if the various types of compounds are intimately mixed in such a way that the *average* atomic composition of a region larger than certain minimum size (for instance, larger than a sphere 50 Å in diameter) remains unchanged, such a region is perceived by X-rays or neutrons in the small angle scattering regime as a uniform *phase*.

Coals may contain several different phases, each of them of different average chemical composition. This simply reflects the fact that coals may contain regions (phases) whose *average* elemental composition varies from one another in a way that is *sensed* by the particular scattering technique used. The word *sensed* is an important qualification: differences in chemical composition alone are not always discernible, or may be perceived by one scattering technique but not another. This is analogous to a colour-blind person being able to see black-and-white *contrast*, but not the red-green one, although red and green are obviously two different colours. A blind person would not see any contrast at all, and some people could see contrast in both cases.

The notion of *scattering contrast* is pivotal in the small angle scattering theory and practice. With no contrast there is no scattering at all. Scattering of X-rays (SAXS) depends on the *electronic density contrast*, scattering of neutrons (SANS) on the *nuclear scattering cross section contrast*, and scattering of light on the *dielectric contrast*. Since by definition there is no scattering contrast within any single phase, all scattering occurs at the *interface* between different phases. This is why small angle scattering signal is sensitive to the shape and extent of this interface and, therefore, can be used to quantify the geometry and size of the internal surface in porous media.

3.2. THEORETICAL CONCEPTS

3.2.1. *Basic results of small angle scattering theory*

Small angle scattering is an exact science and, therefore, different types of contrast are well defined and represented by mathematical formulae via the quantities called scattering length densities. There are many textbooks and review articles which address various aspects of the small angle scattering theory, the ultimate one still being Guinier et al. (1955). A recent review of the application of SAXS and SANS to the microstructural studies of rocks is given by Radlinski et al. (1996).

The contribution of different phases to the total neutron (SANS) scattering cross section can be determined from the values of the corresponding scattering length densities, $\rho_n(r)$. The scattering amplitude for neutrons, $A(Q)$, is related to $\rho_n(r)$ by:

$$dA(Q) = \rho_n(r) \exp(-iQr) \, dV \qquad (1)$$

where Q is the scattering vector and dV the scattering volume element. The scattering length density depends on the coherent scattering lengths of individual nuclei, b_i. These scattering lengths, which are generally well known, contain fundamental information about how the atomic nucleus is interacting with thermal neutrons used in SANS experiments. The exact relationship is:

$$\rho_n(r) = (1/V) \, \Sigma_i \, b_i \, \delta(r-r_i) \qquad (2)$$

where the summation is extended over all the nuclei, i, contained in the volume V. The symbol $\delta(r-r_i)$ means that the neutron-nucleus interaction occurs exactly at the point where the nucleus is located, i.e. at the position $r=r_i$ (bold symbols indicate that nuclei are located in 3D space). For X-rays, the scattering amplitude is:

$$dA(Q) = I_e \rho_e(r) \exp(-iQr) \, dV \qquad (3)$$

where $I_e = (e^2/mc^2) = 2.82 \times 10^{-13}$ cm is the scattering amplitude for a single electron and $\rho_e(r)$ is electronic density (i.e. the number of electrons per unit volume). Comparing relations (1) and (3) one can see that the product $I_e \rho_e(r)$ plays the same role in SAXS as the scattering length density $\rho_n(r)$ does in SANS.

As stated before, coals may contain several macroscopic phases that can be distinguished in the small angle scattering experiments. A single phase is a fine mixture of a number of chemical compounds, labelled with a running index j, each compound containing a number of different atoms (isotopes), labelled with a running index i. The scattering length density for such a single-phase, complex mixture of molar mass M can be calculated if its density and elemental composition are known:

$$\rho_n = (N_A d/M) \, [\Sigma_j \, p_j \, (\Sigma_i \, s_i \, b_i)_j] \qquad (4)$$

where N_A is the Avogadro's number, d is density, s_i is the proportion by number of nucleus i in the compound j and p_j is the proportion by molecular number of the compound j in the mixture. In practical calculations it is convenient to construct a hypothetical 'supra-molecule', the chemical formula of which corresponds to the lowest common denominator derived from the atomic analysis of coal and ash, calculate its molar mass and use the independently measured coal density for the value of d in equation (4). If X-rays are used for scattering, the product $I_e\rho_e$ can by simply calculated as:

$$I_e\,\rho_e = (N_A d/M)\,N_e\,I_e \tag{5}$$

where N_e is the number of electrons per one 'supra-molecule' of composition as in the square brackets in equation (4).

The quantity measured in a small angle scattering experiment is the scattered intensity $I(Q)$, equivalent to the differential cross section $d\Sigma/d\Omega$:

$$I(Q) = A(Q)\,A^*(Q) \tag{6a}$$

which can be also expressed as:

$$I(Q) = C_O \int_V \gamma(\mathbf{r})\exp(-i\mathbf{Qr})\,dV \tag{6b}$$

where $\gamma(\mathbf{r})$ is the density-density correlation function, which represents the geometry of the scattering object. C_O is unity for neutrons and I_e^2 for X-rays. For a multi-phase system the correlation function has the form (Debye et al. 1957, Goodisman et al. 1981):

$$\gamma(\mathbf{r}) = \sum_{i,j} P_{ij}(r)(\,\rho_i-\rho_o)(\,\rho_j-\rho_o) \tag{7}$$

where $\rho_i=\rho_{ei}$ or ρ_{ni} (electronic or nuclear scattering length density of phase i), ρ_o is the volume average of ρ_e or ρ_n over the entire sample, indices i and j indicate separate phases of the system and $P_{ij}(r)$ is the probability that a point at distance $\mathbf{r}$ away from a randomly selected point in phase i happens to be in phase j. For a two phase system, $\gamma(\mathbf{r})$ can be readily determined once the geometry of the scattering object and the scattering length density of the object and the medium in which it is immersed is known. For multi-phase systems, however, one needs to know the exact geometrical distribution of every phase in order to be able to use equation (7). Such detailed information is not readily available and this has been one of the reasons why SANS and SAXS had not been extensively applied to study the geometry of the pore space in rocks and coals in the past.

3.2.2. Small angle scattering, fractals, and the internal surface area

As we said above, all scattering occurs at the *interface* between different phases and, therefore, the small angle scattering signal is sensitive to the shape and extent of the internal surface in porous materials. About fifty years ago, in the pioneering days of small angle scattering, it has been recognised that for *two-phase systems* the form of scattering in the large-Q region should be universal for any scattering system (Porod 1951-1952, Guinier et al. 1955, Debye et al.

1957). This was thought to happen because no matter how convoluted the interface might be, at small enough scale its shape would be made more regular by the granular (molecular) nature of matter, which could be approximated by a locally planar interface. The formal result in the large-Q limit is:

$$(d\Sigma/d\Omega) = 2\pi C_o(\rho_1-\rho_2)^2 Q^{-4}(S/V) \qquad (8a)$$

or

$$S_o = S/V = \lim \{Q^4 (d\Sigma/d\Omega) /[2\pi C_o(\rho_1-\rho_2)^2]\} \qquad (8b)$$

where the limit is taken for large values of the scattering vector Q in the small angle region.

This is the famous *Porod limit*, which importance permeates the entire field of small angle scattering and certainly warrants detailed discussion. It is also very important in the current context of scattering by coals. The quantity $d\Sigma/d\Omega$ on the left hand side of equation (8a) is the scattering cross section measured in absolute units, cm^{-1}, over a range of values of the scattering vector Q. On the right hand side, the quantity C_o is a number which depends on what sort of radiation is being scattered (equation 6b), $(\rho_1-\rho_2)^2$ is formally defined as the scattering contrast (see equation 7), and $S_o= S/V$ is the ratio of the surface area of the interface, S, to the sample volume, V, in which it is embedded.   As explained above, it was expected that for a small enough sample volume (i.e. in the large-Q limit of the small angle scattering region) the value of S/V would reach a plateau and represent the true specific surface area averaged over the sample volume.

It follows that in order to quantify the internal surface area of any system it is sufficient to (1) make sure that the system is two-phase, (2) measure the absolute scattering cross section over a range of Q-values, (3) identify the Q-range over which the scattering cross section has the Q^{-4} functional dependence and (4) calculate the scattering contrast. Then, the specific surface area could be calculated from equation (8b).

Starting from early 1950's, measurements in Porod limit have been used to determine the specific surface area for numerous systems. As the precision of the small angle scattering techniques improved with time, it became apparent that some systems never reach the Porod limit or instead show a power-law scattering with an exponent different from -4. There may be a variety of reasons for such a behaviour, which could be explained by various aspects of the small angle scattering theory and experimental practice, and these were frequently invoked to explain the discrepancy between the Porod formula and the actual experimental findings. Only in early 1980's the connection was made between the scattering results and shared geometrical characteristics of the 'anomalous' systems.

The breakthrough was facilitated by two developments: (1) the realisation that non-Porod exponents can be explained by certain broad distributions of sizes of scattering objects (Schmidt 1982) and (2) the advent of fractal geometry, which showed that such distributions were a special case of deeper structural symmetry of the scattering systems (Mandelbrot 1983). This dilation symmetry (or self-similarity) was shown to be a common feature of mathematical objects called fractals.

Without quite using the terminology, the notion that rocks are fractal on the *macro scale* is second nature to geologists. It is the need to define scale for an intrinsically self-similar (i.e. looking the same in the zoom-in/zoom-out situations) object which places the geological hammers, coins and rulers on the photographs of geological formations. In an important early work, the self-similar (dilation) symmetry has been demonstrated for the *microstructure* of rocks. This was done by processing the electron microscopy images of rocks in a way inspired by fractal geometry (Katz and Thompson 1985). This work was later developed into a comprehensive study of a large number of sedimentary rocks (Krohn and Thompson 1986, Krohn 1988a, Krohn 1988b).

Theoretical calculations of small angle scattering cross sections for two-phase self similar objects soon followed, and one of the first materials studied were coals (Bale and Schmidt 1984) and sedimentary rocks (Wong et al. 1986). It turned out that, not surprisingly, the scattering is dominated by the fractal interface between the grains and the pore space, which could be seen as a crumpled, convoluted surface embedded in the rock. Using ideas that lead to the concept of Porod limit it is easy to see that such a surface would show more and more detail as one zooms in. Therefore, the value of S/V in equation (8) would never plateau, but instead increase steadily as Q increases. One could interpret scattering from such a *surface fractal* as a very extensive Porod limit region with S/V systematically increasing with Q.

There is another type of fractals which are less important for this work: the mass fractals. These are spatially extensive objects with very little volume, like a river system or a system of interlocked cracks in a rock. Details of calculations of small angle scattering from both types of fractals can be found in the original literature cited by Radlinski et al. (1996) and will not be repeated here. It is a general result that small angle scattering from ideal fractals results in a power-law dependence of the scattering cross section, $d\Sigma/d\Omega$, on the scattering vector Q. In keeping with our discussion above, the negative power law exponent is always larger than -4 expected in Porod limit.

Furthermore, for real (i.e. material) two-phase fractal objects the power law scattering is expected within the Q region for which $Ql_{max} \gg 1$ and $Ql_{min} \ll 1$. These two inequalities ascertain that the range of Q is such that the scale of length, r, probed by X-rays or neutrons is smaller than the scattering object's total diameter l_{max} but larger than the size l_{min} of the individual building block. The question of lower size limit effect is related to the size (l_{min}) and shape of the building blocks (or elementary units) of a real fractal object. These can be atoms, molecules, molecular clusters, or larger objects. In the case of coals studied here the building blocks are clusters of a large number of finely dispersed particles of organic and mineral matter. The meaning of the upper-limit cutoff is discussed below.

Quantitatively, the correlation function for real surface fractals with the upper limit ξ for the range of scale invariance (equivalent to l_{max}) was proposed by Mildner and Hall (1986), based on the correlation function of Bale and Schmidt (1984) for perfect surface fractals:

$$\gamma(r) = \exp(-r/\xi) \, [1 - C(r/\xi)^{3-D_s}] \qquad (9)$$

where $C = S_0/(4\phi(1-\phi)V$, V is the sample volume, ϕ is porosity and S_0 a constant of area dimension, like in equation (8) in Porod limit. The quantity ξ (equivalent to l_{max}) is the upper-size cutoff (sometimes called the correlation length), and D_s is the (surface) fractal dimension of the scattering object. The surface fractal dimension can be as small as 2, but is always less than 3. The lower limit corresponds to the usual Euclidean dimension of a smooth surface, for which the Porod limit is expected at large magnifications (large Q-values). As D_s increases, the surface becomes more and more space-filling and would become arbitrarily close to any point in 3D space (although not necessarily solid) at the limit $D_s = 3$. For a smooth interface ($D_s=2$), S_0 is, of course, the interfacial surface area.

The scattering intensity can be obtained from the correlation function following the usual procedure of Fourier transform of (9) (equation 6b):

$$I(Q) = Q^{-1} \, \Gamma(5-D_s) \, \xi^{5-D_s} \, [1+(Q\xi)^2]^{(D_s-5)/2} \, \sin[(D_s-1)\arctan(Q\xi)] \quad (10)$$

For $\xi Q \gg 1$, the limiting behaviour is:

$$I(Q) = Q^{D_s-6} \, \Gamma(5-D_s) \, \sin[(D_s-1)(\pi/2)], \qquad 2 \le D_s < 3 \qquad (11)$$

which results in the scattering intensity being of the form $I(Q) = const \, Q^{6-D_s}$, i.e. a power law that reduces to Porod limit for $D_s=2$, as expected.

Similar calculations have been performed for mass fractals (Freltoft et al. 1986). In this case the scattering intensity also follows the power law, but the exponent is always less than three and is equal to the (mass) fractal dimension.

As discussed above, the fractal surface is self-similar and the surface area per unit volume at a linear scale r, S(r), increases while r decreases. To be able to perform calculations, one needs a quantitative formula describing how S(r) changes with the scale r (Allen 1991):

$$S(r) = S_0(r/\xi)^{2-D_s} \qquad (12)$$

This scaling relationship enables one to calculate the surface area per unit volume at any scale length once the 'smooth' specific surface S_0 is known. In principle, S_O can be always measured at the scale ξ, but for experimental reasons the scattering data may be hard to get in this region if ξ is large (of the order of micrometres or more). However, the value of ξ in equation (12) can be substituted by any scale within the range covered by the small angle scattering experiment at which the scattering is Porod-like, i.e. exhibits the Q^{-4} behaviour. This scale provides the 'anchor' value for S_O (calculated using equation 8b), which can be used in equation (12) to calculate the specific surface at any other scale, r, once the surface fractal dimension, D_S, has been determined from the slope of the scattering curve outside the Porod region using equation (11).

4. Results

4.1. SMALL ANGLE SCATTERING FROM BOWEN BASIN COALS

4.1.1. *Scattering by individual phases and the two-phase approximation*

Five phases in coals. Bowen Basin coals contain five phases of distinctly different chemical composition and spatial continuity: macerals, ash, oil, water and unfilled pores. Unfilled pores could not be identified and, therefore, their volume fraction remains unknown. The dominant maceral in these Permian coals is vitrinite (Beeston et al. 1978). Oil is assumed to be chemically equivalent to extractable organic matter (EOM). The weight fraction for the measurable four phases in Bowen Basin coals studied in this work is shown in figure 2. The composition of coals is clearly dominated by macerals (82% to 95%), the second largest component being ash (5% to 15%). The amount of oil is very small (less than 2%) and systematically decreases with increased maturity, as expected. Water holding capacity varies from 2% to 5%, being generally larger for more mature coals.

Small angle neutron scattering. The neutron scattering length density for the five phases in coals, calculated from their chemical composition and specific density according to equation (4), is presented in figure 3. These values group into three categories: large values for the maceral and ash, small values for water and voids (unfilled pores) and an intermediate value for oil. Neutrons can be scattered by interface between any two phases, and in order to assess whether the two-phase approximation would be valid it is important to consider the relative weight of contribution from the every possible combination of any two phases. The purpose of this exercise is to find out whether there are particular two phases, scattering between which dominates the SANS signal.

According to equations (7) and (8a), the quantity that determines the intensity of scattering is the scattering contrast, defined as square of the difference between the scattering length densities, $(\rho_i - \rho_j)^2$, and the extent of the interface. According to the relative abundance data presented in figure 2, the interface between the ash and maceral should be most extensive. On the other hand, the contrast term (figure 3) varies from nearly zero for the mature coals, through about $0.05c_{max}$ for immature coals to about $0.1\ c_{max}$ for coal 5C, where c_{max} is the maximum contrast possible in the system (assuming full dewatering), i.e. that between the ash and void phase. Therefore, it seems to be a reasonable approximation if maceral and ash are combined into one 'solid supra-phase', water and voids into another 'fluid supra phase', and oil (EOM) is neglected (or added to the fluids) owing to its very small quantity.

The above procedure leads to a 'solid - fluid' two-phase approximation for SANS on coals. This approximations means that the scattering signal is dominated by the scattering on the 'solid-fluid' interface, but there may be a small scattering component originating from the underlying five-phase structure of coals. One method to determine the size of this component is to use a pore-filling fluid whose scattering length density matches that of the 'solid' phase,

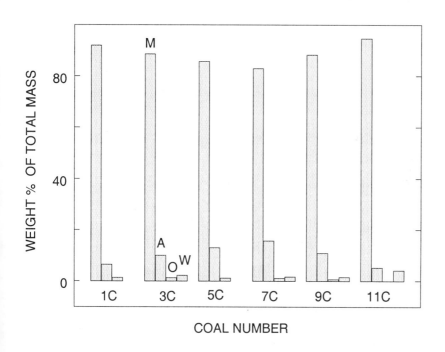

Figure 2. Phase proportion in Bowen Basin coals (in wt %). M - maceral, A - ash, O - oil, W - water holding capacity. Water data are not available for coals 1C and 5C.

and measure the residual scattering signal. Such a contrast-matching technique has been used by Gethner (1986) to study the microstructure of Illinois No. 6 coal and he measured the residual scattering intensity to be about 20% of the total SANS scattering. Contrast-matching data are not available for Bowen Basin coals.

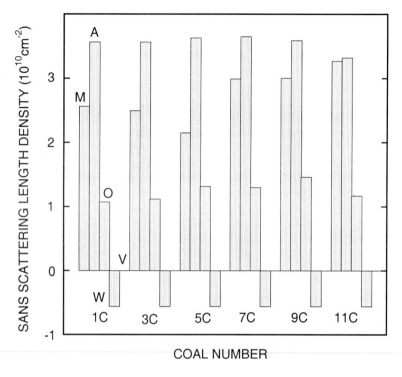

Figure 3. Neutron scattering length density for main phases in coals. M - maceral, A - ash, O - oil, W - water, V - void. Note that SLD for water is negative

Figure 4 presents a comparison of the neutron scattering length density calculated for maceral (coal), total oil and major components of the oil: asphaltenes, polars, aromatics and saturates. These data illustrate that the scattering length density of an oil can change in a very broad range, depending on the oil composition. In the case of Bowen Basin coals the scattering length density for oils does not vary much for coals of various maturity because all the EOMs have similar composition.

Small angle X-ray scattering. The X-ray scattering length density (product $I_e\rho_e$) for the five phases in coals, calculated from their chemical composition and specific density according to equation (5), is presented in figure 5. Unlike for neutrons, there is no grouping of the scattering length densities for most abundant phases into the high and low values, the values for oil, water and maceral being roughly 50% of the value for ash. This is unfortunate, since the scattering from even small amount of ash is accentuated by high contrast. Furthermore, there is evidence that various types of maceral have enough electronic density contrast to be detected by SAXS (see Lin et al., 1978, and references therein).   All this means that the two-phase approximation is much worse for SAXS than for SANS.   This approximation works best for SAXS when one type of maceral is predominant (which is vitrinite in the case of Bowen Basin coals), the ash content is low and the combined amount of oil and water is low, which happens for coals 1C and 11C (figure 2).   On balance, however, the two-phase approximation is much better justified for neutron scattering than for X-ray scattering and, therefore, SANS was selected as the scattering technique predominantly relied upon in this study.

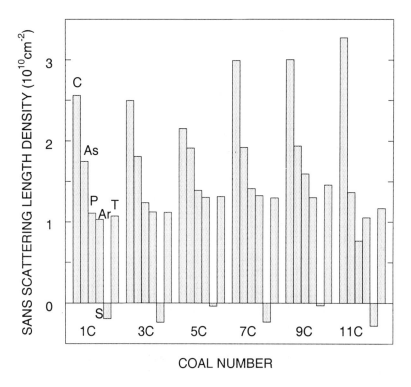

Figure 4.   Calculated neutron scattering length density for the organic components of Bowen Basin coals: the average value for maceral (C - coal), total oil (T), and major components of the oil: asphaltenes (As), polars (P), aromatics (As) and saturates (S). Specific gravities for oil components used in calculations are given in the figure. Specific gravities for coals are listed in table 3.

4.1.2. The microstructure of coal: fractal dimension and the specific surface area

SANS spectra of Bowen Basin coals are presented in figure 6.   The scattering vector covers the range 3×10^{-3} Å^{-1}< Q < 0.13 Å^{-1}, which corresponds to the linear size range of the scattering features (which can be interpreted as pores) 50 Å < r < 2000 Å.   The latter range has been calculated from the former using the usual relationship $r = 2\pi/Q$. The scattering curves have a steep power law dependence in the small-Q range, which on the log-log plot results in a straight line whose slope is the power exponent. In the light of our theoretical discussion in section 3.2.2, such behaviour of the scattering curve is indicative of fractal structure of the scattering object. Fractal character (mass or surface) as well as fractal dimension can be deduced from the slope. We note that the intensity of

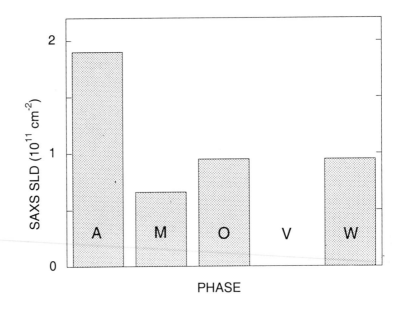

Figure 5. X-ray scattering length density (product $I_e\rho_e$) for the five phases in Bowen Basin coals. A - ash, M - maceral, O - oil, V - void, W - water.

SANS scattering decreases as the coal rank increases, which indicates decreasing specific surface area with maturity.

Another salient feature of the SANS curves is the presence of flat scattering background (of the order of 1 cm^{-1}), which dominates the scattering in the large-Q region. The origin of this background is discussed in the following section. At this stage it is sufficient to state that the background is caused by effects other than fractal scattering and, therefore, needs to be subtracted before the slopes are calculated. Two examples of SANS spectra after background subtraction are shown in figure 7a and 7b for coals 1C and 11C, respectively. These spectra are representative of all coals studied here: the slopes in the large-Q and small-Q regions may be different, indicating possible different architecture of micropores and mesopores. Table 2 lists slopes calculated in the large-Q and small-Q regions and the value of background subtracted from the data.

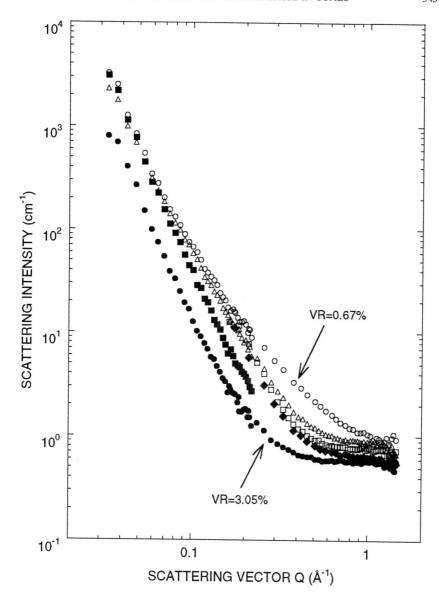

Figure 6. SANS absolute scattering cross section for Bowen Basin coals. The meaning of the symbols is as follows: open circles - 1C (VR=0.67%), open triangles - 3C, open squares - 5C, full squares - 7C, full diamonds - 9C, full circles - 11C (VR=3.05%).

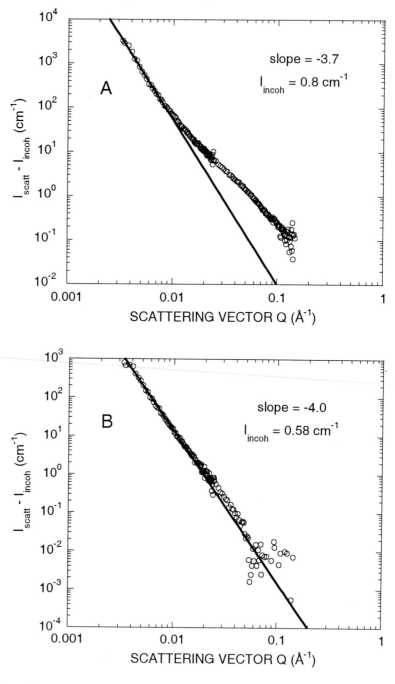

Figure 7. SANS absolute scattering cross section after background subtraction for coals 1C (A) and 11C (B).

Coal name	VR (%)	Slope, $2\pi/Q < 10$ nm	Slope, $2\pi/Q > 100$nm	background (cm^{-1})
1C	0.67	-2.4	-3.7	0.80
3C	0.95	-3.1	-3.4	0.84
5C	1.12	-3.0	-4.05	0.73
7C	1.4		-3.95	0.70 e
9C	1.6			0.595
11C	3.05	-4.0	-3.9	0.58

Table 2. Slopes in the large-Q and small-Q regions of the SANS data and the flat neutron scattering background in absolute units. Symbol 'e' indicates an estimated (not measured) value.

The values of slopes versus vitrinite reflectance are shown in figure 8. There is a general trend, in both small- and large-pore region, for the interface to become smoother with increased coal maturity. This is manifested by the slope decreasing from values close to -3 to the Porod limit value of -4 as vitrinite reflectance increases. The corresponding fractal dimension is $D_s=6+$slope (where the slope is always negative); it decreases from -3 (corresponding to a very rough interface) to -2 (smooth interface). This is consistent with progressive coal crystallisation with increased maturity. One exception is coal 1C, which shows SANS slope of -2.4 in the micropore region. Such slope corresponds to a mass fractal, indicating that at this low level of maturity the individual building blocks forming the maceral matrix are not yet consolidated, resulting in relatively high structural inhomogeneity on the 100-200 Å scale.

If the Porod-like scattering region can be observed, the calculation of specific surface area is straightforward. Two examples of Porod plots (for coals 11C and 9C, respectively) are shown in figure 9A and 9B. From these plots the values of $Q^4I(Q)$ are determined as indicated by arrows and substituted to equation (8b) together with appropriate contrast values. Specific surface area can be calculated at an arbitrary scale r using equation (12), where $\xi = 2\pi/Q^p_{min}$, and Q^p_{min} is the Q-value at the crossover from Porod region to the fractal region. For coal 3C there is only an indication of data turning into the Porod region at the small-Q end of the SANS experimental range. Therefore, $\xi = 2000$ Å was used for this coal. No calculation was done for coal 1C owing to the unknown value of ξ.

Figure 10 illustrates the evolution of specific surface area for Bowen Basin coals with maturity, calculated at two scales: 5 Å and 60 Å. Since the diameter of the methane molecule is about 3 Å, the smaller scale roughly corresponds to about

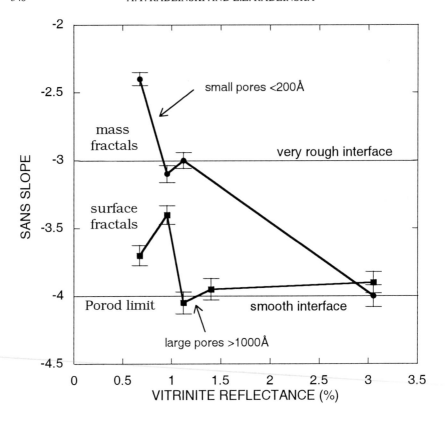

Figure 8. SANS slopes versus vitrinite reflectance for Bowen Basin coals. The slopes were determined from the log-log plots of the SANS absolute scattering cross section versus Q in the small-Q (large pores) and large -Q (small pores) limit.

50% coverage by a monolayer of methane molecules. The specific surface area at this scale decreases from about 200m^2/g for sub-bituminous coal 1C to 3m^2/g for anthracite 11C. At the 60 Å scale, corresponding to the total surface coverage by large asphaltene molecules, this range narrows to 20m^2/g to 3m^2/g. The lower value is the same on both scales since the interface in the anthracite is smooth and, therefore, the surface area is independent of the size of the molecular probe used.

4.1.3. The neutron scattering background and hydrogen content in coals

In addition to SANS signature caused by the microstructure of coal, there is a neutron scattering component caused by anomalously high *incoherent* scattering of thermal neutrons on hydrogen nuclei. The incoherent scattering is Q-independent and, therefore, results in increased flat scattering background. This background can be measured for coals in the large-Q end of the SANS spectrum. The question arises whether the incoherent component of the SANS background can be separated and used to calculate the hydrogen concentration in coals.

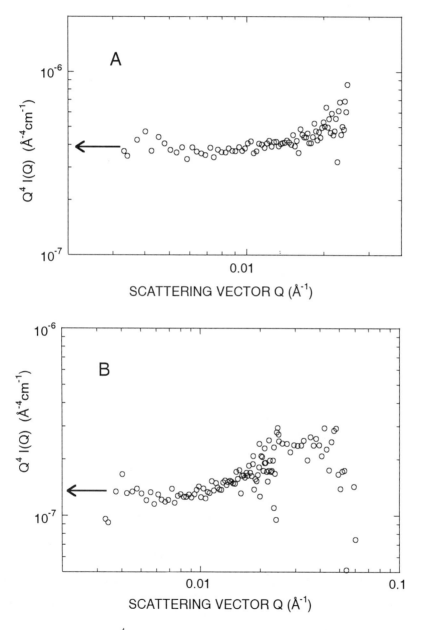

Figure 9. Porod plot ($Q^4 \times I(Q)$ versus Q) for coal 7C (A) and 11C (B). The limiting value indicated by the arrow is used to calculate the specific internal surface area for coal.

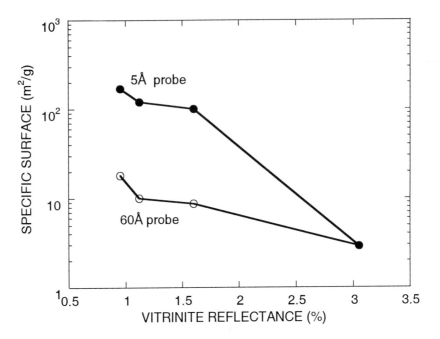

Figure 10.   Specific internal surface area versus maturity determined from SANS measurements for Bowen Basin coals. Based on fractal scaling, the specific surface area can be calculated for a surface-covering probe (molecule) of any arbitrary diameter. Probe size 5 Å corresponds roughly to the size of methane molecule. Probe size 60 Å corresponds to the size of asphaltene molecule.

The incoherent background can be estimated by empirical methods (Dubner et al. 1990). At $\lambda = 4.75$ Å used in our experiments, the reasonable estimate of the incoherent scattering cross section for hydrogen is 8.0×10^{-23} cm^2, which results in the incoherent background being given by

$$d\Sigma_{inc}/d\Omega = 8.0 \times 10^{-23}(N_p/4\pi) \text{ cm}^{-1} \qquad (13)$$

where N_p is the number of protons (hydrogen nuclei) in a cubic centimetre of the scattering sample. This hydrogen concentration in the coal matrix, N_p^c, can be readily calculated from the weight percent of hydrogen in coal obtained from elemental analysis using the following simple formula:

$$N_p^c = (d_{coal} \text{ wt\%H })/(100m_h) \qquad (14)$$

where d_{coal} is coal specific density and $m_h = 1.674 \times 10^{-24}$g is the mass of hydrogen atom. Furthermore, the hydrogen concentration present in water filling the pore space, N_p^w, can be calculated from the value of wt%H_2O, the moisture holding capacity of coal:

$$N_p^w = M_h/m_h = 6.69 \times 10^{20} \, (d_{coal} \, wt\% \, H_2O) \qquad (15)$$

where M_h is the mass of all hydrogen atoms in the molecules of water present in one cubic centimetre of coal.

Thus, the incoherent scattering background can be calculated from expression (13) by combining the contributions from hydrogen present in the coal matrix (equation 14) and in water (equation 15). Numerical data used for these calculations are presented in table 3 and the results are compared with the measured SANS background in figure 11.

Coal name	measured flat background cm^{-1}	hydrogen in coal wt%	water in coal wt%	coal density g/cm^3	incoherent scattering of maceral cm^{-1}	incoherent scattering of water cm^{-1}
1C	0.85	4.5		1.38	0.24	
3C	0.84	5.0	2.3	1.41	0.27	0.014
5C	0.73	5.1		1.31	0.25	
7C	0.70	3.7	1.7	1.42	0.20	0.010
9C.	0.60	4.1	1.6	1.47	0.23	0.010
11C	0.56	3.3	4.2	1.39	0.17	0.025

Table 3. Measured SANS flat background for Bowen Basin coals in absolute units of cm^{-1} (column 2) and calculated incoherent scattering background due to hydrogen present in the maceral (column 6) and water (column 7, assuming full water loading).

It transpires that the incoherent scattering background contributes between 20% to 30% to the total SANS flat background, depending on the origin of coal. The remaining 70 to 80% originates from the scattering on the small-scale structural inhomogeneities. The contribution from moisture retained in coal is much smaller than that from the organic hydrogen and can be neglected in first approximation. The magnitude of the incoherent background is 0.2 to 0.3 cm^{-1} and there seems to be no systematic variation with coal maturity. In contrast to this, the structural homogeneity of coal appears to increase with maturity, which results in the systematic decrease of the total measured SANS background in the maturity series from sub-bituminous coals to anthracites.

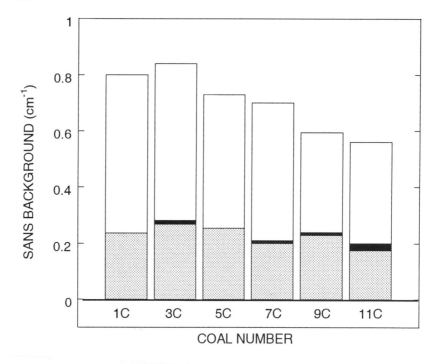

Figure 11. Measured total flat SANS background versus maturity for Bowen Basin coals. Calculated contributions caused by the incoherent scattering on hydrogen nuclei present in coal (gray bar) and in pore water (black bar) are also shown.

4.1.4. SAXS and small scale structural features in coals

Because of the multiphase character of coals and the specific values of the scattering length densities, the SAXS data are in general not as easy to interpret in the framework of fractal scattering as the SANS results. One exception is the small-Q region ($0.1 \text{ Å}^{-1} < Q < 0.6 \text{ Å}^{-1}$), where for experimental reasons only SAXS data are available. Furthermore, in this region the SANS signal is influenced by the incoherent scattering background, which is insignificant for X-ray scattering. As illustrated in figure 12, there is a scattering peak in SAXS signal at $Q_{max} = 0.3 \text{ Å}^{-1}$. This peak becomes more pronounced with increased coal maturity and corresponds to a structural feature spread throughout the sample at an increasing number density. The characteristic dimension of this feature is $R = 2\pi/Q_{max}$, which is about 20Å.

In figure 13 we compare the SAXS and SANS spectra for anthracite (coal 11C). Because of the very low ash content and uniform maceral composition in this coal, both techniques perceive the anthracite as two-phase (maceral and voids) and the two scattering curves are parallel to each other, as expected. The small-scale scattering feature at $Q = 0.3 \text{ Å}^{-1}$ is beyond the range of SANS, but is clearly discernible in SAXS data.

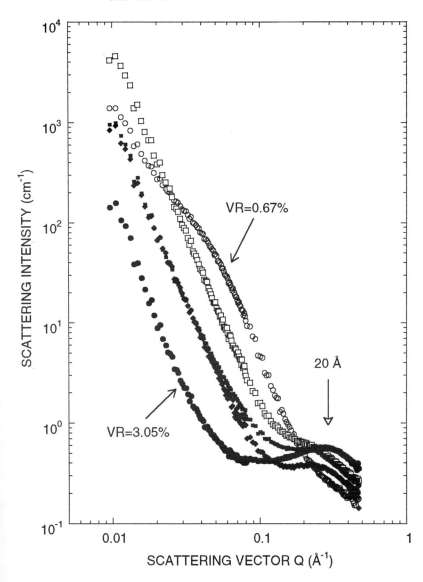

Figure 12. SAXS absolute scattering cross section versus maturity for Bowen Basin coals. Note the development of a broad scattering peak centred at $Q = 0.3$ Å^{-1} for high rank coals. The meaning of the symbols is as follows: open circles - 1C (VR=0.67%), open squares - 5C, full squares - 7C, full diamonds - 9C, full circles - 11C (VR=3.05%).

The full geometry of this structural feature cannot be established based on SAXS data only. It is most likely, however, that the distance of 20Å detected by SAXS is the smallest dimension (width), and the other characteristic dimensions are much larger and, therefore, undetectable against the strong background of fractal scattering in the small-Q region. Since it is known that increasing coal maturity results in increasing aromatisation of the vitrinite and consequent alignment of polyaromatic sheets (Tissot and Welte 1984), it is likely that the distance of 20Å corresponds to the repeat distance of stacked aromatic rings aligned into parallel sheets. Assuming full aromatisation, the specific surface area of these sheets can be readily calculated to be 700m²/g for a coal of specific density 1.4 g/cm³ (Table 3). For partially aromatised (less mature) coals this number would be accordingly smaller.

An immediate consequence of the above finding is a strong possibility that the internal surface area of high rank coals is actually larger than that calculated solely from fractal scaling using equation (12). It is therefore possible that the actual surface area available for methane adsorption can be significantly larger than that presented in figure 12, especially for the more mature coals.

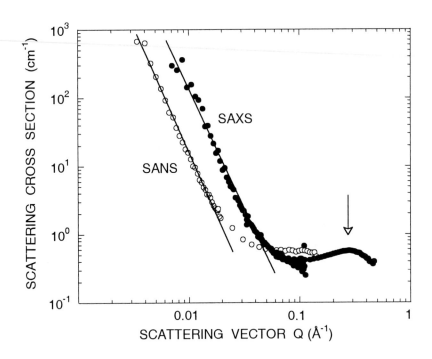

Figure 13. Comparison of SAXS (full circles) and SANS (open circles) absolute scattering cross section for the anthracite 11C. Note the fractal scattering region in the low-Q range and an additional SAXS peak centred at Q = 0.3 Å⁻¹ (indicated by the arrow).

4.2. SCANNING ELECTRON MICROSCOPY

The aim of this aspect of our study has been to illustrate the fractal character of the pore space/coal matrix interface using SEM observations in a wide range of scales. Most of the previous applications of SEM in coal petrology concentrated on the identification of various types of macerals, inorganic intrusions and the classification of pores (see, for example, Gamson and Beamish, 1991). It has been demonstrated, however, that one can infer the fractal dimension of rock microstructure using statistical analysis of SEM images (Katz and Thompson 1985, Krohn 1988b), and that the fractal dimensions obtained from SANS are consistent with the SEM statistical data (Radlinski et al. 1998b). Since we believe that small angle scattering technique provides more direct means to determine a fractal dimension than SEM, in this work we use SEM for illustrative purposes only.

As explained in section 3 above, the small angle neutron scattering technique is insensitive to much of the structural detail in a coal and responds instead to the micro-geometry of the internal boundary between the maceral and the pore space in a very wide range of sizes. In order to access this boundary at various scales by SEM it is necessary to use a wide range of magnifications, which in this study extended from x500 to x500,000. Only a small percentage of the several hundreds of SEM images recorded in the course of this work are presented here.

Figure 13 illustrates the predominant morphology of the pore space recorded at magnification x10,000. As the vitrinite reflectance increases from 0.67% to 3.05%, the appearance of coals at this magnification changes from granular to crystalline-like. Furthermore, as illustrated below, at increased magnification some new, smaller-scale structural details are revealed for coals of any maturity. This phenomenon constitutes the defining feature of fractal objects: fractals have no characteristic scale and, therefore, they look similar at any magnification. For mathematical fractals there is no limit to such behaviour, but all the real objects, including coals, are physically constrained from above by the sample size and from below by the size of chemical molecules and, therefore, can be self-similar only within a limited range of scales. This scale-independent self-similarity is statistical in nature and may not be quite well reflected by the SEM technique for two reasons. Firstly, the SEM resolution deteriorates at high magnifications due to the decreasing depth of focus and increasing problems with correcting for the imperfections of the electron optics. Secondly, with SEM one can only examine a small fragment of the sample volume at a time and the human brain is not well suited for taking the statistical average over a large number of visual images.

Figure 14 shows a series of SEM micrographs of pore walls for the sub-bituminous coal 1C, taken in the magnification range x500 to x500,000. Although the quality of images deteriorates for magnifications above x100,000, there is no doubt that the internal pore surface never becomes flat even at the highest magnifications. The same effect is illustrated for the bituminous coal 7C in figure 15 and for anthracite 11C in figure 16.

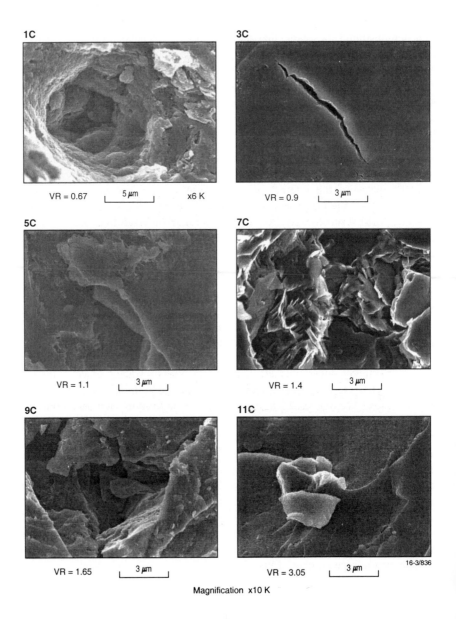

Figure 14. The morphology of structural features analysed by SEM in six Bowen Basin coals. Magnification x6,000 (coal 1C) and x10,000 (coals 3C to 11C).

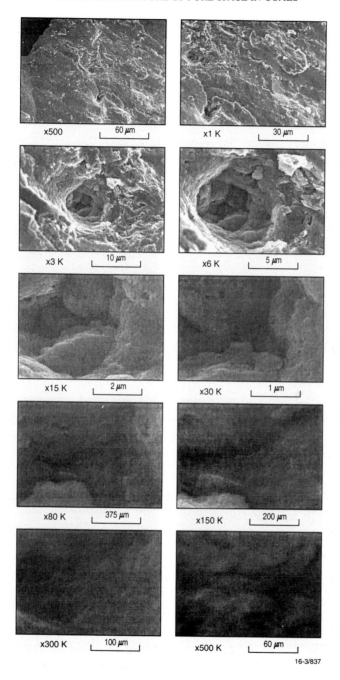

Figure 15. SEM images of the pore/maceral interface in a magnification range of x500 to x500,000 for coal 1C. Although the quality of SEM micrographs deteriorates at high magnifications, the interface never appears to be smooth.

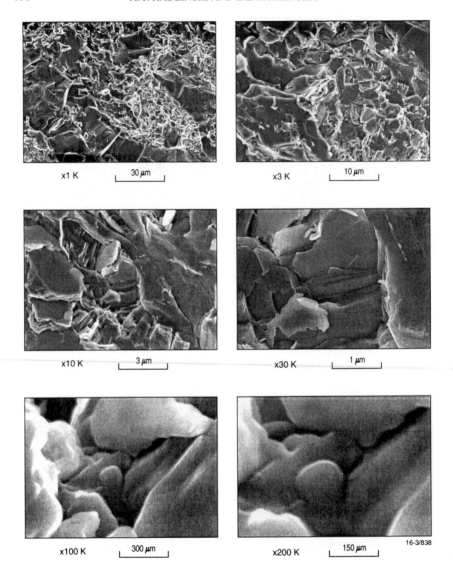

x1 K   30 μm

x3 K   10 μm

x10 K   3 μm

x30 K   1 μm

x100 K   300 μm

x200 K   150 μm

16-3/838

Figure 16. SEM images of the pore/maceral interface in a magnification of x1,000 to x200,000 for coal 7C. Note new structural features appearing at each increased magnification.

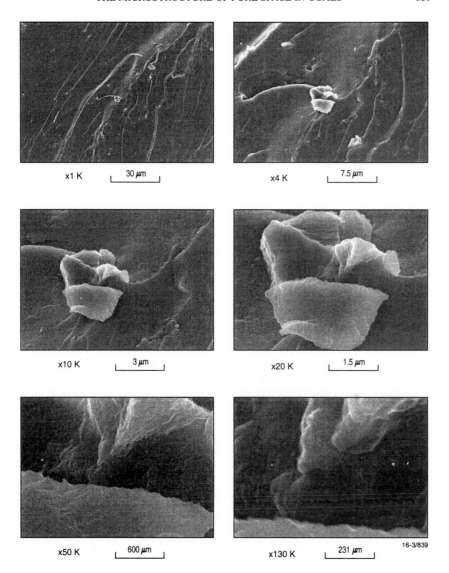

Figure 17. SEM images of the pore/maceral interface in a magnification range of x1,000 to x130,000 for the anthracite 11C. Note new structural features appearing at each increased magnification.

The small angle scattering (SANS and SAXS) data cover the size range 1.3 nm to 200 nm, and the SEM images have resolution roughly in the range 60 nm to 100 μm. From the analysis of SANS spectra discussed above, the fractality of the pore/coal interface in the corresponding size range is evident. The structural features giving rise to the SANS signal reported in this work can be best seen on SEM images of magnification x100,000 and larger.

5. Discussion

This work constitutes a comprehensive small angle scattering and SEM study of the micro-geometry of pore space in a natural maturity series of coals originating from the Bowen Basin, Queensland, Australia. The pore size region probed by SAXS and SANS is 1.3 nm to about 200 nm, and by SEM from about 50 nm to about 5μm. The study has been focussed on the smallest-size micropore region, which contributes most to the internal surface area of coals.

Small angle scattering proved to be a very useful and artefact-free method for microstructural research in many different fields. Therefore, we have discussed in detail various theoretical and practical aspects of SANS and SAXS which are relevant to the study of coal microstructure. For completeness, we have covered a range of issues including (i) the theoretical foundations of small angle scattering, (ii) the multi-phase nature of coals and how it is perceived by neutrons and X-rays, (iii) the scattering background and the role of hydrogen incoherent scattering, (iv) the fractal scattering on the rock-pore interface, and (v) how to determine the internal surface area of a coal from the small angle scattering data.

In the context of coal seam gas, the particularly useful application of small angle scattering is for the determination of internal surface area available for gas adsorption. This information is important since in the high-pressure underground conditions the majority of the gas molecules are adsorbed to pore walls in a quasi-liquid state. Traditionally, the internal surface area in coals has been determined using the adsorption and desorption isotherms for methane, neon, carbon dioxide and other gases (Berkowitz, 1979; for a recent summary see, for instance, Yee et al., 1993). The sorption method provides information which depends both on the internal surface area accessible to gas molecules and the physico-chemical affinity between these molecules and the pore walls. The well-known consequence is that the results are molecule-specific. The results are also rock-specific: for instance, the specific internal surface area of organic-poor shales is not much different from its counterpart in coals (Radlinski et al. 1996), but shales are totally inert in respect to gas sorption (Yee et al. 1993).

In contrast to the isotherm method, SAXS and SANS can be used to directly measure in a non-specific way the extent of the internal surface area, both open and closed-off to the gas molecules present in the pore space. Therefore, the SAS-determined values should lie at the upper limit of the range provided by the other types of measurement. Figure 10 shows the variation of the specific surface area with maturity determined from SANS data for four Bowen Basin coals. As discussed above, the result depends on the size of the molecular probe. For sizes of the order of 5 Å, typical for simple gases, the area decreases from about 200m²/g for subbituminous coals to 100m²/g for bituminous coals, in quantitative accordance with the long-accepted trend observed for a large number of coals (Berkowitz 1979). This result indicates that SANS can be reliably used to determine the internal surface area in coals using the experimental and interpretative procedures outlined in this paper.

The specific surface area obtained for the anthracite C11 is only 3m²/g. This is much less than the value of over 100m²/g expected from the above mentioned trend (Berkowitz 1979). Although our result is based on data for one particular type of anthracite only, it has been long known that anthracites scatter much less than other types of coals and, therefore, the outcome is not unexpected. Although more specific studies of anthracites involving both SANS/SAXS and adsorption isotherms are needed, based on the present SAXS results it is likely that the discrepancy for high rank coals may be explained by extra surface area available due to formation of stacked layers of polyaromatic sheets of the characteristic repeat distance 20 Å. For a fully aromatised coal the specific surface area of these sheets would be about 700m²/g (calculated using specific coal density of 1.4 g/cm³), which is more than required to compensate for the decreasing contribution from the fractal sites.

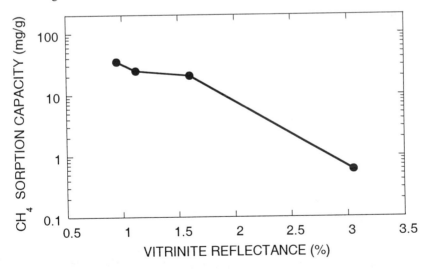

Figure 18. Estimated adsorption capacity for methane for Bowen Basin coals. The numbers were calculated from the specific surface area data presented in figure 10 assuming 50% surface coverage by methane molecules. Gas adsorption between polyaromatic sheets is not included.

The specific surface areas calculated for Bowen Basin coals have been translated into the amounts of methane that could be adsorbed per unit weight of coal at saturation (figure 18). These numbers refer only to gas molecules adsorbed on the fractal component of coal microstructure. A roughly linear decrease with VR, from 40mg/g for subbituminous coals to 6 mg/g for anthracite, is observed. Gas adsorption between the polyaromatic sheets in high rank coals was not taken into account in these calculations and would result in markedly increased gas adsorption capacity with increasing coal maturity, possibly even above the value for subbituminous coals.

6. Summary

We used small angle neutron scattering (SANS), small angle X-ray scattering (SAXS) and Scanning Electron Microscopy (SEM) to analyse the pore space for a series of Bowen Basin coals of various maturity, ranking in the range VR = 0.7% to VR = 3.1%. SANS and SAXS methods are non-invasive and provide data from which main microstructural characteristics averaged over the sample volume can be readily quantified. The linear pore size range studied in these experiments was 1.3 nm to 200 nm. Using fractal analysis of the small angle scattering data we concluded that for the lower rank coals (VR less than about 1%) the pore-coal interface is rough in the pore size range 3 nm to 200 nm. For coals with VR larger than 1%, the interface gradually becomes smooth on scales larger than 100 nm, but remains rough at smaller scales. The degree of roughness is decreasing with the increased coal rank and for anthracites the coal/pore interface becomes smooth in the entire scale range 1.3 nm to 200 nm.

These findings are quantified by the calculated specific area of the internal pore-coal interface. Owing to the fractal character of the pore space, the specific surface area can be obtained from measured scattering spectra using the scaling procedure. The numerical result depends on the size of measuring yardstick, which can be interpreted as the mean diameter of molecules adsorbed on the pore surface. For the particular case of coals used in this study, when measured with the 5 Å yardstick, the specific area decreases from about 200 m^2/g for low rank coals to 3m^2/g for anthracites. For a 60 Å yardstick, these limits are 20m^2/g and 3m^2/g, respectively.

Apart from the fractal micro-architecture at scales above 100 Å, we observed an additional microstructural feature of the characteristic smallest dimension of 20Å, which develops in coals as their maturity increases. This feature, identified by SAXS, is most pronounced for anthracites. The full geometry of this feature could not be established at this time, but it is likely to be related to the increased aromatisation of the coal structure and its consequent organisation into aligned polyaromatic sheets . As a consequence, the specific internal surface area of mature coals may be significantly larger than that calculated solely from the fractal scaling. The upper-limit estimate of this additional surface area for anthracites is $700m^2/g$.

SEM observations were performed on fragments of samples previously analysed by SAXS and SANS. Internal surfaces of relatively large pores (of the order of 1 μm across) opening to the exposed surface were observed at magnifications from x500 to x500000. Coals of all ranks appeared to have microstructural features in the size range 200 Å to about 5μm. For low rank coals, however, structural features down to the size of 50 Å were observed. The wide size range of observed structural detail is consistent with the fractal microstructure of coals determined using the small angle scattering technique.

7. Acknowledgments

We thank Sally Stowe and other staff members of the Australian National University Electron Microscopy Unit for their expert assistance and advise with the Hitachi 4500 scanning electron microscope. We greatly benefited from discussions with. J.-S. Lin and G. D. Wignall at Oak Ridge National Laboratory. We thank Chris Boreham for the critical reading of the manuscript. A. P. Radlinski publishes with permission of the Director, Australian Geological Survey Organisation. The experimental research at Oak Ridge was supported by the Division of Materials Sciences, U.S. Department of Energy under Contract No. DE-AC05-84OR21400 with Martin Marietta Energy Systems Inc.

References

M. Agamalian, G. D. Wignall and R. Triolo, 1997, "Optimization of a Bonse-Hart ultra-small-angle neutron scattering facility by elimination of the rocking-curve wings", Journal of Applied Crystallography **30**, 345-352.

A. J. Allen, 1991, "Time-resolved phenomena in cements, clays and porous rocks", Journal of Applied Crystallography **24**, 624-634.

H. D. Bale, M. L. Carlson, M. Kalliat, C. Y. Kwak and P. W. Schmidt, 1984, "Small-angle X-ray scattering of the submicroscopic porosity of some low-rank coals", *The Chemistry of Low-Rank Coals*, ACS Symposium Series vol. **264**, Washington, 78-94.

H. D. Bale and P. W. Schmidt, 1984, "Small-angle X-ray scattering investigation of submicroscopic porosity with fractal properties", Physical Review Letters **53** (6), 596-599.

J. W. Beeston, A. G. Galligan and N. J. Zillman, 1978, "The relationship between subsidence, depth of burial and coal quality in the German Creek Formation, central Queensland", Coal Geology **1** (2), 39-50.

N. Berkowitz, 1979, "An introduction to coal technology", Academic Press, New York, 345 pp.

P. Debye, H. R. Anderson and H. Brumberger, 1957, "Scattering by an inhomogeneous solid. II. The correlation function and its application", Journal of Applied Physics **28** (6), 679-683.

W. S. Dubner, J. M. Schultz and G. D. Wignall, 1990, "Estimation of incoherent backgrounds in SANS studies of polymers", Journal of Applied Crystallography **23**, 469-475.

T. Freltoft, J. K. Kjems, and S. K. Sinha, 1986, "Power-law correlations and finite-size effects in silica particle aggregates studied by small-angle neutron scattering", Physical Review B, **33** (1), 269-275.

P. D. Gamson and B. B. Beamish, 1991, "Characterisation of coal microstructure using scanning electron microscopy", Proceedings of Queensland Coal Symposium, Brisbane 29-30 August 1991, 9-21.

J. S. Gethner, 1986, "The determination of the void structure of microporous coals by small-angle neutron scattering: Void geometry and structure in Illinois No. 6 bituminous coal", Journal of Applied Physics **59** (4), 1068-1085.

J. Goodisman, H. Brumberger and R. Cupelo, 1981, " Determination of surface areas for supported metal catalysts from small-angle scattering", Journal of Applied Crystallography **14**, 305-308.

J. A. Gray and W. H. Zinn, 1930, "New phenomena in X-ray scattering", Canadian Journal of Research **2** , 291-293.

A. Guinier, G. Fournet, C.B. Walker and K.L. Yudowitch, 1955, "Small-angle scattering of X-rays', John Wiley and Sons, New York, 259 pp.

A. J. Katz and A. H. Thompson, 1985, "Fractal sandstone pores: implications for conductivity and pore formation", Physical Review Letters **54** (12), 1325-1328.

C. E. Krohn, 1988a, "Sandstone fractal and Euclidean pore volume distributions", Journal of Geophysical Research **93** (B4), 3286-3296.

C. E. Krohn, 1988b, "Fractal measurements of sandstones, shales and carbonates", Journal of Geophysical Research **93** (B4), 3297-3305.

C. E. Krohn and A. H. Thompson, 1986, "Fractal sandstone pores. Automated measurements using scanning-electron-microscope images", Physical Review B, **33** (9), 6366-6374.

J. Lambard and Th. Zemb, 1991, "A triple-axis Bonse-Hart camera used for high-resolution small-angle scattering", Journal of Applied Crystallography **24**, 555-561.

J.-S. Lin, R. W. Hendricks, L. A. Harris and C. S. Yust, 1978, "Microporosity and micromineralogy of vitrinite in a bituminous coal", Journal of Applied Crystallography **11**, 621-625.

B. B. Mandelbrot, 1983, "The fractal geometry of nature", W. H. Freeman and Co., New York, 468 pp.

D. F. R. Mildner and P. L. Hall, 1986, "Small-angle scattering from porous solids with fractal geometry", Journal of Physics D: Applied Physics **19**, 1535-1545.

G. Porod, 1951-1952, "Die Roentgenkleinwinkelstreuung von dichtgepackten kolloiden Systemen I, II", Kolloid-Zeitschrift, **124**, 83-114; **125**, 51-57; 109-122.

A. P. Radlinski, C. J Boreham, G. D. Wignall and J.-S. Lin, 1996, "Microstructural evolution of source rocks during hydrocarbon generation: A small-angle scattering study", Physical Review B, **53** (21), 14152-14160.

A. P. Radlinski, C. J. Boreham, P. Lindner, O. G. Randl, G. D. Wignall and J. M. Hope, 1998a, "Small angle neutron scattering signature of oil generation in artificially and naturally matured hydrocarbon source rocks", submitted to Organic Geochemistry, 26 pp.

A. P. Radlinski, E. Z. Radlinska, M. Agamalian, G. D. Wignall, P. Lindner, and O. G. Randl, 1998b, "The fractal geometry of rocks", submitted to Physical Review Letters.

P. W. Schmidt, 1982, "Interpretation of small-angle scattering curves proportional to a negative power of the scattering vector", Journal of Applied Crystallography **15**, 567-569.

B. P. Tissot and D. H. Welte, 1984, "Petroleum formation and occurrence", Springer-Verlag, Berlin, p. 237-238.

P.-z. Wong, J. Howard and J.-S. Lin, 1986, "Surface roughening and the fractal nature of rocks", Physical Review Letters **57** (5), 637-640.

D. Yee, J. P. Seidle and W. B. Hanson, 1993, "Gas sorption on coal and measurement of gas content", in: Hydrocarbons from Coal, AAPG Studies in Geology No. 38, B. E. Law and D. D. Rice eds, 203-218.

COALBED METHANE CHARACTERISTICS OF THE MIST MOUNTAIN FORMATION, SOUTHERN CANADIAN CORDILLERA: EFFECT OF SHEARING AND OXIDATION

S. J. VESSEY and R.M. BUSTIN
Department of Earth and Ocean Sciences, The University of British Columbia, Vancouver, B.C., Canada, V6T 1Z4

1. Abstract

The coalbed methane potential of the Jurassic-Cretaceous Mist Mountain Formation was investigated to determine if shearing and oxidation of coal could account for the low volumes of methane encountered in the formation. The coal is of suitable rank and composition to host significant coalbed methane reserves, yet tests to date indicate lower than the expected volumes of methane are present.

Adsorption isotherms indicate that Mist Mountain Formation has good to excellent reservoir capacity (Langmuir volumes 13 to 30 cc/g daf bases). Shearing does not significantly affect the methane adsorption capacity of the coal; any effect shearing may have on adsorption capacity is overshadowed by the effect of variation in maceral content and oxidation. Adsorption tends to increase with increasing vitrinite content, irrespective of whether or not the coal is sheared. Oxidation results in decreased methane adsorption capacity in the Mist Mountain Formation coals. However, the decreased volume of methane adsorbed due to oxidation alone is unlikely to result in the sub-economic volumes of methane that desorb from the coal.

It appears that shearing and oxidation have facilitated leakage of methane resulting in under saturation: both shearing and oxidation enhance the permeability of coal and therefore have facilitated the diffusion and pressure dependent flow of methane from the coal to groundwater.

2. Introduction

The Jurassic-Cretaceous Mist Mountain Formation of the southern Canadian Cordillera (Fig. 1) contains up to 1975 megatonnes of measured coal reserves ranging in rank from high volatile bituminous to semi-anthracite (Smith, 1989). Up to 17 major coal seams occur many of which are thick (up to 20 metres), laterally continuous, rich in

367

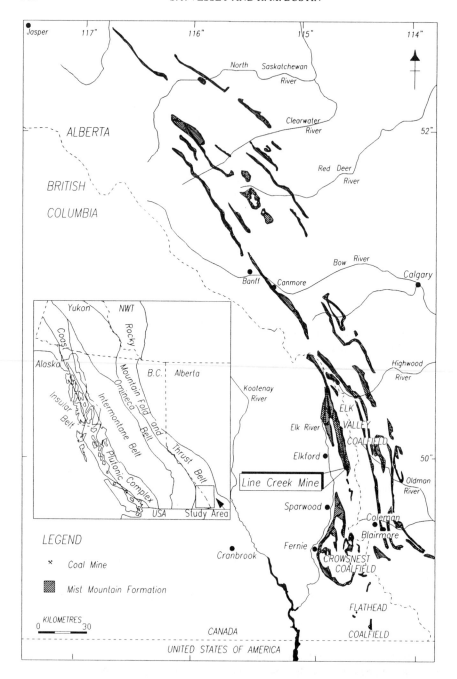

Figure 1. Location map of the study area showing outcrop of the
Mist Mountain Formation (after Gibson, 1985)

vitrinite and have variable ash contents (4-45%) (Smith, 1989). The Mist Mountain coals are thus of suitable rank and composition to host significant quantities of coalbed methane. Mist Mountain Formation coals contain much less methane (5 to 15 cc/g) then anticipated compared to coals of similar rank and composition in other basins (Feng et. al., 1984; Johnson and Smith, 1991; Dawson and Clow, 1992). The reason for the low methane content of the coal is unclear. Possibilities include escape of methane along shear zones that are common in Mist Mountain Formation coal (Feng et. al., 1984) and an apparent under-saturation of the coal due to changes in temperature and pressure because of uplift and erosion since gas generation.

In order to resolve the reasons for the low gas content, a series of coals from the Mist Mountain Formation in the Line Creek area of south-eastern British Columbia were investigated. Series of samples were collected to determine if shearing, oxidation, or coal composition could be responsible for the low volumes of methane.

2.1 FACTORS AFFECTING THE GAS STORAGE CAPACITY OF COAL

Methane can be retained in coal in one of three states: as adsorbed molecules on the internal surfaces of coal; as free gas within pores and fractures; and as a solute in pore fluids. Adsorption in micropores is the primary mechanism of methane retention in coal; excess methane exists as free gas which may then be dissolved into solution (Rightmire, 1984). The amount of gas that coal can adsorb is dependant on the internal surface area of the coal, which in turn, is dependant on pore size distribution and porosity (Gan et. al., 1972).

Previous studies have attributed variable gas contents mainly to coal rank and composition (Kim, 1977; Meissner, 1984; Ayers and Kelso, 1989; Levine, 1992, 1993; Schraufnagel and Schafer, 1996). Coal rank plays an important role in determining the gas storage capacity of coal because it directly influences porosity and pore size distribution (Lamberson and Bustin, 1993). Coal porosity decreases to a minimum at high volatile bituminous rank, then increases between the ranks of low volatile bituminous and anthracite (Gan et al., 1972). The predominant pore size in coal also changes with rank; low rank coal is mainly macroporous, while coal above high volatile bituminous rank is typically microporous (Mahajan and Walker, 1978). Maceral composition of coal also directly influences gas storage capacity through porosity and pore size distribution (i.e. Lamberson and Bustin, 1993; Clarkson and Bustin, 1996; 1997). Porosity is greatest in inertinite, intermediate in vitrinite and least in liptinite (Harris and Yust, 1976). Pore size distribution also varies between macerals; inertinite is mesoporous, liptinite is macroporous, and vitrinite contains both meso- and micropores. Although vitrinite is less porous than inertinite, it adsorbs a greater volume of gas because it contains a higher proportion of micropores (Clarkson and Bustin, 1996). Lamberson and Bustin (1993) demonstrate that maceral content is at least as important as rank in determining the gas storage capacity of some Canadian coals. Mineral matter adsorbs very little methane, therefore a high ash content reduces the quantity of gas a coal is able to store (Meissner, 1984).

The effect of shearing and oxidation on the gas storage capacity of coal is not well understood and has received very little study. Shearing increases the number of microfissures and reduces the grain size of coal (Bustin, 1982), features which might be expected to increase the gas storage capacity of coal. However, our unpublished studies at The University of British Columbia have shown that subjecting coal to up to 30% strain does not significantly affect the micropore capacity or distribution, but does increase the coal's permeability. Feng et al. (1984) noted that sheared coal desorbs more methane at a faster rate than unsheared coal. Oxidation results in oxidation rims, microfissures and a change in surface chemistry of coal (Chandra, 1982). The effect of these oxidation features on gas storage capacity is not well documented. Clarkson (1992) found that laboratory oxidation resulted in a slight reduction in the gas storage capacity of coal.

2.2 STRATIGRAPHY AND STRUCTURE OF THE MIST MOUNTAIN FORMATION

The Late Jurassic-Early Cretaceous Mist Mountain Formation is the major coal-bearing stratum of the southern Canadian Cordillera. In this study coals from the Line Creek Coal Mine (Fig. 1) were selected for study because of accessibility. Here the Mist Mountain Formation is in excess of 650 metres thick and contains up to 17 major and numerous minor coal seams (Fig. 2) of high to low volatile bituminous rank.

In the Line Creek vicinity, the Mist Mountain Formation can be divided into two units based on the abundance and lateral continuity of sedimentary facies. The lower unit extends from the base of the formation to the top of 7 seam (Fig. 2) and consists of alternating thick, laterally continuous coal seams and channel sandstones interspersed with crevasse splay and floodplain facies (Vessey, 1998). Coal seams in the lower unit average 4.5 metres thick and 26% raw ash content. Above seam 7, both coal seams and channel sandstone decrease in thickness and lateral continuity and sandstones are replaced by a greater proportion of crevasse splay and floodplain facies (Vessey, 1998). In the upper unit coal seams average 2.5 metres thick, have 28% raw ash content, and were deposited in a fluvial-alluvial floodplain environment.

The Mist Mountain Formation strata are faulted and folded into a series of northwest trending synclinal structures bounded by high angle reverse and normal faults (Pearson and Grieve, 1980). Coal seams are cut by numerous high to low angle reverse faults and are commonly sheared (Bustin, 1982).

3. Coalbed Methane in the Mist Mountain Formation

The rank and thickness of Mist Mountain Formation coals suggest they should contain large CBM resources (Johnson and Smith, 1991). Gas has been noted in the Mist Mountain Formation coal since mining began in the late 1800's. Early underground mines experienced gas blowouts and consequently drilling to control gas pressure was

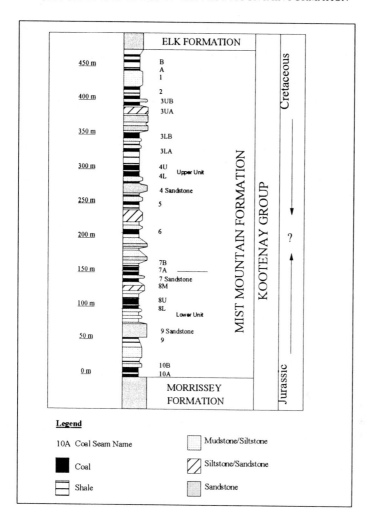

Figure 2. Generalised stratigraphic section of the Mist Mountain Formation at Line Creek showing coal seam nomenclature.

common (Johnson and Smith, 1991). All mines currently in operation are open pits; hence gas build-up does not affect mining. Interest in the CBM potential of the Mist Mountain Formation coal developed with increased CBM production in the USA in the 1980's. At least ten CBM exploration wells have been drilled into the Mist Mountain Formation but none have intersected economically viable CBM reserves (Dawson, 1995).

Feng et. al. (1984) noted a change in stored methane related to depth of burial of the Mist Mountain Formation while testing methane desorption from three drill holes in the Elk Valley Coalfield (Fig. 1). Coal seams less than 200 metres deep appear to have degassed naturally and contain less than one quarter the amount of methane of seams at a depth of 300 metres. Feng et. al. (1984) also noted that sheared coal seams released more methane than unsheared seams during desorption, and that less methane was retained in sheared coal once desorption ceased.

Dawson and Clow (1992) found that the adsorption capacity of Mist Mountain Formation coal was at least an order of magnitude more than the gas content of the coal, indicating the coal is under-saturated. They suggest that the low volumes of methane encountered during desorption is due to natural degassing of the coal because drilling occurred close to outcrop of the seams tested. Dawson and Clow (1992) also suggest that differences in adsorption between seams may be due to the extent of shearing of the coal.

3.1 METHODS

Fourteen samples were obtained from the Line Creek Coal Mine, south-eastern British Columbia and were supplemented by an additional sample of higher rank Mist Mountain Formation coal from Canmore, Alberta for comparative purposes. Three pairs of samples were collected to test the effect of shearing on methane adsorption. In each pair, a sheared and unsheared sample of similar ash content was selected. Two sets of samples were collected to test the effect of oxidation on methane adsorption. The first set was collected at progressively increasing distances from the face of a road cut. The second set was divided in two, half of which was left to oxidise under laboratory conditions at room temperature for a period of one year. A third suite of samples were collected to evaluate the effect of coal lithotype on methane adsorption capacity. The sample did not contain sufficient fusain to test for methane adsorption.

Samples were ground to pass through a 250-micrometre mesh and the moisture, ash content and free swelling index (FSI) were determined using standard ASTM procedures. Maceral composition of the coal was determined based on a 300-point count (mineral matter free) of a polished, crushed particle pellet according to standard procedures (Bustin et. al., 1985). Mean random vitrinite reflectance (R_o) was also determined for each sample using standard techniques (Bustin et. al., 1985), based on a minimum of 30 reflectance measurements per pellet. High-pressure methane adsorption was performed on approximately 100 grams of each of the samples at 30°C, at equilibrium moisture and in a high-pressure volumetric adsorption analyser similar to that described in Mavor et. al. (1990). The volume of methane adsorbed by each sample at nine successively higher pressures (0.4 to 11 MPa) was measured over a period of up to five days. Results were fitted to the Langmuir equation and recalculated to an ash free basis.

3.2 RESULTS

3.2.1. Coal Composition and Rank

Ash, moisture, FSI and random reflectance data for all samples are presented in Table 1. The ash content of the samples ranges from 4 to 44% and FSI values range from 0 to 4. Random reflectance values for the Line Creek samples range from 0.7% to 1.27% which correspond to high to medium volatile bituminous ranks. The Canmore sample has a low volatile bituminous rank and random reflectance of 1.73%.

The relative percentage of each maceral was calculated on an ash free basis (Table 2 and Figure 3). Vitrinite content of the samples ranges from 27.6 to 84.5% and is primarily in the form of unstructured detrovitrinite. The coal contains between 13 and 57% semifusinite and other inertinite accounts for up to 16% of the total maceral content. Liptinite is a minor component of all samples.

Table 1 – Ash, Equilibrium Moisture (EQ), Free Swelling Index (FSI) and Reflectance (Ro%). data for studied samples.

Sample	Seam	Ash Content (%)	EQ Moisture (%)	FSI	Ro%
Sheared suite					
96-28-346	6 – unsheared	43.04	3.02	1.5	1.09
96-28-347	6 – sheared	36.84	2.50	0	1.26
97-06-038	10A – sheared	18.05	2.02	1	1.20
97-06-039	10A – unsheared	15.72	2.16	3	1.07
97-07-074	10A – sheared	44.34	2.71	1	1.27
97-07-075	10A - unsheared	38.66	2.08	1	1.17
Oxidised Suite					
96-23-268	4L – unoxidised	13.84	8.09	0	0.70
96-23-268a	4L – lab oxidised	13.84	6.27	0	0.90
97-03-026	8U – oxidised	35.05	2.30	0	1.12
97-03-027	8U – partly oxidised	6.81	3.37	0	1.06
97-03-028	8U - unoxidised	5.69	4.35	0	1.13
Lithotype Suite					
98-01-002	8U – durain	5.10	2.16	4	1.07
98-01-003	8U – clarain	20.68	1.17	1.5	1.13
98-01-004	8U – vitrain	7.52	1.06	3.5	1.11
Canmore					
98-01-001	Canmore anthracite	4.32	1.46	0	1.73

Table 2 – Maceral Composition Data for all Samples. Telo.-telocollinite, Detro. - detrovitrinite, Inerto.- inertodetrinite, Semifu.- semifusinite, Lip.- liptinite.

Sample	Seam	Vitrinite (%)		Inertinite (%)				Lipt. (%)
		Telo.	Detro.	Fusinite	Macrinite	Inertodet	Semifu.	
Sheared Suite								
96-28-346	6 – unsheared	2.6	42.3	18.0	1.0	0.0	36.0	0.0
96-28-347	6 – sheared	1.3	46.0	8.0	0.3	0.0	44.0	0.3
97-06-038	10A – sheared	2.3	40.3	5.0	0.3	0.0	52.0	0.0
97-06-039	10A – unsheared	4.6	43.6	6.6	0.6	0.0	44.3	0.0
97-07-074	10A – sheared	3.0	54.0	8.6	1.3	0.0	33.0	0.0
97-07-075	10A - unsheared	2.3	49.0	4.3	1.3	0.0	43.0	0.0
Oxidised Suite								
96-23-268	4L – unoxidised	0.0	42.0	1.0	0.0	0.0	57.0	0.0
96-23-268a	4L – lab oxidised	2.3	48.3	8.0	0.3	0.0	41.0	0.0
97-03-026	8U – oxidised	2.3	46.6	10.0	0.6	0.0	40.0	0.3
97-03-027	8U – partly oxidised	6.6	47.0	10.3	2.0	0.0	34.0	0.0
97-03-028	8U - unoxidised	3.6	54.6	7.6	0.0	0.3	33.6	0.0
Lithotype Suite								
98-01-002	8U – durain	1.6	47.0	9.3	0.6	0.0	41.0	0.3
98-01-003	8U – clarain	0.6	27.0	14.0	1.6	0.0	56.6	0.0
98-01-004	8U – vitrain	3.0	47.0	11.6	0.3	0.0	38.0	0.0
Canmore								
98-01-001		22.0	62.5	2.5	0.0	0.0	13.0	0.0

3.2.2. Methane Adsorption

Methane adsorption isotherms for all 14 samples are illustrated on an ash and moisture free basis in figure 4. The methane monolayer capacities range from 11.5 to 27 cc/g for the Line Creek samples and 30.3 cc/g for the Canmore sample (Table 3). The volume of methane adsorbed by the Mist Mountain Formation coal increases with rank (Fig. 5).

3.2.3. The Effect of Maceral Content on Methane Adsorption

Small changes in the adsorption capacity of the Mist Mountain Formation coal can be related to changes in maceral content. Within the lithotype suite of samples, adsorption increases with increasing brightness (vitrinite content; Figs. 6 and 7).

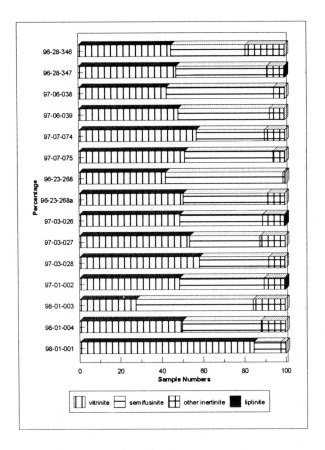

Figure 3. Maceral composition of studied samples, mineral matter free basis

3.2.4. The Effect of Oxidation on Methane Adsorption

Natural and laboratory oxidation affects the methane adsorption of coal in different ways. Natural oxidation led to a small decrease in adsorption with increased oxidation (Fig. 8). In contrast, laboratory oxidation increased the adsorption capacity of the coal (monolayer capacity from 11.5 to 13.5 cc/g; Fig. 8). We obtained similar higher

Table 3 – Methane Monolayer Volumes for all Samples.

Sample	Seam	Methane Monolayer Capacity (cc/g @ STP, DAF)
Sheared Suite		
96-28-346	6 – unsheared	24.2
96-28-347	6 – sheared	13.0
97-06-038	10A – sheared	20.6
97-06-039	10A – unsheared	21.8
97-07-074	10A – sheared	25.2
97-07-075	10A - unsheared	21.1
Oxidised Suite		
96-23-268	4L – unoxidised	11.5
96-23-268a	4L – lab oxidised	13.5
97-03-026	8U – oxidised	17.8
97-03-027	8U – partly oxidised	18.3
97-03-028	8U - unoxidised	18.8
Lithotype Suite		
98-01-002	8U – durain	25.7
98-01-003	8U – clarain	23.0
98-01-004	8U – vitrain	27.0
Canmore		
98-01-001		30.3

adsorption capacities following laboratory oxidation of low rank coals. These results verify that laboratory oxidation is not equivalent to the natural oxidation process (Huggins et. al., 1983; Huggins and Huffman, 1989). It is possible that without the influence of water, the changes in surface chemistry that occur in natural oxidation cannot be reproduced in the laboratory.

3.2.5. The Effect of Shearing on Methane Adsorption

Interpretation of the effect of shearing on methane adsorption is complicated by compositional variation between sheared and unsheared samples. The sheared suite of samples collectively does not show a consistent trend in methane adsorption (Fig. 9). In two sets of samples, (10A and 6U seams) the sheared coal adsorbed less methane than the unsheared coal, but the third sample set (also from 10A seam) had increased methane adsorption capacity in the sheared sample (Fig. 8). The adsorption capacity of the two sets of samples from 10A seam can be related to composition of the coal. Samples with the highest vitrinite content adsorb the most methane; irrespective of whether or not the coal is sheared (Tables 1 and 3; Figs. 9A and 9B).

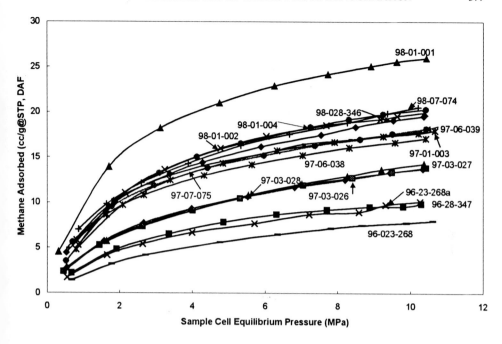

Figure 4. Methane adsorption isotherms for all samples.

The adsorption capacity of the samples from 6U seam is not related to the maceral content of the coal; vitrinite content of the samples are similar, at 47% in the sheared and 45% in the unsheared sample. However, the unsheared coal adsorbed a much greater volume of methane than the sheared sample; monolayer volumes are 24.2 cc/g and 13 cc/g respectively (Fig. 9). The difference in adsorption capacity is probably not due to shearing alone since with other coals the difference between sheared and unsheared coals is minimal. It is more likely that the differences in adsorption capacity are due to oxidation: the sheared samples do not fuse on a FSI test whereas the unsheared sample has an FSI of at least 1.5,

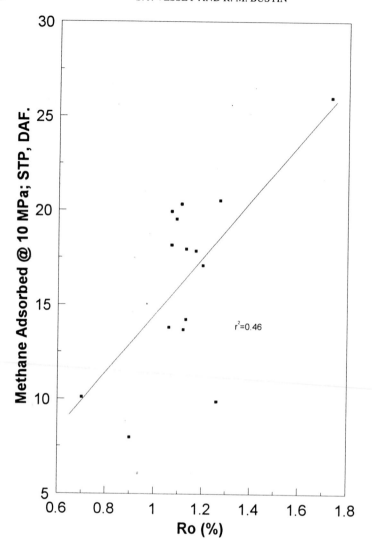

Figure 5. Cross-plot of volume of methane adsorbed and sample rank.

4. Discussion

Adsorption isotherm analyses demonstrates that coals of the Mist Mountain Formation have moderate to high reservoir capacities. Shearing does not influence the methane adsorption capacity of the Mist Mountain Formation coal enough to account for their low volumes of retained methane. Maceral content is a more important control on methane adsorption capacity than shearing. Vitrinite contains a large proportion of micropores, and accordingly, is capable of storing greater volumes of methane than inertinite or liptinite (Lamberson and Bustin, 1993). Small changes in vitrinite content in the Mist Mountain Formation coals appear to mask any influence of shearing on

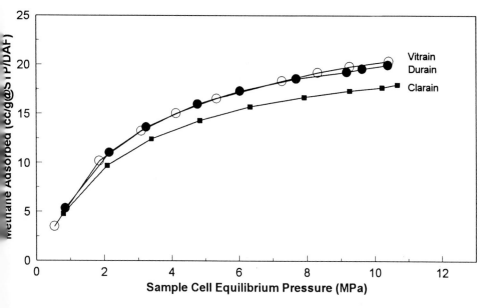

Figure 6. Methane adsorption isotherms of the lithotype suite of samples.

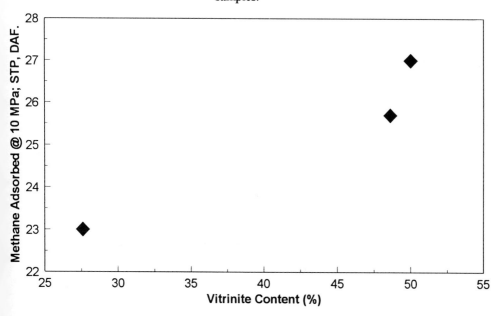

Figure 7 Cross-plot of volume of methane adsorbed and vitrinite content of the lithotype suite of samples.

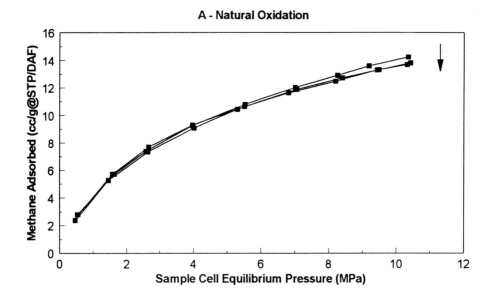

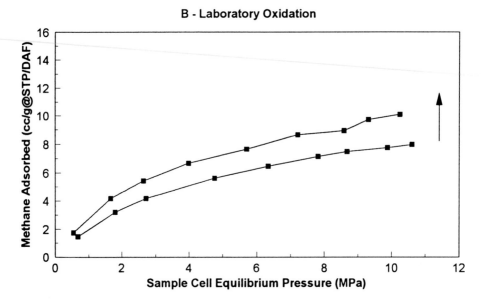

Figure 8 - Methane adsorption isotherms of the oxidation suite of samples
Arrows indicate increasing oxidation as determined by FSI

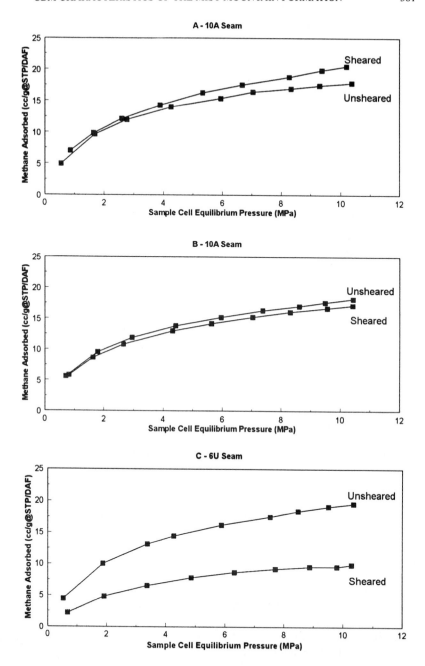

Figure 9. Methane adsorption isotherms for the sheared suite of samples

methane adsorption capacity. Natural oxidation decreases the gas storage capacity of coal, but it is unlikely that the decrease due to oxidation alone is enough to result in the sub-economic volumes of methane that desorb from the Mist Mountain Formation. Results from this study do not indicate whether oxidation or maceral content has a greater influence on methane adsorption capacity.

The rank and composition of the Mist Mountain Formation coal suggests large volumes of methane were produced during coalification which, when combined with the moderate to high measured reservoir capacity indicates gas leakage must account for the low measured gas contents. Decrease in formation temperature due to erosion and uplift may have increased the methane adsorption capacity of the coal subsequent to gas generation (Mavor et. al., 1990). However uplift alone can not account for the low gas contents. Levy et. al. (1997) found that each 1°C increase in temperature results in only a 0.12 mLg^{-1} decrease in methane adsorption capacity in Bowen Basin coals of Australia. Such small decreases in methane adsorption capacity with increased temperature can not account for the order of magnitude difference between capacity and gas content reported (Dawson and Clow, 1992) Alternatively we suggest that shearing and oxidation have facilitated leakage of methane resulting in the marked degree of under saturation. Shearing and oxidation is pervasive throughout much of the Mist Mountain coal and greatly enhances the permeability of coal. The highly sheared coals are commonly aquifers, which in turn promotes oxidation of the coal and facilities diffusion and pressure dependent flow of methane from the coal to groundwater.

5. Conclusions

The thickness, rank and depth of burial of coal seams within the Mist Mountain Formation suggest they may be a favourable reservoir for significant CBM reserves. Data presented herein confirms that the coal is theoretically capable of storing economic quantities of methane. However, desorption data indicate that lower volumes of methane than expected are present in the coal (Dawson, 1995).

Shearing and oxidation enhance the permeability of coal (Bustin, 1982) and therefore may facilitate the escape of methane from coal reservoirs. Shearing does not significantly affect the methane adsorption capacity of the coal. Any effect shearing may have on adsorption capacity appears to be overshadowed by the effect of variation in maceral content and oxidation. Adsorption tends to increase with increasing vitrinite content, irrespective of whether or not the coal is sheared. Natural oxidation results in decreased methane adsorption capacity in the Mist Mountain Formation coals. However, the decreased volume of methane adsorbed due to oxidation alone is unlikely to result in the sub-economic volumes of methane that desorb from the coal. It is probable that a combination of decreased adsorption due to oxidation and reduced retention of the methane in the coal due to shearing and oxidation account for the low volumes of methane in the Mist Mountain Formation.

6. References Cited

Ayers, W.B., and Kelso, B.S. (1989) Knowledge of methane potential for coalbed resources grows, but needs more study. Oil and Gas Journal, 87: 64-67.

Bustin, R.M. (1982) The effect of shearing on the quality of some coals in the South-eastern Canadian Cordillera: CIM Bulletin, v. 75, p. 76-83.

Bustin, R.M., Cameron, A.R., Grieve, D.A. and Kalkreuth, W.D. (1985) Coal petrology, its principals, methods, and applications, 230 p.

Chandra, D. (1982) Oxidized coals, in Stach, E., Mackowsky, M.-T., Teichmuller, M., Taylor, G.H., Chandra, D. and Teichmuller, R. Stach's textbook of coal petrology: Berlin, Gebruder Borntraeger, p. 198-205.

Clarkson, C.R. (1992) Effect of low temperature oxidation on micropore surface area, pore distribution and physical and chemical properties of coal, UBC Department of Geological Sciences.

Clarkson, C.R. and Bustin, R.M. (1996) Variation in micropore capacity and size distribution with composition in bituminous coal of the Western Canadian Sedimentary Basin: Fuel, v. 75, p. 1483-1498.

Clarkson, C.R. and Bustin, R.M. (1997) Variation in permeability with lithotype and maceral composition of Cretaceous coals of the Canadian Cordillera: International Journal of Coal Geology, v. 33, p. 135-151.

Dawson, F.M. and Clow, J.T. (1992) Coalbed methane research: Elk Valley Coalfield, In the Canadian Coal and Coalbed methane geoscience forum: Parksville, BC., p. 57-71.

Dawson, F.M. (1995) Coalbed methane: a comparison between Canada and the United States: Geological Survey of Canada Bulletin, v. 489, p. 60.

Feng, K.K., Cheng, K.C. and Augsten, R. (1984) Preliminary evaluation of the methane production potential of coal seams at Greenhills Mine, Elkford, British Columbia: CIM Bulletin, v. 77, p. 56-61.

Gan, H., Nandi, S.P. and Walker, P.L. Jr. (1972) Nature of the porosity in American coals: Fuel, v. 51, p. 272-277.

Gibson, E. (1985) Stratigraphy, sedimentology and depositional environments of the coal-bearing Jurassic-Cretaceous Kootenay Group, Alberta and British Columbia. Geological Survey of Canada Bulletin 357, 108 pp.

Gregg, S.J. and Sing, K.S.W. (1982) Adsorption surface area and porosity: New York, Academic Press.

Harris, L.A. and Yust, C.S. (1976) Transmission electron microscope observations of porosity in coal: Fuel, v. 55, p. 233-236.

Huggins, F.E., Huffman, G.P. and Lin, M.C. (1983) Observations on low-temperature oxidation of minerals in bituminous coals: International Journal of Coal Geology, v. 3, p. 157-182.

Huggins, F.E. and Huffman, G.P. (1989) Coal weathering and oxidation the early stages in Nelson. C.R., ed, Chemistry of coal weathering, Elsevier, Amsterdam, 230 p.

Johnson, D.G.S. and Smith, L.A. (1991) Coalbed methane in Southeast British Columbia, British Columbia Geological Survey, p. 19.

Kim, A.G. (1977) Estimating methane content of bituminous coalbeds from adsorption data. U.S. Bureau of Mines Report of Investigations 8245: 22 pp.

Lamberson, M.N. and Bustin, R.M. (1993) Coalbed methane characteristics of Gates Formation coals, northeastern British Columbia: effect of maceral composition: AAPG Bulletin, v. 77, p. 2062-2076.

Levine, J.R. (1992) Influence of coal composition on coal seam reservoir quality: a review. Coalbed Methane Symposium, Townsville Australia, 1992: I to XXVIII.

Levine, J.R. (1993) Coalification: the evolution of coal as source rock and reservoir. In Law, B.E. and Rice, D.D. (Editors), Hydrocarbons From Coal. American Association of Petroleum Geologists, AAPG Studies in Geology. 38: 39-77.

Levy, J.H., Day, S.J. and Killingly, J.S. (1997). Methane capacities of Bowen Basin coals related to coal properties: Fuel, v. 74, p. 1-7.

Mahajan, O.P. and Walker, P.L. Jr. (1978) Porosity of coal and coal products, *in* Karr, C. Jr. ed. Analytical methods for coal and coal products, Volume 1: New York, Academic Press, p. 125-162.

Mavor, M.J., Owen, L.B. and Pratt, T.J. (1990) Measurement and evaluation of isotherm data, Proceedings of 65th annual technical conference and exhibition, Society of Petroleum Engineers, Volume SPE 20728, p. 157-170.

Meissener, F.F. (1984) Cretaceous and lower Tertiary coals as sources for gas accumulations in the Rocky Mountains area, *in* Woodward, J., Meissener, F.F. and Clayton, J.L. eds. Source rocks of the Rocky Mountain region, Volume 1984 Guidebook, Rocky Mountain Association of Geologists, p. 401-431.

Pearson, D.E. and D.A. Grieve (1980) Elk Valley Coalfield, Volume 1980-1, British Columbia Geological Survey Paper, p. 91-96.

Rightmire, C.T. (1984) Coalbed methane resources, *in* Rightmire, C.T., Eddy, G.E. and Kirr, J.N. eds. Coalbed methane resources of the United States, Volume 17, AAPG studies in geology series, p. 1-13.

Schraufnagel, R.A. and Schafer, P.S. (1996) The success of coalbed methane. In: Saulsberry, J.L., Scahfer, P.S., Schraufnagel, R.A. (Editors) A Guide to Methane Reservoir Engineering Gas Research Institute (GRI Reference Number GRI-94/0397), Chicago, Illinois. P. 1.1- 1.10.

Smith, G.G. (1989) Coal resources of Canada, Geological Survey of Canada, Paper 89-4, 146 p.

Vessey, S. (1998) Quality of the Jurassic-Cretaceous Mist Mountain Formation coals, sotuern Canadian Cordillera: relationships to sedimentology and coalbed methane potential. Unpublished M.Sc. thesis, The University of British Columbia, 147 pp.

DECREASE OF DESORPTION INTENSITY OF COALBED METHANE DUE TO HYDRAULIC FRACTURING

Z. WEISHAUPTOVÁ[1], J. MEDEK[1] and J. NĚMEC[2]
[1]*Institute of Rock Structure and Mechanics,*
Academy of Sciences of the Czech Republic
182 09 Prague 8, Czech Republic
[2]*EUROGAS Ltd.,*
272 04 Kladno 4, Czech Republic

1. Introduction

During the prospecting extraction of methane from coal seams in Ostrava district by the method of hydraulic fracturing in many cases it has been observed that the gas releasing does not follow the assumed pattern of continuous flow with gradually decreasing intensity as it is known from the degassing of drilling cores by canister test. On the contrary, after a short initial outburst, the gas stream quickly decreases to a minimum. Due to the fact that in the pumped water a high sludge content was present, it was assumed that the sludge formed during the disintegration of the coal seam by the shock wave of pressurized water may restrict the gas release owing to clogging of transport pores by very fine particles. The overpressure of water towards gas present in coal introduces these particles into the orifices of transport pores, blocking them partially and thus reducing the gas escape.

The aim of presented study was to simulate the effect of pressurized water and sludge on alternations of the primary porous system of coal, and to determine their influence of these alterations on possible changes in methane desorption kinetic.

2. Restriction of gas permeability through transport pore by sludge particles

According to our informations acquired under laboratory conditions [1], the processes associated with the blockade of pores by sludge particles may be simulated in two different ways: a) the sludge suspension can be directly forced into coal pores of the respective size, or b) the sludge suspension can be formed by the pressure water impact on a mixture of mineral matter and coal. In the first case, the contact of porous system of coal with an already available suspension is modeled; in the second case the effect of a suspension in the moment of its formation is observed. The crucial blocking factor is the relation between pore orifice size and the particle radius, as shown in Fig. 1.

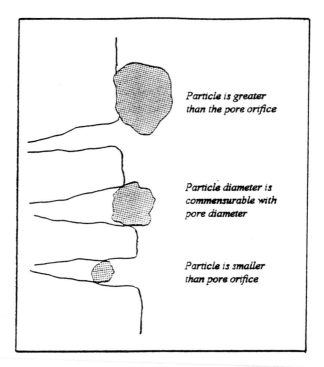

Figure 1. Clogging of transport pores by sludge particles

When the particle size exceeds the size of the largest pore will not be caught in the pore orifice and therefore it behaves only as an indifferent neighbour touching the external surface of coal, and does not affect the media motion in the porous system

If the particle size is commensurate with the pore orifice size, it closes the pore like a plug. This plug is not leak-proof, as both surfaces are irregular. If the particle is not firmly wedged into the pore such a plug can be released, when external or internal liquid media are forcefully moving.

If the particle is smaller than the pore orifice, it penetrates into such a depth where the pore cross section decreases so far enough that a particle gets seized in this point.

Due to the surface irregularities of both the particle and pore, the complete sealing off of the pore does not occur; only its effective cross section is reduced, which causes a subsequent local shift of the pore size distribution. At a considerable reduction, the braking effect appears becoms evident not only at the blocking site but, due to the pore network, also in the pore region connected with the clogged pore.

Among the transport pores are included the mesopores, macropores and coarse pores. Of this group, the course pores with effective size given by radii exceeding 7,5 μm, can be most affected by the presence of suspension particles obstructing the entrance into a porous system. Owing to the fact that according to granulometric analysis the particle size in the suspension attains diameters under 5 μm (Table 2), macropores with radii exceeding about 1 μm should be included into the pore set, which can be also affected by the presence of solid particles. Suspension particles within this size range can be considered active for the considered action.

Hydraulic fracturing is accompanied with the disintegration of coal bodies which renders the main part of the porous system accessible. Due to the fact that during the hydraulic fracturing the intensity and rate of degassing depend as well on the extent of the disclosed porous system, the effect of the coal grain size has been included in the study of this problem.

3. Experimental

3.1. COAL SAMPLE

All measurements have been made on the bituminous gas-bearing coal from Ostrava district obtained from a bore hole at 923 m depth. The coal was characterized by following values: moisture $W^a = 0.8$ wt % , ash in dry basis $A^d = 2.2$ wt%, volatile matter in dry ash-free basis $V^{daf} = 29.2$ wt% and specific density $d_r = 1.472$ g/cm^3.

The coal was used in three grain size fractions: 1 - 2 mm, 0.5 -1.0 mm and 0.2 - 0.5 mm. For porosimetric measurements, the grain size 1 - 2 mm was chosen as optimum. Smaller grains distort the pore distribution by intergranular spaces which are of such a size that the device indicates them as pores. At a large grain, the communication within the porous system is restricted.

Samples used in the study are denoted as follows.
For porometric measurements:
a - original sample, b - sample saturated with water, c - sample with sludge suspension
For the high-pressure adsorption/desorption:
A - original sample, B - sample saturated with water, C - sample with sludge suspension

3.2. EXPERIMENTAL METHODS

Measurements of pore distribution changes due to the action of both the water and the sludge suspension were carried out using a high-pressure mercury porosimeter (POROSIMETER 2000, Carlo Erba).

For porosimetric measurements, a coal sample was partly saturated with water (Sample b), and partly contacted with the sludge suspension (Sample c). As it was documented experimentally [1], the presence of water in pores does not influence the pore volume filling with mercury and the results reflect with high probability the state of the porous system after contact with both of the mentioned media. For comparison, the standard porosimetric measurements of the dry sample (Sample a) have also been carried out.

To determine the effect of the pressurized water (Sample B) and pressurized sludge suspension (Sample C) on the desorption kinetics of methane, a high-pressure sorption device [2] has been used, adapted for introduction of pressurized water and sludge into water free coal saturated by methane till the sorption equilibrium state at

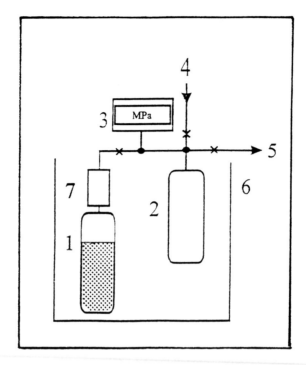

(1) Sample cell; (2) reference cell; (3) electronic digital transducer;
(4) methane pressure bottle; (5) vacuum pump; (6) constant temperature air bath;
(7) container for glass ampoule

Figure 2. Schematic diagram of high-pressure apparatus

pressure P_{ads}. For this purpose, the device (Fig.2) was provided with a container for a small glass ampoule (volume 17 cm³), filled with water, whose breakage was controlled from outside by an external screw. The temperature of experiment was 25 °C, and the sample weight was about 25 g. Pressure conditions were derived from measurements on a dry sample (Sample A), where the value of P_{ads} near to 4.2 MPa was chosen for the adsorption equilibrium. During the subsequent desorption, the equilibrium was attained at about P_{des} = 1.85 MPa. These conditions were considered standard, and all further measurements with both the pressurized water and sludge were controlled according to them. For the determination of the restricting effect of pressure water, the coal in sorption equilibrium at the pressure P_{ads} was completely wetted with water which was subsequently forced into pores by methane under pressure of about 11 MPa. After 1 min, the pressure was reduced to the original equilibrium P_{ads} and the desorption was carried out till the equilibrium pressure P_{des}. In the study of the sludge effect, the coal sample was mixed, prior to the introduction into the measuring chamber, with 20 wt% of dust particles of the dried sludge, and further measurements followed the above procedure. The only difference was that, after breaking the glass ampoule, the water with dust particles formed a sludge which was then forced into coal pores.

In the constant volume device the desorbed gas volume V_{des} is a function of pressure $V_{des} = f. P$, where f is the constant involving both the conditions of state and experimental conditions. The recorded pressure is directly proportional to the desorbed quantity of gas, and under constant experimental conditions, it is an adequate quantity for characterizing the desorption kinetics in coordinates pressure P versus time τ. If, under a high pressure, the first desorption step is carried out into vacuum, the pressure increase is so fast that it cannot be recorded reliably and therefore illustrated in the diagrams. Further points illustrate the changes of kinetics quite clearly.

4. Results

Changes of volumes and pore distribution due to the grain size of the original coal (sample a) are shown in Table 1.

TABLE 1. Effect of the grain size on the open pore volume V of coal **a**

Grain size [mm]	V_{mm} [mm³/g]	V_{meso} [mm³/g]	[%]	V_{macro} [mm³/g]	[%]	V_{macro}/V_{meso}
1.0 - 2.0	21.15	10.14	48.0	10.91	52.0	1.08
0.5 - 1.0	31.10	12.33	39.7	18.77	59.3	1.52
0.2 - 0.5	37.48	13.94	37.2	23.54	62.8	1.69

Note:

V_{mm} is the total volume of meso- and macropores (from 2.0 to 7500 nm), V_{meso} is volume of mesopores (from 2.0 to 35 nm), and V_{macro} the volume of macropores (>35 nm).

The quoted values are the means of 5 measurements.

Changes of the porous structure caused by forcing the water (sample b) or sludge (sample c) into coal, measured by mercury porosimeter are shown in Table 2 for sample with grain size 1 - 2 mm.

The last column in Table 2 illustrates the ratio of macropores volume V_{macro} to the volume of mesopores V_{meso} which characterizes the change of distribution of the studied pore categories.

The restriction of pore volume was confirmed by changes of surface area of pores. In the original coal and coal containing solid particles in pores the standard BET method of N_2 isotherm at -196 °C was used and the surface area of micropores S_{micro} was determined from the CO_2 isotherm at 25°C [3]. With the content of 0.025 g particles/1g of coal, the surface area S_{BET} changed from original 8 m³/g to 6 m³/g and the area of micropores decreased from 120 m³/g to 110 m³/g.

Results of high-pressure measurements are graphically illustrated in Fig.3, which demonstrates the time dependence of changes of desorption pressure for individual grain size fractions.

TABLE 2. Effect of presence of water and sludge on the pore distribution in coal

Sample	Grain	Water / Particles	V_{mm}	V_{meso}	V_{macro}	V_{macro} / V_{meso}
	[mm]	[g/g of coal]	[mm³/g]	[mm³/g] (%)	[mm³/g] (%)	
a	1 - 2	0.00/ 0.000	21.15	10.14 (48.0)	10.91 (52.0)	1.08
b	1 - 2	0.46/ 0.000	20.37	12.66 (62.0)	7.71 (38.0)	0.61
c	1 - 2	0.41/ 0.036	17.97	13.05 (74,0)	4.92 (26.0)	0.35

Note:
a - original water-free coal, b - coal after water saturation, c - coal after saturation with sludge;
the sludge contains: 0.23 % of silty clay, finally banded coal to coal dust; granulometric composition of solid particles in sludge:
10 - 30 μm = 15%, < 10μm = 55%, <5μm = 30%, clusters of smaller particles prevailing, isolated particles < 30 μm only exceptionally

5. Discussion

5.1. EFFECT OF WATER AND SLUDGE ON THE CHANGES OF POROUS STRUCTURE

The increasing volume of meso- and macropores V_{mm} with decrease in a grain size according to porosimetric measurements suggest that the coal fracturing is accompanied with exposure of further inlet orifices into its porous network (Table 1). This, as it has been shown by desorption measurements illustrated in coordinates P versus time τ in Figure3 a,b,c affects the desorption kinetics, as more gas is released from smaller grains due to the larger quantity of transport pores with simultaneous shortening of escape paths.

The porosimetry was also used to prove the effect of presence of sludge particles on the volume of the porous system. It is evident from Table 2, that the water alone does not affect substantially the overall pore volume, the changes in their distribution for the benefit of mesopores being caused by the contraction of macropores due to coal swelling. In contrast, the presence of sludge particles causes the decrease of the overall pore volume V_{mm} and further increase of the volume of mesopores V_{meso}. The difference between the ratio V_{macro} / V_{meso} in Samples b and c can be attributed, above all, to the reduction of the volume V_{macro} by the volume of sludge particles. The presence of sludge particles in porous systems was also confirmed by the reduction of the surface area S_{BET} and area of micropores S_{micro}, caused by the restriction of communication with the system of transport pores. The restriction of large pores prevents also the access to a part of micropores.

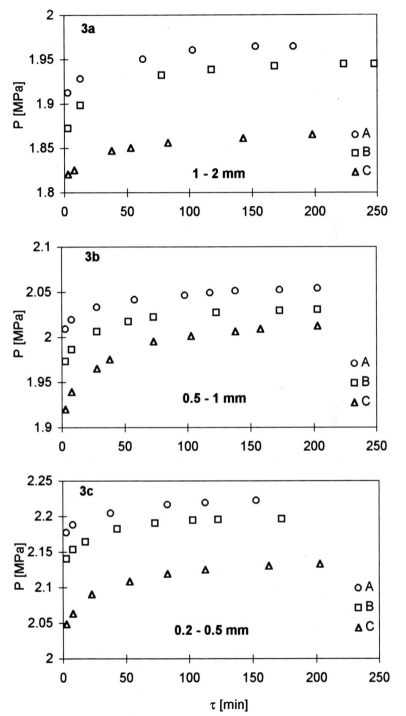

Figure 3. Time dependence of methane desorption from coal samles with various grain size.

A - dry sample, B - sample with water, C - sample with sludge suspension

5.2. EFFECT OF WATER AND SLUDGE ON THE DESORPTION COURSE.

The diagrams in the Fig.3a,b,c illustrate clearly the differing time behaviour of desorption of tested samples, depending on their character and grain size. During the same time interval, the largest quantity of methane expressed by increase of methane pressure is released from the original Sample A, a smaller quantity from Sample B with pressurized water, and the smallest one from Sample C with pressure-forced sludge. This sequence corresponds with the action of individual barriers as they were determined by porosimetry in analogous Samples a, b and c without adsorbed methane. The course of the desorption kinetics has been expressed by the function

$$P = \tau / (a + b. \ \tau) \ , \tag{1}$$

where P (MPa) is pressure, τ (min) is time, and a, b are constants. The transformation into linear form in the coordinates τ/P versus τ enables the constant b (MPa^{-1}) resp. its reciprocal value v_{glob} (MPa) to be determined. Both values shown in Table 3, can be used as differentiation parameters, characterizing the effective degassing rate expressed by the pressure change. The highest degassing rate is found in original samples with unblocked porous system. It increases with decreasing grain size due to shortening of the degassing path within the porous system. The dependence on the grain size is maintained also in samples with water and sludge, the rate decreasing again according to the degree of blocking of the outlet pores. Most remarkable rate differences are evident for the largest grain size, for which the dependence of quantity τ/P on time τ in linearized form is shown in Fig.4.

TABLE 3. Constant b and rate parameter v_{glob} during the pressure sorption

Sample	Grain size [mm]	Constant b [MPa^{-1}]	Parameter v_{glob} [MPa]
A	1.0 - 2.0	0.500	2.000
	0.5 - 1.0	0.487	2.053
	0.2 - 0.5	0.450	2.222
B	1.0 - 2.0	0.572	1.748
	0.5 - 1.0	0.493	2.028
	0.2 -0.5	0.455	2.198
C	1.0 - 2.0	0.634	1.577
	0.5 -1.0	0.497	2.012
	0.2 - 0.5	0.468	2.137

A-original coal, B-coal with pressurized water, C-coal with pressure-forced sludge

6. Conclusion

- It follows, from results of the high pressure porosimetry and high pressure desorption, simulating selected conditions of the hydraulic mining, that the pressurized water or sludge cause changes in the porous structure of coal, affecting thus the rate and extent

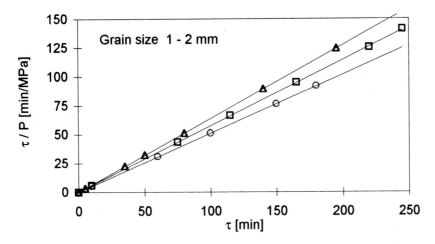

Figure 4. Relationship between desorption parameter τ/P and time τ in linear coordinates
o (A) - original, □ (B) - water, Δ (C) - sludge

of desorption of methane contained in the coal substance. In the case of water, swelling can be suggested, narrowing down the transport paths, while for the sludge, the cause consists in the formation of steric barriers, which are blocking the coarse pores and a part of macropores. The highest reduction of the desorbed quantity, compared with the original sample, was established for coal with pressure-forced sludge.

The desorption of methane, adsorbed at increased pressure, proved a dependence on the coal grain size, its rate increasing with the decreasing grain size.

Acknowledgements

This work was supported by EUROGAS Ltd under Contract 620/77

7. References

1. Weishauptová,Z. and Medek,J. (1998) Change of pore distribution due to different water content, *Acta Montana*, in press
2. Weishauptová,Z. and Medek,J. (1998) Bound forms of methane in the porous system of coal, *Fuel*, 77, 71-76
3. Medek,J. (1977) Possibility of micropore analysis of coal and coke from the carbon dioxide isotherm, *Fuel*, 56,131- 133

COAL SEAM GAS EMISSIONS FROM OSTRAVA - KARVINA COLLIERIES IN THE CZECH REPUBLIC DURING MINING AND AFTER MINES CLOSURE

Georges Takla, Zdenek Vavrusak
OKD, DPB PASKOV, akciova spolecnost
Rude armady 637
739 21 Paskov
Czech Republic

Abstract

The Ostrava-Karvina part (OKR) of the Upper Silesian coal basin is the most important hard coal basin in the Czech Republic. This basin extends over an area of 1,600 km². The Carboniferous basin fill contains 255 seams with a net coal thickness of 150 m. In the north-western part of the basin, the coal seam gas contains methane and carbon dioxide. But in the remaining part it contains only methane. The gas content in the coal is estimated between 4.4 to 20 m^3/t. The seams of the western and south collieries are prone to outbursts of coal and gas.

In the active mines, a system of methane drainage is operating to ensure mine safety. Other safety measures are taken in the mines which exploit the seams susceptible to outbursts of coal and gas to predict and to prevent their occurrence.

Some of the collieries are already closed or being closed. Coal mines closure does not mean that methane liberation into the underground abandoned areas would be stopped. After closure water level in the mines would rise. During this stage some mine gas may leak uncontrollably to the surface via old mine shafts and natural migration channels. This migration is greatly influenced by rapid drops in atmospheric pressure. A risk of explosion or fire would occur in the basements of buildings or industrial plants in the densely populated city of Ostrava. A complete system of gas extraction, gas monitoring and safety measures in the buildings is being built to prevent the gas emissions to the atmosphere and to ensure the inhabitants safety.

1. Introduction

The hard coal reserves in the Czech Republic are found in three basins: Upper Silesian coal basin (Czech part), Lower Silesian coal basin (Czech part) and Central Bohemian coal basins (Figure 1).

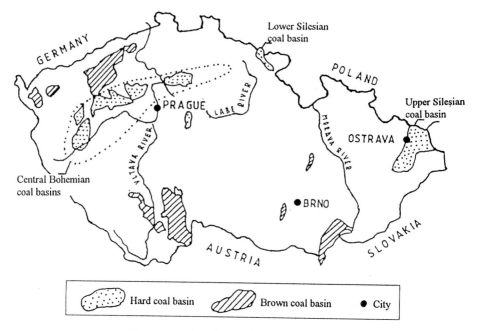

Figure 1. Location of coal basins - Czech Republic

The Ostrava-Karvina Coal Basin (OKR) contains the most significant Czech coal deposits. It covers an area of about 1,600 square kilometres (about 0.4 million acres). This area constitutes to about 20 % of the Upper Silesian Coal Basin, which straddles the Polish / Czech frontier. Within that area, about 335 km^2 belong to mining claims of individual collieries.

Hard coal production in the OKR accounts for more than 95 % of the total hard coal production in the Czech Republic. More than 99.8 % of methane emissions from mines in the Czech Republic are related to mining operations in the OKR.

Coal in OKR is extracted by two mining companies, OKD, a.s., Ostrava and CMD, a.s., Kladno.

2. The Ostrava-Karvina coal basin: brief geologic description

Coal bearing sediments of the OKR range in age from Namurian A to Westphalian A. The OKR contains two formations: the Ostrava formation and the Karvina formation [3].

The Ostrava Formation was deposited under coastal plain conditions, with marine transgressions and regressions. The Ostrava Formation contains 168 coal seams with an average coal seam thickness of 0.73 m (2.4 ft). The coal seams of the Ostrava Formation are characterised by a relatively high dispersion of the volatile matter values.

The Karvina Formation was formed after a short hiatus under entirely non-marine conditions. This resulted in a substantially greater coal seam thickness, with less dispersion in coal rank values. The Karvina Formation contains 87 coal seams with an average coal seam thickness of 1.2 m (3.94 ft).

Both formations contain productive layers with a total of 255 coal seams, 120 of which are considered as workable. The total net coal thickness is approximately 150 metres (492 feet).

The coal bearing sediments are overlain by Miocene clays in the northern part of the OKR, and by hard rocks of the Silesian and Upper Silesian Secondary and Tertiary nappes in the south-eastern part. The overburden thickness varies between 220 and 1,000 meters (722-3,280 ft).

In the OKR, coals ranging from high volatile bituminous coal rank to anthracite are mined. As an average, OKR coal contains 0.6 % sulphur, 15 % ash, 3.5 % moisture, and 23 % volatile matter. The average heating value in situ conditions is 25.5 MJ/Kg (corresponding to 10,965 Btu/lb.). Approximately 73 % of the coal is of coking grade.

The structural style of the OKR changes across the Orlova - Michalkovice folded and faulted zone (Figure 2). The structural differences between the two sides of the tectonic feature are also expressed in the hydrogeology of these areas. Furthermore, this tectonic boundary separates two areas with different rates of gas emanations in coal mining operations. Coal mines in the western area tend to release higher amounts of gas during mining, combined with a higher water production, compared to the ones on the eastern side of the tectonic feature [1].

According to the present state of knowledge, and also in comparison to the Polish part of the Upper Silesian coal basin, the coal of the OKR contains an average of 4.4 to 20 cubic metres of methane per metric ton of coal (equivalent to 141-640 cft/t) [8]. In the north-western part of the basin, the coal seam gas contains methane and carbon dioxide. But in the rest part it contains only methane. The seams of the western and south collieries are prone to outbursts of coal and gas [2].

The coal seams in the OKR are generally considered as dry. The primary source of water in the coal mines is the Miocene detritus. The water in the mines is highly saline, with a high amount of dissolved chlorides and sulphates.

3. Brief history of gas emissions in OKR

Coal mining in the OKR started about 200 years ago. In the course of industrial mining development, difficulties were experienced from the high methane content of

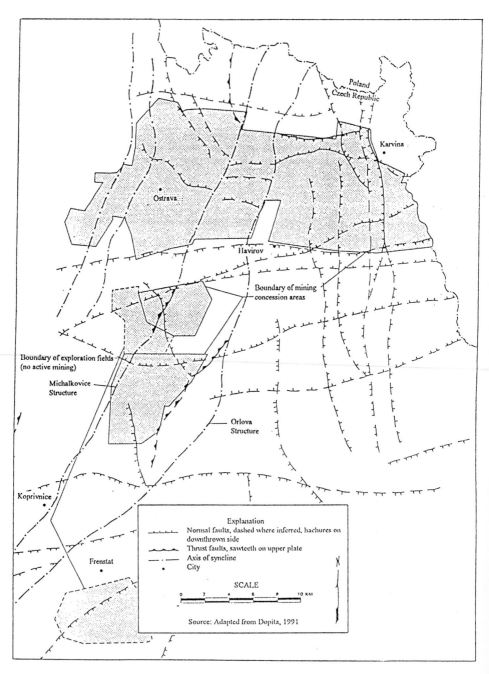

Figure 2. Tectonic map of the Upper Silesian coal basin (Ostrava-Karvina coal field)

the coal. Since the middle of the last century, a total of 68 gas explosions was recorded
from the mines operated, with many miners killed and severe economic losses. At the
beginning of this century, total coal production per year in the OKR exceeded five
million tons. The oldest recording of the quantity of methane drained by mine
ventilation is known from the year 1910 [6]. In the course of this year the coal
production increased to 7.67 million metric tons (8.45 million short tons) per year,
accompanied by a release of methane averaging 589,000 cubic meters per day (20.8
million cubic feet per day). From the years 1905-1908, first methane outbursts became
known from exploration drilling in coal beds of new coal fields south of the mining
areas.

Because of the advancement in mine mechanisation and increase of coal mining
intensity, it became necessary to introduce a system of methane drainage in the mines,
for safety reasons. Following the experience in other European coal mine districts, a
system was designed to insulate and drain part of the methane liberated during coal
mining.

During the 35 years of operation of mine methane drainage, an efficiency of about
30 % of all the methane released was reached. This means that about 30 % of the total
amount of methane released from mining operations in the whole mining district is
collected and drawn into the methane drainage plants. In some longwalls, it is possible
to reduce the amount of methane released into the atmosphere by up to 50 % or more
(Figure 3).

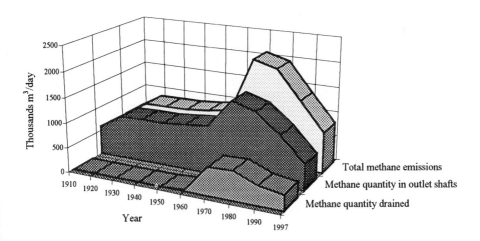

Figure 3. Methane emissions in Ostrava - Karvina coal basin
Source [4, 6]

The methane drainage system was designed for mine safety in the first place, so in the first period of operation all drained gas was vented into the atmosphere. But within a short period of time, after further development of the mine gas drainage network, some mines heating plants were modified to enable them to use mine gas for heating and for the production of hot water.

4. Overview of the present state of gas emissions

At present, the mine gas drainage system in the OKR consists of 21 individual mine gas drainage plants with a total of 115 vacuum pumps installed. An extensive underground pipeline network connects about 5,000 mine gas sources, mainly mine boreholes drilled in the roof of longwalls, to this system. In addition to these boreholes, abandoned mine and „oldman" workings are connected to the system as well, separated from active mine workings by seals. From individual mine gas drainage plants, pipelines feed into the main pipeline system. The gas mixture collected is then delivered to consumers in local industry. The gas delivered consists of a mixture of methane with air, with a methane content of about 50-55 %. The equipment that burns that gas has to be designed in a way to take such gas composition into account. The typical gas mixture consists of 50-55 % of methane, 2-4 % of carbon dioxide, 1-3 % of oxygen, and 47-38 % of nitrogen. The heating value of such gas is about 19-20 MJ per cubic metre (510-537 Btu/cft) [4].

The present gas pipeline network forms a system of about 92 km (57 miles) total length. In view of the very low operating pressure, the diameter of the pipe used is relatively large: 200 up to 500 mm (7.8-19.7 in). This is due to the technical characteristics of the vacuum pumps used, which produce a pressure of only 45-50 kPa (6.5-7.25 psi) on the delivery side.

The captured gas is used by the collieries themselves for heating and hot water production and by the local steelworks and heating plants. A small amount of the captured gas is vented to the atmosphere because of low content of methane usually less than 40% or because of technological reasons (Figure 4).

Due to the recent decline in coal mining industry (Figure 5) and improvements in the efficiency of mine ventilation systems, the importance of mine gas drainage for the safety of mine operations is gradually decreasing. In view of this development, our company has worked out new methods for collecting methane from active mines.

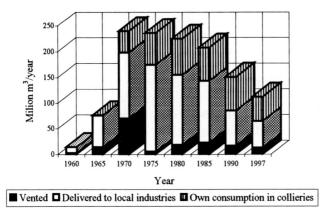

Figure 4. Methane production from mine gas drainage in Ostrava - Karvina collieries and its utilization
Source [4]

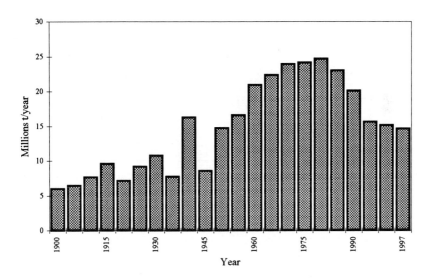

Figure 5. Coal production in Ostrava - Karvina coal basin
Source: Statistical data of OKD, a.s.

Based on a detailed study of geological data, new locations and rock sequences where methane could have accumulated in greater quantities were explored. At such locations specially designed mine boreholes are drilled, leading to increases in total coalbed methane production. Locations favourable for methane production could be found in the upper part of mine claims, in weathered Carboniferous strata, in faulted intervals, and also in areas with unconsolidated rocks covering coal mine areas. The development of such opportunities considerably increases mine gas production from gas drainage plants in active mines. This activity so called supplementary mine gas drainage started in 1993. The 87 boreholes of total length 17, 028 m have been drilled. The total production reached a quantity of about 24 millions m^3 (Figure 6).

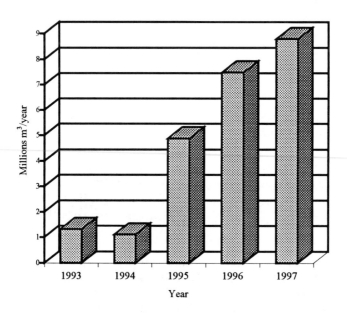

Figure 6. Methane recovery by supplementary gas drainage
Source: [4]

In addition to conventional gas drainage in coal mines, our company has also been involved in a fundamental development program for exploration, production and utilisation of coalbed methane from abandoned coal mines.

5. Gas emissions from abandoned coal mines

Due to the political and social changes in the Czech Republic, which started in 1989/1990, coal production decreased as the coal industry was restructured and unprofitable coal mines were closed down. In the Ostrava part of the mining district of the OKR, four collieries were shut down. Mine claims of these collieries cover an area of more than 100 square kilometres (more than 24,700 acres) extending partly over very densely populated quarters of Ostrava town. After closing operations, the collieries were liquidated and hoisting and outlet shafts filled up in steps. Individual coal mines are connected to one underground whole which is connected to the surface. The number of known old shafts exceeds 300, and there are about 60 old adits/galleries there. The water level in the mines would raise as predicted, along with a gradual flooding of the mine workings to the level expected. During this stage some mine gas may leak out to the surface and out of control via old mine shafts and other potential leaks in their vicinity, and also via natural migration channels, such as fissures and tectonic faults. In compliance with gained knowledge in shafts closing and gas volume measurement in them and in degassing boreholes already drilled as it will be stated below we can also state the fact that barometric pressure fluctuation - even as a short-term acting factor, influencing gas pressure in the underground areas - will have the main influence upon gas pressure and its possible leakage to the surface. A risk of explosion or fire would occur due to such gas possibly finding its way into basements of buildings or industrial plants. This risk would be combined with the risk of deteriorating the quality of the air by releasing methane into the atmosphere. To prevent from and/or limit the unchecked mine gas leakage to the surface a system of active and passive degassing protection measures and gas monitoring has been created and is being implemented.

At three former mine sites, mine gas drainage systems and plants had been in operation at the surface before the decision to close the mines was taken. The authors' company conducted a study, showing the necessity to maintain mine gas drainage from shut-down mines even after the termination of mining in the OKR. This was the reason why the company bought the surface equipment of complex mine gas drainage plants of two mines before their closure.

Before the mine shafts were filled up, boreholes were drilled at mine locations characterised by main concentrations of methane. These boreholes were connected to an existing pipeline. Precautions had do be taken to avoid damage of the mine gas pipeline while the mine shafts were filled up. The abandoned shafts of these two mines connected to their gas drainage plants were adapted as gas production shafts, by filling in their upper section only and leaving the bottom parts open and connected to former mine working levels, so that those could act as a reservoir for mine gas. In the upper filled up part of such shafts, pipes had to be installed so that the mine gas collected could be pumped into the mine gas drainage plant (Figure 7).

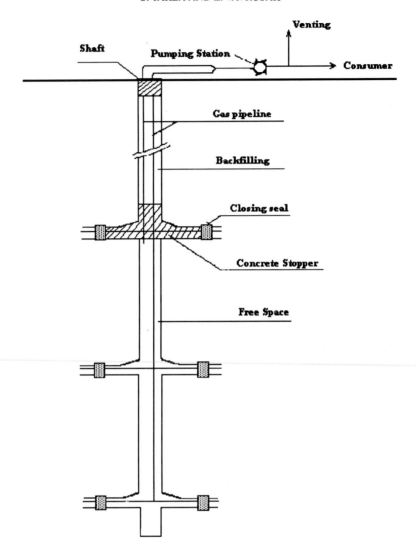

Figure 7. Mine gas pumping scheme in shafts
Source: OKD, DPB PASKOV, a.s.

A similar situation as in the case described above incited our company (DPB) to start the implementation of a drilling surface gob wells program in the area of the abandoned coal mines with the same expected good quality of gas.

The first two surface gob wells verified:

- the possibility of recovering gas from abandoned coal mines area,
- the technical and technological requirements for drilling, completing and testing such wells and
- the possibility of commercial use of this gas.

These wells also demonstrated the feasibility of gob gas extraction from abandoned coal mines area. (Figure 8 demonstrates the gas production test from gob well V-3). Last year DPB drilled two more wells.

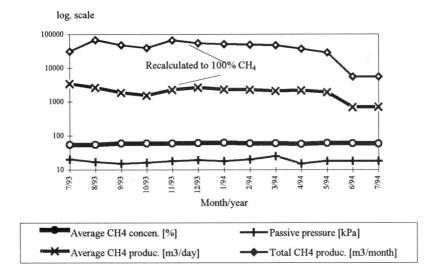

Figure 8. V-3 gob well gas production
Source: OKD, DPB PASKOV, a.s.

DPB believes that mine gas of a quality sufficient for industrial use can be pumped out from closed down mines in the north-eastern part of the Ostrava mining district. The methane content of the gas mixture produced varies from 40 to 80 %, but an increase in methane content can be expected once the filling up of the abandoned shafts is completed.

Five other shafts were selected to be adapted in the same manner as gas production shafts and connected to recently constructed gas pumping stations to form permanent passive pressure within the underground areas of the rest closed mines district where the mine gas contains less than 30 % of methane and a high quantity of carbon dioxide.

The described system of gas drainage plants and gas pumping stations could be realised only on the shafts which were in operation when the decision of implementation of such system has been taken. The assumed areas to be influenced by each station is determined in compliance with the possible underground connection of the shafts and the nearest old communication with the surface. This communication may be because of existence of some old shaft fulfilled in an unknown way or of old surface wells.

The passive protection measures are taken in the area which is not influenced by the active measures and where the thickness of the Miocene overburden over the exploited Carboniferous coal layers is less than 50 m. This limit has been determined in accordance with verified foreign experience, with regard to specific conditions of the OKR, and without any regard to local Tertiary and Quaternary layers composition.

In such areas in all the old liquidated shafts boreholes have to be drilled to the Carboniferous layers and to be equipped with 2,5 m high chimneys enabling the gas releasing to the atmosphere.

In the same time, a network of more than 50 degassing boreholes was set. The direction of the boreholes is towards the highest places of the worked - out coal seams areas. The boreholes would be equipped with the same chimneys like the shafts. The boreholes and chimneys equipment is shown in the scheme (Figure 9).

This passive measure would serve for gas positive pressure elimination in old workings and for monitoring of gas pressure depending upon barometric pressure changes and water level rising in the underground areas.

A mobile pumping station already constructed is an integral part of this passive subsystem. It would be able to pump gas from any borehole and / or old shaft when necessary.

The function of the whole system - the pumping stations, the old shafts and the boreholes requires to be both measured and evaluated. To equip the individual integral parts with measuring devices seems to be different according to the type and the importance of the part's position. It shows to be necessary to concentrate the gained information in one place - in a central control room, to form a necessary data bank with the main aim not only to evaluate the past or the present state but to forecast the possibility of occurrence of all the extraordinary situations.

6. The outbursts of coal of rock and gas prediction and prevention in OKR

The outbursts of coal or rock and gas are very well known phenomena during the coal mining in OKR. In the period of the years 1894 to 1997 about 480 outbursts have been registered [5,6] . Only three cases were outbursts of rock (sandstone) and gas and the others were outbursts of coal and gas (methane or mixture of methane and carbon dioxide). 318 cases were initiated by blasting works, i.e. about 66 % of the total number. The biggest quantity of burst coal was estimated to 540 t and of burst rocks was estimated to 1,200 t. The biggest volume of released gas was estimated to 100,000 m^3 of methane, [5] .

An important feature of the outbursts is that they occur in the faces just in operation, it means in direct connection with the disturbance of the primary state of stress of the rock massif. Therefor the majority of the outbursts happens in the untouched area of the mine field.

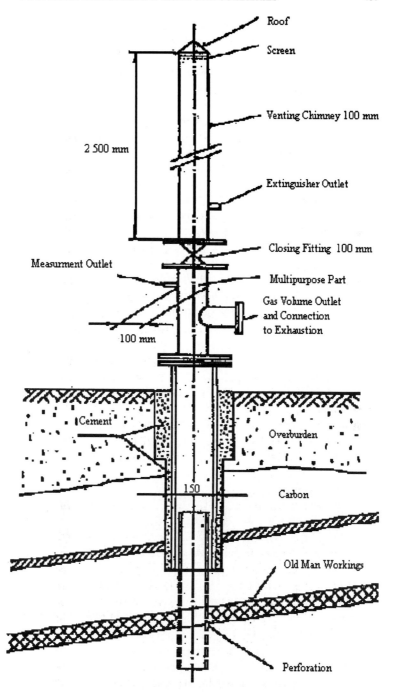

Figure 9. Monitoring and degassing borehole equipment scheme
Source: OKD, DPB PASKOV, a.s.

    To eliminate the outburst events, many prognostic methods and manners of implementing the preventive measures were developed. A continuous measurement of the gas pressure and of the rate of gas desorption is used successfully. When the limit values of these prognosis are measured a shock blasting operation is used as a passive measure to initiate an outburst when the miners are not present at the face.

    As for the active methods and measures of prevention, mostly blasting is applied to drive the workings and the water infusion in the coal pillar by means of long boreholes from the headings is used as the main method. Sometimes destressing drilling is also used.

    Five mines of OKR were prone to occurrence of outburst of coal and gas, only in one of them outbursts of rock (sandstone) and gas occurred. In another coal mine the released gas during the outburst was a mixture of methane and carbon dioxide. Three of these mines are now closed. The exploitation of coal in one of now existing two mines would be finished in the end of 1998. So in the near future only one mine will be prone to outbursts occurrence. The mentioned mines were closed or would be closed because of the unprofitable coal exploitation in them.

    The number of occurred outbursts has been significantly diminished in OKR not only due to reducing of mining activity but mainly due to the improved efficiency of the applied measures of outbursts prediction and prevention. (Figure 10).

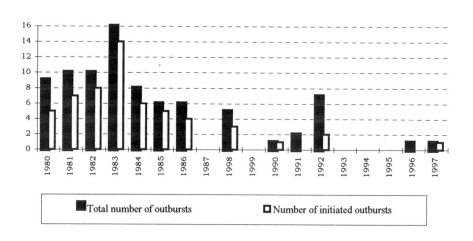

Figure 10. Number of outbursts occurred in Ostrava - Karvina collieries
Source [5]

9. References

1. Dopita, M., and Havlena,V.: Geology and Mining in the Ostrava-Karvina coal field, Moravske tiskarske zavody, Ostrava, (1972).
2. Dopita, M. ET AL.: Geologie ceske casti hornoslezske panve (Geology of the Czech part of the Upper Silesian basin). Ministry of environment of Czech Republic, Prague, (1997).
3. Havlana, V.: Geologie uhelnych lozisek 2, (Geology of Coal Deposits 2), SNTL, Prague, (1964).
4. OKD, DPB Paskov, akciova spolecnost: Statistical data of gas emissions and drainage in Ostrava-Karvina collieries unpublished, (1960-1997).
5. OKD, DPB PASKOV, akciova spolecnost, Statistical data of geodynamic phenomena in Ostrava-Karvina collieries, unpublished, (1975-1997).
6. Ostrava Bureau of Mines, unpublished protocols, Archives.
7. Rakowski, Z., Lat., L., Hruzik, B. and Dvoracek., J.: Nove poznatky v problematice prutrzi uhli a plynu v OKR (New knowledge about the problems of outbursts of coal and gas in OKR), SNTL, Prague, (1983).
8. Raven Ringe Resources, Inc.: Assessment of the Potential for Economic Development and Utilization of Coalbed Methane in Czechoslovakia, U.S.EPA, Washington, D.C., (1992).

COUNTERMEASURES AND RESEARCHES FOR PREVENTION OF METHANE EMISSION INTO THE ATMOSPHERE IN A JAPANESE COAL MINE

K. Ohga[1], S. Shimada[2], Higuchi[1] and G. Deguchi[3]

1) Hokkaido University, Sapporo, 060 JAPAN

2) University of Tokyo, Tokyo,113 JAPAN

3) Japan Coal Research Center, Tokyo,001 JAPAN

Abstract

In order to prevent methane emission from underground coal mines into the atmosphere, various technologies are required.

In this paper, prediction methods of methane emission from the longwall panel and methane drainage methods are specifically described .

1. Introduction

In Japan, the coal mining industry is in a severe situation because policy on it has changed. Therefore, some coal mines had been closed. We have only two coal mines, one is Taiheiyo Coal Mine in Hokkaido and the other is Ikeshima Coal Mine in Kushu. Both coal mines are situated under the sea. It is difficult to increase the air flow rate for ventilation. In the design of mining panels and methane drainage boreholes, it is an important thing to estimate the methane emission from the mining panel during mining of coal .

The evaluation of the various prediction methods of methane emission from the new mining panel at the Taiheiyo Coal Mine are attempted. One is a statistical method, which is estimation by the statistical analysis of measured data of methane emission from the longwall panel during mining of coal. Another method is estimating the methane emission from the measured data of methane content of coal. Still another method is estimation of methane emission from mining panels by calculation using simulation program of methane emission from the surrounding mining panels. The last one is estimation from the results of measured methane emission during driving of gate road. In this paper, details of these estimation methods and results of evaluation are described.

2. Taiheiyo Coal Mine

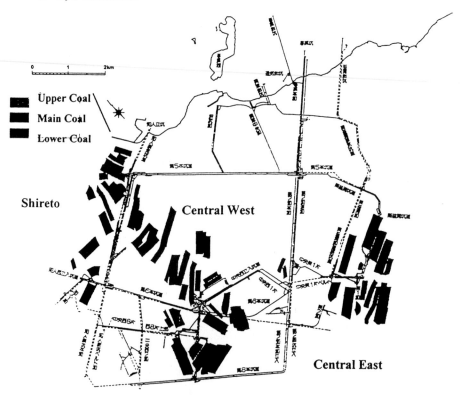

Fig.1 Mining Areas and Mining Panels in Taiheiyo Coal Mine

Taiheiyo Coal mine is in eastern part of Hokkaido. Coal production is about 6000 t/ day . Most of coal are produced by longwall mining method , width of longwall is more than 250 m. Average mining depth is about 600 m below the sea. Gas Make per clean coal is about 20 m3 /t. There are plans for a higher production longwall panel, width of which is more than 300m and production of coal per day is 10,000 t/day. Average mining depth will be 700m at that time. In this coal mine, mining areas are divided into three areas, which are Shireto Area, Center West and Center East Area as shown Fig.1.

3.   Methane Drainage Methods

Methane drainage from virgin coal seam is difficult due to the low permeability of coal. Therefore, methane drainage from the upper relaxed areas using three different length boreholes is usually used.   Fig.2 -A shows the outline of methane drainage by long-length boreholes with length of more than 500 m. These boreholes are drilled from the drainage station excavated in the rock seam above the mining coal seam. The number of these boreholes are usually 3 or 4.

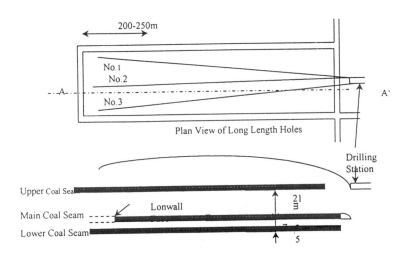

Plan View of Long Length Holes

Cross Section of Long Length Holes(A-A')

Fig.2-A   Methane Drainage by Long-length Boreholes

Fig.2-B shows the outline of methane drainage by middle-length boreholes with length from about 200m to 300m. These boreholes are drilled from the drainage station in rock seam as shown in this figure.

Furthermore, short length boreholes with a length of less than 100m are drilled from the tail gate of mining panel. Usually, combination of the mentioned three length boreholes are used.

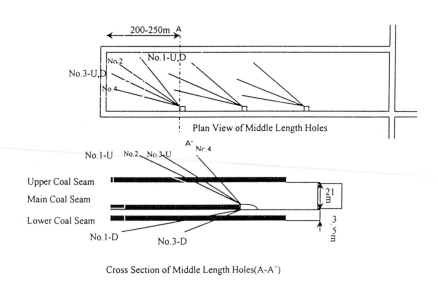

Plan View of Middle Length Holes

Cross Section of Middle Length Holes(A-A')

Fig.2-B Methane Drainage by Middle-length Boreholes

4. Estimation of Methane Emission from the Mining Panel

At this coal mine, we have tried to predict the methane emission from the mining panel using four methods; One is the method by statistical analysis. Second one is based on measurement results of methane content of coal in the mining panel. Third one is estimation by using a simulation program of methane flow rate from the mining panel. Last one is estimation from the methane emission during driving gate road of the mining panel.

4.1. Statistical Analysis of Methane Emission during Coal Production

Statistical data used in this analysis were measured and calculated as follows.

Amount of methane drainage is calculated from methane concentration, methane flow rate and pressure in the drainage pipe. These are measured at each drainage stations and the measured values are transmitted to the computer on the surface and calculated.

Methane emission into the ventilation is calculated from methane concentration of intake and return air and air flow rate. These data also is transmitted to the computer and calculated.

Fig.3 shows the location of each sensor in a mining panel. The circle marks show the location of methane concentration sensors in the ventilation. Square mark is the flow rate measurement device in the ventilation air. Triangle mark is the location of the sensors to measure the methane concentration and flow rate in the drainage pipe at the drainage station.

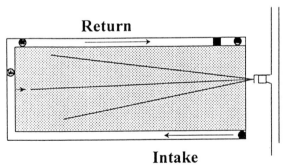

Return

Intake

● **Methane Concentration**
■ **Flow Rate**
□ **Methane Concentration and Flow Rate in the Drainage Pipe**

Fig.3 Location of Each Sensors

We carried out correlation analysis among total gas emission, gas drainage, gas make, gas emission rate, gas drainage efficiency, length of panel, width of longwall, total coal production and average coal production rate in each mining area. The results of the analysis is shown in Table 1. + marks indicate the strength of relationship between each factors. +++ marks indicate that there is strong relationship between factors and - marks indicate that there is no relationship between two factors.

In this Shireto Mining Area, the relation between total coal production and total gas emission, between gas drainage rate and average coal production rate, between gas make rate and average coal production, and between gas emission rate and average coal production rate are strong.

Fig.4 shows the relation between total gas emission and total coal production in each mining area. In each mining area except for CW (Central West Mining Area) a strong relationship is indicated. Distribution on CW can be divided into two point sets which are upper and lower groups. The mining panels in the upper group have not been affected by anything and the mining panels in the lower group have been affected by mining adjacent mining panel, wherein , methane emission is found to decrease.

Table 1 Results of Correlation Analysis

	Length of Panel (m)	Width of Longwall (m)	Total Coal Production (t)	Average Coal Production Rate (t/day)
Total Gas Emission	+	+	+++	+
Gas Drainage	-	+	-	+++
Gas Make	-	+	-	+++
Gas Emission Rate	-	-	-	+++
Gas Drainage Efficiency	-	-	-	-

Fig.5 shows the relation between total gas emission and length of panel. There is no substantial relation between these factors. Fig.6 shows the relation between average coal production rate and gas emission rate in each mining area. There are strong relations between these values in each mining area. The lines in Fig.6 are regression lines of each mining area. By using the regression lines, we will be able to estimate the methane emission from the mining panel in each mining area.

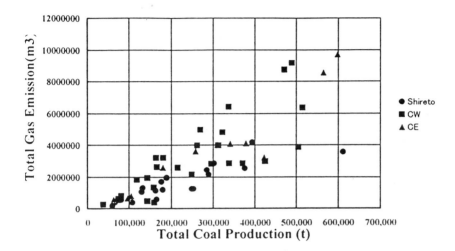

Fig.4 Relation between Total Gas Emission and Total Coal Production

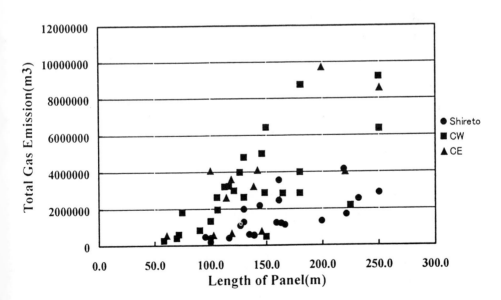

Fig.5 Relation between Total Gas Emission and Length of Panel

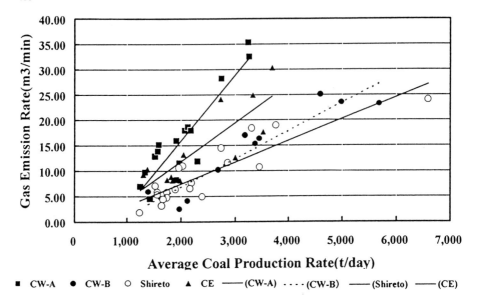

Fig.6 Relation between Average Coal Production Rate and Gas Emission Rate

4.2. Estimation from Methane Content of Coal

In this method, methane content of coal seam in the mining panel must be determined initially. Therefore, the coal samples are collected from the mining panel by drilling machine as cuttings and desorbed gas from the coal cuttings is measured by special measurement device which we developed (Ohga k. *et al.*,1991).   The determination method is basically same as the UDBM method. This instrument has 4 separate vessels and sensors to measure temperature and atmospheric pressure. Each vessel has a pressure sensor and the increase of pressure in each vessel caused by gas emission from the coal sample is automatically measured and recorded at fixed time intervals. After measurement the data is transmitted to a personal computer via RS 232 port. The amount of emission gas is calculated from the measured pressure in   each vessel by the personal computer. In this measurement, about 40g of coal cuttings size of which is between 4 and 16 mesh is used for coal sample.

A rough emission gas from the mining panel can be estimated considering the

measured results and historical data on methane emission from the mining panel
directly.

4.3   Estimation by using a simulation program

Emission gas is calculated by using a numerical model on methane gas flow rate from
the mining panel(Ohga K. *et al.*,1993). Fig.7 shows the numerical model used in this
estimation. The flow rate of methane from the main coal seam which is mined is
calculated by using a model of the plan view. The flow rate from the underlying coal
seam and overlying coal seam is calculated by a section view model .

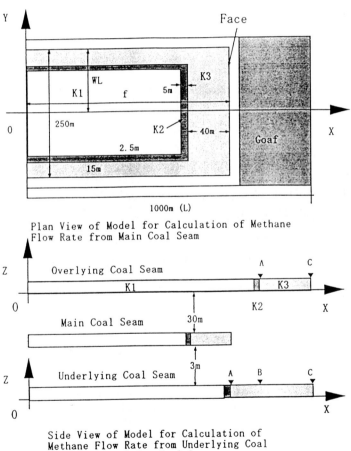

Plan View of Model for Calculation of Methane
Flow Rate from Main Coal Seam

Side View of Model for Calculation of
Methane Flow Rate from Underlying Coal
Seam and Overlying Coal Seam

Fig.7 Numerical Model

The gas permeability in the large part of mining panel is K1, the permeability of the highly stressed zone is K2, and the permeability of disturbed zone adjacent to the gate road and the working face is K3.

Fig.8 shows a result of this calculation. Methane flow rate from the working coal seam is from 10.7 to 13.2 m3/min. Methane flow rate from the lower coal seam into the working area through the gob area is from 4.6 to 6.9 m3/min. From the lower coal seam through the gob area to the upper disturbed area is from 3.8 to 4.8 m3/min. From the upper coal seam emission gas is about 0.8 m3/min. Therefore, we can capture the methane from 4.6 to 5.6 m3/min by cross measure boreholes. To reduce the methane emission from the working coal seams into the working area, drilling of the horizontal methane drainage boreholes for working coal seams are needed.

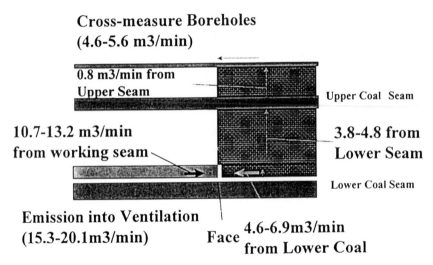

Fig.8 Results of Calculation (Central East Mining Area)

4.4 Estimation from Measured Methane Emission during Driving Gate Road

We attempted to estimate from the measurement results of methane emission during the driving of gate road. We could not use the methane derectors because the change of methane concentration in the ventilation during the advancing of gate road is very little. Therefore, we collected intake and return air samples and analyzed them by gas-chromatography.

Fig.9 shows a result of measurement of methane emission during the driving of gate road and coal production. Square marks show the results of gas make during driving gate road and the curve shows the measurement results of gas make during mining of panels. In the beginning of coal production, there is no big difference between gas make during advancing gate road and producing coal. These measurement has been carried out few times. Therefore, it is not enough to evaluate this estimation method.

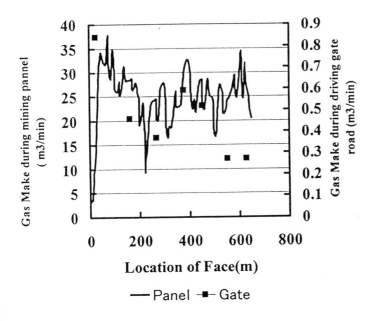

Fig.9   Measurement Results

4.   Conclusion

1) In order to make a long range mining plan, statistic analysis is a convenient method.

2) To make a short range mining plan and design of methane drainage boreholes, a numerical method using measurement results of methane content of coal and the prediction method based on measurement of methane emission during the driving

of gate roadway are effective.

3)   Combination of the mentioned estimation methods is the best way to make an accurate prediction of the methane emission from new mining panels.

4)   To reduce methane emission from the working coal seams, drilling of methane drainage boreholes in the mining coal seams is required as based from the results of numerical method. However, it is difficult to drill horizontal long-length boreholes into the coal seams and drain the methane from the coal seams effectively. Therefore, development of drilling techniques of the horizontal boreholes in coal seams and maintaining techniques of drainage boreholes are required.

References

1.   Ohga,K., Higuchi, K. & Hucka, V.J.(1991), A NEW TECHNIQUE OF DETERMINING THE METHANE CONTENT IN-SITU, Proc. Of 5th US Mine Ventilation Symposium, pp.494-499

2.   Ohga, K.,Higuchi, K. & Deguchi, G.(1993), Countermeasures and Prediction Gas Emission from Longwall Panel, Proc. Of 6th US Mine Ventilation Symposium , pp.99-104,

MODELING THE HYDROTHERMAL GENERATION OF COALS AND COAL-SEAM GAS

D.L. LOPEZ(1), M. COBB(1), S.D. GOLDING(2), AND M. GLIKSON(2)

(1) Department of Geological Sciences, Ohio University, Athens, Ohio 45701
(2) Department of Earth Sciences, University of Queensland, Brisbane, QLD 4072, Australia

Abstract

A two-dimensional computer program (HYDROMAT) for modeling the maturation of coal in sedimentary basins affected by hydrothermal circulation has been constructed. The program considers the effect of fluid flow on the thermal regime through time. The program contains: 1) a geodynamic module to simulate compaction, erosion, deposition, and tectonic uplift and subsidence of the sediments, 2) a fluid flow and thermal module to solve the coupled fluid flow and heat transfer equations, and 3) a chemical kinetic module to simulate the maturation of organic matter. The effect of magmatic intrusions during the evolution of a basin is considered. Coordinates fixed to the strata (Lagrangian coordinates) are used to follow the evolution of the coal generating units. The kinetic module uses the distribution of activation energies for parallel Arrhenius-type first order rate expressions to calculate the extent of chemical transformations over the thermal history of the basin. The program has been used to assess the effect of advective and/or convective fluid circulation on the generation of hydrocarbons. Our initial results suggest that for basins affected by high heat flow and fluid circulation, HYDROMAT can reproduce better the maturation of organic matter than conventional conductive models.

1. Introduction

This paper presents the scientific basis for the construction of the computer program HYDROMAT, a program that simulates the maturation of organic matter produced by the circulation of hydrothermal fluids in sedimentary basins. Hydrothermal fluids can play an important role in maturing organic matter in sedimentary basins that have had a complex thermal history, such as foreland basins at plate boundaries. The Bowen Basin in Australia (Glikson et al., 1995, Golding et al., 1996) is a good example of this kind of basin, its complex thermal history is associated with the separation of Australia from

Gondwana (Faraj, 1995). Other examples of sedimentary basin showing evidence of magmatic and hydrothermal activity are the Manus back-ar basin, Papua New Guinea (Both et. al., 1986) and the Pacific Northwest (Summer and Verosub, 1989).

The chemical reactions resulting in the transformation of organic matter into coal, oil or gas can be produced by the increased temperature in the earth crust as sediments are buried or by the intense heat generated at relatively shallow depths when the regional heat flow is high and/or magmatism occurs. Hydrocarbons generated by burial mechanisms alone are believed to form within the temperature range from ~60 to 150° C, and hydrocarbons matured by hydrothermal fluids at a temperature range from ~60 to >400° C (Simoneit, 1984). One striking difference between the two physical-geochemical paths is that the reaction rates for the conversion of organic matter to petroleum in hydrothermal system are several orders of magnitude faster than for hydrocarbons formed by burial alone (e.g. Peter et al., 1991). Laboratory work by Seewald (1994) demonstrates that in hydrothermal systems, water participates directly in the redox reactions involving species such as ethane, ethene, and inorganic minerals. Water is an important source of aqueous hydrogen gas, therefore it is an important reactant in hydrocarbon redox reactions.

The effect of rapid heating by hydrothermal fluids on the maturation of organic matter has been documented in the lower Jurassic of NW-Germany (the Lias δ shales, Altebaumer et al., 1993). These shales are surrounding the igneous intrusion called the Bramsche Massif. The hydrocarbons from this site were compared with the source rock-type shales of the Douala Basin in Cameroon, which produced hydrocarbons by burial mechanisms. Strong differences between the generation of soluble organic matter in the two systems were found. Mean vitrinite reflectance show a clear shift of hydrocarbon generation maximum from 0.95% Ro in the Douala Basin series towards 1.5% R_o around the Bramsche Massif. The rapid heating rate provided by the intrusion of the "Bramsche Massif" shifted the oil generation window towards higher vitrinite reflectance values. High heating rates affect more the chemical reactions and structural changes which produce increasing vitrinite reflectance than the generation rates of hydrocarbons. Altebaumer et al. (1993) work shows the differences between the generation of hydrocarbons by hydrothermal and burial mechanisms and the need for different approaches in the modeling of both phenomena.

In areas affected by magmatic intrusions, the high temperature of the intrusion promotes hydrothermal circulation. Advection can be the more important heat transfer mechanism. As a consequence, the temperature field is different from the temperature field in a purely conductive regime. Regions of the system were the fluid is moving downward represent zones of cold water recharge that bends the isotherms downward. On the other hand, regions with water moving upward (carrying with it the heat) are usually hotter. The isotherms are bent upwards. This circulation generates a different temperature evolution through time affecting the maturation reactions.

Modeling the generation of hydrocarbons by burial mechanisms usually considers only the conductive heat flow (e.g. Ungerer et al., 1990; Welte and Yukler, 1981). For modeling the hydrothermal organic matter in hydrothermal systems, consideration should be given to the transient heat flow coupled with the fluid flow equations. The combination of geological, geophysical, and geochemical models in hydrocarbon research to simulate the formation, accumulation, and migration of hydrocarbons usually includes several models (Ungerer et al., 1984): backstripping and geodynamics, thermal, maturation, and diphasic fluid flow models (migration model). These models are usually organized as separated modules of a larger program. Because oil and gas migration is not modeled in our program, it contains only three basic modules: a geodynamic module, fluid flow and heat transfer module, and maturation module.

"Backstripping models" use simple porosity versus depth relationships to model the decompaction of the sediments restoring the past geometry of the rock strata (Ungerer et al., 1990). Backstripping allows the evaluation of the structural origin of the basin through the characterization of the subsidence rates. The porosity reduction with depth is assumed to be produced by the expulsion of water. The effect of mineral diagenesis and mass transport is neglected. Backstripping allow us to model the initial conditions of the basin from which we start the simulation of the different physical and chemical processes that produced the final geometry of the basin. "Geodynamic models" are used for basins formed by extension to reconstruct the heat flow history. A significant increase of subcrustal heat flow is associated with lithosphere extension during rifting, followed by thermal relaxation and subsidence (MacKenzie, 1978). Sedimentation rates to fill the basin are considered in these models. In our program, we refer to the "Geodynamic Module" as the program which incorporates the backstripping, deposition, compaction, erosion and subsidence for the reconstruction of the stratigraphic two-dimensional section through time. This program also generates the grid used in the solution of the coupled fluid flow and heat transfer equations.

The common models used to simulate the evolution of the temperature field within the basin are called "thermal models" because they consider only conductive heat effects. Temperature evolution depends on the basal heat input, and heat transfer and redistribution within the sediments. Basal heat flow from the basement is considered to have two components: a radiogenic contribution of the crust and a subcrustal heat flow (Ungerer, 1990). These models neglect the advective transport of heat (e.g. Welte and Yukler, 1981; Ungerer et al., 1990), and cannot be used in the modeling of the hydrothermal maturation of coal. Circulation of hydrothermal fluids within the basin play an important role in the evolution of the temperature field. Water circulation can deflect the direction of the heat flow vectors discharging heat and heating up the regions of water discharge, and cooling down the regions of water recharge (Forster and Smith, 1989; López and Smith, 1995, 1996). The effect of fluid circulation on the thermal regime is more intensive if convective cells are generated within the system.

In some basins (e.g. the Bowen Basin, Golding et al., 1996) magmatic events produced igneous intrusives that cross-cut the coal. The occurrence of magmatic events within the basin, during the maturation of organic matter, imply that the sediments and organic matter were subjected to temperatures as high as 800-1000° C. The simulation of this type of thermal regime requires a multiphase computer program that can handle these temperature and pressure conditions. In P-T coordinates, near the critical point the partial derivatives of the properties of water (e.g. density, heat capacity) are undetermined. According to Hayba and Ingebritsen (1994), the solution of the heat flow and heat transfer equations in pressure and enthalpy space offers a better solution for the regions of pressure and temperature close to the critical point. In P-H space the two phase conditions are represented as a region rather than as a curve and the partial derivatives of the properties of waters are well defined (Hayba and Ingebritsen, 1994). However, most programs solve the fluid flow and heat transfer equations in pressure (or water head) and temperature space (e.g. TOUGH, Pruess, 1991; Shafault, López and Smith, 1995). The program HYDROTHERM (Hayba and Ingebritsen, 1994) solves the fluid flow and heat transfer equations in enthalpy and pressure space and then obtains the temperature field. However, it is a finite difference program that cannot handle the irregular grids generated during the evolution of sedimentary basins.

"Maturation models" simulate the generation of total hydrocarbons, oil, coal, and gases in sedimentary basins. The formation of hydrocarbons is a function of the temperature of the sediments as well as the initial composition of the reacting organic matter. The more frequently used maturation models are kinetic models of hydrocarbon formation that incorporate information from laboratory simulations such as pyrolysis and hydropyrolysis (Winters et al., 1983) and the thermal history of the basin. Hydrocarbon generation (mass rate of hydrocarbon formation) can be found at each time step during the simulation of the history of the basin. The temperature field at the beginning and end of each time step is used for this purpose. These models describe coal, oil and/or gas formation through multiple parallel reactions governed by first-order kinetics and the Arrhenius law (Ungerer et al., 1990; Hermanrud, 1993: Sweney, 1990). This approach is used in HYDROMAT to model the evolution of vitrinite reflectance, and the generation of gas and oil.

The different numerical models for maturation of hydrocarbons in sedimentary basins use either Eulerian coordinates or Lagrangian coordinates (e.g. Bethke et al., 1985). Lagrangian coordinates follow the sediments during compaction. Eulerian coordinates are fix coordinates and do not follow the strata. Most recent computer programs use Lagrangian coordinates rather than Eulearian because they seem to handle better the changes in position of the compacting sediments (e.g. Grigo et al., 1993). Lagrangian coordenates are used in HYDROMAT. The grid space is adjusted after each time step to preserve uniform lithologies within each cell.

2. The program HYDROMAT

The physical and chemical processes involved in the maturation of organic matter are complex. As the organic matter is heated up either by burial and/or by hydrothermal fluids, the maturation reactions occur. Heating and chemical reaction occur simultaneously. However, computer models can only simulate these processes in an approximated way because time and space are discretized. The heating process is assumed to occur uniformly within each cell of strata at every time intervals. The initial and final temperatures at every cell of the discretized strata are used to calculate vitrinite reflectance and hydrocarbon generation for the time interval. The different modules of the program are run one after another at every time interval until the present time of the basin is obtained. Figure 1 illustrates the way that the program flows.

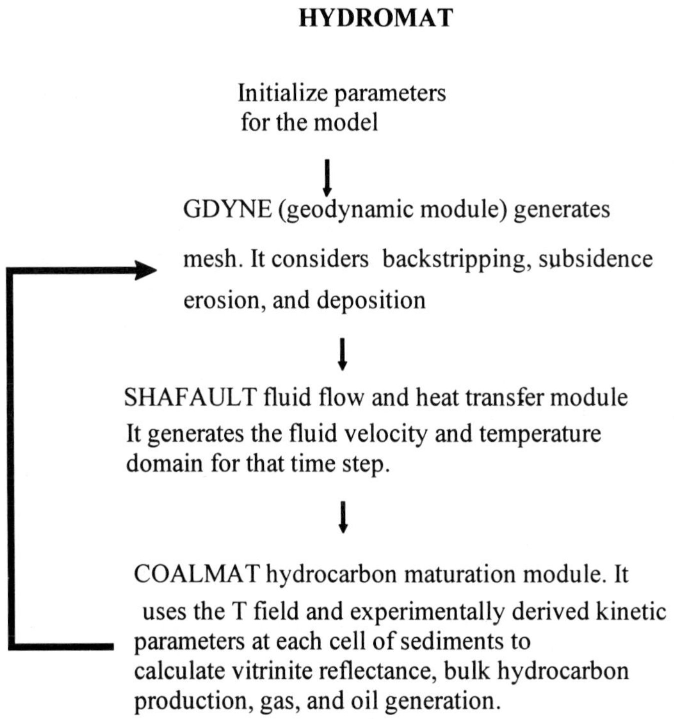

HYDROMAT

Initialize parameters
for the model

$\downarrow$

GDYNE (geodynamic module) generates

mesh. It considers backstripping, subsidence

erosion, and deposition

$\downarrow$

SHAFAULT fluid flow and heat transfer module
It generates the fluid velocity and temperature
domain for that time step.

$\downarrow$

COALMAT hydrocarbon maturation module. It
 uses the T field and experimentally derived kinetic
parameters at each cell of sediments to
calculate vitrinite reflectance, bulk hydrocarbon
production, gas, and oil generation.

Figure 1. Diagram showing the different modules of HYDROMAT and the way they are run at every time step.

Brief descriptions of each module are as follows:

2.1 GEODYNAMIC MODULE

In order to model the geodynamics of a basin we must be able to describe the deposition/erosion and subsidence/uplift of the basin through time. The geodynamic module (GDYNE) of HYDROMAT generates a 2-dimensional mesh of sediment boxes representing the sedimentary layers of the basin at each point of its evolution. In order to generate this mesh, well log data must first be generated from stratigraphic sections at various locations in the basin. The sedimentary layers of each well are then further subdivided into layers that correspond to depositional periods. It is assumed that the basin has an impermeable basement and that the wells only go as deep as the basement. We require that the sedimentary layers of each well have the same number of subdivisions in order to create the mesh. We interpolate the basin between pairs of wells by assuming a straight line connects the corresponding subdivisions of each well. The deposition of a single layer occurs across the entire basin at the same time (each layer is an isochrone). This forces the deep regions of the basin to have higher rates of deposition than the shallow regions.

The layers of each well are backstripped (or decompacted) by numerically solving an expression for the z-coordinates of their top and bottom as the layers above them are removed (Allen and Allen, 1990). The thickness of a layer is determined at each stage of its burial and from this information the history of each well can be reconstructed. A basin mesh can then be created for any level of deposition using the backstripped wells. The porosity of the decompacting sediments is assumed to be exponentially decreasing with depth. Porosity is calculated using the equation:

$$\phi = \phi_o \, e^{-cy} \qquad (1)$$

(Athy, 1930; as cited by Allen and Allen, 1997) where ϕ is the porosity of the sediments at depth y, ϕ_o is the porosity of the sediments at the surface at the time of deposition and c is the compacting factor.

The permeability of the sediments is calculated using the Kozeny-Carman equation (Bear, 1972) that gives the permeability as a function of the porosity:

$$k = C_o(\phi^3/(1- \phi)^2)/M_s^2 \qquad (2)$$

where M_s is the specific surface with respect to a unit volume of solid and C_o is the Kozeny constant. C_o is 0.15 and $M_s = 1.0 \times 10^6 \, m^{-1}$ in our model.

The effects of subsidence and uplift are modeled shifting vertically the positions of the wells. Uplift and erosion are modeled simultaneously. As the sedimentary layers are uplift, the upper layers above sea level are eroded. The sediments are not allowed to decompact after they are buried to a maximum depth; porosity and permeability maintain their minimum values during uplift and subsequent burial. The thickness of the eroded sediments does not appear in the present day stratigraphy. This thickness is estimated from geological maps of the area and the uplift and subsidence history of the basin. Backstripping of the present day sediments assumes that their porosity and permeability were acquired at maximum depth (which could be much deeper than that observed in the present).

2.2 FLUID FLOW AND HEAT TRANSFER MODULE

If the intrusives occurring in the basin are several kilometers deep, we can model fluid flow and heat transfer in the basin using a finite element code Shafault developed by López and Smith (1995). This program can handle the geometry of the basin and the intrusions in a special way. Shafault (López and Smith, 1995) is a transient 3D finite element code that solves the coupled fluid flow and heat transfer equations. It was written to investigate the effect of gravitationally driven groundwater flow on the fluid flow and heat transfer regime at fault zones. It overlaps a 3D domain representing the permeable country rock and a 2D permeable domain representing the fault. In the same way that the two dimensional domain can represent a fault, we have used it to represent a narrow low permeability, high temperature domain representing a dike.

2.3 MATURATION MODULE

Classical reaction kinetics cannot be applied in organic matter maturation because the precursors of petroleum for the petroleum generation reactions cannot be rigorously defined (Schenk et al., 1997). To solve this problem petroleum geochemists define pseudo-kinetics concepts combining the reactions that have similar activation energy into one pseudo-reaction. They have found that most chemical reactions converting organic matter into coal, gas, or oil can be represented by first order kinetics with an activation energy between 40 and 80 kcal/mol. Following the commonly accepted models for the maturation of organic matter (e.g. Sweeney, 1990; Welte et al., 1981; Sweeney and Burnham, 1990), we assume that the different chemical reactions for the generation of oil and gas, and evolution of vitrinite reflectance follow a first order Arrhenius law:

$$- \frac{dm_r}{dt} = \frac{dm_p}{dt} = A e^{\left(-\frac{E}{RT}\right)} m_r \tag{3}$$

At constant temperature Equation 3 predicts a decrease in mass of reactant m_r with time as the mass of the product m_p increases. A is the frequency or pre-exponential factor in s^{-1}, E is the activation energy in Kcal/mol, and R is the gas constant.

The composition of the initial organic matter can be quite variable. Many different reactions can be happening at the same time, every one with a different initial organic matter component and a different activation energy. We have extended the data point approach of Sweeney (1990) to a 2-dimensional domain. We solve the differential equation and find the fraction of unconverted organic matter, vitrinite reflectance, and the generated gas or oil.

The maturation module COALMAT can be used to evaluate conversion of organic matter into total hydrocarbons, oil, gas, or the evolution of vitrinite reflectance. For every case (total, oil, gas, vitrinite) different input files for the activation energies and the stoichiometric or weighing coefficients are needed. The input files (preexponential factor A, weighing factors for a discrete set of activation energies) are generated using the experimental results of pyrolysis experiments. Kinetic programs that calculate the preexponential factor A, and the weighting factors corresponding to different activation energies use a least-square iteration method that compares the measured hydrocarbon formation in the pyrolysis experiments with calculated data at different temperatures until the error function is minimized (Schenk et al., 1997).

3. Application of the program

HYDROMAT has been applied to model a section of the Bowen Basin, Eastern Australia. A complete description of the modeled section and discussion of the results can be found in Cobb et al. (these procceedings). Data for the model is based in previous work in coal petrology, isotopic composition, and alteration mineralogy (e.g. Saxby et al., 1996; Golding et al., 1996; Tonguc et al., 1997). In this paper, we present only some of the results to illustrate the application of the program and the effect of fluid circulation on the organic matter maturation as compared with maturation produced only by conductive heat. Figure 2 shows the fluid flow and thermal regime along the modeled section after 34 MY of sediment deposition. For the conditions modeled in this example, fluid circulation transfer heat from the bottom of the domain to the discharging regions (upflow limbs of the convective cells). The cooler recharging water has also an effect on the temperature field pulling down the isotherms. Throughout the history of the basin, the circulation of water affects the maturation of organic matter. Figure 3 shows the modeled vitrinite reflectance along the German Creek coalseam of the Bowen Basin in the modeled section. Two simulations are presented in this diagram. The curve presenting lowest vitrinite reflectance (open circles) shows the results for convective case and the open triangles represent the results for a low permeability (10^{-22} m^2) domain or conductive case. The diamonds represent values of vitrinite reflectance reported for wells in the area. For the thermal history modeled in this example, the conductive results show that the sediments should have maturities higher than the observed values. For the convective case, the basal heat flow is transferred to the circulating water as well as to the sediments affecting the maturation process. The convective case appears to describe the observed vitrinite reflectance much better then the purely conductive model.

4. Conclusions

Our initial applications of the program HYDROMAT suggest that in basins affected by high heat flow, hydrothermal circulation can play an important role in controlling the maturation reactions. Modeling of fluid circulation (advective or convective heat) as well as conductive heat describes better the observed data, such as thermal gradients, vitrinite

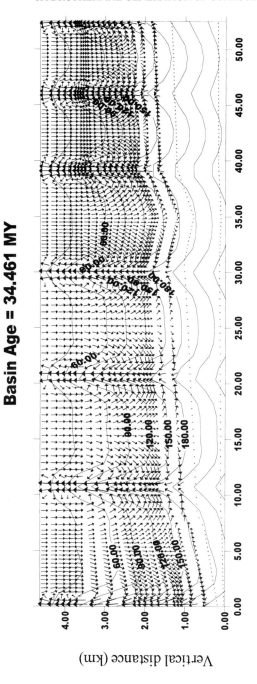

Figure 2. Fluid velocity and thermal regime for the a section of the Bowen Basin, Australia. Convective cells are formed affecting the temperature field and organic matter maturation process.

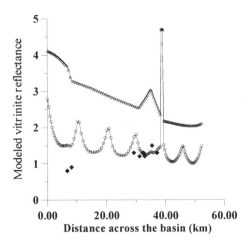

Figure 3. Values of vitrinite reflectance along the German Creek Coal Seam in a section of the Bowen Basin, Australia. Open circles represent the convective case, triangles the conductive case, and solid diamonds measured vitrinite reflectance.

reflectances, alteration mineralogy, and isotopic composition of minerals and fluids (Cobb et al., these proccedings).

Further work is need to obtain a better understanding of the role of hydrothermal fluids in basins with complex thermal and tectonic history, such as basins located close to plate margins. Many of them are present targets of hydrocarbon exploration (e.g. the Northland Basin in New Zealand, PESA News, 1998). Some aspects that need to be investigated further include the effect of the permeability on the thermal regime and maturation of organic matter, the effect of width of the intrusion, the effect of the mineralogy (temperature) of the intrusion, a comparison between the effect of the basal heat flow and the heat generated by the intrusion on the maturation of organic matter, the minimum distance for the intrusion to perturb the thermal regime and affect the maturation reactions, and the effect of topographically driven flow in the basins presenting margins with high topographic relief at some times of their history.

We also need to improve the computer code to solve the coupled fluid flow and heat transfer equations. Shafault can model the fluid flow and thermal regime of a system affected by high basal heat flow and deep intrusives. However, for systems affected by shallow intrusives, a computer program that can solve the fluid flow and heat transfer equations in pressure-enthalpy space is needed to avoid the computational problems occurring close the critical point in pressure-temperature space.

5. References

Allen, P.A., and Allen, J.R. (1990) *Basin Analysis: principles and applications*, Blackwell Science, Cambridge, Massachusetts.

Altebäumer, F.J., Leythaeuser, D., and Schaefer, R.G. (1983.) Effect of geologically rapid heating on maturation and hydrocarbon generation in lower Jurassic shales from NW-Germany, *Advances in Organic Geochemistry*, 80-86.

Bear, J. (1972) *Dynamics of Fluids in Porous Media*. Dover Publications, Inc.

Bethke, C.M. (1985) A numerical model of compaction-driven groundwater flow and heat transfer and its application to the paleohydrology of intracratonic sedimentary basins, *J. Geophys. Res.* **90**, 6817-6828.

Both, R., Crook, K., Taylor, B., Brogan, S., Chapbell, B., Frankel, E., Liu, L., Sinton, J., and Tiffin, D. (1986) Hydrothermal chimneys and associated fauna in the Manus back-arc basin, Papua New Guinea. *EOS* **67**, no. 21, 489-490.

Faraj, B. S. (1995) *Cleat mineralogy of late Permian coal measures, Bowen Basin, Queensland, Australia*. Ph. D. thesis, University of Queensland.

Forster, C., and Smith, L. (1989) The influence of groundwater flow on thermal regimes in mountainous terrain: a model study, *J. Geophys. Res.* **94**, 9439-9451.

Glikson, M., Golding, S.D., Lawrie, G., Szabo, L., Fong, C., Baublys, K., Saxby, J.D., and Chatsfield, P. (1995) Hydrocarbon generation in Permian coals of Queensland, Australia: source of coalseam gases, *Bowen Basin Symposium*, Mackay, 205-216.

Golding, S.D., Glikson, M., Collerson, K.D., Zhao, J.X., Bublys, K., and Crossley, J. (1996) Nature and source of carbonate mineralization in Bowen Basin coals: implications for the origin of coalseam gases. In: *Mesozoic Geology of the Eastern Australia Plate Conference, Geological Society of Australia Extended Abstracts* **43**, 489-495.

Grigo, D., Maragna, B., Arienti, M.T., Fiorani, M., Parisi, A., Marrone, M., Sguazzero, P., and Uberg, A.S. (1993) Issues in 3D sedimentary basin modeling and application to Haltenbanken, offshore Norway. In: A. G. Doré et al. (Editors), *Basin Modelling: Advances and Applications. Norwegian Petroleum Society (NPF), Special Publication* **3**. Elsevier, Amsterdam, 455-468.

Hayba, D.O., and Ingebritsen, S.E. (1994) The computer model Hydrotherm, a three-dimensional finite difference model to simulate ground-water flow and heat transport in the temperature range of 0 to 1,200° C, *U.S. Geological Survey, Water-Resources Investigations Report* **94-4045**.

Hermanrud, C. (1993) Basin modelling techniques -an overview. In: A. G. Doré et al. (Editors), *Basin Modelling: Advances and Applications. Norwegian Petroleum Society (NPF), Special Publication* **3**. Elsevier, Amsterdam, 1-34.

López, D.L., and Smith, L. (1996) Fluid flow in fault zones: Influence of hydraulic anisotropy and heterogeneity on the fluid flow and heat transfer regime, *Water Resour. Res.* **32**, 3227-3235.

López, D. L., and Smith, J. L. (1995) Fluid flow in fault zones: Analysis of the interplay of convective circulation and topographycally-driven groundwater flow, *Water Resour. Res.* **31**, 1489-1503.

MacKenzie, D. (1978) Some remarks on the development of sedimentary basins, *Earth and Planetary Science Letters* **40**, 25-32.

PESA News (1998) Queensland, Australia, April/May, p. 53.

Peter, J.M., Peltonen, P., Scott, S.D., Simoneit, B.R.T., and Kawka, O.E. (1991) Carbon-14 ages of hydrothermal petroleum and carbonate in Guaymas Basin, Gulf of California -implications for oil generation, expulsion and migration. *Geology* **19**, 253-256.

Pruess, K. (1991) *TOUGH2-A general purpose numerical simulation for multiphase fluid and heat flow*, Lawrence Berkeley Laboratory, Report number **LBL-29400**.

Saxby, J.D., Glikson, M. and Szabo, L.S., (1996) Maturation of Queensland Mesozoic coals: thermal and microscopic analysis of petroleum and gas generation. In: *Mesozoic Geology of the Eastern Australia Plate Conference, Geological Society of Australia Extended Abstracts*, **43**, 489-495.

Schenk, H.J., Horsfield, B., Krooss, B., Schaefer, R.G., and Schwochau, K. (1997) Kinetics of petroleum formation and cracking. In: Welte D.H., Horsfield B., and Baker D.R. (eds) *Petroleum and Basin Evolution*, Springer, 231-269.

Sweeney, J.J. (1990) BASINMAT, Fortran program calculates oil and gas generation using a distribution of discrete activation energies. *GEOBYTE*, April 1990, 37-43.

Sweeney, J.J., and Burnham, A.K. (1990) Evaluation of a simple model of vitrinite reflectance based on chemical kinetics. *AAPG Bulletin* **74**, 1559-1570.

Seewald, J.S. (1994) Evidence for metastable equilibrium between hydrocarbons under hydrothermal conditions, *Nature* **370**, 285-287.

Simoneit, B.R.T. (1984) Hydrothermal activity and its effects in sedimentary organic matter, *Organic Geochemistry* **6**, 857-864.

Summer, S., and Verosub, K.E. (1989) A low-temperature hydrothermal maturation mechanism for sedimentary basins associated with volcanic rocks, In: *Origin and Evolution of Sedimentary Basins and Their Energy and Mineral Resources, Geophysical Monograph 48, IUGG* **3**, 129-136.

Tonguç, U., Golding, S., and Glikson, M. (1997) Carbonate and clay mineral diagenesis in the late Permian coal measures of the Bowen Basin, Queensland, Australia: Implications for thermal and tectonic histories, *EUG 9 abstracts, European Union of Geosciences Meeting, 23-27 March 1997*, Strasbourg, France, 572.

Ungerer, P., Burrus, J., Doligez, B., Chénet, P.Y., and Bessis, F. (1990) Basin evaluation by integrated two-dimensional modeling of heat transfer; fluid flow, hydrocarbon generation, and migration, *AAPG Bulletin* **74**, No. 3, 309-335.

Ungerer, P., Bessis, F., Chénet, P.Y., Durand, B., Nogaret, E., Chiarelli, A., Oudin, J.L, and Perrin, J.F. (1984) Geological and geochemical models in oil exploration: principles and practical examples, *in* G. Demaison (ed.), *Petroleum geochemistry and basin evolution:AAPG Memoir* **35**, 53-77.

Welte, D.H., and Yukler, M.A. (1981) Petroleum origin and accumulation in basin evolution -A quantitative model, *AAPG Bulletin* **65**, 1387-1396.

Winters, J.C., Williams, J.A., and Lewan, M.D. (1983) A Laboratory study of petroleum generation by hydrouspyrolisis, *Advances in Organic Geochemistry*, 524-533.

SIMULATING THE CONDUCTIVE AND HYDROTHERMAL MATURATION OF COAL AND COAL SEAM GAS IN THE BOWEN BASIN, AUSTRALIA

M. COBB(1), D.L. LOPEZ(1), M. GLIKSON(2), AND S.D. GOLDING(2)
(1) Department of Geological Sciences, Ohio University, Athens, Ohio 45701
(2) Department of Earth Sciences, University Queensland, Brisbane, QLD 4072, Australia

Abstract

We model the hydrothermal maturation of hydrocarbons in the Bowen Basin of Australia with the program HYDROMAT (López et al., these proceedings). HYDROMAT is used to determine the vitrinite reflectance and the generation of bulk hydrocarbons, hydrocarbons with one to four carbons, and hydrocarbons with five carbons or more during the evolution of the Bowen Basin. The uplift, subsidence, erosion, deposition, fluid flow and heat transfer, and thermal history of the Bowen Basin are reconstructed in order to model the maturation history of the hydrocarbon producing sediments. The effect of high temperature Cretaceous intrusions on the Bowen Basin is also modeled. We focus on the German Creek and Pleiades coal seams and find good agreement between our results and actual data for vitrinite reflectance. Our results support previous experimental studies, which suggested that hydrothermal fluid circulation and high temperature sills and dikes created the high ranking coals and methane gas deposits found in the Bowen Basin. According to our modeling results the regional Triassic high heat flow event was responsible for generating the largest fraction of the total hydrocarbons produced during the entire history of the Bowen Basin. This is consistent with the coexistence of bitumen and minerals of hydrothermal alteration in the Bowen Basin sediments.

1. Introduction

A deeper understanding of the formation of coal, oil, and coal seam gases in a sedimentary basin can be obtained through a model of sufficient complexity. Although any model of a real sedimentary basin must be greatly simplified, valuable insight into the relationships between deposition, erosion, subsidence, uplift, basal heat flow, fluid and heat transfer, and hydrocarbon production can still be obtained. This paper will focus on modeling the hydrothermal maturation of coal and generation of coal seam oil and gases in the Bowen Basin of Australia using a new model, HYDROMAT, described elsewhere (López et al., these proceedings). Recent work by Golding et al. (1996) demonstrates that a purely conductive model of hydrocarbon formation in the Bowen Basin is inconsistent with petrographic observations, mineralogy, and isotopic information, which suggest the influence of hydrothermal fluid circulation. We have simulated a section of the Bowen basin with sediments similar to those of the actual basin in order to determine the effect of fluid flow on hydrocarbon maturation.

435

The Bowen Basin of Australia (Fig. 1) is Permo-Triassic in age and it is the northernmost structural element of the Bowen-Gunnedah-Sydney Basin System. The Bowen Basin was formed through back-arc extension and foreland subsidence before the break up of Gondwana. It is filled with up to 10,000 m of marine and non-marine saliciclastic sediments intercalated with coal seams and accumulations of natural gas (Golding et al. (1996)). Figure 1 shows the location of the Bowen Basin and the area that we have modeled.

The coals of the Bowen Basin are extremely mineralized with pyrite, hematite, barite, calcite, ankerite, siderite and dolomite (Golding et al., 1996) showing that the basin has been affected by hydrothermal circulation. Carbonate mineralization mixed with bitumen indicates simultaneous emplacement of the carbonates with hydrocarbon generation. Isotopic work in the carbonate veins show that the composition of $^{87}Sr/^{86}Sr$ suggests a magmatic component for the fluids precipitating the carbonates, rather than a radiogenic crustal source. In addition, Nd isotopic ratios ($^{143}Nd/^{144}Nd$) of the carbonates have values similar to the isotopic ratios observed by the Whitsunday-Cumberland Islands Cretaceous volcanic-plutonic province and central Queensland Cainozoic alkali basalts. Golding et al. (1996) also investigated the isotopic composition of C in calcite ($\delta^{13}C$) and found that the values reflect deposition of calcite from fluids having compositions close to the equilibrium between CO_2 and CH_4 as found in hydrothermal systems. They found a steep slope between the C and O isotope correlation trend suggesting that the carbonates were deposited either in a narrow temperature range or at high temperatures. Composition of $\delta^{13}C$ in coal seam methane indicates a predominantly thermogenic origin.

The vitrinite reflectance of coals from the Bowen Basin have been studied extensively (Saxby et al., 1996) and have been found to have erratic depth profiles. In the Northern part of the basin, inverted profiles with depth are not uncommon. This erratic behavior is attributed to hydrothermal circulation playing an important role in the maturation of the organic matter in the basin. Saxby et al. (1996) also found a strong correlation between gas and bitumen content, the gas content increasing with the increasing rank of the coal.

Tonguç et al. (1997) have dated the Triassic cleat-fill illites, the Late Triassic illite-smectites, the Cretaceous intrusions, and the Tertiary flows and intrusions in the Bowen Basin using K-Ar techniques. Considering their reported results and more recent work (S.D. Golding and U. Tonguç, personal communication), the thermal, subsidence, and uplift history of the basin can be summarized as follows:

i)   Triassic Hunter Bowen Orogeny (235- 245 Ma). Possibly two separate contractional pulses with magmatism to east in the New England Fold Belt. Subsequent rapid uplift occurred (Faraj, 1995).

ii)   Late Triassic regional meteoric hydrothermal event (215-210 Ma). It is recorded in the illite/smectite K-Ar ages in the central and northern Bowen Basin and at depth in the southern Bowen Basin. Extension and magmatism occurred to the east in the New England Fold Belt.

iii) Cretaceous Magmatism (145-100 Ma). Separate intrusive pulses occurred at 145-130 Ma and 110-100 Ma with a possible third intrusive event at about 80 Ma. It is recorded in K-Ar ages of bimodal but dominantly mafic dikes and sills and contact aureole illites. Early hydrothermal activity (145-140 Ma) recorded in illite/smectite K- Ar ages in shallow parts of bore holes in the southern Bowen Basin indicates as well that Cretaceous magmatism occurred. This data could indicate the early stage breakup of Gondwana.

iv)   Tertiary magmatism. 30-25 Ma. Basaltic magmatism occurred with lesser felsic plugs; mainly in the south west of the basin.

Based on vitrinite reflectance and alteration mineralogy, Golding and Tonguç (Personal Communication) have estimated the geothermal gradients to be around 70 °C/km for the Late Triassic thermal event, and 30-40 °C/km for the Cretaceous event. For the compressional periods of the basin, a default gradient of 20-25 °C/km was estimated. These values and the subsidence history of the basin as reported by Faraj (1995) have been used to construct a generalized subsidence, uplift, and thermal history of the basin as displayed in Fig. 2. A generalized thermal conductivity of 2.5 watts/°C-m was used to calculate the basal heat flow history of our Bowen Basin model.

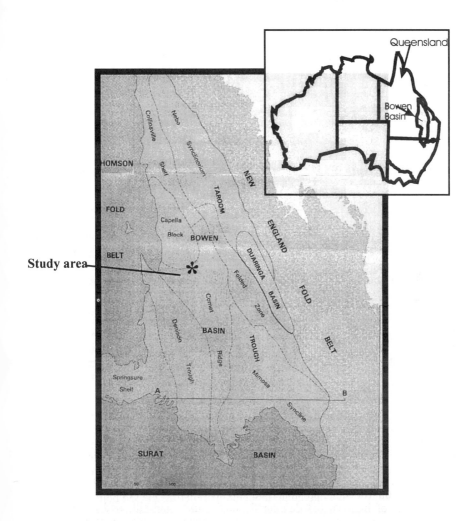

Figure 1. Location of the Bowen Basin in Australia and study area. After Balfe et al. (1985).

In order to model the complex processes involved in the maturation of coal in the Bowen Basin we needed a region that presented less complexity but still contained the basic features discussed previously concerning the effects of hydrothermal circulation. We selected the region close to German Creek (Fig. 3) for this purpose because it was less intruded and has a reasonable number of studied wells.

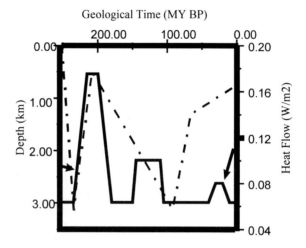

Figure 2. Heat flow and subsidence history of the Bowen Basin.

2. Modeled Section

The section AA' shown in Fig. 3 is presented in Fig. 4. The main coal seams are shown in Fig. 4: Pleiades, Aquila, Tigris, Corvus, and German Creek. The section is complex close to the wells DDH251 and DDH247, and was probably affected by a fault. However, it is not computationally possible to model the section in all its complexity. We have simplified the section in this diagram for the purpose of our model. We have also simplified the geology and included only the upper and lower seams (Pleiades and German Creek) as well. In addition, because much of the basin sediments have eroded, we estimated that there was an additional 4 km of overlaying sediments when the basin was at its greatest thickness (Fig. 5). We associate the uplift and subsidence seen in Fig. 2 with the erosion and deposition of the upper sediments. The physical parameters used for the modeled sediments are listed in Table 1.

During the history of the basin, the New England Belt was directing gravitationally driven fluid towards the basin. This implies a greater head to the east of the modeled domain and a higher topography of the sea bottom to the east and lower to the west. The topographic relief of the sea bottom is difficult to know because the only information we have is generalized data about the depositional environments in the basin through time (Table 2, A. Falkner, personal communication). Person et al. (1993) have modeled the Cooper-Eromanga basins. They used a maximum topographic relief for the sea bottom close to 0.5 m/km. For this initial work, we have used a topographic relief ranging from 0.0 to 0.16 m/km. These values were chosen because the available information does not show significant changes in deposition environment along the transect at a given time. However, more investigation is needed to determine better the paleowater depths at different points of the basin. This topographic relief is important because it affects the fresh water heads in the modeled domain. According to the work of López and Smith (1995), high topographic gradient along the water table produces advective fluid movement (no convective cells are formed) and low gradient produces convective circulation. At the fault zone modeled by

López and Smith (1995), a gradient of 3 m/km was the limit between advective and convective flow regime for a fault having a permeability equal to 10^{-12} m². A gradient of 3m/km is higher than the observed topographic relief in most sedimentary basins, suggesting that convection could happen in some sedimentary basins.

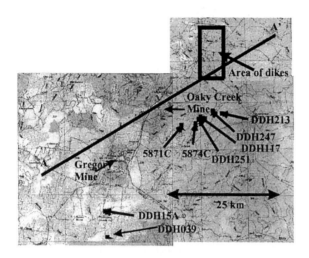

Figure 3. Modeled cross-section showing the location of wells in the area.

In this first model, we have not considered topographically driven fluid that could be transferred from regions to the East, outside the modeled domain. More information about the geometry of that region through time is needed and will be incorporated later. Initially, we included the coal as a separate layer. However, the coal had a very high porosity and permeability at the time of deposition and it compacted very fast as it was buried in our model. This high compaction produced an unrealistic low permeability region that did not allow fluid circulation. The extremely high porosity during the early history of the coal layer was also problematic. High porosity produced high permeability and fluid velocities creating numerical problems in the solution. For these reasons, and because coal is usually associated with shale layers, we have modeled the coal and the shale as a single unit of sediments. The permeability of the shales was modeled using the Kozeny-Carmen formula (Specific surface area of 1.0×10^6 m^{-1}, Kozeny constant 0.15) (Bear 1972) and was given a lower bound of 5×10^{-16} m² because there is evidence (cleat in coal, fractures in shales) that the sediments were permeable during the history of the basin. Upper limits for the porosity and permeability were used as well. Effective porosity for the flow was limited to values lower than 30% and permeabilities to values lower than 5×10^{-12} m². High permeabilities and porosities are likely to occur only in the upper parts of the domain in contact with air or ocean water.

For the maturation of hydrocarbons, Boreham et al. (1998) have determined the frequency distribution of activation energies for the production of light gas (C1-C4, one to four carbons), hydrocarbons with five or more carbons (C5+), and bulk hydrocarbons for different coal samples in the Bowen Basin. We use these kinetic data in our model to simulate the generation of gases, oil, and bulk hydrocarbons. The kinetic parameters used to model the maturation of vitrinite in our model are those reported in Sweeney and Burnham

(1990). Table 3 shows actual values of vitrinite reflectance for the wells in the study area. One intrusive dike around 110 Ma was modeled in the region of the cross section close to the heavily intruded region (Fig. 3). The dike was modeled as thin, hot, low permeability slab that intruded the domain. The width of the dike was 10 m and its initial temperature was 800° C.

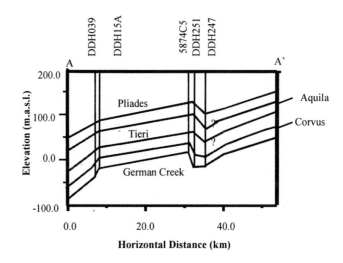

Figure 4. Coal seams present in the German Creek region, Bowen Basin, Australia.

Table 1. Simulation parameters

Parameter	Definition	shale	sandstone	shaley sandstone
θ	surface porosity	0.63	0.49	0.56
λ	solid thermal conductivity (W/m-°K)	2.5	2.5	2.5
ρ	solid density (kg/m³)	2800	2650	2680
C	solid specific heat (KJ/kg °C)	1.01	1.01	1.01
c	coefficient for slope of the porosity versus depth curve (1/m)	0.70	0.3	0.43

H	basal heat flow	variable through the basin's history	
T_{ref}	reference surface temperature	5° C	
Pref	reference surface pressure	1 atm	

3. Discussion of Results

Figures 6, 7, 8, and 9 show the results obtained in our modeling work. Figure 6 shows the temperature and water velocity fields at several times during the history of the basin. With the topographic relief of the sea-sediment interface considered in our model, convective circulation occurs along the cross-section. Multiple cells are formed which significantly alter the deeper isotherms. In the downward regions of the convective cells, the isotherms are pulled downward. This behavior corresponds to the cooling effect of the recharging water. In comparison, at the upward limbs of the convective cells, temperature increases because heat is transported from the bottom of the domain heating the rocks of the discharge regions. This behavior is observed through the history of the basin. The thickness of the modeled domain changes with time as the sediments are buried, compacted, uplifted, and eroded. However, the number of convective cells formed in the 52.5 km long domain remains almost constant throughout the history of the basin; 12 convective cells are generally formed and in some cases 16.

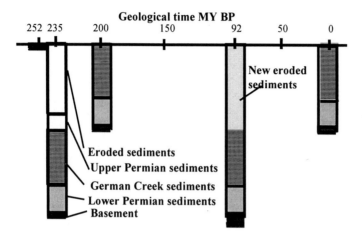

Figure 5. Modeled burial, subsidence, uplift and erosion history of the Bowen Basin.

Table 2. Depositional environments and thickness for the sediments found in the German Creek region of the Bowen Basin.

Sedimentary Unit	Depositional Environment	thickness (m)
Tertiary sediments	allvial plain	0-50 m
Jurassic sediments	alluvial plain, lacustrine	0-?
Triassic sediments	alluvial plain, lacustrine	0-2000
Rangal coal measures	alluvial plain, peat mire	100-200
Burngrove Formation	alluvial plain	100-250
Fairhill Formation	alluvial plain, peat mire	120-350
Mc Millan Formation	marine	0-100
German Creek Formation upper coals	lower delta plain, peat mire, delta mouth bar	110
German Creek Formation lower	shallow marine	160
Undiferentiated Back Creek Group	shallow marine	90-180
Blair Athol	graben fill	0-250
Reid Dome Beds	graben fill	0-400
Basement		

The position of the upward and downward limbs of the convective cells changes slightly, shifting the position of the isotherms a few hundred meters. A change in this cell structure occurs when the intrusion is emplaced. The high temperature around the intrusive is very localized and is dissipated after a few hundred years. Only the region located a few hundred meters from the intrusive is affected. Strong convective currents are formed close to the intrusive. This result is in agreement with modeling work around intrusive bodies by

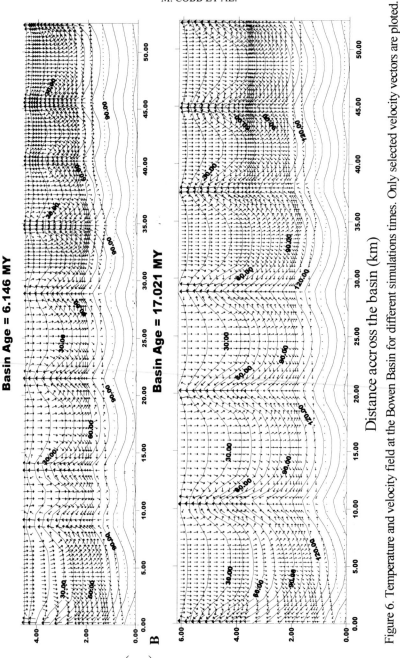

Figure 6. Temperature and velocity field at the Bowen Basin for different simulations times. Only selected velocity vectors are ploted.

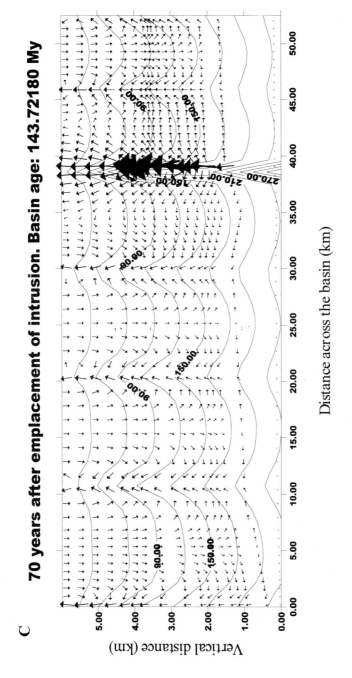

Figure 6. Continuation.

Cathles and Erendi (1997). According to these authors, a dike with the dimensions modeled in our problem should have decreased its temperature to 25% of the initial temperature after a few hundred years. However, the Bowen Basin has been subjected to multiple intrusive events, as shown in the area enclosed by a rectangle in Fig. 3. The dikes observed in that region could be the remnants of longer dikes and even the roots of sills that have been eroded. These multiple intrusive events should be the source of continued heat release to the basin affecting the maturation of hydrocarbons across the basin.

Table 3. Representative values of vitrinite reflectance (mode) for wells in the German Creek area. Some names represent several coalseams (at close depths) instead of a single one.

Well	Horizontal Distance along AA' (km)	Pleiades coalseam Ro	Aquila Coalseam Ro	Tieri Coalseam Ro	Corvus Coalseam Ro	German Creek Coalseam Ro
DDH039	6.8					0.8
DDH15A	8.3	0.7	0.7	0.8		0.9
5871C	29.3					1.3
5874C	31.2					1.2
DDH251	32.5	1.0	1.1	1.2		1.3
DDH117	32.9					1.2
DDH247	35.4				1.4	1.5
DDH213	37.1					1.3

It is important to note that in homogeneous-isotropic domains, hydrothermal circulation can affect the temperature field, bending the isotherms. However, the temperature increases with depth if the source of heat is at the bottom of the domain (basal heat flow). Inverted temperature fields (higher at the bottom than at the top of the domain) can be produced locally if we have a heterogeneous permeability field (López and Smith, 1996). Another way to produce an inverted temperature profile, as those observed in the Northern Bowen Basin, is by having a source of heat at the top of the domain. Such is the case when a sill is emplaced. This situation is observed in the Northern Bowen Basin (M. Glikson, personal communication). Further work is needed to model these scenarios of multiple intrusive events and different locations of the heat source (dikes and sills).

The effect of the complex evolution of the temperature field on the maturation of organic matter in the basin can be observed in Figs. 7 and 8, and in Figure 3 of the paper describing the program HYDROMAT (López et al., these proceedings). In Figure 7, vitrinite reflectance is plotted against distance along the basin for Pleiades coal seam at the present time. A similar graphs is presented for the German Creek coalseam in Figure 3 (López et al., these proceedings). Both Figures show that observed values of vitrinite reflectance can be better described by convective heat circulation. Regions that have been subjected to higher temperature (upward limbs of the convective cells) show a higher maturation than regions of lower temperature (downward flows). The maturation gradient for the convective case (difference in maturation between the two seams) is not uniform through the basin.

The maturation profile follows the oscillatory behavior promoted by a spatially periodic temperature field. In Figure 8, the bulk fraction of generated hydrocarbons, C1-C4 gases, and C5+ hydrocarbons are plotted against distance along the basin for the permeable and highly impermeable German Creek coal seam at the present time. A sharp spike is observed in around the intrusive for vitrinite reflectance, bulk hydrocarbons, gas C1-C4, and C5+

hydrocarbons. The intrusive body has a deep effect on the maturation reactions for the region close to the intrusive increasing vitrinite reflectance to values higher than 4 and releasing almost instantaneously gases like methane. This is consistent with values of vitrinite reflectance found by M. Glikson in regions close to dikes (personal communication) and the correlation between methane gas content and increasing coal rank. The cluster of data points from Table 3 in the vicinity of our results indicates that our model approaches the actual hydrocarbon maturation in some regions of the basin.

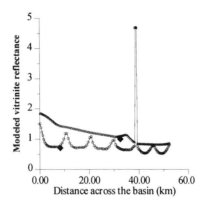

Figure 7. Values of vitrinite reflectance along the Pleiades Coal Seam. Open circles represent the convective case, triangles the conductive case, and solid diamonds measured vitrinite reflectance.

The measured vitrinite reflectance values are closer to the convective case than to the conductive case (Fig. 7), suggesting that fluid circulation played an important role in the maturation process. For the observed thermal gradients, conduction alone would produce overmaturation of the sediments. Topographic differences produce the decreasing general trend in maturity in Figures 7 and 8, and the spike around 35-40 km from the origin.

Maturity increases with depth except in the vicinity of the intrusive. In both cases, the intrusive has the same effect on hydrocarbon generation in the surrounding sediments. The thermal and subsidence history of the basin indicates that the basin was subjected to uplift and intensive heating (up to 175 mW/m²) during the late Triassic. Our modeled results (Fig. 9) indicate that oil generation was accomplished during this heating event, especially in the deeper seam. Figure 9 shows the maturation of the organic matter versus time (generated fraction of bulk hydrocarbons, methane C1-C4, and hydrocarbons C5+). Two points were selected, one for well DDH15A and the other for well DDH251. The effects of the different high heat flow events are clear from Fig. 9. Maturation increases with the onset of the first thermal pulse during the late Triassic and reaches the hydrocarbon window during that time, increasing the release of bulk hydrocarbons. During this time, we have the highest basal heat flow and hydrothermal convection consistent with Tonguç et al. (1997) observations that the bitumen and generation of oil coexisted with the deposition of hydrothermal carbonates in the basin. According to our modeling results and the work of Golding et al. (1996), Saxby et al. (1996), and Tonguc et al. (1997), uplift of the basin sediments, high heat flow, hydrothermal convection, deposition of minerals from hydrothermal alteration, and generation of bulk hydrocarbons coexisted.

It should also be observed that hydrocarbon production is higher (except near the intrusive) in the low permeability case (deeper seam) indicating that the seams where subjected to higher temperatures. For the convective case, it should be noted also (Figures 9B and 9C) that the generation of oil (C5+) precedes the generation of gases (C1-C4). Oil generation reaching between 10% and 100% (and C1-C4 gases only 1%) during the Triassic heating event. During the Cretaceous heating event maximum gas generation occurs and there is a significant increase in the generation of other hydrocarbons as well.

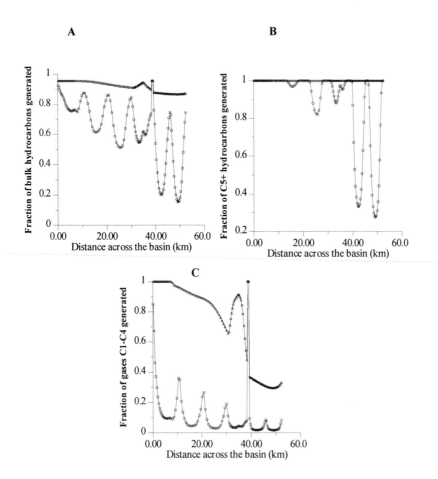

Figure 8. Values of modeled fraction of generated bulk hydrocarbons (A), C1-C4 gases (B), and C5+ hydrocarbons (C), along the German Creek Coal Seam of the Bowen Basin. Symbols as in Figure 7.

4. Conclusions

This initial modeling work in the Bowen Basin generated values of vitrinite reflectance for the German Creek and Pleiades coal seams close to the observed values. This strongly suggests that hydrothermal processes where a significant mechanism for heat transport in the Bowen Basin during its history. In addition, our results indicate that during the Triassic heat pulse the largest fraction of bulk hydrocarbons (in particular heavier hydrocarbons) was produced. It was during this time of high heat flow that the hydrocarbon producing sediments entered their hydrocarbon window. Although this does not account for the very high values of vitrinite reflectance found in some Bowen Basin coals and the methane gas associated with these coals. Our results for a single intrusive dike clearly show that dikes and sills can generate these high vitrinite reflectance values and high concentrations of methane gas. This effect is consistent with the Cretaceous intrusions of the Bowen Basin and the high ranking coals which are found in their vicinity (Golding et al., 1996).

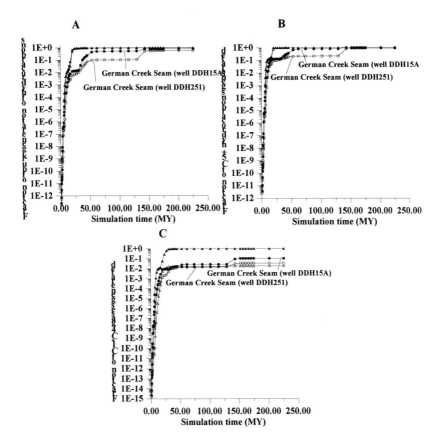

Figure 9. Simulated evolution of fraction of bulk hydrocarbons (A), C1-C4 gases (B), and C5+ hydrocarbons (C) generated during the modeled history of the Bowen Basin. Circles represent the convective case, triangles the conductive case. Open symbols represent well DDH251 and filled symbols well DDH15A. In Figures A and B the conductive case for both wells overlap.

Kinetic studies for coal samples in the Bowen Basin indicate that coals with different vitrinite reflectance are characterized by different activation energy frequency distributions for the generation of methane, other gases, and oil (Boreham et al., 1998). This behavior suggests that the results of our simulations should differ from the real values because the model does not approach the complexity of the maturation processes. Traditional kinetic models cannot predict accurately hydrocarbon generation because they use single frequency distributions (e.g. Sweeney, 1990) during the simulation of the organic maturation reactions through time. New approaches for the maturation module are needed which consider the variations in the distribution of activation energies for hydrocarbon generation as vitrinite reflectance increases.

The results of this research project concerning the maturation of organic matter in the Bowen Basin are encouraging because they reproduce the sequence of events observed in the results of different types of experimental studies (e.g. coal petrology, geochemistry of alteration minerals, and isotopic studies). This is our first modeling work on the maturation of organic matter by hydrothermal fluids and we realize that many of the assumptions of our model need to be improved in order to obtain results that are more realistic. However, it is clear that fluid circulation plays an important role in the maturation reactions and explains some of the physical and chemical characteristics of the coals in this basin. More research is needed to explore other hydrothermal scenarios such as the emplacement of multiple dikes and sills, to determine the effect of parameters such as permeability of the sediments in the maturation process, and the effect of the location of heat sources and topographic relief of the water table.

5. References

Balfe, P.C., Draper, J.J., Scott, S.G., and Belcher, R.C. (1985) Bowen Basin Solid Geology map 1:500,000, *Queensland Department of Mines*.

Bear, J. (1972). *Dynamics of Fluids in Porous Media*. Dover Publications, Inc., 764 pp.

Boreham, C.J., Horsfield, B., and Schenck, H.J. (1998). Predicting the quantities of oil and gas generated from permian coals, Bowen Basin, Australia using pyrolytic methods. *Marine and Petroleum Geology*, in press.

Cathles, L.M., Erendi, A.H.J., and Barrie, T. (1997). How long can a hydrothermal system be sustained by a single intrusive event? In: A special issue on the timing and duration of hydrothermal events. *Economic Geology*, vol. **92**, pp. 763-771.

Faraj, B. S. (1995). Cleat mineralogy of late Permian coal measures, Bowen Basin, Queensland, Australia. Ph. D. thesis, University of Queensland, pp. 216.

Golding, S.D., Glikson, M., Collerson, K.D., Zhao, J.X., Bublys, K., and Crossley, J. (1996). Nature and source of carbonate mineralization in Bowen Basin coals: implications for the origin of coalseam gases. In: *Mesozoic Geology of the Eastern Australia Plate Conference, Geological Society of Australia Extended Abstracts* No. **43**, pp. 489-495.

López, D.L., and Smith, L. (1996). Fluid flow in fault zones: Influence of hydraulic anisotropy and heterogeneity on the fluid flow and heat transfer regime, *Water Resour. Res.*, vol. **32**, pp. 3227-3235.

López, D. L., and Smith, J. L. (1995). Fluid flow in fault zones: Analysis of the interplay of convective circulation and topographically-driven groundwater flow. *Water Resour. Res.* vol. **31**, pp. 1489-1503.

Person, M., Toupin, D., Weick, J., Eadington, P., and D. Warner (1993). Hydrological constraints in petroleum generation within the Cooper and Eromanga Basins, Australia: I Mathematical Modeling. *Geofluids '93 Extended Abstracts*, pp. 264-267.

Saxby, J.D., Glikson, M. and Szabo, L.S., (1996) Maturation of Queensland Mesozoic coals: thermal and microscopic analysis of petroleum and gas generation. In: *Mesozoic Geology of the Eastern Australia Plate Conference, Geological Society of Australia Extended Abstracts*, **43**, 489-495.

Sweeney, J.J. (1990). BASINMAT, Fortran program calculates oil and gas generation using a distribution of discrete activation energies. *GEOBYTE*, April 1990, pp. 37-43.

Sweeney, J.J., and Burnham, A.K. (1990). Evaluation of a simple model of vitrinite reflectance based on chemical kinetics. *AAPG Bulletin* **74**, pp. 1559-1570.

Tonguç, U., Golding, S., and Glikson, M. (1997). Carbonate and clay mineral diagenesis in the late Permian coal measures of the Bowen Basin, Queensland, Australia: Implications for thermal and tectonic histories. *EUG 9 abstracts, European Union of Geosciences Meeting*, 23-27 March 1997, Strasbourg, France, pp. 572.

MODELLING OF PETROLEUM FORMATION ASSOCIATED WITH HEAT TRANSFER DUE TO HYDRODYNAMIC PROCESSES

Hydrodynamic heat transfer enables rapid petroleum formation

R. H. BRUCE [1], M. F. MIDDLETON [2], P. HOLYLAND [3],
D. LOEWENTHAL [4] and I. BRUNER [5]

1. *Department of Minerals and Energy of Western Australia, East Perth, Western Australia (formerly of School of Physical Sciences, Curtin University of Technology, Perth, Western Australia)*
2. *Department of Geology, Chalmers University of Technology, Gothenberg, Sweden*
3. *Consultant, Brisbane, Queensland (formerly of Perth, Western Australia)*
4. *Department of Geophysics and Planetary Sciences, Tel Aviv University, Israel*
5. *Institute for Petroleum Research and Geophysics, Holon, Israel*

1. Abstract

This paper presents the results of modelling the formation of petroleum in the framework of rapid hydrodynamic processes in sedimentary basins. The study is motivated by the observation that many Pb-Zn deposits occur within hydrocarbon prone sedimentary basins, and that expulsion of hot basinal brines, over a geologically short period, may contribute significantly to the generation of hydrocarbons. Modelling of the generation of bitumen by hydrous pyrolysis shows that peak generation can occur after approximately 1000 years at a temperature of 200 °C. Hydrothermal modelling of advective heat transfer in a shaly formation above a hot porous layer, that permits (relatively) rapid fluid flow, was carried out. The results of this modelling indicated that with advective flow velocities in the vicinity of 0.315 m/yr (which is a reasonable value within the petrophysical constraints of the model) can significantly alter the temperature at a distance of some 200 m above the hot layer after 1000 years. It is concluded that a source bed approximately 200 m above such a hot layer, and buried 3 km deep within a basin, can attain 200 °C, and thus peak generation, in the vicinity of 1000 years.

2. Introduction

Gradualism is generally assumed as the norm for sedimentary basin processes. That is, it is assumed that it takes millions of years for the generation of a sedimentary basin, and the petroleum and mineral deposits that it contains. However, laboratory studies demonstrate that petroleum can form in only a few years at appropriately high temperatures (Saxby *et al.*, 1986), or under appropriate catalytic conditions, such as hydrous pyrolysis (Lewan, 1992).

Experimental hydrous pyrolysis (Lewan, 1991) provides evidence for simultaneous generation and expulsion of oil. Kerogen decomposes into a tarry soluble bitumen, which in turn partially decomposes into a liquid oil as thermal stress increases. The net volume increase generated by decomposition results in bitumen expansion along bedding fabric of the rock matrix, and the lack of pore volume in the bitumen-impregnated rock matrix results in expulsion of the generated oil. In such source shales, fracture permeability is a far more effective for primary petroleum migration than matrix permeability. Thus, under appropriate thermal and hydrous conditions, petroleum may form rapidly within a sedimentary basin. This paper compares the conditions under which hydrocarbons have been observed to form rapidly in the laboratory to realistic geological conditions where similar processes may occur.

3. Hydrous pyrolysis model

Hydrous pyrolysis is one of the more realistic ways of generating petroleum in the laboratory, and such results have been extensively published (Lewan, 1992; Lewan, 1991; Lewan, 1985; Barth *et al.*, 1989). These experiments are generally carried out in the temperature range 200 to 370 °C, and residence times are commonly in the vicinity of several days. Maximum bitumen yields are found in the vicinity of 310-320 °C after 72 hours for Kimmeridge oil shale (Barth *et al.*, 1989). This process has been modelled according to kinetic theory by Nielsen and Barth (1991). Their model may be used to predict peak bitumen yields for other heating times.

The model of Nielsen and Barth (1991) describes the bitumen concentration (B) in terms of a distribution of n activation energies:

$$B = K_0 \ \Sigma_i \ [W_i \ F_i] \tag{1}$$

where K_0 is the total amount of bitumen that can be generated, W_i = constant . exp[-$(E_{ai} - E_{ak})^2/2\beta^2]$, i = 1 ... n, F_i is a function of reaction rates as described by the Arrhenius equation, E_{ai} is the ith activation energy in the gaussian distribution of the appropriate activation energy distribution, E_{ak} is the central value of the activation energy distribution, and β is the width of the distribution. The constant is determined to satisfy $\Sigma_i \ W_i = 1$.

They suggest an optimum model, which reproduces both (1) the bitumen yield versus pyrolysis temperature data (Barth *et al.*, 1989) for 72 hours heating time and (2) expected values of activation energy for the thermal breakdown of oil-prone kerogen (Quigley and Mackenzie, 1988). Their optimum model has the following parameters: ln (K_o) = 4.35 mg/g shale, E_{ab} = 223.92 kJ/mol, E_{ak} = 238.73 kJ/mole and β = 15 kJ/mol, where E_{ab} is the activation energy of gas generation from the bitumen. Figure 1 shows the bitumen yield versus temperature derived from the above model for a variety of heating times from the laboratory situation of 72 hours to 1000 years.

The peak of bitumen generation moves towards lower temperatures with increasing time of heating (Figure 1). Thus, after 1000 years, generation of significant bitumen can occur within realizable temperatures in sedimentary basins. Indeed, significant bitumen can be generated even after several days for temperatures in the vicinity of 300 °C - this may occur near igneous intrusions (Buchardt and Lewan, 1990; Reekmann and Mebberson, 1984).

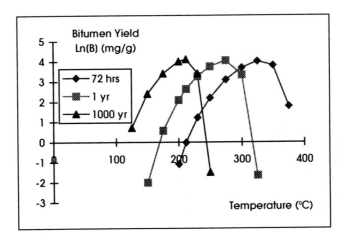

Figure 1.   Bitumen yield versus temperature for various times of heating based on the kinetic model of Nielsen and Barth, 1991 (equation 1).

Thus, the kinetic theory of hydrous pyrolysis indicates that significant volumes of petroleum may be generated at temperatures in the vicinity of 200 °C, if maintained for the order of 1000 years.

4. Temperatures in sedimentary basins

It is widely known that average geothermal gradients in sedimentary basins occur generally in the range of 20 to 30 °C/km (Kappelmeyer and Haenel, 1974). This geothermal gradient range is essentially for (1) sedimentary basins with little or no circulation of water, and (2) an average thermal conductivity of the sedimentary sequence of approximately 2.0 W/mK. Under these conditions, assuming a surface temperature of 10 °C, a temperature of 150 °C occurs in the range 7.0 to 4.7 km. A source bed, therefore, at 2 km depth of burial with a geothermal gradient of 30 °C/km will have a temperature of 70 °C, and according to the bitumen generation model in equation (1) will not generate petroleum within geologically short periods (less than 1 million years).

A thermal conductivity of 2.0 W/mK is typical for shales (Middleton, 1994), and thus also a quite reasonable value for a shale-rich sedimentary basin. In such a basin, the existence of fluid pathways is significantly reduced by the predominance of shales, which are aquitards with very low matrix permeability. Thus, one might not expect to find much heat transfer due to convective processes.

Nevertheless, it is a valuable exercise to investigate how much fluid and heat transport might occur in such a basin. This is justified on the basis that many sedimentary basins have associated Pb-Zn mineralization. This mineralization, on the basis of current models of genesis of stratiform, sediment-hosted type deposits (Clendenin and Duane, 1990; Wallace, 1994), suggests that fluids move freely through sedimentary basins, and that these fluids from time to time originate from deep within the sedimentary basin. Thus, at a time of upward, or advective (movement of fluid in one direction) movement of deeper fluids, temperatures may be elevated in parts of the sedimentary basin near the fluid pathways, and the pressure within such formations may also be elevated slightly above hydrostatic pressure. Figure 2 shows the conceptual model of such fluid flow in a sedimentary basin.

Data on the permeability of various rocks within hydrothermal systems (Holyland 1988) is presented in Figure 3. Based on this classification, we will look at hydrothermal migration through shale dominated rocks with permeability in the vicinity of 0.05 mD, or 5×10^{-17} m², which is within the range of permeability for clay dominated rocks (see Figure 3).

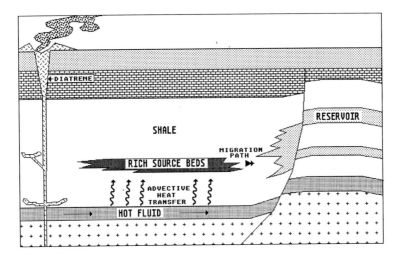

Figure 2. Schematic diagram showing conceptual model of hydrocarbon generation in a hydrothermally effected basin.

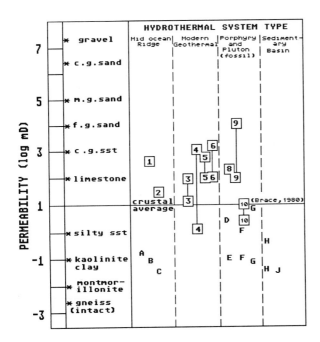

Figure 3. Permeability data for various types of rocks and various hydrothermal system types. 1 to 10 are actual observed system permeabilities. A to J are modelled permeabilities. After Holyland (1988).

5. Hydrothermal model

For the purpose of investigating the possibility of advective heat and fluid transport through a shale dominated basin, a simple thermal model is proposed based on the theory presented by Carslaw and Jaeger (1959). Figure 4 shows the model geometry in schematic form.

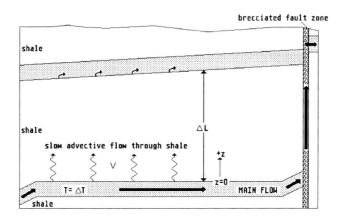

Figure 4. Schematic diagram of the hydrothermal model used to estimate the temperature disturbance, based on equation (2), in a shaly formation above a hot porous layer. The distance ΔL is the distance above the hot porous layer where the slow vertically moving fluid can escape into another high permeability layer.

The temperature T(z,t) at time t and distance z within the shale bed overlying a thin porous fluid pathway, with temperature ΔT above the stable temperature regime, is given by:

$$T(z,t) = 0.5 \, \Delta T \, \{ \, \mathrm{erfc}[(z\text{-}Vt)/(4\kappa t)^{1/2}] + e^{Vz/\kappa} \cdot \mathrm{erfc}[(z\text{-}Vt)/(4\kappa t)^{1/2}] \, \} \qquad (2)$$

where V is the velocity of vertical advective flow within the shaly formation, κ is thermal diffusivity, and erfc is the complementary error function.

In geological terms, this model essentially describes the case where hot fluid is flowing, under pressure and from a deep source, through a thin very porous and permeable layer, which is overlain by a thick and less permeable shaly layer. The main fluid flow passes through the porous layer and then upward along a brecciated fault zone; such flow of hot brines is well known to exist in sedimentary basins (Eugster, 1985; Lindblom, 1982) and is believed to be involved in Pb-Zn formation (Figure 4). The flow that we are interested in is the much slower flow through the shaly unit above the hot permeable layer. This flow will occur over a greater rock volume, and also thermally disturb a greater rock volume, albeit not to such a great degree as the zone of main (faster) flow.

Equation (2) can now be used to estimate the temperature disturbance within an organic rich layer within the shaly formation above the anomalously hot porous layer. For the purpose of these calculations, the source rock layer is assumed to be 200 m above the hot layer, the anomalous temperature of the hot layer is assumed to be 100 ° C, and other parameters as in Table 1. The temperature disturbance within the shaly formation above as a function of advective velocity (V) and time is shown in Figure 5.

TABLE 1. Assumed modelling parameters

thermal diffusivity:	$\kappa = 0.01$ cm²/s
anomalous temperature:	$\Delta T = 100$ °C
water viscosity (150 °C):	$\mu = 0.25$ cP

Formation under abnormal pressure

anomalous pressure difference:	$\Delta P = 25$ bar (363 p.s.i.)
thickness of shale formation:	$\Delta L = 500$ m
permeability:	$k = 0.05$ mD
porosity:	$\phi = 2\%$

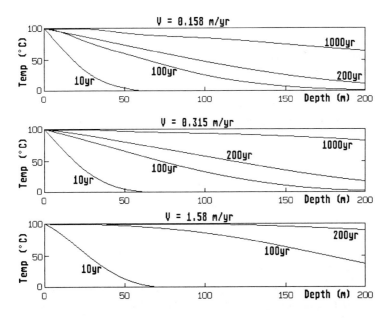

Figure 5. Plots of temperature versus distance from the hot layer, based on equation (2), for various times and advection velocities. Temperature disturbance in the overlying shale is greater closer to the hot layer, for increased advection velocity and for longer heating times.

It is seen from Figure 5 that an advective velocity of approximately 0.158 m/yr will cause a region 200m above the anomalously hot layer (fluid pathway) to rise approximately 50 percent of the anomalous temperature after about 1000 years. Furthermore, if the advective velocity is greater than 0.315 m/yr, a region 200 m above the anomalously hot zone will be at least 80 percent of the anomalous temperature for heating longer than 1000 years.

Thus, if the flow is maintained in the anomalously hot layer for approximately 1000 years, and the advective velocity through the shaly formation above the hot zone is in the range 0.158 to 0.315 m/yr, then the temperature of a region (perhaps source rock) 200 m from the fluid flow zone will be appreciably increased. Figure 5 also indicates that this region will be almost at the anomalous temperature after only 200 years, if the advective velocity is approximately 0.315 m/yr.

An estimate of the appropriate order of magnitude of the velocity (V) can be obtained by considering the anomalous pressure difference that may develop in the shaly formation in conjunction with the permeability assumed above (0.05 mD). The velocity (V) of advective flow may be determined from Darcy's Law (Bear, 1972) :

$$V\varphi = -(k/\mu) \{\Delta P/\Delta L\} \tag{3}$$

where φ is porosity of the shaly formation, k is permeability, μ is dynamic viscosity, $\Delta P/\Delta L$ is the abnormal pressure gradient above hydrostatic in the shale.

Assuming the values of the parameters given in Table 1, the advective velocity (V) is calculated from equation (3) to be 1.58 m/yr. This calculation shows that advective velocities with similar values to those in Figure 5 are found for advective flow in shale dominated sedimentary basins.

With respect to the assumed parameters in the above calculation, the permeability is obtained from Figure 3, shale porosity from Rieke and Chilingarian (1974) and water viscosity from published charts (Cosse, 1993). The anomalous pressure difference can vary considerably depending on the pressure of the fluid in the hot permeable layer below, the advection velocity and the duration that the fluid has been flowing through the shaly formation. A value of 25 bar (363 p.s.i.) is assumed, which is a conservative estimate of abnormal pressures commonly observed in sedimentary basins.

We must also at this point consider the prospect of overpressure within the kerogen-rich source rock itself, due to generation of hydrocarbons. This can be considered in two stages: (i) pre-peak oil generation and (ii) post-peak oil generation. In the pre-peak oil generation phase, the thermal process modelled above should proceed with only minor decrease of flow velocity due to overpressure from oil generation. During the post-peak oil generation phase, overpressure should build in the source-rock layer, and the velocity of flow will diminish with increasing temperature. If the overpressure in the source-rock layer becomes equal to the excess pressure driving the hot fluid upward, then the advective flow process will stop. Thus, the generation process is "self-limiting", and oil generation due to this mechanism may not pass beyond the peak-generation phase.

The process may proceed if a strong anisotropy in permeability exists within the source-rock layer. Thus, if the horizontal permeability (either natural layering or as fractures induced by oil generation) is greater than the vertical permeability, fluid flow (oil migration) after the peak-generation phase can occur horizontally. In this study, however, we are concerning ourselves more with a process that brings the source rock into the oil-generation window, rather than the complexities of secondary migration out of the oil kitchen.

6. Petroleum generation considerations

It is apparent from the modelling (Figure 5), if an advective flow of velocity of 0.315 to 1.58 m/yr can be maintained within a shaly formation for approximately 1000 years, then an anomalous temperature of approximately 100 °C can be attained as far from the hot porous bed as 200m. Thus, if a 20 m thick petroleum source bed is buried at 3 km with the stable thermal gradient is 30 °C/km and is 200 m above the hot porous bed, then the temperature of the source bed after 1000 years of experiencing advective heat transfer will be approximately 200 °C. The temperature of the source bed will be maintained at this temperature, as long at the temperature anomaly (i.e. the hot fluid flow) is maintained in the porous layer below.

Reference to Figure 1 shows that if the source bed is maintained near 200 °C for periods of greater than 1 year then significant volumes of hydrocarbons can be generated. Indeed, the model (see Figure 1) shows that for a heating period of 1000 years with a temperature of 200 °C, the bitumen concentration is at its peak value. Thus, economic volumes of hydrocarbons can be generated in short periods of time (ca. 1000 yrs) in sedimentary basins that have experienced hot fluid flow for similar durations and advective velocities in the range 0.315 to 1.58 m/yr. The widespread occurrence of Pb-Zn mineralization in hydrocarbon-prone basins (Rickard et al., 1979; Lindblom, 1982) provides strong coincidental support for the hypothesis that significant hydrocarbon generation can occur in geologically short periods of time, due to hot fluid flow (Cathles and Smith 1983) in such basins.

7. Discussion

In some sedimentary basins in geologically active areas a low-temperature (ca. 200 °C) hydrothermal mechanism has been described for the maturation of sediments (Summer and Verosub, 1989). This involves heat transport away from deep faults or volcanic centres by laterally flowing aquifers. In the United States San Juan Basin, thermal fluids may explain the regional maturation pattern of increasing maturation towards the volcanic centre and also results in the highest maturation levels being found in sediments that were not the site of deepest burial.

In the Swiss and French Molasse Basin, second order variation in vitrinite reflectance has been related to a regional discharge of warm fluids controlled by tectonic structures (Schegg and Moritz, 1993). In the German Molasse Basin, generally gas fields are closer than oilfields to the Alpine front (Bachmann and Muller, 1991). Similar observations in the Appalachian Trough (Oliver, 1986) possibly reflect the temperature distribution of fluids as they flowed into the foreland basin.

The Australian Cooper-Eromanga Basin is not known to be associated with active volcanic centres. However, mathematical modelling of the basin indicates that the subsurface thermal regime and petroleum generation have been markedly affected by a topography-driven groundwater flow system (Person et al., 1993; Eadington et al., 1993).

Studies of the Western Canada Sedimentary Basin (Jessop and Majorowicz, 1993) indicate that the assumption of conductive steady state heat flow from the Precambrian basement does not predict vertical temperature profiles that match records from drilling records and logs. A conclusion is that the temperature field within the sedimentary strata has undergone hydrological distortion.

Thermal conduction has been a major parameter in quantitative maturation studies. However, it would be more realistic to take into account permeable sediments. These provide excellent pathways for convective (and advective) heat transfer by flowing fluids and enable preferential heating of regions at rates much higher than that provided by conduction alone (Lerche, 1990).

Permeability values 3 to 6 orders of magnitude lower than average crustal permeabilities, or observed permeabilities of present day analogues, have been commonly used in modelling of hydrothermal systems (Holyland, 1988) (refer Figure 3). Consequently rates for hydrothermal ore formation and hydrothermal reaction rates described in the literature have been lower by the same magnitude.

Convection dominates over conduction when permeability exceeds 0.05 mD - this is less than the permeability of kaolinite and greater than that of montmorillonite clay (Cathles, 1977). It would thus be unusual to have a purely conductive heating system geologically.

Laboratory measurements of the permeabilities of rocks of hand specimen scale give values much lower than the effective permeabilities found in actual field reservoirs where larger fractures are found (Holyland, 1988). The northward and northeastward increase of heat flow in the Alberta Basin may demonstrate that regional permeabilities are larger than those estimated from laboratory measurements, because of the existence of fractures (Jessop and Majorowicz, 1993).

Thus convective and advective heat flow has probably been present in more sedimentary basins than previously recognised.

8. Conclusions

The modelling in this study shows that peak hydrocarbon generation can occur at temperatures in the vicinity of 200 °C after exposure times of only 1000 years.

Hydrodynamic and petrophysical considerations show that fluid flow velocities through shaly formations of approximately 1.58 m/yr are not unreasonable.

Hydrothermal modelling, using velocities in the range 0.315 to 1.58 m/yr, shows that rocks up to 200 m from a porous layer with flowing hot fluid can experience sufficient thermal disturbance (200 °C) to allow the onset of peak hydrocarbon generation in times of the order of 1000 years.

Rapid hydrocarbon generation may, therefore, be more widespread in sedimentary basins than is currently believed.

9. Acknowledgements

The authors gratefully acknowledge the permission of the Petroleum Exploration Society of Australia (PESA) to reprint this paper as part of these Proceedings. This paper was first published in 1996 in PESA Journal No. 24.

The principal author thanks the University of Queensland for the invitation to give a presentation (based on this research) at the International Conference on Coal Seam Gas and Oil, Brisbane, March 1998.

10. References

Bachmann, G.H. and Muller, M. (1991) The Molasse basin, Germany: evolution of a petrolifierous foreland basin. In: Spencer, A.M. (ed.) *Generation, Accumulation and Production of Europe's Hydrocarbons.* Special Publication of the European Association of Petroleum Geoscientists No. 1, Oxford University Press, Oxford, 263-276.

Barth, T., Borgund, A.E. and Hopland, A.L. (1989) Generation of organic compounds by hydrous pyrolysis of Kimmeridge Oil Shale - Bulk results and activation energy calculations, *Organic Geochemistry*, **14**, 69-76.

Bear, J. (1972) *Dynamics of Fluids in Porous Media,* Elsevier, New York.

Buchardt, B. and Lewan, M.D. (1990) Reflectance of vitrinite-like macerals as a thermal maturity index for Cambrian-Ordovician Alum Shale, Southern Scandinavia, *American Association of Petroleum Geologists Bulletin,* **74**, 394-406.

Carslaw, H.S. and Jaeger, J.C. (1959) *Conduction of Heat in Solids,* Oxford University Press, Inc., 510 pp.

Cathles, L.M. (1977) An analysis of the cooling of intrusives by ground-water convection which includes boiling, *Economic Geology,* **72**, 804-826.

Cathles, L.M. and Smith, A.T. (1983) Thermal constraints on the formation of Mississippi Valley lead-zinc deposits and their implications for episodic dewatering and deposit genesis, *Economic Geology,* **78**, 983-1002.

Clendenin, C.W. and Duane, M.J. (1990) Focused fluid flow and Ozark Mississippi Valley-type deposits, *Geology,* **18**, 116-119.

Cosse, R. (1993) *Basics of Reservoir Engineering, Oil and Gas Field Development Techniques,* Institut Francais du Petrole Publications, Editions Technip, 342 pp.

Eadington, P.J., Hamilton, P.J., Lisk, M., Toupin, D., Person, M. and Warner, D. (1993) Hydrologic Controls on Petroleum Generation within the Cooper and Eromanga Basins, Australia. II Fluid Inclusion, isotopic, and geothermometric data., *GEOFLUIDS '93* Extended Abstracts. Torquay, England, 338-341.

Eugster, H.P. (1985) Oil shales, evaporites and ore deposits, *Geochimica et Cosmochimica Acta*, **49**, 619-635.

Holyland, P. (1988) *Structure and hydrodynamics of Renison Tin Mine*, Ph.D. Thesis, University of Queensland, unpublished.

Jessop, A.M. and Majorowicz, J.A. (1993) Heat flow from the basement of the Western Canada Sedimentary Basin and its redistribution in the sedimentary succession., *GEOFLUIDS '93* Extended Abstracts. Torquay, England, 92-95.

Kappelmeyer, O. and Haenel, R. (1974) *Geothermics with special reference to application*, Geoexploration Monograph Series 1, No.4, Berlin-Stuttgart, Gerbruder Borntraeger, 238 pp.

Lerche, I. (1990) *Basin Analysis: Quantitative Methods, Vol. 1*, Academic Press, 7.

Lewan, M.D. (1985) Evaluation of petroleum generation by hydrous pyrolysis experimentation, *Philosophical Transactions of the Royal Society of London*, **A, 315**, 123-134.

Lewan, M.D. (1991) Generation and expulsion of oil as determined by hydrous pyrolysis, *American Association of Petroleum Geologists Bulletin*, **75**, 620.

Lewan, M.D. (1992) Water as a source of hydrogen and oxygen in petroleum formation by hydrous pyrolysis, *American Chemical Society, Division of Fuel Chemistry*, **37/4**, 1643-1649.

Lindblom, S. (1982) *Fluid inclusion studies of the Laisvall sandstone lead-zinc deposit, Sweden*, Meddelanden fran Stockholms Universitets Geologiska Institution, Doctoral Thesis, Stockholm University.

Middleton, M.F. (1994) Determination of matrix thermal conductivity from dry drill cuttings, *American Association of Petroleum Geologists Bulletin*, **78**, 1790-1799.

Nielsen, S.B. and Barth, T. (1991) An application of least-squares inverse analysis in kinetic interpretations of hydrous pyrolysis experiments, *Mathematical Geology*, **23**, 565-582.

Oliver, J. (1986) Fluids expelled tectonically from orogenic belts: Their role in hydrocarbon migration and other geologic phenomena, *Geology*, **14**, 99-102.

Person, M., Toupin, D., Wieck, J., Eadington, P. and Warner, D. (1993) Hydrologic Constraints on Petroleum Generation within the Cooper and Eromanga Basins, Australia: I Mathematical Modeling, *GEOFLUIDS '93* Extended Abstracts. Torquay, England, 264-267.

Quigley, T.M. and Mackenzie, A.S. (1988) The temperature of oil and gas formation in the subsurface, *Nature*, **333**, 549-552.

Reekmann, S.A. and Mebberson, A.J. (1984) Igneous intrusions in the north-west Canning Basin and their impact on oil exploration, in *The Canning Basin, W.A.*, edited by P.G. Purcell: Geological Society of Australia and Petroleum Exploration Society of Australia, 389-399.

Rickard, D.T., Willden, M.Y., Marinder, N.-E. and Donnelly, T.H. (1979) Studies of the genesis of the Laisvall Sandstone Lead-Zinc Deposit, Sweden, *Economic Geology*, **74**, 1255-1285.

Rieke, H.H. and Chilingarian, G.V. (1974) *Compaction of Argillaceous Sediments*, Elsevier, 424 pp.

Saxby, J.D., Bennett, A.J.R., Corcoran, J.F., Lambert, D.E., and Riley, K.W. (1986) Petroleum generation: Simulation over six years of hydrocarbon formation from torbanite and brown coal in a subsiding basin, *Organic Geochemistry*, **9**, 69-81.

Schegg, R. and Moritz, R. (1993) Implications for Paleogeothermal Anomalies in the Molasse Basin (Switzerland and France), *GEOFLUIDS '93* Extended Abstracts. Torquay, England, 96-99.

Summer, N.S. and Verosub, K.L. (1989) A Low-Temperature Hydrothermal Maturation Mechanism for Sedimentary Basins associated with Volcanic Rocks, *Geophysical Monograph 48*, IUGG **3**, 129-136.

Wallace, M.W. (1994) Burial diagenesis, Pb-Zn sulphides and hydrocarbon maturation in the Devonian reefs of the Lennard Shelf, W.A., Geological Society of Australia, Abstracts Number 37, *12th Australian Geological Convention*.

FLORAL INFLUENCES ON THE PETROLEUM SOURCE POTENTIAL OF NEW ZEALAND COALS

J.NEWMAN
Newman Energy Research Ltd
2 Rose Street, Christchurch 8002, New Zealand

C.J. BOREHAM
AGSO Isotope and Organic Geochemistry
GPO Box 378, Canberra 2601, Australia

S.D. WARD
Ian R. Brown and Associates
PO Box 24-147, Wellington, New Zealand

A.P. MURRAY
Woodside Offshore Petroleum
1 Adelaide Terrace, Perth WA 6000,
Australia.

A.A. BAL
Z & S Asia Pty Ltd
46 Ord Street, West Perth, WA 6005, Australia

Abstract

New Zealand coals range in age from Cretaceous to Miocene and represent a wide variety of mire floral communities. Chemical and petrographic analysis demonstrate considerable variability in coal properties, and particularly vitrinite chemistry. Although some chemical variability can be attributed to environmental controls such as mire drainage and marine influence, other trends correspond with age, climate and floral assemblage.

Mire flora is primarily reconstructed using palynological analysis. Cretaceous and Eocene coals are dominated by gymnosperm and angiosperm pollen respectively. Paleocene sequences exhibit large fluctuations in mire flora and provide ideal sample sets for investigation of potential floral controls on petroleum source potential.

Proximate analysis, specific energy, sulphur, palynology and Rock-Eval data presented for 36 coals demonstrate a general increase in petroleum source potential with increasing angiosperm palynomorph dominance. Notable exceptions are coals from immediately beneath the Cretaceous/Tertiary boundary at Greymouth Coalfield, which have moderate to high Rock-Eval yield despite very low angiosperm palynomorph abundance.

Pyrolysis-gcms of 3 Paleocene and 2 Cretaceous samples shows generally good oil generative potential, particularly for perhydrous coals. Stepwise pyrolysis-GCMS and bulk kinetic parameters indicate that angiosperm derived coals may generate oil at slightly lower temperatures than gymnosperm derived coals. This difference in generation temperatures is small, however there is evidence that current kinetic models do not account for rank variation in the immature analogues used to predict generation histories. Consequently, actual generation temperatures may be significantly underestimated, and the difference in generation temperatures for gymnosperm vs angiosperm derived coals may be more significant, than current models suggest.

1.0 Introduction

New Zealand coals span a wide range of geological ages and ranks, and are known to have sourced oil and gas in at least the Taranaki and West Coast Basins (Cook, 1987; Collier & Johnston, 1991; Czochanska, 1988). Both stratigraphy and geochemistry indicate that hydrocarbons in these western areas are sourced from a Cretaceous to Eocene sequence of terrestrial and paralic coals and sediments. Traditional coal analysis and petroleum geochemistry demonstrate considerable variability in petroleum source potential of the coals, resulting from differences in vitrinite chemistry (Newman et al., 1997). Some of this chemical variability can be attributed to original peat mire oxygenation and the effects of marine influence. However, there are indications that coal chemistry may also be influenced by mire floral community, and the possible implications for petroleum source potential are the subject of this paper.

Previous work shows that New Zealand oils can sometimes be correlated with coals of particular ages and ranks using biomarker geochemistry (Czochanska et al. 1987 & 1988; Collier & Johnston, 1991; Killops et al. 1995). Gymnosperms and ferns were the dominant element in Cretaceous mires, and oils sourced from Cretaceous strata contain distinctive gymnosperm derived diterpanes. Angiosperms became increasingly successful mire colonists during the early Tertiary, resulting in oils containing oleananes. However, environmental interpretations and oil-source correlations based on oleanane abundance may require consideration of diagenetic conditions (Murray et al. 1997).

The present investigation directly assesses the floral origin and petroleum geochemistry of specific coal samples, to determine whether there is a relationship between mire flora, hydrocarbon yield and the kinetics of hydrocarbon generation.
This is the first stage of longer term research. Biomarker data are not reported here but will be published later.

Mire flora has been reconstructed primarily from coal palynology (Warnes 1992 , Ward et al. 1995, Moore 1996), which confirms the results of earlier wood tissue analysis (Shearer & Moore 1994a&b). Comparing the palynological assemblage of coals with that of associated sediments demonstrates that mire floral communities commonly have lower diversity than associated regional floras. Quantitative palynological analysis reveals that coals are usually dominated by just a few palynomorph types. These observations endorse palynological analysis as a means to reconstruct mire floras without undue contamination from transported pollen and spores. Palynological analysis has been undertaken at the University of Canterbury by senior postgraduates and staff of CRL Energy Research & Testing and Newman Energy Research Ltd.

Petroleum source potential of coal samples has been assessed using traditional proximate and specific energy analysis (CRL Energy Research & Testing), Rock-Eval pyrolysis, pyrolysis-GCMS and stepwise pyrolysis-GCMS (AGSO).

2.0 Method

2.1 PALYNOLOGY

Coal samples are ground to less than 1mm before oxidation by the wet Schulze process (Traverse, 1988), modified to suit the material. Each sample was reacted with concentrated $HNO_3/KClO_3$ for 10 – 15 minutes, followed by a wash step and the addition of 10% KOH solution for 5 minutes. All samples were passed through a 100μm seive , and unwanted fine material was removed by the addition of 10% bleach solution for 10 minutes, followed by "short centrifuging", whereby the samples are accelerated to 2000 r.p.m. over 30s, repeatedly. When necessary, inertinite and mineral matter were removed by separation in a saturated solution of potassium iodide (specific gravity 1.6).

All slides were examined with a Zeiss KM microscope, using a scan interval of 0.6mm and x300 magnification. Higher magnifications were used to aid identification where needed. Approximately 250 pollen and spore grains were recorded for each sample, and where slides were only partially completed upon reaching this total, a further 10 tracks were scanned for the presence of key species. For the purposes of this paper the palynological results are presented as total abundance of gymnosperms, angiosperms and spores. However, each of these classes was subdivided into numerous individual taxa and the detailed count data will be published later.

2.2 TRADITIONAL COAL ANALYSES

Traditional analyses referred to in this paper include proximate analysis, specific energy and total sulphur. Proximate analysis expresses coal composition in terms of moisture

(ISO 5068-1983), ash (ISO 1171-1981), volatile matter (ISO 562-1981) and fixed carbon. Volatile matter is the hydrocarbon yield during pyrolysis at over 900°C, and fixed carbon is the organic matter remaining after pyrolysis. Volatile matter measurement has excellent reproduceability and it is a useful parameter against which other properties, such as Rock-Eval S1 + S2 yield, can be assessed (Newman et al. 1997).

Volatile matter and specific energy are affected by levels of moisture and inorganic material in a coal sample. Both parameters are therefore corrected to a dry, mineral matter and sulphur free basis (dmmSf, Newman, 1997). This correction is implicit wherever volatile matter and specific energy are referred to in this paper.

2.3 BULK PYROLYSIS

Rock Eval and (where measured) total organic carbon (TOC) content of the powdered coal samples were measured using standard procedures (Boreham et al., 1994). Because coal is predominantly composed of organic material, hydrocarbon yield is expressed as total S1+S2 (ash free basis), and not as a proportion of total organic carbon (as for Hydrogen Index), which can introduce errors (Newman et al. 1997; Section 3.0, this paper).

2.4 PYROLYSIS-GAS CHROMATOGRAPHY MASS SPECTROMETRY (PYGCMS)

The coal (15mg) was sandwiched between a quartz wool plug inside a quartz insert and transferred to the Quantum MSSV split injector (GC^2 Chromatography, UK) held at 300°C. Nine pyrolyses, each of either 20 or 30°C steps at 5°C/min (300-330°C, 330-350°C, etc) were performed by programming (Newtronics Micro 96 controller) the injector (under 12 psi He pressure, 50ml/min split flow controlled by a Brooks electronic flow controller placed at the end of the split line) between 300°C and 540°C. After achieving the final temperature for each step the injector was cooled 50°C and held at this temperature over the time the pyrolysis products were analysed by gcms. Over the course of the pyrolysis step the C_{2+} pyrolysate was trapped within a 1m megabore pre-column (HP5, 0.88μm film thickness) by immersing a 10cm length of column in liquid nitrogen. Methane was not retained and passed into the analytical column. After 10 minutes, the trap was removed and the pyrolysis products separated on a 25m widebore fused silica capillary column (HP1, 0.33mm ID, 0.52μm film thickness) under gas chromatograph temperature programming conditions from 30°C to 310°C at 6°C/min and detected using full scan mode (VG 70E, 15-500 dalton scan range at 0.5 sec/decade). An external standard gas mixture of CH_4 and n-butane (0.5 ml injection of 2000ppm of each gas component) was used to quantify the amount of compound classes in the pyrolysate. The methane response factor was based on the m/z=16 signal while the total ion current response of n-butane in the standard gas was used for the C_{2+} components. A Pt/Rh calibrated thermocouple (3mm o.d.; J.C. Instruments , UK) was placed at the same position occupied by the sample inside the quartz tube in order to best represent the true temperature at the site of pyrolysis. Throughout temperature programming from 300-540°C the temperature of the calibrated thermocouple was within ±2°C of the Newtronics digital temperature readout which measures the temperature (Pt thermocouple) positioned in the middle of the injector heater block.

2.5   CHEMICAL KINETICS FROM OPEN SYSTEM PYROLYSIS

A Rock-Eval II instrument equipped with a kinetics module supplied by Humble Scientific (USA) was used to determine bulk kinetic parameters. Green River Shale standard AP22 (E_a=54 kcal/mole carries 95% of the product potential, A=7.3x10^{14} sec^{-1}) was employed to convert the measured Rock-Eval Tmax temperature (ie. outside the pyrolysis oven; Table 1) to that experienced by the sample at the reaction site. An addition of 26°C to the measured Tmax value at 25°C/min pyrolysis rate was needed to obtain the required correlation. Thus, the instrument was considered calibrated when the temperature adjustment gave the required Ea, and A to within ±20% of the quoted value for the Green River standard. This error in A translated to a ±1°C shift in the Tmax value for hydrocarbon generation at a geological heating rate of 5.3°C/Ma (10^{-11}°C/min). For the coals, rates of hydrocarbon evolution were determined at 4 heating ramps of 5, 15, 25 and 50°C/min over a temperature range of 300-550°C. Discrete distribution of activation energies, their reaction potentials and a common frequency factor were then obtained using the KINETICS^{TM} software package (Humble Scientific, USA).

Chemical kinetics for gas (C_1-C_5) and oil (C_{6+}) generation have been determined by a simplified method that combines bulk kinetic parameters with compositional data from stepwise py-gcms in order to derive semi-quantitative compositional kinetic parameters (Boreham et al., in press).

2.6 SAMPLE SELECTION

Samples were chosen to represent the widest possible range in floral composition, with relatively small variation in other parameters which could influence hydrocarbon generation potential. For example, coals which were clearly marine influenced during deposition were generally avoided, except for one sample which is included as a member of a larger sample set. Differences in burial history (rank) were avoided where possible, and corrected for when necessary (Section 3.0).

3.0 Management And Interpretation Of Bulk Data

3.1 SCREENING DATA FOR THE EFFECTS OF MINERAL CONTAMINATION, WEATHERING AND OTHER ANOMALIES.

Traditional coal analyses, Rock-Eval, and palynological data appear in Table 1. It is necessary to correct for inorganic contamination, and to screen data for anomalies, before attempting to relate floral composition to hydrocarbon potential. The procedures used will be demonstrated here using the Pakawau and Drillhole 698 datasets, and DH633. Volatile matter (Section 2.2) is fundamental to these procedures, and is discussed in more detail in Section 3.3. The reasons for using Rock-Eval S1+S2 instead of the more commonly used Hydrogen Index, HI (100*S2/TOC) are discussed in Section 3.4.

Figure 1 shows volatile matter related to raw Rock Eval S1+S2 (hereafter referred to as Rock Eval yield), and demonstrates considerable scatter in the Pakawau fraction. When Rock Eval yield is corrected to ash free basis (Figure 2) the relationship improves, except for 4 Pakawau samples which have anomalously low yield. Petrological examination in fluorescence has established that three of these outcrop samples have been chemically altered by weathering, and this is endorsed by their unusually high moisture content (Table 1). These samples are therefore omitted from future discussion. The anomalously low yield of the remaining sample (49/363) does not result from either weathering or high mineral content, and may be a consequence of experimental error. Figure 3 shows a good linear relationship between volatile matter and Rock Eval yield (ash free) for all other members of the Pakawau/DH698/DH633 set.

The examples in Figures 1 – 3 show that traditional coal analysis provides a useful method for screening Rock-Eval data for anomalous and unreliable results. Relating volatile matter to Rock Eval yield also indicates whether Rock-Eval results from different laboratories, or from different analysis runs at the same laboratory, represent coherent datasets. In the present example Pakawau results were obtained from Geotech and DH698 results from AGSO, but the uniform relationship with volatile matter (Figure 2) indicates good correspondence.

3.2 CORRECTING THE ROCK-EVAL YIELD OF IMMATURE COALS FOR SUPPRESSION AT LOW RANK

When utilising sample sets with different burial history it is important to define rank variation. Usual measures of coal rank, such as vitrinite reflectance, are strongly affected by isorank variability in the vitrinite chemistry of many New Zealand coals (Newman & Newman, 1982; Newman et al. 1997). This effect is exemplified by samples from DH698 and DH1494 (Table 1 & Figure 4). Rank is constant within each of these sample suites, because the samples were obtained in close stratigraphic proximity. However, there is a significant rank difference between the two suites. Samples from Drillhole 698 are at the subbituminous/high volatile bituminous C rank boundary, and

Table 1. Proximate, specific energy, sulphur, Rock-Eval and palynological data for Cretaceous to Eocene New Zealand coals. Sulphur values were not available for the Waikato and Ohai samples shown here. However, sulphur is invariably low (0.2-0.4%) in these coals.

Location	Sample type	CRL No.	Age	Unit	Proximate analysis			Sp.E. MJ/kg a.d.	S % a.d.	VM dmmSf %	Tmax (deg. C)	Rock-Eval					Palynology		
					M a.d.	A a.d.	VM a.d.					S1 (mg/g)	S2 (mg/g)	S3 (mg/g)	TOC (wt%)	HI	Angio-sperms	Gymno-sperms	Pterido-phytes
Waikato	Face	46/630	L. Eocene	Taupiri	17.0	2.8	34.5			43.0	425	0.7	53.9	29.7	54.6	99	76.8	18.4	4.8
Waikato	Face	46/610	L. Eocene	Taupiri	15.6	2.9	37.2			45.6	428	1.3	86.9	27.6	58.3	149	68.0	23.2	8.8
Waikato	Face	46/613	L. Eocene	Taupiri	14.6	3.0	40.7			49.4	429	5.6	171.3	23.4	59.9	286	72.4	13.2	14.4
Waikato	Face	46/614	L. Eocene	Taupiri	15.1	9.6	35.0			46.5	432	0.3	73.9	28.7	52.5	141	65.2	1.2	33.6
Waikato	Face	46/621	L. Eocene	Taupiri	15.1	4.6	37.3			46.5	428	1.1	94.3	25.0	57.2	165	66.8	7.2	26.0
Ohai	DH375	42/090	L. Cret.	Morley	13.8	7.5	32.9			41.8	428	1.8	82.9	22.2	56.9	146	19.5	66.5	14.0
Ohai	DH375	42/106	L. Cret.	Morley	9.5	45.2					429	1.0	54.9	14.6	30.5	180	29.9	36.2	33.9
Ohai	DH375	42/104	L. Cret.	Morley	14.2	5.1	33.1			41.0	429	1.2	80.8	22.4	58.7	138	15.8	44.6	39.6
Ohai	DH375	42/101	L. Cret.	Morley	13.8	3.1	35.2			42.4	426	1.3	88.3	24.5	60.5	146	22.1	31.4	46.5
Ohai	DH375	42/129	L. Cret.	Morley	6.2	67.3					432	0.5	26.2	8.9	13.9	189	39.4	39.1	21.5
Greymouth	DH698	52/172	Paleocene	Dunollie	9.1	9.7	39.5	25.89	1.04	48.0	424	8.4	173.7	9.8	63.2	275	68.2	29.0	2.8
Greymouth	DH698	52/174	Paleocene	Dunollie	7.5	5.7	45.7	28.33	1.28	52.4	421	22.7	249.7	8.8	69.2	361	81.4	5.5	13.1
Greymouth	DH698	52/175	Paleocene	Dunollie	11.2	4.9	36.2	26.51	0.67	42.8	424	6.7	153.0	18.5			14.3	51.0	34.7
Greymouth	DH698	52/177	Paleocene	Dunollie	12.4	2.6	35.2	26.69	0.28	41.2	429	3.0	130.0	18.3			33.1	58.9	8.0
Greymouth	DH698	52/178	Paleocene	Dunollie	11.7	1.8	36.1	27.20	0.27	41.6	427	4.8	117.3	12.8	70.4	167	38.8	48.3	12.9
Greymouth	DH698	52/179	Paleocene	Dunollie	11.8	2.8	35.3	27.04	0.25	41.1	427	2.5	148.0	19.7			42.4	47.0	10.6
Greymouth	DH755	59/657	Paleocene	Dunollie	10.6	4.6	38.9	27.37	0.78	45.5	423	2.7	140.9	10.5	55.4	254	33.3	64.3	2.4
Greymouth	DH755	59/658	Paleocene	Dunollie	9.1	21.9	31.0	22.05	0.34	43.1	427	1.6	112.3	9.5	53.5	210	40.4	59.2	0.4
Greymouth	DH755	59/659	Paleocene	Dunollie	10.8	4.1	38.2	27.29	0.89	44.6	424	3.0	144.3	10.3	62.7	229	39.4	54.4	6.2
Greymouth	DH712	50/645	L. Cret.	Rewanui	3.0	4.9	41.7	30.74	0.34	45.0	428	3.3	216.0	14.9			4.7	71.7	23.6
Greymouth	DH628	26/376	L. Cret.	Rewanui	9.9	3.2	37.2	29.14	0.29	42.6	429	10.5	166.1	10.7	74.2	224	4.3	74.0	21.7
Greymouth	DH633	26/662	L. Cret.	Rewanui	9.3	5.8	41.8	28.20	0.59	48.9	427	11.9	198.2	11.3	52.5	377	8.4	78.2	13.4
Pakawau	Outcrop	53/383	Paleocene	Puponga	11.4	22.5	29.1	18.12	0.48	42.0	427	1.6	58.3	43.8	48.0	121	51.3	16.5	32.2
Pakawau	Outcrop	53/381	Paleocene	Puponga	6.5	33.8	26.9	18.50	0.38	41.7	432	2.0	113.1	8.2	46.0	244	55.6	25.6	18.8
Pakawau	Outcrop	49/368	Paleocene	Puponga	7.3	23.8	32.2		0.41	44.8	420	3.8	142.1	14.9	53.0	268	37.3	19.4	43.3
Pakawau	Outcrop	54/890	Paleocene	Puponga	7.2	41.1	24.6		0.31	43.0	426	0.6	88.7	6.9	39.0	229	58.7	22.2	19.1
Pakawau	Outcrop	54/891	Paleocene	Puponga	9.7	28.7	28.5		0.33	44.0	430	1.8	68.3	31.3	44.0	155	27.7	38.3	34.0
Pakawau	Outcrop	49/363	Paleocene	Puponga	5.3	26.3	34.1		1.17	47.8	422	3.1	95.7	36.2	51.0	186	35.8	56.8	7.4
Pakawau	Outcrop	53/382	Paleocene	Puponga	11.4	21.8	33.2	17.71	0.96	48.0	427	3.0	71.2	64.4	47.0	151	43.6	38.5	17.9
Pakawau	Outcrop	49/365	Paleocene	Puponga	4.4	43.0	25.8		0.67	44.4	431	2.0	92.3	10.2	34.0	246	48.1	39.8	12.1
Pakawau	Outcrop	52/888	Paleocene	Puponga	5.7	21.6	37.4		1.84	50.0	426	7.4	199.2	2.8	55.0	364	36.4	60.2	3.4
Buller	DH1494	46/739	L. Eocene	Brunner	3.0	2.4	47.2	32.42	4.39	49.8	429	20.0	259.9	0.0	74.3	350	95.2	2.4	2.4
Buller	DH1494	46/742	L. Eocene	Brunner	3.8	1.2	44.2	32.44	3.88	46.3	432	16.9	238.1	0.0	77.1	309	86.9	5.5	7.6
Buller	DH1494	46/745	L. Eocene	Brunner	4.9	0.4	40.0	32.42	1.42	42.1	436	12.3	225.5	0.3	77.6	291	94.0	2.0	4.0
Buller	DH1494	46/748	L. Eocene	Brunner	4.6	0.3	40.2	32.54	1.08	42.1	437	10.0	227.9	0.8	76.8	297	94.7	0.7	4.6
Buller	DH1494	46/749	L. Eocene	Brunner	4.7	2.6	39.2	31.77	1.02	42.0	437	11.7	227.6	0.5	75.5	301	96.4	1.0	2.6

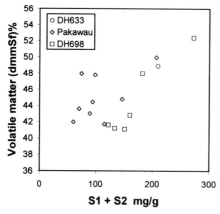

Figure 1. Volatile matter related to
S1+S2 for DH's 698 & 633, and
Pakawau samples (Table 1).

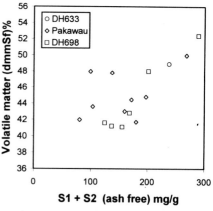

Figure 2. Data as for Fig. 1, with
S1+S2 corrected to ash free basis.

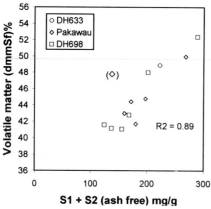

Figure 3. Data as for Figs 1 and 2,
omitting weathered samples, and
excluding anomalous value from the
correlation of volatile matter and S1+S2.

those from Drillhole 1494 are
high volatile bituminous B. Each
suite exhibits a range of
reflectance values, due to
variability in vitrinite chemistry.
Samples with highest reflectance
(and lowest volatile matter)
have normal chemistry, while
samples with low reflectance (and
high volatile matter) are
perhydrous. This chemical
variability causes the reflectance
ranges for the two sample sets to
overlap, even though the
reflectance of coals with normal
chemistry in DH1494 is 0.20%
higher than in DH698.

Figure 5 shows an alternative
method of rank assessment, which
is based closely on Suggate rank (Suggate 1959; Sykes et al. 1994). In this example,
data are related on a chart of volatile matter and specific energy. Rank and type trends
are illustrated, as are the approximate positions of ASTM coal rank boundaries. Rank
increases from left to right, and the "normal type line" represents the path followed by a
coal of normal vitrinite chemistry as it increases in rank. Perhydrous coals follow paths

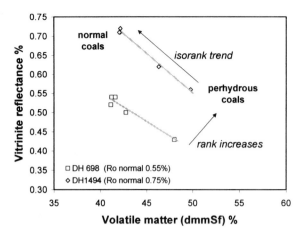

Figure 4. Rank and type effects on vitrinite reflectance, exemplified by samples from DH698 and DH1494. Ro normal is the reflectance of vitrinite with normal chemistry.

higher on the chart. It is important to note that the normal type line shown in Figure 5 is approximate and not intended to exactly correspond to Suggate's "average type line" (Suggate 1959; Sykes et al. 1994). It is clear from Figure 5 that there is both type and rank variation within the Table 1 dataset. Rank ranges from subbituminous B/A to high volatile bituminous B, i.e. from very immature to (arguably) marginally mature. Coal type – which is largely controlled by vitrinite chemistry – ranges from normal to strongly perhydrous.

Figure 3 illustrated a good linear correlation between volatile matter and Rock Eval yield for an isorank dataset. However, more complex trends emerge when additional data from Table 1 are plotted (Figure 6). The samples in Figure 3 display a simple relationship because they are the same rank. Figure 6 clearly shows that, for immature coals of normal type, Rock Eval yield (S1+S2) increases as rank increases. The hydrocarbon yield from pyrolysis of normal type coals is therefore strongly rank dependant at the relatively low temperatures employed by Rock-Eval analysis (<600°C). Similar observations have been made by other workers (Teichmuller & Durand, 1983; Suggate & Boudou, 1993). The serial sample (isorank) trends in Figure 6 also show that Rock-Eval yield is strongly type dependant, especially at low ranks.

In order to compare the Rock-Eval yield of immature coals of different ranks it is necessary to apply a rank correction to S1+ S2. Figure 6 shows representative lines of equal type for normal coals (those lowest on the chart) and for progressively more perhydrous coals. Clearly, coals of normal chemistry require the most correction. For the purpose of this paper, all data are corrected to the rank of DH698 and Pakawau samples, i.e., for each sample, Rock-Eval yield is adjusted to that of similar type coals at the subbituminous/high volatile bituminous rank boundary.

3.3 VOLATILE MATTER
The high temperature process by which volatile matter is measured (>900°C) produces a much larger pyrolysis yield than Rock-Eval (Newman et al. 1997). The products of

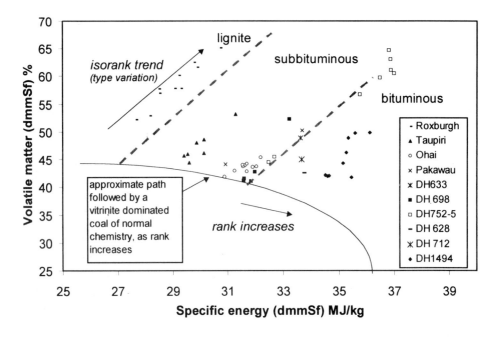

Figure 5. New Zealand coals, including those in Table's 1 and 2, shown on a modified "Suggate rank" diagram. The coalification path of normal type coals is only approximate, and does not equate exactly to the "average type" line of Suggate (1959). ASTM rank boundaries cannot be precisely depicted on this diagram, but their approximations provide a useful indication of the ranks represented in Tables 1 and 2.

coal pyrolysis >600°C are dominated by methane, carbon dioxide and carbon monoxide (Boreham et al., 1998), and an exact correlation between volatile matter and Rock-Eval yield is therefore unlikely. However, it is clear that volatile matter yield is systematically related to the amount of hydrocarbon released during pyrolysis, based on the generally good relationship between (1) volatile matter and specific energy (Figure 5), and (2) volatile matter and Rock-Eval yield (Figure 3), for coals of equal rank. Volatile matter has an advantage in being a direct gravimetric measure of pyrolysis yield, whereas Rock-Eval S1 + S2 can show varying amounts of suppression because detection depends on the concentration of heteroatom species in pyrolysates (Boudou et al. 1994).

Volatile matter is of course highest at low rank, when coals contain a high proportion of thermally labile compounds. Volatile matter declines progressively as increasing rank binds organic carbon in forms which are thermally stable. This is the only sense in which volatile matter is rank dependant, and it is possible to correct for this rank effect when using volatile matter to compare the source potential of coals of differing rank (Newman, 1989). Therefore, for immature coals, volatile matter is in some respects

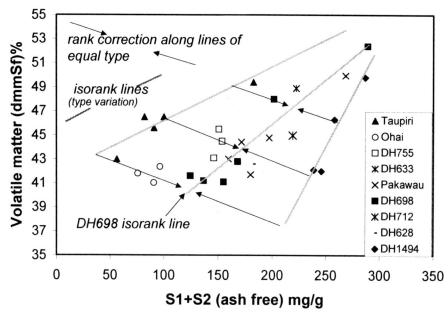

Figure 6. A plot of volatile matter and S1+S2 for coals ranging in rank from sub-butiminous B/A (Taupiri) to high volatile bituminous B (DH1494). Isorank lines drawn for Taupiri, DH698 and DH1494 samples illustrate the influence of varying type and rank on S1+S2. Arrows follow lines of approximately equal type and show the method for correcting S1+S2 values to the rank of DH698, 633 and Pakawau samples.

superior to Rock-Eval as an indicator of petroleum source potential, particularly in view of the strongly depressed Rock-Eval yield of low rank coals (Fig. 6).

3.3 ROCK EVAL S1+ S2 (ASH FREE AND RANK CORRECTED) vs HI

Hydrogen index (HI) is commonly used as an indicator of the source potential of immature coals. Expressing S2 as a ratio with TOC allows the organic matter of samples with different mineral content to be compared without correcting S2 to an ash free basis. However, S1+S2 (ash free) should describe the petroleum source potential of coals more precisely than HI. This is partly because S1 in coals represents free hydrocarbons (bitumen) and not generated hydrocarbons (Teichmuller & Durand, 1983; Bostick & Daws, 1994). Also, decarboxylation reactions are a major process within this low maturity range (Durand and Paratte, 1983), resulting in the HI showing 'erratic' behaviour because of TOC loss (Hydrogen Index denominator). HI also suffers because it does not fully accommodate rank influences on volatile matter yield within the immature range (obviously HI varies legitimately, due to hydrocarbon expulsion, once samples attain adequate maturity).

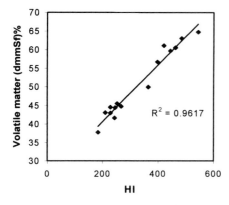

Figure 7. A plot of volatile matter
and HI for an isorank set of coals
from Table 1 and 2. The correlation
is good.

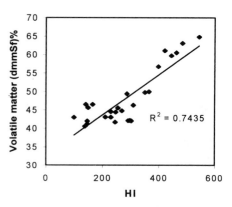

Figure 8. A plot of volatile matter
and HI for coals of varying rank
fromTable 1 and 2. The correlation
is poor compared with that in
Figure 7.

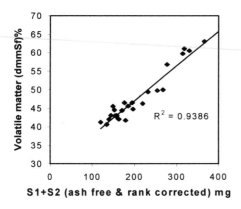

Figure 9. A plot of volatile matter and
S1+S2 (ash free and rank-corrected)
for the same coals as in Figure 8,
showing a superior correlation.

Figure 7 relates HI to volatile
matter for a subset of isorank
samples for which HI is available
(i.e., where TOC has been
measured). These are Pakawau and
DH755 samples from Table 1, and
additional samples from Table 2,
included to give a larger HI range.
Figure 7 demonstrates a strong
relationship, which shows that HI
can be a good source potential
indicator for immature coals when
rank is held constant. However, HI
performs relatively poorly when
immature coals of variable rank (at
and below high volatile bituminous
B) are plotted (Figure 8). For this
reason S1+S2, ash and rank corrected, is preferable to HI when comparing the source
potential of immature coals (Figure 9).

4.0   Relationship Between Mire Flora And Petroleum Source Potential

It has previously been observed that New Zealand coals of different ages have different vitrinite bulk chemistry (Newman 1989). Lower to middle Eocene coals in particular tend to have higher volatile matter, at any given rank, than most Cretaceous and Paleocene coals. Both palynological (Warnes 1992 , Ward et al. 1995, Moore 1996) and wood tissue analysis (Shearer & Moore, 1994a&b) indicate that the older coals are gymnosperm dominated, whereas the lower to middle Eocene coals are angiosperm dominated. In addition, paleobotanical studies of leaf physiognomy have identified a cool temperate climate in the Late Cretaceous in contrast to a sub-tropical to tropical climate in the lower to middle Eocene (Daniel et al. 1992).

The objective of this paper is to assess whether different floral origins result in different petroleum source potential for coals which achieve maturity in New Zealand's petroleum basins. Although total proportions of angiosperm, gymnosperm and pteridophyte (fern) palynomorphs may seem crude indicators of floral origins for coal, they represent a valid first step in evaluating trends. These total proportions summarise a large body of more detailed palynological data which will be added to and subjected to multivariate analysis in the next phase of the research.

Figure 10 relates total angiosperm palynomorph abundance to Rock-Eval yield (ash free and rank corrected) for all unweathered samples in Table 1. The samples range in age from Late Cretaceous (Ohai) to Late Eocene (Taupiri), and represent a wide range of floral communities and climatic conditions. Nevertheless, there is a surprisingly strong trend whereby Rock-Eval yield increases with the proportion of angiosperm palynomorphs.

Exceptions to this trend include some Eocene coals which exhibit considerable variation in hydrocarbon yield despite a uniformly angiosperm dominated flora. As indicated in Figure 10, the Eocene samples with relatively low yield originated in a raised mire environment, in which peat was flushed persistently by meteoric water (Newman, 1987). This regime leached mineral matter and depleted organic matter of hydrogen. It is significant that the ash content of these samples is less than 0.5% (Table 1), except for the basal sample which has a higher ash content due to inclusion of small amounts of dirty coal at the seam floor.

Other samples depart from the general trend in Figure 10 by having very high Rock-Eval yield in relation to angiosperm palynomorph abundance. One of these is marine influenced, which is considered likely to enhance source potential (Baskin & Peters 1992; Sinninghe Damste et al. 1989). This is the only coal in Table 1 which is considered likely to have been marine influenced. Coals in the upper part of the DH1494 seam have high sulphur, but this was introduced well after initial burial. Analysis of a large geological and chemical database for this coal deposit demonstrates that sulphur has not enhanced hydrogen content as it would if introduced during or soon after peat accumulation (Newman 1991; Newman et al 1997). For example, high sulphur also

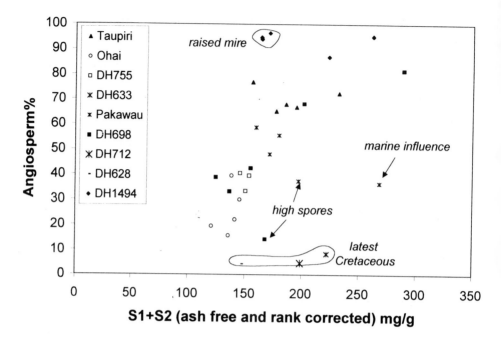

Figure 10. Abundance of angiosperm palynomorphs related to S1+S2 (ash free and rank corrected) for the coals in Table 1.

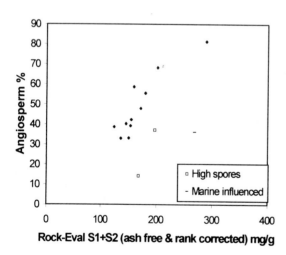

Figure 11. Abundance of angiosperm palynomorphs related to S1+S2 (ash free and rank corrected) for all Paleocene samples in Table 1

occurs towards the seam floor in some locations, in coal which has the relatively low volatile matter (hence low S1+S2 yield) attributed to raised mire accumulation. The upward increase in Rock-Eval yield for this seam is attributed to a progressively rising water table and oxygen depletion, without marine influence.

Two other samples, one each from DH698 and the Pakawau sample set, have high Rock-Eval yield relative to angiosperm palynomorph abundance. Both of these samples

Table 2. Additional data for Figures 7 to 9.

| Location | Sample type | CRL No. | Proximate analysis | | | Sp. E. MJ/kg | S % | VM dmmSf | Rock-Eval | | | | | |
			M a.d.	A a.d.	VM a.d.	a.d.	a.d.	%	Tmax deg. C	S1 (mg/g)	S2 (mg/g)	S3 (mg/g)	TOC (wt%)	HI
Greymouth	DH752	59/651	4.4	11.6	49.8	27.91	8.54	59.7	428	15.8	262.3	5.4	59.1	444
Greymouth	DH753	59/652	3.1	21.6	48.8	25.56	4.97	64.8	427	22.3	295.2	5.0	54.3	543
Greymouth	DH753	59/653	4.5	14.2	50.6	26.92	9.21	63.1	427	23.5	290.7	5.1	60.1	484
Greymouth	DH754	59/654	4.7	13.3	49.6	27.23	9.17	61.1	427	18.2	257.7	5.5	61.4	420
Greymouth	DH754	59/655	5.3	15.2	47.7	26.16	9.76	60.5	425	15.4	264.7	8.4	57.3	462
Greymouth	DH755	59/656	6.1	10.7	47.2	27.50	7.10	56.8	428	13.4	234.8	6.9	59.1	397

contain exceptionally high proportions of fern spores (Table 1). Although other data indicate that the ferns themselves are unlikely to directly enhance source potential, their presence may indicate unusual mire conditions.

Finally, coals from immediately below the Cretaceous/Tertiary boundary at Greymouth Coalfield have high Rock-Eval yield relative to angiosperm abundance. This productive chemistry sets them apart from other gymnosperm dominated coals (Ohai, DH698), for reasons which are not yet understood.

The interaction of environmental and floral controls on coal properties are potentially complex, and there may be other complicating factors such as differential rates of pollen/spore production by different plants, and differential preservation potential for different plant tissues. It would therefore be surprising if a diverse set of Late Cretaceous to Late Eocene coals, such as that in Table 1, demonstrated a consistent relationship between angiosperm palynomorph abundance and petroleum source potential. For this reason it is desirable to restrict the age range of samples, and to isolate floral influence from other variables as far as possible. The Paleocene period provides a good opportunity for this more specific study. The rank of Paleocene samples in Table 1 is uniform, and their depositional environment is predominantly fluvial floodplain. The Paleocene also provides valuable diversity in mire floral assemblage, related to expansion of angiosperms into environments previously dominated by gymnosperms and ferns.

Figure 11 relates angiosperm palynomorph abundance to Rock-Eval yield for Paleocene samples in Table 1. These exhibit a consistent trend whereby Rock-Eval yield increases as angiosperm palynomorph abundance increases. The three coals within this dataset which have elevated source potential relative to angiosperm palynomorph abundance have been discussed above.

5.0 Influence of Floral Content on the Timing of Gas and Oil Generation

5.1 INTRODUCTION

Five samples were selected for a more detailed investigation of oil generation potential and compositional kinetics (Boreham et al. 1998). These include 3 Paleocene samples from Drillhole 698 (AGSO #8370, #8373 and #8369), representing high, low and intermediate angiosperm abundance and S1+ S2 yield. Two Cretaceous samples were also included (AGSO #8376 and #8377). These were selected because they have moderate to high S1+S2 despite being gymnosperm dominated (Figure 10), and therefore represent a potentially important departure from the overall angiosperm/Rock Eval relationship.

The DH698 samples and #8377 (26/662) are isorank, at the sub-bituminous-high volatile C bituminous rank boundary. Sample #8376 (26/376) is slightly higher rank, but still high volatile bituminous C (Figure 5). All the coals are below the threshold for hydrocarbon generation. Bulk chemical data for all 5 samples, which will be referred to here by their AGSO laboratory number, appear in Table 3. It should be noted that, despite the stated intention to isolate floral influences from other potential controls, there is a direct relation between sulphur and angiosperm abundance in the Paleocene sample suite. The sulphur levels are low in comparison with known marine influenced coals, and there is no additional evidence of marine influence. Nevertheless, future work on a new sample set aims to more completely separate floral and sulphur influence on petroleum geochemistry.

Table 3. Bulk properties of coals discussed in Section 5. These coals also appear in Table 1, where they are referenced by CRL number, shown here with AGSO numbers.

AGSO No.	CRL ID	Drillhole	Age	Rank from Fig. 5	Angio-sperm %	TOC %	S1 mg/g	S2 mg/g	S3 mg/g	Tmax deg. C	HI
8369	52/172	698	Paleocene	subbit/bit	68.2	63.2	8.35	173.70	9.80	424	275
8370	52/174	698	Paleocene	subbit/bit	81.4	69.2	22.70	249.70	8.80	421	361
8373	52/178	698	Paleocene	subbit/bit	38.8	70.4	4.80	117.30	12.75	427	167
8376	26/376	628	Latest Cret.	bituminous	4.3	74.2	10.50	166.05	10.65	429	224
8377	26/662	633	Latest Cret.	subbit/bit	8.4	52.5	11.90	198.15	11.25	427	377

5.2 PY-GCMS

Molecular composition of the pyrolysate was analysed by full scan gcms in order to determine yields of individual components or classes of compounds. Figure 12 depicts the full scan traces for the New Zealand coals pyrolysed in a single step from 300 to 540°C. At the final pyrolysis temperature the maturity of the low rank coals has been raised to a rank equivalent of approximately 2% Ro and encompasses the entire oil window and the initial phase of dry gas generation (Boreham et al., 1997). The high bulk pyrolysis yield (S2) shown by coals #8370 and #8377 is reflected in the enhanced C_{15+} alkene/alkane pairs compared to the lower molecular weight linear hydrocarbons, and this is seen as an indication of their good oil potential (Horsfield, 1989). In contrast,

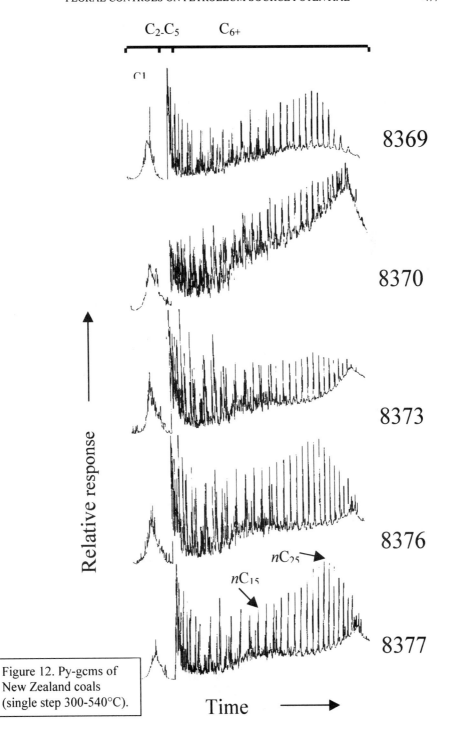

Figure 12. Py-gcms of New Zealand coals (single step 300-540°C).

coal #8373 shows higher relative contents of the low molecular weight components consistent with its potential for mainly gas.

Py-gcms over smaller temperature increments (stepwise pyrolysis of 20 to 30°C increments) provides more detailed information on the change in pyrolysate composition as the coal is artificially matured from an immature to overmature stage. As an example, Figure 13 depicts the traces for coal #8370. The increase in vitrinite reflectance of an artificially matured coal under the same experimental conditions has previously been 'calibrated' against the final pyrolysis temperature using a Permian coal from the Bowen Basin, Australia (Boreham et al., in press). There, the vitrinite reflectance increased between 0.05 to 0.15% for each 20°C step up to a final vitrinite reflectance of Ro = 1.7% at 510°C (Boreham et al., in press). Furthermore, the Ro difference between successive pyrolysis steps progressively increases as the temperature increases. This calibration curve (Boreham et al., in press) is used to define the Ro level at the end of each pyrolysis step for the artificially matured coals.

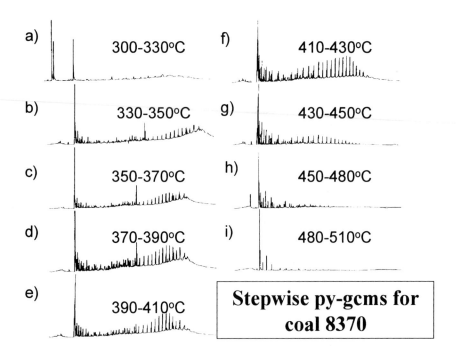

Figure 13. Stepwise py-gcms for coal #8370.

The value of the stepwise approach can be seen in the systematic changes in various compositional features. The highest methane yields are consistently shown by coal #8373 (Fig. 14a), concomitant with the lowest yields of wet gas (Fig. 14b) and liquid hydrocarbons (Fig. 14c) over all the temperature-incremented intervals. This confirms

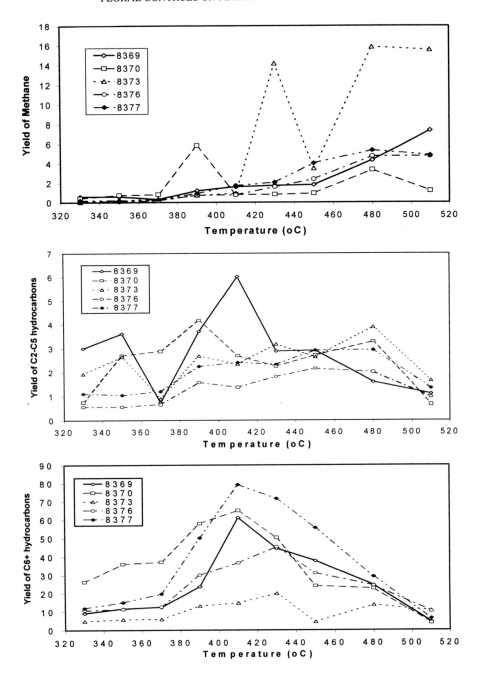

Figure 14. Plot of yield (mg/gTOC) of (a) methane (b) wet gas (C_2-C_5) (c) oil (C_6+) versus final temperature of individual pyrolysis steps.

the gas-proneness of the coal as suggested by its low HI. The two coals #8370 and #8377 show the highest liquid hydrocarbon yields (Fig. 14c) and are considered oil-prone in accord with their high HI values (Table 3). However, higher yields of liquid hydrocarbons occur at lower temperatures (<390°C) for the angiosperm-dominant coal (#8370) compared to the gymnosperm-dominant coal (#8377). Extrapolation to subsurface processes may imply that the onset of oil generation occurs at a lower maturity for the angiosperm dominated coal. Indeed, kinetic analysis strongly supports this view (see below).

These compositional differences are further expressed in two important ratios, the C_1/C_2-C_5 ratio (DWR or dry-to-wet gas ratio of Figure 15a) and the C_1-C_5/C_{6+} ratio (GOR or gas-to-oil ratio of Figure 15b). The DWR for the coals remains fairly constant within the range 0.2 to 0.4 during pyrolysis from 300 to 430°C (final Ro = 1.0%). Over this temperature range, corresponding to the oil window, GOR also remained uniform within each individual coal. Coal #8373 shows the lowest DWR even though it has the highest GOR. Both characteristics signify its enhanced potential for wet gas generation. The major oil potential shown by coals #8370 and #8377 is mainly expressed in their lower GOR. The progressive increase in oil yields up to peak oil generation is also seen at the molecular level in the n-alkene/alkane contents (Figs. 13 a-e). Furthermore, high molecular weight n-alkenes/alkanes are initially generated followed by a progressive increase in the lighter homologues.

Beyond peak oil generation (430°C; 1.0% Ro), n-alkene/alkane yields are on the rapid decline and low molecular weight homologues dominate (Figs. 13 f-g). At temperatures >450°C (>1.15% Ro), n-alkene/alkanes are no longer generated (Figs. 13 h-i). Here, DWR and GOR steadily increase as wet gas and oil generation become exhausted, together with the commencement of the main phase of methane generation (peak methane generation should occur at maturities higher than those obtained here; Boreham et al., 1997). Within this overmature zone, #8373 now shows the highest DWR representing its limited liquid leg. On the other hand, the two oil-prone coals #8370 and #8377 have the lowest DWR representing their extended liquid hydrocarbon (condensate?) leg.

5.3 CHEMICAL KINETICS

5.3.1 *Bulk Kinetics*
Table 4 lists the bulk kinetic parameters for conversion of coal to bulk pyrolysate (gas and oil). These kinetic parameters were derived assuming that the conversion of coal to total pyrolysis products can be modelled by a set (n) of parallel reactions ($\Sigma R(i)$ i=1 to n), each obeying the first order rate law (Schenk et al. 1996 and references therein):

$$d[R(i)]/dt = -k(i)[R(i)] \qquad (1)$$

where $R(i)$ is the ith reactant of the initial coal and $k(i)$ is its associated rate constant. The initial $[R(i)]$ is assigned a reaction potential. The sum of all the reactions potentials is equal to HI.

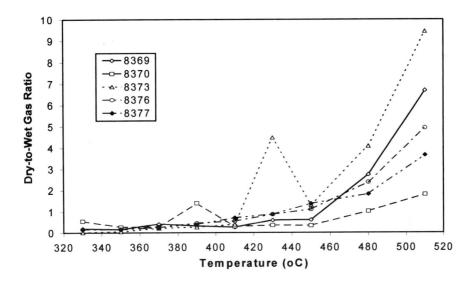

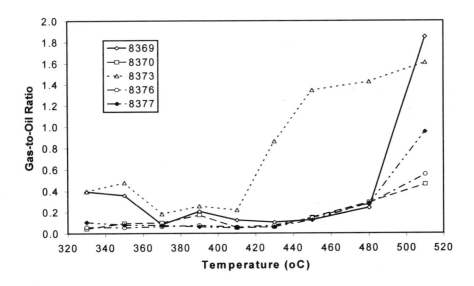

Figure 15. Plot of (a) dry-to-wet gas ratio (DWR) (b) gas-to-oil ratio (GOR) versus individual stepwise pyrolysis final temperature.

In general, the rate of reaction increases with temperature and the temperature dependency of the rate constant obeys the empirical relationship (termed the Arrhenius equation):

$$\ln k(i) = \ln A - Ea(i)/RT \qquad\qquad (2)$$

where A is the frequency factor (or pre-exponential factor), T is the temperature, R is the universal gas constant, and Ea(i) is the activation energy of the ith reactant. The activation energy defines the minimum energy that reactants must obtain in order to form products, and the frequency factor is the fraction of collisions that have sufficient energy to lead to reaction and thus relates to the probability that the reaction will occur.

All coals show a rather broad distribution in activation energy (Table 4) reflecting the many explicit reactions associated with pyrolytic release of hydrocarbons which, in turn, is related to the heterogeneous composition of coal.

A clearer picture for the timing of hydrocarbon generation arises when the kinetic parameters are applied to a common temperature history, especially one which is geologically relevant. Extrapolation of the laboratory data over ten orders of magnitude to the geological time/temperature regime incorporates uncertainties (Schenk et al., 1996). Most significant are assumptions that the mechanism of reaction remains unchanged and that the process of petroleum generation is a simple conversion of coal to petroleum (oil and gas) where each reaction step behaves independently. However, acceptance of this kinetic model allows the entire petroleum generation process to be predicted from an analysis of the immature analogue.

Although kinetic models appear to work well for marine organic matter, coals show considerable divergence of predicted generation histories from coals within a natural maturity sequence ranging in rank from high volatile bituminous C to anthracite (Boreham et al., 1998). In the natural maturity sequence, Rock Eval yields are higher and kinetic parameters show higher temperatures at peak hydrocarbon generation compared to that predicted from the immature analogue. Re-incorporation of generated bitumen with the coal matrix has been suggested as a possible explanation (Boreham et al., 1997; 1998).

Despite the potential shortcomings recognised for the above kinetic model, the approach remains the most popular 'link' between experimental and field observations (Schenk et al., 1996). Combination of the bulk kinetic parameters in Table 4 with a simple linear heating rate of 1°C/Ma results in the calculated cumulative hydrocarbon yield (represented as a transformation ratio) and rate of hydrocarbon generation plots of Figure 16 and Figure 17, respectively. These indicate that petroleum generation generally begins, peaks and then ends at lower maturity for those coals with higher angiosperm contents and confirms the results of the compositional analysis from stepwise pyrolysis.

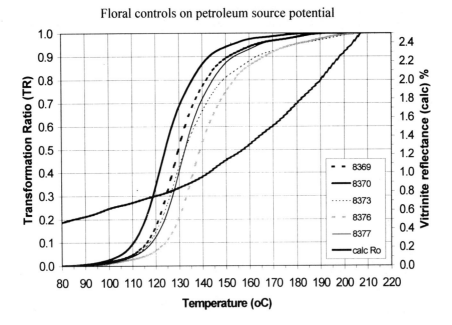

Figure 16. Plot of (a) transformation ratio (b) vitrinite reflectance versus temperature at 1°C/Ma. The reflectance curve is based on the model of Burnham & Sweeney (1989).

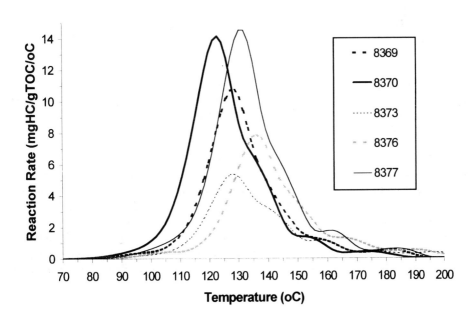

Figure 17. Plot of calculated reaction rate versus temperature at 1°C/Ma.

Also plotted on Figure 16 is the interpreted Ro derived using the kinetic parameters for vitrinite maturation (Burnham and Sweeney, 1989; Sweeney, 1990). These kinetic parameters represent an 'average' for coals of different ages and thus should be considered only as a guide to calculated sub-surface maturity levels for these specific coals. For the same amount of coal conversion (identical TR) the present model indicates at maximum a 15°C difference between the most labile and least labile coal corresponding to an approximately 0.15% difference in Ro.

5.3.2 Compositional Kinetics

Compositional kinetic parameters have most commonly been determined using the open system approach for the bulk kinetics with continuous trapping and measurement of individual pyrolysis products for at least three linear heating rates (Ungerer, 1990; Tang and Stauffer, 1994). However, this approach is labour and time intensive. A simplified method is offered here to derive individual kinetic parameters for both gas and oil generation which combines bulk kinetic parameters with compositional data from stepwise py-gc. The pseudo-rate profile from open-system stepwise pyrolysis (Fig. 18a) can be 'aligned' with that generated from bulk kinetics (Fig. 18b). Individual activation energies are assigned a GOR value (Table 4) depending on where they are positioned under the stepwise pyrolysis evolution curve. This semi-quantitative compositional kinetic approach, much simpler in design to the 'fully computational' compositional kinetics, requires identical A values for both oil and gas kinetics over the same range of activation energies but with differing reaction potentials at each individual activation energy. This results in GOR's being rather insensitive to heating rate. However, this level of complexity in the hydrocarbon generation model may be quite adequate under 'normal' conditions where the quality of the maturation model is usually low due to poor constraints on the geological inputs.

Table 4 lists the derived kinetic parameters for both gas and oil generation and is presented in a form that is suitable for input into most PC-based geohistory modelling software packages. Figure 19 shows the relative proportions of the transformation ratio (TR, proportion of coal converted to hydrocarbons) for gas and oil, together with the rate of gas and oil generation at 1°C/Ma. Many of the features expressed in this figure have been discussed in detail under the Bulk Kinetic and Py-gcms headings. Clearly, liquid hydrocarbon generation is dominant over gas generation, gas generation occurs throughout all of the 'oil window', and the timing order for hydrocarbon generation and the evolution in hydrocarbon composition (GOR) is readily derived over the entire maturity range up to a rank of approximately 2.0% Ro. For example, at geologically significant heating rates the onset of oil generation (defined as 10% conversion of the oil potential; TR=0.1 on a 0 to 1 scale and proportioned accordingly on Fig. 19) for coal #8370 corresponds to a temperature and Ro of 108°C and 0.65%, respectively, whereas they are at 123°C and 0.75% Ro for coal #8376. At peak oil generation (TR= 0.5 on a 0 to 1 scale and proportioned accordingly on Fig. 19) for coal #8370 the temperature and vitrinite reflectance are at 125°C and 0.8% Ro, respectively, whereas for coal #8376 they are at 137°C and 0.92% Ro, respectively. However, see the discussion in the following section.

#8377 stepwise py-gcms + bulk kinetics

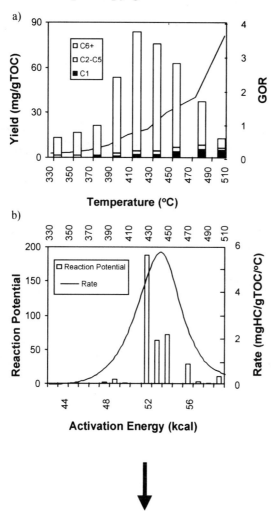

Compositional kinetics

Figure 18. Integration of stepwise py-gcms with bulk kinetics for coal 8370 using (a) bar chart of yields of methane, C2-C5 and C6+, and GOR (b) activation energy distribution, reaction potential and rate of hydrocarbon generation at 5°C/min.

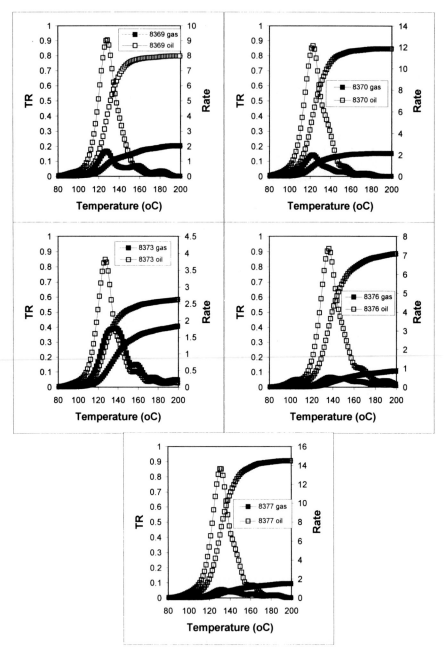

Figure 19. Plot of transformation ratio and rate (mg/gTOC/°C) for gas and oil generation at 1°C/Ma.

Table 4. Bulk and compositional kinetic parameters for New Zealand coals

	Ea (kcal/mol)	Bulk Kinetics		Compositional kinetics		
				Gas (C1-C5)	Oil (C6+)	GOR
		Proportion	Reaction potential mgHC/gTOC	Reaction potential mgHC/gTOC	Reaction potential mgHC/gTOC	
8369	A = 1.121E+13 s-1					
	46	0.0120	3.30	0.87	2.43	0.36
	47	0.0028	0.77	0.20	0.57	0.36
	50	0.4823	132.63	23.02	109.61	0.21
	51	0.2196	60.39	6.47	53.92	0.12
	52	0.1459	40.12	4.62	35.51	0.13
	53	0.0316	8.69	1.68	7.01	0.24
	54	0.0601	16.53	10.73	5.80	1.85
	55	0.0090	2.48	1.61	0.87	1.85
	57	0.0366	10.07	6.53	3.53	1.85
8370	A = 7.244E+12 s-1					
	45	0.0038	1.37	0.12	1.25	0.10
	46	0.0087	3.14	0.29	2.86	0.10
	48	0.0122	4.40	0.40	4.00	0.10
	49	0.5013	180.97	26.29	154.67	0.17
	50	0.1703	61.48	8.02	53.46	0.15
	51	0.2055	74.19	9.68	64.51	0.15
	52	0.0123	4.44	0.86	3.58	0.24
	53	0.0605	21.84	6.88	14.96	0.46
	55	0.0094	3.39	1.07	2.32	0.46
	56	0.0160	5.78	1.82	3.96	0.46
8373	A = 1.173E+13 s-1					
	46	0.0065	1.09	0.35	0.73	0.48
	47	0.0080	1.34	0.43	0.90	0.48
	50	0.3810	63.63	12.31	51.31	0.24
	51	0.1641	27.40	12.67	14.73	0.86
	52	0.2245	37.49	21.47	16.02	1.34
	53	0.0086	1.44	0.85	0.59	1.43
	54	0.1106	18.47	10.87	7.60	1.43
	56	0.0400	6.68	4.12	2.56	1.61
	57	0.0142	2.37	1.46	0.91	1.61
	59	0.0426	7.11	4.39	2.73	1.61
8376	A = 3.904E+13 s-1					
	48	0.0208	4.66	0.26	4.40	0.06
	49	0.0027	0.60	0.03	0.57	0.06
	52	0.4250	95.20	5.39	89.81	0.06
	53	0.1858	41.62	2.72	38.90	0.07
	54	0.1767	39.58	5.16	34.42	0.15
	55	0.0305	6.83	1.49	5.34	0.28
	56	0.0666	14.92	3.26	11.66	0.28
	57	0.0382	8.56	1.87	6.69	0.28
	59	0.0341	7.64	2.74	4.90	0.56
	60	0.0122	2.73	0.98	1.75	0.56
	61	0.0073	1.64	0.59	1.05	0.56
8377	A = 2.426E+13 s-1					
	47	0.0064	2.41	0.18	2.23	0.08
	48	0.0172	6.48	0.48	6.00	0.08
	49	0.0016	0.60	0.04	0.56	0.08
	51	0.4976	187.60	10.62	176.98	0.06
	52	0.1685	63.52	3.60	59.93	0.06
	53	0.1926	72.61	7.78	64.83	0.12
	55	0.0760	28.65	6.27	22.38	0.28
	56	0.0076	2.87	0.63	2.24	0.28
	57	0.0031	1.17	0.57	0.60	0.95
	58	0.0294	11.08	5.40	5.68	0.95

5.3.3 Are the Results of Kinetics Studies on Immature Coals Complicated by Initial Sample Rank?

Section 3.2 demonstrates that Rock-Eval yield is strongly influenced by coal rank through the immature range. It is therefore desirable to consider whether the processes by which kinetics parameters are determined using immature samples may also be influenced by their initial rank. Such a dependence would affect the accuracy of temperatures and reflectances attributed to oil generation. However, kinetics studies can still provide important information on the relative timing of generation from coals of different character, provided the samples investigated have the same initial rank.

In the present sample set there is a good relationship between modelled generation temperature and floral origin, for coals which are the same rank (#8369, #8370, #8373 in DH698 and #8377 in DH633 , Figure 5). That is, despite differences in total yield, gymnosperm dominated samples 8373 and 8377 report similar generation temperatures, while increasing angiosperm dominance corresponds to progressively lower generation thresholds (Figure 16). However, despite similar floral origins to 8373 and 8377, sample 8376 reports significantly higher generation temperatures. This could be due to the higher rank of coal 8376 (DH628 in Figure 5).

Figure 20 compares the kinetics of gymnosperm dominated coals in this study with a high volatile bituminous B coal from Buller Coalfield (DH1494 in Figure 5). This coal, from the low sulphur interval in DH1494 (Table 1), has vitrinite of normal chemistry despite being angiosperm dominated. Kinetics studies on this Buller coal report higher generation temperatures than for the significantly lower rank coals in the present study (Figure 20). It is unlikely that this difference can be attributed to significant oil generation by the Buller coal prior to analysis (Newman et al. 1997). The relatively high generative temperatures modelled for the Buller sample may simply be due to its higher rank. This implies that increasing rank has lead to an internal structural rearrangement that is not accounted for in the current kinetic models.

5.3.4 TIMING OF OIL EXPULSION FROM COAL

Of relevance to this discussion is work on Buller coals by Newman, Price and Johnston (1997). This empirical study assessed angiosperm dominated coals ranging in rank from high volatile bituminous B to medium volatile bituminous. Each isorank sample suite featured coals ranging from perhydrous to relatively hydrogen depleted. Rock-Eval data indicated that oil expulsion did not commence before ranks of about high volatile A bituminous (c. 1.0% Ro). Saxby et al. (1982) has also suggested similar maturity levels are required before expulsion can occur from Australian coals. Thus a maturity-lag after peak oil generation is implicated before expulsion of oil from coal.

Although a reduction in pore space with increased maturity will undoubtedly assist oil mobility, it is likely that the main driving force for oil expulsion is the associated gas content as indicated by GOR and to a lesser extent DWR. Primary migration is likely to occur past peak oil generation where GORs are high (Boreham and Powell, 1994). The role of methane during petroleum migration by gaseous solution was discussed by Price

Floral controls on petroleum source potential

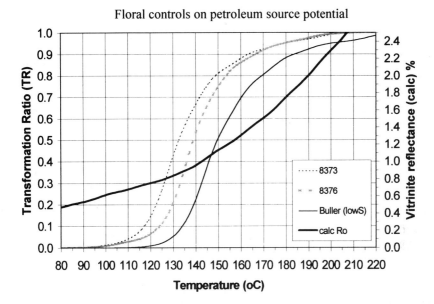

Figure 20. Plot of (a) transformation ratio (b) vitrinite reflectance versus temperature at 1°C. This subset of samples represents immature coals which are high volatile B bituminous (Buller), high volatile C bituminous (8376), and at the sub-bituminous/bituminous rank boundary (8373). Differences in modelled generation temperatures may reflect the different ranks of the initial samples. The reflectance curve is based on the model of Burnham & Sweeney (1989).

et al. (1983) and Price (1989), and related to delayed oil expulsion from New Zealand coals by Newman et al. (1997), who proposed microfracturing during methanogenesis as an additional mechanism aiding oil expulsion. With respect to the current kinetics model, there is a gradual change in GOR over the oil window, resulting in no definite maturity 'cutoff' between the end of oil generation and beginning of gas generation (Boreham et al., in press). Thus, the kinetic parameters alone make it difficult to predict a GOR-expulsion threshold. Furthermore, within the gaseous hydrocarbons, the change in DWR is not accounted for within the 2-component compositional approach (gas and oil) presented here. Multi-component compositional models (Behar et al., 1997) can better address the change in DWR over the oil window. Additionally, these complex kinetic models can more rigorously account for the increase in the light liquids compared to the heavy liquids at elevated maturation levels which would result in an extension of the end of the oil window to slightly higher maturities with the associated generation of a more readily expellable phase.

6.0 Conclusions

This first stage of research into floral controls on coal chemistry has identified a general trend whereby petroleum source potential increases in direct relation to angiosperm abundance in the original peat mire. This trend is almost entirely independent of sulphur variation. Some latest Cretaceous coals from Greymouth Coalfield have high oil generative potential despite low angiosperm abundance, for reasons which have not yet been indentified.

Immature coals ranging in rank from sub-bituminous to high volatile bituminous B were included in the study in order to investigate a wide variety of floral origins. Rank controlled variations in Rock-Eval yield necessitate rank correction of S1+S2 before the source potential of all coals can be directly compared. Hydrogen index does not adequately accommodate rank variations in the immature range. Consequently volatile matter is graphed with S1+S2 (ash free basis) to construct a diagram which displays rank and type variation in Rock-Eval yield. This diagram allows the Rock-Eval yield of coals which vary in rank to be adjusted to an equal rank basis.

Correlations of Rock-Eval yield with volatile matter also provide a method for screening data to identify analyses which are anomalous due to outcrop weathering or experimental error. Volatile matter is also a suitable reference against which the compatability of different Rock-Eval datasets can be evaluated.

Full scan py-gcms confirms that high Rock-Eval S1+S2 yield corresponds to enhanced oil generation (C15+ alkane/alkene pairs), whereas samples with lower S1+S2 are relatively gas prone. Stepwise py-gcms, which illustrates the relative timing of generation onset and peak generation, shows that angiosperm derived coals in the present data set attain the onset of generation and peak generation earlier than gymnosperm derived coals, regardless of total generative potential. This trend cannot be definitively attributed to floral controls, due to parallel variations in sulphur content, and further work is planned to resolve this ambiguity.

Bulk and compositional kinetics applied to immature samples of different rank indicate that the results of present models may not be independent of the rank of coals used for kinetics studies. This inference is based on an observed increase in predicted generation temperature with increasing rank of the immature analogue, and implies that results for immature coals of differing rank are not comparable, and may significantly underestimate the temperatures and reflectances actually required for oil generation. This effect is additional to the widely recognised maturity lag between petroleum generation and expulsion for coals.

However, where the rank of immature analogues in a sample set is uniform, the results of kinetic models are expected to accurately represent the relative timing of petroleum generation for different coals. Four samples in the present study have a common rank, at the sub-bituminous/high volatile bituminous rank boundary. Sample #8370, which is angiosperm dominated and has high oil generation potential, is predicted by the present model to achieve onset and peak generation at 108°C/0.65%Ro and 123°C/0.8%Ro respectively. Sample #8377, which also has high generation potential but is gymnosperm dominated, is predicted to achieve onset and peak generation somewhat later, at 118°C/0.74%Ro and 130°C/0.85Ro respectively. As discussed above, this difference in peak generation temperatures may be underestimated by present kinetic models. However, if oil expulsion is delayed until after peak generation the geological implications of earlier generation from angiosperm derived coals are unclear. Therefore, the most important conclusions of the work to date are (1) that angiosperm derived coals appear to have generally greater petroleum source potential than gymnosperm derived coals, and (2) that existing kinetic models may underestimate the temperatures required for petroleum generation from coals in general.

References

Behar, F., Vandenbroucke, M., Tang, Y., Marquis, F. and Espitalié, J. (1997) Thermal cracking of kerogen in open and closed systems: determination of kinetic parameters and stoichiometric coefficients for oil and gas generation. Org. Ceochem. 26, 321-339.

Boreham, C.J. and Powell, T.G. (1993) Petroleum source rock potential of coal and associated sediments: qualititative and quantitiative aspects, in B.E.Law and D.D. Rice (eds.), Hydrocarbons from coals. *AAPG Studies in Geology* **38**, 133-158.

_____, Summons, R.E., Roksandic, Z., Dowling, L.M. and Hutton, A.C. (1994) Chemical, molecular and isotopic differentiation of organic facies in the Tertiary lacustrine Duaringa oil shale deposit, Queensland, Australia, *Organic Geochemistry* **21**, 685-712.

_____, Golding, S. and Glikson, M. (1997) Factors controlling gas generation and retention in Bowen Basin coals, Australia, *18th International Meeting on Organic Geochemistry, Maastricht, The Netherlands.* Abstracts Part 1. Forschungszentrum, Julich, Germany, 41-42.

_____, Golding, S.D. and Glikson, M. (1998) Factors controlling the origin of gas in Australian Bowen Basin coals. Org. Geochem. In press.

_____, C.J., Horsfield, B. and Schenk, H.J. (in press) Predicting the quantities of oil and gas generated from Permian coals, Bowen Basin, Australia using pyrolytic methods, *Marine and Petroleum Geology.*

Bostick, N.H. and Daws, T.A. (1994) Relationships between data from rock-Eval pyrolysis and proximate, ultimate, petrographic and physical analysis of 142 diverse US coal samples. *Org. Geochem.* **21**, 35-49.

Boudou, J.P., Espitalié, J., Bimer, J. and Salbut, P.D. (1994) Oxygen groups and oil suppression during coal pyrolysis. Energy and Fuels 8, 972-977.

Burnham, A.K. and Sweeney, J.J. (1989) A chemical kinetic model of vitrinite maturation and reflectance, *Geochim. et Cosmochim. Acta* **53**, 2649-2657.

Collier, R.J. and Johnston, J.H. (1991) The identification of possible hydrocarbon source rocks, using biomarker geochemistry, in the Taranaki Basin, New Zealand, *Journ. Southeast Asian Earth Sciences* **5**, 231-239.

Cook, R.A. (1987) *The geology and geochemistry of crude oils and source rocks,* New Zealand Geological Survey Petroleum Report 1250, Institute of Geological and Nuclear Sciences, Lower Hutt.

Czochanska, Z., Sheppard, C.M., Weston, R.J. and Woolhouse, A.D. (1987). A biological marker study of oils and sediments from the West Coast, South Island, New Zealand. *New Zealand Journ. Geol. Geophys.* **30**, 1-17.

_____, Gilbert, T.D., Philp, R.P., Sheppard, C.M., Weston, R.J., Wood, T.A. and Woolhouse, A.D. (1988) Geochemical application of sterane and triterpane biomarkers to a description of oils from the Taranaki Basin, New Zealand. *Org. Geochem.* **12**, 123-135.

Daniel, I.L., Kennedy, E.M., Lovis, J.D., Moore, T.A., Newman, J., Shearer, J.C. and Warnes, M.D. (1992) Reconstruction of terrestrial paleoclimate during the Cenozoic, using evidence from coal constituents, paleobotany, and palynology, Abstract, *4th Australia-New Zealand Climate Forum,* Cairns, Australia.

Durand, B. and Paratte, M. (1983) Oil potential of coals: a geochemical approach. In: *Petroleum Geochemistry and Exploration of Europe, Geological Society Special Publication No. 12* (ed J. Brooks). Oxford, Blackwell Scientific Publications, 255-265.

Horsfield, B. (1989) Practical criteria for classifying kerogens; some observations from pyrolysis-gas chromatography. *Geochim. et Cosmochim. Acta* **53**, 891-901.

Killops, S.D., Raine, J.I., Woolhouse, A.D. and Weston, R.J. (1995) Chemostratigraphic evidence of higher-plant evolution in the Taranaki Basin, New Zealand, *Organic Geochemistry* **23**, 429-445.

Moore, N. (1996) Seam identification, correlation and coal quality prediction using in-seam variations in key palynomorph abundances, *Proceedings of the 1996 AusIMM (NZ Branch) Conference, Greymouth.*.

Murray, A.P., Sosrowidjojo, I.B., Alexander, R., Kagi, R.I., Norgate, C.M. and Summons, R.E. (1997) Oleananes in oils and sediments: Evidence of marine influence during early diagensis? *Geochim. et Cosmochim. Acta* **61**, 1261-1276.

Newman, J. (1989) Why are some high rank Tertiary coals more peculiar then others? Some thoughts on climate and floral assemblage, *Proceedings of 3rd Coal Research Conference*, Wellington, New Zealand.

_____ (1991) Controls on the distribution, timing and effects of diagenetic sulphur enrichment on some New Zealand coals, *Proceedings of 4th Coal Research Conference,* Wellington, New Zealand.

_____ and Newman, N.A. (1982) Reflectance anomalies in Pike River coals: evidene of variability in vitrinite type, with implications for maturation studies and "Suggate rank", *NZ Journ. Geol. Geophys* **25**, 233-243.

_____, Price, L.C. and Johnston, J.H. (1997) Hydrocarbon source potential and maturation in Eocene New Zealand vitrinite rich coals; Insights from traditional coal analyses, and Rock-Eval and biomarker studies, *Journal of Petroleum Geology* **20**, 137-163.

Newman, N.A. (1987) High alumina ash West Coast coals. *Proceedings of the 2nd Coal Research Conference,* Wellington, New Zealand.

Saxby, J.D. (1982) A reassessment of the range of kerogen maturities in which hydrocarbons are generated J Petroleum Geol 5 117-128.

Schenk, H.J., Horsfield, B., Krooss, B., Schaefer, R.G. and Schwochau, K. (1996) Kinetics of petroleum formation and cracking, in D.H. Welte, D.R. Baker and B.Horsfield (eds.), *Petroleum and Basin Evolution* , Springer Verlag, Heidelberg, pp231-269.

Shearer, J.C. and Moore, T.A. (1994a) Grain size and botanical analysis of two coal beds from the South Island of New Zealand, *Paleogeog., Paleoclim., Paleoecol.* **80**, 85-114.

_____ (1994b) Botanical control on banding character in two New Zealand coal beds, *Paleogeog., paleoclim., Paleoecol.* **110**, 11-28.

Suggate, R.P. (1959) New Zealand coals; their geological setting and its influence on their properties, *New Zealand Dept of Sci. and Ind. Res. Bull* **134**.

_____ and Boudou, J.P. (1993) Coal rank and type variation in Rock-Eval assessment of New Zealand coals. *Journ. Petrol. Geol.* **16**, 73-88.

Sweeney, J.J. (1990) Basinmat: Fortran program calculates oil and gas generation using a distribution of activation energies. *Geobyte* **April** 37-43.

Sykes, R.P., Suggate, R.P. and King, P.R. (1993) Timing and depth of maturation in southern Taranaki Basin from reflectance and rank (S). *Proceedings of the New Zealand Petroleum Symposium,* Christchurch, 373-389.

Tang, Y. and Stauffer, M. (1994) Multiple cold trap pyrolysis gas chromatography: a new technique for modelling hydrocarbon gneration, in N. Telnaes, G. van Grass and K. Oygard (eds.) *Advances in Organic Geochemistry 1993, Organic Geochemistry* **22**, 863-872.

Teichmuller, M. and Durand, B. (1983) Fluorescence microscopical rank studies on liptinites and vitrinites in peat and coals, and comparison with results of the Rock-Eval pyrolysis. *Int. Journ. Coal Geol.* **2**, 197-230.

Traverse, A. (1988) *Paleopalynology*, Unwin-Hyman, Boston.

Ungerer, P. (1990) State of the art in kinetic modeling of oil formation and expulsion, *Organic Geochemistry* **16**, 1-25.

Warnes, M.D. (1992) Interpretations of coal forming vegetation of the Morley Coal Measures, Ohai Coalfield, *Geological Society of New Zealand miscellaneous publication* **63a**:

Ward, S.D., Moore, T.A. and Newman, J. (1995) Floral assemblage of the "D" coal seam (Cretaceous): implications for banding characteristics in New Zealand coal seams, *N Z Journ. of Geol. and Geophys.* **38**, 283-297.

THE INFLUENCE OF DEPOSITIONAL AND MATURATION FACTORS ON THE THREE-DIMENSIONAL DISTRIBUTION OF COAL RANK INDICATORS AND HYDROCARBON SOURCE POTENTIAL IN THE GUNNEDAH BASIN, NEW SOUTH WALES

LILA W. GURBA AND COLIN R. WARD

School of Geology
University of New South Wales
Sydney NSW 2052 Australia

Abstract

Three-dimensional modelling of vitrinite reflectance has been used to enhance the understanding of lateral and vertical rank variations in the Permian coals of the Gunnedah Basin, New South Wales, Australia. The level of organic maturity of the coals has been investigated using both petrographic (vitrinite reflectance and fluorescence) and chemical methods (proximate and ultimate analyses, and electron microprobe data). The coal is of high-volatile bituminous rank, with a mean maximum vitrinite reflectance of between 0.56 and 1.1%. In addition to maturation-induced trends, a significant influence of depositional environment has been identified on vitrinite reflectance and other coal rank indicators in different parts of the sequence.

Lower than normal vitrinite reflectance is developed in several parts of the Permian sequence, where marine strata overlie the coal-bearing interval or where lower delta plain facies are present. The coals in these intervals have a perhydrous character, increased fluorescence intensity and contain framboidal pyrite, that combine to make them distinctive in petrographic studies. When plotted against depth all vitrinite reflectance values in these parts of the sequence are shifted to the lower side of the more "normal" depth/reflectance regression line. Such anomalies can be recognised at equivalent horizons over wide areas, suggesting basin-wide marine flooding events. If not allowed for in some sections rank, as expressed by vitrinite reflectance or volatile matter content, would appear to decrease instead of increase with depth.

Coals in other parts of the section have anomalously high vitrinite reflectance values, and contain hydrogen-poor material described elsewhere as 'pseudovitrinite'. Data from such coals plot to the right of the regression line in vitrinite reflectance profiles.

Chemical and petrographic studies show that the different vitrinite types follow separate coalification tracks, and hence both high and low-value anomalies need to be taken into account when interpreting maturation patterns. The depositional controls and the rank trends both have implications to maturation studies, and to prospectivity mapping for coalbed methane and petroleum generation.

1. Introduction

An important parameter in assessing coalbed methane or petroleum potential of a sedimentary basin is degree of thermal maturation (coal rank). Being a concept rather than a property (Diessel, 1992a) rank cannot be measured, but it can be assessed from those physical and chemical coal properties that change most during coalification. A wide variety of methods have been used to study thermal maturation including vitrinite reflectance, volatile matter content, moisture content and calorific value. Since these properties change at different rates during the coalification process, some rank indicators are more appropriate than others for particular rank stages (Teichmüller and Teichmüller, 1982). The appropriate rank parameters for use in a particular study are those which exhibit the highest sensitivity, i.e. the greatest rate change with depth over the part of the rank range under investigation. (Rank indicators commonly used for dispersed organic matter are not discussed in the present paper).

With the growth of coal petrology, vitrinite reflectance has now become the most commonly used parameter to indicate organic maturation. However, representation of rank by a single parameter may for some purposes be an oversimplification. Vitrinite reflectance can respond to other effects in addition to true maturity (Price and Barker, 1985; Teichmüller, 1987; Gentzis and Goodarzi, 1994, Mukhopadhyay and Dow, 1994), and the level of maturation of organic matter can be misinterpreted if such effects are not taken into account. It is also desirable to use several overlapping rank indicators, so that the various parameters supplement each other in the evaluation of basin-fill successions.

Many geological factors control coalbed methane (CBM) formation, as discussed for example by Creedy (1988), Bailey (1995) and Fails (1996). Depth of burial and coal rank are commonly used as approximate indicators of coalbed methane potential. However, classic concepts involving increasing coal rank and methane contents with depth as a dominant factor in CBM have become increasingly irrelevant. The highest rank areas of a basin do not necessarily contain the highest coalbed methane concentrations. Not all coal rank indicators, moreover, follow Hilt's Rule and indicate increasing rank with depth, especially in the high-volatile bituminous range.

The purpose of this paper is to discuss the use of different parameters to indicate coal rank and rank trend in the Gunnedah Basin, New South Wales. The Gunnedah Basin has for many years been the subject of petroleum and (more recently) coal bed methane exploration. The coal is of high-volatile bituminous rank, with a mean maximum vitrinite (telocollinite) reflectance of between 0.56 and 1.1 %. Recent work by Gurba and Ward (1997, 1998) and Gurba (in preparation) has shown that there is a strong influence of depositional factors on the coal rank indicators in this basin that may lead to faulty rank interpretation if not taken into account.

2. Geological Setting

The Gunnedah Basin is part of the Sydney-Bowen foreland basin system of Eastern Australia (Figure 1). It contains up to 1200 m of marine and non-marine Permian and Triassic sediments, resting unconformably upon Early Permian (and possibly Late Carboniferous) silicic and mafic volcanics (Tadros, 1995). Much of the sediment is continuous with that of the adjoining Bowen Basin in the north, but the connection is concealed beneath the thick Jurassic and Cretaceous cover of the overlying Surat Basin. To the south the Permian sequence is continuous with that of the Sydney Basin, although the intervening area in this case is covered by extensive Tertiary basalt flows.

The study area for the present paper is located in the central part of the Gunnedah Basin bounded by the Rocky Glen Ridge in the west and Boggabri Ridge in the east (Figure 1). Around 120 fully cored boreholes have been drilled in this area by the New South Wales Department of Mineral Resources; all of them have penetrated at least the upper part of the Permian succession.

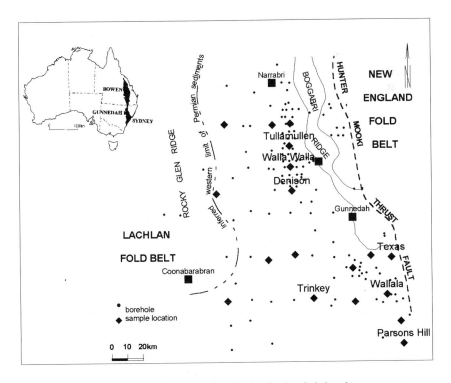

Figure 1. Map of the Gunnedah Basin showing borehole locations.

2.1. DEPOSITIONAL ENVIRONMENTS

The Gunnedah Basin contains two coal measure sequences of Permian age (Tadros, 1993; 1995): a lower succession embracing the Leard and Maules Creek Formations, and a sequence referred to as the Black Jack Group at the top of the Permian succession (Figure 2).

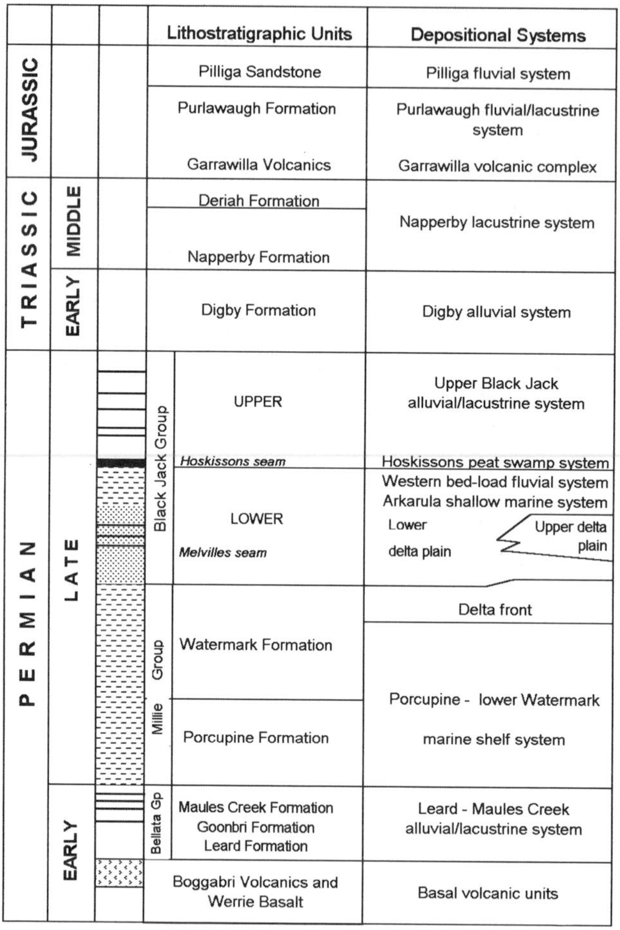

			Lithostratigraphic Units	Depositional Systems
JURASSIC			Pilliga Sandstone	Pilliga fluvial system
			Purlawaugh Formation	Purlawaugh fluvial/lacustrine system
			Garrawilla Volcanics	Garrawilla volcanic complex
TRIASSIC	MIDDLE		Deriah Formation	Napperby lacustrine system
			Napperby Formation	
	EARLY		Digby Formation	Digby alluvial system
PERMIAN	LATE	Black Jack Group	UPPER	Upper Black Jack alluvial/lacustrine system
			Hoskissons seam	Hoskissons peat swamp system
			LOWER	Western bed-load fluvial system / Arkarula shallow marine system / Lower delta plain — Upper delta plain
			Melvilles seam	
		Millie Group	Watermark Formation	Delta front
			Porcupine Formation	Porcupine - lower Watermark marine shelf system
	EARLY	Bellata Gp	Maules Creek Formation / Goonbri Formation / Leard Formation	Leard - Maules Creek alluvial/lacustrine system
			Boggabri Volcanics and Werrie Basalt	Basal volcanic units

Figure 2. Stratigraphic units in the Gunnedah Basin (after Tadros, 1995).

The Permian sequence of the Gunnedah Basin was deposited in three major essentially terrestrial depositional episodes (Figure 2), separated by two marine transgressive/regressive events (Hamilton and Beckett, 1984; Tadros, 1993; 1995). The first terrestrial episode was dominated by colluvial and alluvial environments and formed the Leard and Maules Creek Formations. The first marine transgression, which probably came from the south (Beckett *et al.*, 1983), changed this alluvial sedimentation to coastal fan development and formed the lower part of the Porcupine Formation. The upper part of the Porcupine Formation and the lower part of the overlying Watermark Formation were deposited with continuation of this transgression in a marine shelf environment (Hamilton and Beckett, 1984).

The top of the lower Watermark Formation represents the maximum extent of the first marine transgression. A large deltaic system then prograded across the basin from the north-east, depositing the thick prodelta and delta-front sequences of the upper Watermark Formation. The second terrestrial depositional episode resulting from this progradation is represented by the lower delta plain deposits which form the lower part of the Black Jack Group (Figure 2).

The second marine transgression inundated the lower Black Jack delta system and resulted in widespread, shallow marine conditions. It led to deposition of the Arkarula Sandstone, a distinctive marker horizon within the shallow marine lithofacies succession (Beckett *et al.*, 1983). The third terrestrial depositional episode followed deposition of the Arkarula Sandstone. Termination of the marine conditions allowed the development of a vast peat swamp system, which is represented by the Hoskissons Coal Member. The peat swamps that formed this seam were terminated by a lacustrine system in the east and a westerly-sourced fluvial sandstone system in the west (Tadros, 1993, 1995). These deposits were then overlain by coal-barren alluvial fan and braidplain deposits, sourced initially from the east, to form the Triassic Digby Formation (Jian and Ward, 1993).

The coals in the lower part of the Black Jack Group and in the upper part (and in some cases the whole) of the Maules Creek Formation show evidence of marine influence during or shortly after deposition (Gurba and Ward, 1998). Igneous intrusions also occur in the sequence at a number of localities, and these have locally affected the rank of the coal in surrounding parts of the Permian succession. Vitrinite reflectance anomalies due to igneous intrusions are not discussed in the present paper.

3. Experimental

Approximately 300 samples of core have been collected from numerous boreholes in the Gunnedah Basin for rank evaluation (Gurba, in preparation). The samples were taken mainly from coal seams, with additional material from carbonaceous shales and coaly fragments in non-coal rocks. They were prepared for microscopical examination as polished coal grain mounts or as epoxy-impregnated blocks according to Australian Standards. The polished samples were examined using a Zeiss Axioskop reflected-light microscope, fitted with both white (50W halogen) and blue-violet (HBO) light sources. Maximum reflectance in oil of vitrinite (telocollinite) was measured on the particulate samples. The number of reflectance measurements per sample ranged from 25 to 100, with standard deviations from 0.02 to 0.05.

Polished sections of selected samples were also analysed using an electron microprobe to investigate the elemental composition of vitrinite and other maceral components. A Cameca SX-50 model electron microprobe analyser was used in this study. The analytical routine followed was essentially that described by Bustin et al. (1993, 1996). This technique provided direct quantitative data on the percentages of carbon, oxygen, nitrogen and sulphur in the individual macerals, on spots approximately 5 μm wide. For each point the percentages of iron, silicon and aluminium were also determined, to identify any anomalies due to contamination of the spot being analysed by mineral matter.

Complementary chemical analysis data on many of the coals (proximate and ultimate analyses and sulphur content) were provided for the project by the New South Wales Department of Mineral Resources.

4. Coalification Profiles

Rank changes with depth observed in exploration boreholes in the Gunnedah Basin have been illustrated for the present study in the form of coalification profiles. Coalification profiles were compiled for about 20 locations through the Permian sequence (Figure 1), based on maximum vitrinite (telocollinite) reflectance, moisture content (ad), volatile matter content (daf), carbon and hydrogen contents (daf) and H/C ratio. Along with coal rank data the stratigraphy, depositional environment, intrusions intersected and maceral compositions (for selected boreholes) were all carefully studied. The stratigraphic interval covered by the data ranges from 300 to 600 m. Stratigraphic control, including seam identification, was provided for the boreholes by the New South Wales Department of Mineral Resources.

Optical properties (maximum reflectance of telocollinite) and chemical indices of coal rank (proximate, ultimate and electron microprobe analysis data) were correlated with independent information on the sedimentary environments of the seams and associated strata (Beckett et al., 1983; Hamilton and Beckett, 1984; Hamilton, 1985; 1991; Tadros, 1993; 1995). This was partly with the aim of understanding the variation of vitrinite reflectance in the sequence and partly to identify, through the distribution of different types of reflectance anomalies, some of the environmental factors that have affected the coal-bearing sequence. Such an understanding is important if the three-dimensional pattern of rank distribution within the basin is to be fully delineated.

Profiles of vitrinite (telocollinite) reflectance and other rank indicators in the Gunnedah Basin show several different types of variations, including anomalously low and anomalously high values due to environmental factors (e.g. marine influence), maceral associations and localised geothermal (igneous intrusion) events (Gurba and Ward, 1996, 1998). These are overprinted on a general increase in maturation with depth.

4.1. VITRINITE REFLECTANCE

Plots of vitrinite (telocollinite) reflectance against depth for the Permian coals of the Gunnedah Basin often show significant scatter from the general regression line of increasing reflectance with depth (Figure 3). Considering that the reflectance data were collected from an apparently homogeneous vitrinite population (telocollinite only was measured) and that small standard deviations were observed (0.02 to 0.05), the scatter in reflectance values cannot be explained by geothermal factors alone.

Several different forms of reflectance profiles are shown in Figure 3, including:
- An apparent rank decrease (rather than increase) with depth (Figures 3a and 3b);
- Trends for which several different regression lines can be fitted to the observed data (Figure 3c);
- A large scatter in the data with no evident trend at all (Figure 3d).

Allowance can be made within such profiles for anomalies in the reflectance trend due to igneous intrusions in the sequence. A heat-affected sample identified in DM Parsons Hill DDH 1 (depth = 800 m, $R_V max$ = 2.15%), for example, has been excluded from the graph presented in Figure 3c.

An extreme example of the problem is illustrated in Figures 3a and 3b, where the maximum telocollinite reflectance data show an apparent decrease with increasing depth, rather than an increase. If taken at face value this might for example be interpreted as a reversed rank profile in overturned strata. Such a reversed reflectance profile is reported from the Lower Cretaceous coal measures of Alberta by Kalkreuth and McMechan (1984). In a borehole drilled into overturned strata, the measured reflectance decreased with physical depth but increased with stratigraphic depth. However, the strata in DM Denison DDH 1 and DM Tullamullen DDH 1 (Figure 3) are not overturned and the maximum reflectance decreases with both physical depth and stratigraphic depth. Since overthrusting is ruled out in these boreholes, other explanations need to be considered.

4.2. EFFECT OF DIFFERENT VITRINITE TYPES

Detailed petrographic examination of the samples presented in Figure 3 and the relation to depositional environment reveal that the coal seams lying within the influence of marine stratigraphic intervals (Porcupine and Arkarula transgressions) and also the lower delta plain sediments display lower $R_{TC}max$ values (Figure 4). These coals contain perhydrous vitrinite and abundant framboidal pyrite, and show an increase in vitrinite fluorescence. Measured maximum telocollinite reflectance on these samples is lower that it would be expected from the regional and vertical rank trend.

This depression of Rmax values in the Gunnedah Basin coincides generally with the

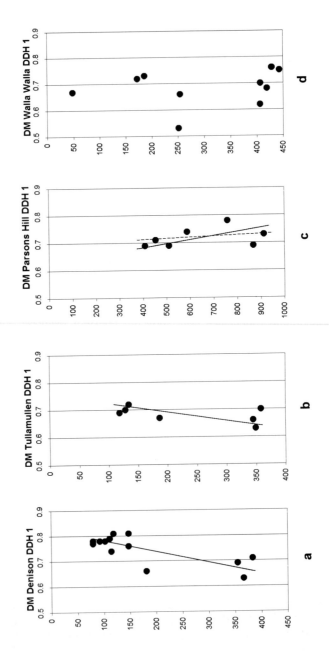

Figure 3. Raw data profiles of mean maximum vitrinite (telocollinite) reflectance against depth for selected boreholes in the Gunnedah Basin. (a) and (b) represent boreholes where a negative coalification gradient (decrease in rank with depth) might be indicated; (c) represents a well for which several different regression lines could be used to fit the observed reflectance data. (d) represents a borehole where a large scatter in reflectance data prevents the drawing of a regression line.

occurrence of marine-influenced sediments in the coal measures, and is further discussed by Gurba and Ward (1998) and Gurba (in preparation). The distribution of these lower reflectance values has thus been used to provide a re-interpretation of the down-hole reflectance profiles as shown in Figure 4. Two parallel trends of downward increase in telocollinite reflectance can be drawn through the reflectance data in each instance, with the reflectances for marine-influenced coals offset by 0.1 to 0.2% on the low-value side of the coalification profile for coals containing non-marine influenced vitrinite components.

Similar low reflectance anomalies due to marine influence have been described in other coalfields (Stach et al., 1982; Newman and Newman, 1982; Price and Barker, 1985 and Diessel, 1990, 1992a,b). Lower than expected reflectance values and related coal properties can be attributed to a variety of factors, and these are summarised by Price and Barker (1985) and Mukhopadhayay and Dow (1994). The effect of abnormal maceral compositions has also been cited as a evidence for anomalous reflectance values. Studies by Hutton and Cook (1980), Kalkreuth (1982), Kalkreuth and McMechan (1984) Raymond and Murchison (1991) and Mastalerz et al., (1993) show that vitrinite reflectance in coal or sediments may be anomalously low in the presence of abundant liptinitic material.

As another factor affecting coalification profiles in the Gunnedah Basin, Gurba and Ward (1996, 1998) have identified a type of vitrinite that resembles material described by Benedict et al. (1968) as 'pseudovitrinite'. This material has a higher reflectance than other telocollinites in the same coal sample and a distinctive slit or microfracture pattern, but exhibits the chemical characteristics of other vitrinite components (Gurba and Ward, 1997, 1998: Mastalerz *et al.*, 1994).

The presence of coal containing 'pseudovitrinite' provides some difficulties in using vitrinite reflectance data as the basis for precise coal rank evaluation, at least in the high-volatile bituminous range. Except for the higher reflectance and the slit pattern, it resembles other telocollinite sufficiently closely to be grouped with that component in most maceral classifications, and its reflectance is therefore measured along with that of more normal telocollinite if standard procedures are followed. Mean vitrinite reflectance values can thus be expected to be higher, for coal of a given rank, where significant proportions of 'pseudovitrinite' are present.

Gurba and Ward (1997) have shown that separate regression lines against depth can be delineated for desmocollinite, telocollinite (excluding 'pseudovitrinite') and 'pseudovitrinite' in the Gunnedah Basin sequence (Figure 5). The maximum reflectance of 'pseudovitrinite' in such cases can be up to 0.15 or 0.20% higher than that of the more normal telocollinite components. If not recognised and allowed for, it will provide reflectance values that plot to the right (high-value side) of the more normal regression line.

The increase in reflectance of the different vitrinite components with rank can be seen

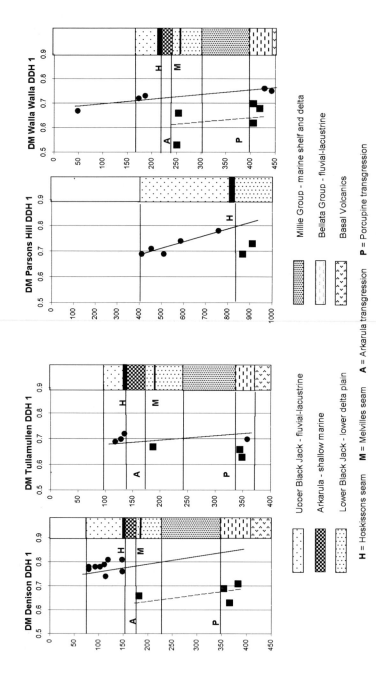

Figure 4. Profiles of mean maximum vitrinite (telocollinite) reflectance against depth for selected boreholes in the Gunnedah Basin, interpreted in relation to depositional environment and vitrinite type. Squares indicate coals with perhydrous vitrinite. The Porcupine and Arkarula marine transgressions are also shown. HS = Hoskissons seam; MS = Melvilles seam.

from Figure 5, including the separate trends due to burial and to intrusive effects. A more substantial increase is associated with an igneous intrusion in the centre of the section studied. The significant differences noted for individual samples are maintained throughout the rank range involved, marked by a general consistency in separation of the data points at each sample depth.

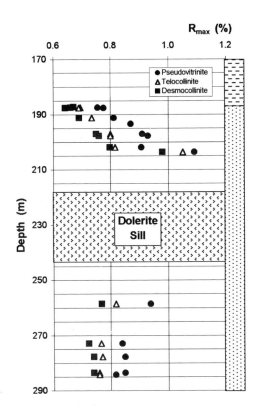

Figure 5. Plot of mean maximum vitrinite reflectance (desmocollinite, telocollinite excluding 'pseudovitrinite', and 'pseudovitrinite') against depth in DM Texas DDH 1 (from Gurba and Ward, 1997).

As with the other vitrinites, the higher-reflecting 'pseudovitrinite' shows an increase in reflectance with depth, with its reflectance data following a trend sub-parallel to those of the telocollinite. This behaviour indicates that the 'pseudovitrinite' has matured thermally in a fashion similar to telocollinite. There is a distinct and relatively constant separation between the reflectance values of up to about 0.15% Rmax (absolute) for high-volatile bituminous coals. At higher rank level the difference in reflectance between telocollinite and 'pseudovitrinite' diminishes. Benedict et al. (1968) have noted that the reflectance of 'pseudovitrinite' is different from that of other telocollinite up to a rank corresponding to around $R_v max = 1.5\%$; at higher ranks it is indistinguishable from other vitrinite types.

4.3. CHEMICAL RANK PARAMETERS

It is very well known that marine-influenced coals of bituminous rank possess unusual properties such as higher sulphur, higher than normal hydrogen content, elevated nitrogen and lower softening temperatures during carbonisation relative to coals of similar rank that are not marine influenced. Their volatile-matter yields are also much higher than would be expected from their degree of geochemical coalification (Teichmüller and Teichmüller, 1982). The relative proportions of different macerals, which are themselves influenced by the depositional environment, can also affect the volatile matter and hydrogen contents (Teichmüller and Teichmüller, 1982; Diessel, 1992a).

The combined effect of depositional environment and maceral associations on chemical rank parameters is illustrated in Figure 6. Coals of the upper Black Jack Group, which are usually inertinite rich (Tadros, 1993, Gurba and Ward, 1998), typically have lower volatile matter and hydrogen (daf) proportions relative to the high-vitrinite seams of the lower Black Jack Group (lower delta plain facies) in the same vertical section. The correlation coefficient of volatile matter (daf) with depth in DM Wallala DDH 1 is very strong ($r = 0.76$), but at the same time the trend indicates that the volatile matter (daf) contradicts Hilt's rule and increases with depth (Figure 6).

In the example illustrated in Figure 6 the plotted points for volatile matter content (daf), hydrogen content (daf) and vitrinite reflectance data fall into two distinct groups, one representing the upper Black Jack Group coals and the other the coals of the lower Black Jack succession (lower delta plain facies). Carbon content (dry, ash-free) and moisture content (air-dried) seem to be the only rank indicators independent of the depositional environment affecting the coal seam.

Electron microprobe data (Gurba, in preparation) also indicate that, for the range of coals studied, the elemental carbon content of telocollinite does not vary significantly between marine and non-marine influenced coals. The scatter in the carbon content (daf) data in Figure 6, which were obtained by conventional ultimate analysis, may therefore be due to different maceral compositions in the samples analysed.

The generally higher volatile matter (daf) yield of the lower Black Jack Group coals is attributed to a combination of their higher vitrinite content and the marine-influenced environment of deposition. Within the vitrinite macerals there is also a greater proportion of desmocollinite in the lower Black Jack Group (Gurba, in preparation). Desmocollinite (vitrinite B) has higher volatile matter and a higher hydrogen content than telocollinite (vitrinite A) in coals at the same rank level (Brown et al., 1964).

Histograms showing the distribution of volatile matter (daf) between the major coal-bearing sequences of the Gunnedah Basin are presented in Figure 7. The frequency distribution for volatile matter in the lower Black Jack coals is highly skewed towards the high values; it is mostly higher than in the coals of the upper Black Jack Group,

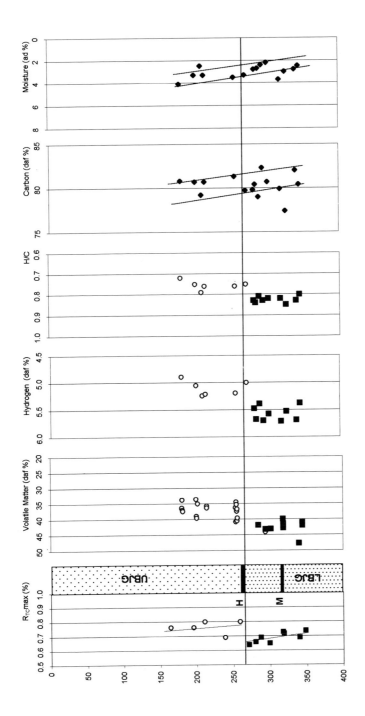

Figure 6. Coalification profiles for DM Wallala DDH 1 based on different rank indicators. Changes in vitrinite reflectance, volatile matter content, hydrogen content and H/C ratio are clearly seen in the lower Black Jack Group coals (lower delta plain facies). HS = Hoskissons seam; MS = Melvilles seam; LBJG = Lower Black Jack Group; UBJG = Upper Black Jack Group.

which are in turn lower in rank. The skewness is again due to the marine-influenced environment of deposition and the associated enrichment in vitrinite in the lower Black Jack Group coals. The extended tail towards lower values in both the upper and lower Black Jack Group is a result of heat-affected samples due to igneous intrusions.

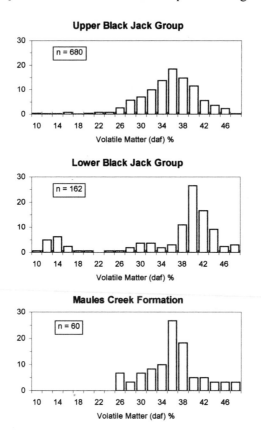

Figure 7. Frequency distributions of volatile matter (dry, ash-free) content in major coal-bearing units of the Gunnedah Basin.

The distribution of volatile matter in the Early Permian coals of the Bellata Group has a broad scatter embracing both Black Jack distributions. This is mainly due to enrichment in desmocollinite and liptinite, and to mixed populations of marine-influenced and non-marine coals in the sample suite.

The depositional environment and the related maceral associations are therefore the main factors (apart form maturity) controlling the volatile matter and hydrogen contents in the Gunnedah Basin coals. The presence of anomalies in maturity values coinciding with marine influence on the coal seams suggests a strong link between the maturity indicators in high-volatile bituminous coals and the depositional environment of the seams.

The general validity of Hilt's Rule in the Gunnedah Basin has been accepted, but it could be questioned in many instances. If the influence of coal type or depositional environment on coal rank indicators is not taken into account (especially in high-volatile bituminous coals), rank may appear to decrease instead of increase with depth (as in Figures 3 and 6). The properties of coal do not change uniformly with rank, and consequently not all coal rank indices are suitable indicators for this sequence (as discussed above). Only when the appropriate parameters are used it is possible to study the relationship between coal rank and depth.

5. Implications for Thermal History Modelling

Figure 8 shows the maximum reflectance of telocollinite (including 'pseudovitrinite') against depth in DM Trinkey DDH 1 (upper Black Jack Group) (for location see Figure 1). Rank as expressed by maximum reflectance increases steadily with depth, with most of data not deviating significantly from the regression line.

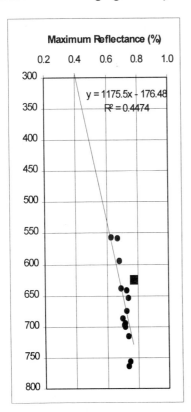 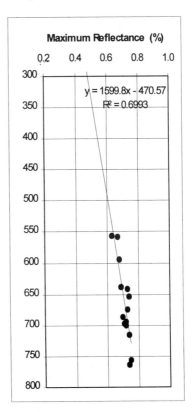

Figure 8. Plot of mean maximum vitrinite (telocollinite) reflectance against depth in DM Trinkey DDH 1 for coals of the upper Black Jack Group. Square indicates a sample containing a high proportion of 'pseudovitrinite'; projected correlation lines with and without this material are also shown.

The sample at a depth of 623.47 m has a slightly higher reflectance than the other samples, a feature which in this case cannot be attributed to igneous intrusive effects. Petrographic examination of this sample shows that the majority of the telocollinite is in the form of the higher-reflecting 'pseudovitrinite'. Computer-drawn regression lines for all data (left) and for data excluding this sample (right), shown in Figure 8, are each characterised by a slightly different slope. If the influence of 'pseudovitrinite' had not been recognised, such an interpretation of the data would have implied a different paleogethermal gradient and, by projection of that gradient, a different thickness of strata removed from the geologic section.

Similar misinterpretations of palaeogeothermal gradient and erosion history may arise from unrecognised inclusion in maturation profiles of anomalously low reflectance values, or anomalies in other rank indicators such as volatile matter, due to the presence of marine-influenced coals in the stratigraphic sequence.

As shown by Gurba and Ward (1997), desmocollinite, telocollinite and the so-called 'pseudovitrinite' follow separate coalification tracks within the Gunnedah Basin sequence. Even reflectance measurements taken exclusively on telocollinite need to be separated into those involving marine-influenced perhydrous vitrinite and those involving the 'pseudovitrinite' component, as well as those based on more normal telocollinite material. If such discrimination is not possible in the data-gathering process, allowance may be made for such effects by considering separately the maturation profiles of stratigraphic intervals formed under different environmental conditions.

6. Coalification Maps

Based on the coal rank indicators discussed above coalification maps have been drawn for three coal-bearing units:
*   The fluvio-lacustrine sequence of the upper Black Jack Group (based on the Hoskissons seam) – Figure 9;
*   Lower delta plain facies of the lower Black Jack Group (based on the Melvilles seam and its equivalents) – Figure 10;
*   The Early Permian mainly fluvio-lacustrine Bellata Group. Separate maps of the upper and lower parts of this unit are incorporated in Figure 11.

6.1. UPPER BLACK JACK GROUP

In order to evaluate the regional coalification pattern and to relate the coal rank pattern to geological structure, a common stratigraphic datum is useful to eliminate differences in maturity due to vertical position. For the present study, a laterally continuous coal seam (the Hoskissons seam) was studied in detail to determine the relationship between coal rank as indicated by mean maximum vitrinite (telocollinite) reflectance ($R_{TC}max$), air-dried moisture content (M ad), and carbon content (C daf) to the geological structure of the basin. The Hoskissons seam was selected as the reference horizon for the upper Black Jack Group because it has a broad extent and because it serves as a genetic stratigraphic sequence boundary in the Permian strata of the Gunnedah Basin (Hamilton and Tadros, 1994).

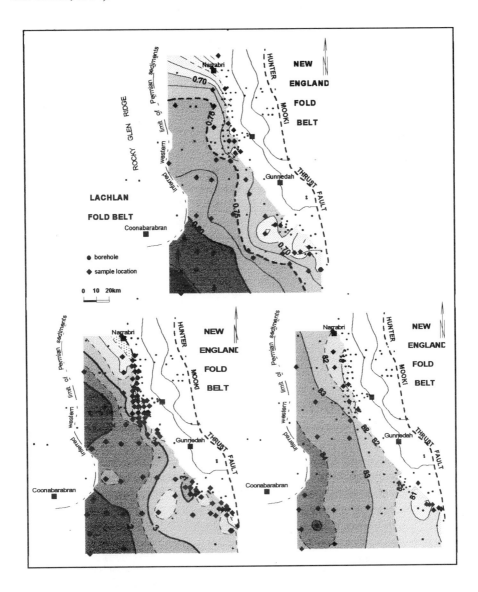

Figure 9. Lateral rank variation in the Hoskissons seam. Top: Lateral variation of mean maximum vitrinite (telocollinite) reflectance (per cent). Lower left: Lateral distribution of air-dried moisture content. Lower right: Lateral distribution of carbon content (dry-ash-free). (Modified from Gurba and Ward, 1995).

The maximum vitrinite (telocollinite) reflectance of the Hoskissons seam varies from 0.65% in the east to more than 0.85% in the west (Figure 9a). This is consistent with a higher coal rank in the west than in the east. The air-dried moisture content of the Hoskissons seam also shows a systematic pattern of lateral rank variation across the Gunnedah Basin area (Figure 9b), decreasing from more than 6% in the eastern part of the basin to below 2% in the south-western part. The carbon content (dry, ash-free) of the Hoskissons seam ranges from 81% in the eastern edge to more than 84% carbon in the south-west (Figure 9c). This corresponds closely to the trends seen in the maximum vitrinite (telocollinite) reflectance and air-dried moisture content.

6.2. LOWER BLACK JACK GROUP

The pattern of coal rank variation for the Melvilles seam (Figure 10) appears to reflect lateral changes in coal rank across the basin similar to those exhibited by the Hoskissons seam. The vitrinite reflectance values for the Melvilles seam, however, are lower than those of the Hoskissons seam, due to marine influence on this part of the succession. If such environmental factors are not taken into account, the coal rank as expressed by vitrinite reflectance would appear to decrease for the lower Back Jack Group coals, relative to those of the upper Black Jack interval.

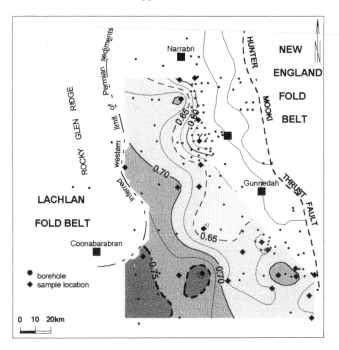

Figure 10. Lateral rank variation of mean maximum vitrinite (telocollinite) reflectance (per cent) in the Melvilles seam.

6.3. BELLATA GROUP

Coals of this sequence, although deposited in essentially non-marine conditions, have been subjected to marine influence from the overlying Porcupine Formation. At the top and up to a distance of about 20 metres below the top of the sequence measured reflectance data deviate towards low values from the "normal" regression line (Gurba and Ward, 1998). The values can be up to 0.2 % lower than these for unaffected coals. In some cases, such as DM Denison DDH 1 (Figure 3a), all coals in the sequence seem to be affected by marine conditions.

A coalification map drawn on the top of the Bellata Group, expressed by vitrinite reflectance, would indicate a lower rank than a similar map drawn on the upper Black Jack Group if depositional effects of the Porcupine transgression on coals were not taken into account. A coalification map drawn on the base of the Bellata Group, after eliminating any reflectance values representing marine influenced vitrinites, would correspond closely to the trends seen in the maximum vitrinite (telocollinite) reflectance for the Hoskissons seam.

6.4. PERSPECTIVE MAPS

In coal rank studies of a sedimentary basin the rank trend is commonly shown by maps drawn on the tops of the relevant stratigraphic units. The differences between Figures 9 and 10, however, indicate that such maps cannot be easily applied to the Gunnedah Basin sequence.

A generalised coalification pattern showing the vertical changes of vitrinite reflectance through the Permian succession of the Gunnedah Basin is given in Figure 11. The vitrinite reflectance shows an overall increase with depth from the top of the Black Jack Group to the base of the Bellata Group. There is, however, a marked departure from this trend at several horizons, with anomalously low values in the lower Black Jack (lower delta plain facies) and uppermost Bellata Group (affected by marine Porcupine transgression) successions.

7. Conclusions

Although maturation apparently exerts the predominant influence over much of the Gunnedah Basin succession, vitrinite reflectance is also affected by the depositional environment of the coal, particularly the occurrence of marine influence.

Comprehensive evaluation of coalification data as described above has led to a completely new interpretation of the coalification pattern. In addition to establishing a slight increase in coalification towards the south-western part of the basin (Gurba and

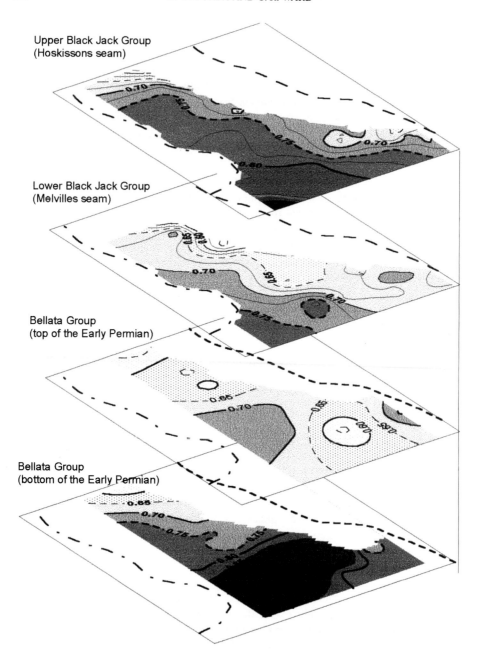

Figure 11. Perspective diagram showing variation in thermal maturity with depth in the Gunnedah Basin (based on maximum telocollinite reflectance).

Ward, 1995), many of the coalification indicators (vitrinite reflectance, volatile matter, hydrogen content) show a clear correlation with depositional environment, especially marine influence.

The lower reflectance values associated with marine influence are not a localised feature; they can be recognised through entire sections of strata and at the same horizons over wide areas of the basin. The horizons at which the relevant changes occur can be mapped, to assist in sequence stratigraphy investigations. For the range studied (high-volatile bituminous coals), vitrinite reflectance is only of value as a rank indicator in intervals with a similar depositional environment.

In general a good correlation has been shown between vitrinite reflectance and volatile matter content (daf) (Gurba, in preparation; see also McCartney and Teichmüller, 1972; Briel and Savage, 1973; Freudenberg et al., 1996). However, these maturity parameters both depend on depositional environment, and both respond in the same manner to marine conditions. This is especially evident for high-volatile bituminous coals, as has been shown above for the Gunnedah Basin succession.

Where marine-influenced coals are present in the lower part of the sequence and fluvial coals in the upper part, as is the case with the Black Jack Group, the resulting reflectance profile, if not carefully analysed, may give the appearance of an abnormally low rate of rank increase with depth, with resulting errors in interpreting geothermal gradients or burial history. In extreme cases a decrease in rank with depth may appear to be present, as in parts of the Gunnedah Basin, where low base-line reflectance gradients are involved.

The presence of different types of telocollinite (marine-influenced, normal telocollinite and material resembling 'pseudovitrinite') has implications in basin modelling and for coalbed methane and petroleum prospectivity. Higher temperatures are apparently required to produce a given reflectance in perhydrous (marine-influenced) vitrinites than in subhydrous vitrinites (e.g. 'pseudovitrinite'), and possibly also to generate hydrocarbon from the respective materials. Price and Barker (1985) have stated that the maturity of the sediment can be miscalculated if either hydrogen-rich or oxygen-rich kerogen is used without reference to the overall kerogen type.

Levine (1992) has pointed out, for example, that some workers use a maturation pathway representing vitrinite-rich, hydrogen poor coal, typical of Northern Hemisphere coals of Carboniferous age, in calculating methane yields. Other coals, such as inertinite-rich Permian Gondwana coals, follow maturation paths that are significantly different. These coals also differ in their volatile matter content. Such a situation applies to the Gunnedah Basin, where the usually inertinite rich and often subhydrous coals of the upper Black Jack Group follow different coalification tracks to the vitrinite-rich and generally perhydrous coals of the lower Black Jack Group. A different coalification track yet again has been distinguished for the 'pseudovitrinite' component (Gurba and Ward, 1997).

8. Acknowledgements

Thanks are expressed to the New South Wales Department of Mineral Resources and to Pacific Power for assistance in the provision of samples and other information used in the project. The personal assistance of Carl Weber, Jeff Beckett, Julie Moloney, Vic Tadros, Adrian Hutton and Harold Read, among others, is gratefully acknowledged.

The study was funded in part by Pacific Power and in part under the Small Grants Scheme of the Australian Research Council.

9. References

Bailey, H.E., Glover, B.W., Holloway, S. and Young, S.R. (1995) Controls on coalbed methane prospectivity in Great Britain, in: M.K.G. Whateley and D.A. Spears (eds), *European Coal Geology*, Geological Society Special Publication 82, pp. 251-265.

Beckett, J., Hamilton, D.S. and Weber, C.R. (1983) Permian and Triassic stratigraphy and sedimentation in the Gunnedah-Narrabri-Coonabarabran region, *New South Wales Geological Survey Quarterly Notes* 51, 1-16.

Benedict, L.G., Thompson, R.R., Shigo, J.J.III and Aikman, R.P. (1968), Pseudovitrinite in Appalachian coking coals. *Fuel* 47, 125-143.

Briel, J.M. and Savage, H.D. (1973) Properties of vitrinite concentrates of South African coals. *Fuel* 52, 32-35.

Brown, H.R., Cook, A.C. and Taylor, G.H. (1964). Variations in the properties of vitrinite in isometamorphic coal, *Fuel* 43, 111-124.

Bustin, R.M., Mastalerz, M. and Raudsepp, M. (1996) Electron-probe microanalysis of light elements in coal and other kerogen, *International Journal of Coal Geology* 32: 5-30.

Bustin, R.M., Mastalerz, M. and Wilks, K.R. (1993) Direct determination of carbon, oxygen and nitrogen content in coal using the electron microprobe, *Fuel* 72: 181-185.

Creedy, D.P. (1988) Geological controls on the formation and distribution of gas in British coal measure strata, *International Journal of Coal Geology* 10, 1-31.

Diessel, C.F.K. (1990) Marine influence on coal seams, *Proceedings of 24th Symposium on Advances in the Study of the Sydney Basin*, University of Newcastle, Newcastle, NSW, 33-40.

Diessel, C.F.K. (1992a) *Coal-Bearing Depositional Systems*, Springer-Verlag, Berlin.

Diessel, C.F.K. (1992b) The problem of syn- versus post-depositional marine influence on coal composition. *Proceedings of 26th Symposium on Advances in the Study of the Sydney Basin*, University of Newcastle, Newcastle, NSW, 154-163.

Fails, T. (1996) Coalbed methane potential of some Variscan foredeep basins, in R. Gayer and I. Harris (eds), *Coalbed Methane and Coal Geology*, Geological Society Special Publication 109, 13-26.

Freudenberg, U., Lou, S., Schluter, R., Schutz, K. and Thomas, K. (1996) Main factors controlling coalbed methane distribution in the Ruhr District, Germany, in R. Gayer and I. Harris (eds), *Coalbed Methane and Coal Geology*, Geological Society Special Publication 109, 67-88.

Gentzis, T. and Goodarzi, F. (1994) Reflectance suppression in some Cretaceous coals from Alberta, Canada, in P.K. Mukhopadhyay and W.G. Dow (eds), *Vitrinite Reflectance as a Maturity Parameter: Applications and Limitations*, ACS Symposium Series 570, American Chemical Society, Washington, pp. 93-110.

Gurba, L.W. (in preparation) Depositional and thermal factors affecting the three-dimensional coalification pattern in the Gunnedah Basin, New South Wales. PhD Thesis, University of New South Wales, Sydney.

Gurba, L.W. and Ward, C.R. (1995) Coal rank variation in the Gunnedah Basin, in R.L. Boyd and G.A. McKenzie (eds), *Proceedings of 29th Symposium "Advances in the Study of the Sydney Basin"*, Department of Geology, University of Newcastle, N.S.W., 180-187.

Gurba, L.W. and Ward, C.R. (1996) Reflectance anomalies in Permian coals of the Gunnedah Basin -

implications for maturation studies, in R.L. Boyd and G.A. McKenzie (eds), *Proceedings of 30th Symposium "Advances in the Study of the Sydney Basin"*, Department of Geology, University of Newcastle, N.S.W., 69-76.

Gurba, L.W. and Ward, C. R. (1997) Chemical composition and coalification paths for vitrinite types in the Gunnedah Basin, New South Wales, *Proceedings 7th New Zealand Coal Conference*, Coal Research Limited, Wellington, New Zealand, 478-489.

Gurba, L.W. and Ward, C.R. (1998) Vitrinite reflectance anomalies in high-volatile bituminous coals of the Gunnedah Basin, New South Wales, Australia, *International Journal of Coal Geology* 36, 111-140.

Hamilton, D.S. (1985) Deltaic depositional systems, coal distribution and quality, and petroleum potential, Permian Gunnedah Basin, New South Wales, Australia, *Sedimentary Geology* 45, 35-75.

Hamilton, D.S. (1991) Genetic stratigraphy of the Gunnedah Basin, N.S.W., *Australian Journal of Earth Sciences* 38, 95-113.

Hamilton, D.S. and Beckett, J. (1984) Permian depositional systems in the Gunnedah region, *Geological Survey of New South Wales Quarterly Notes* 55, 1-19.

Hamilton, D.S. and Tadros, N.Z. (1994) Utility of coal seams as genetic stratigraphic sequence boundaries in nonmarine basins: an example from the Gunnedah Basin, Australia, *AAPG Bulletin* 78(2), 267-286.

Hutton, A.C. and Cook, A.C. (1980) Influence of alginite on the reflectance of vitrinite from Joadja, NSW, and some other coals and oil shales containing alginite, *Fuel* 59, 711-714.

Jian, F.X. and Ward, C.R. (1993) Triassic depositional episode, in N.Z. Tadros (ed), *The Gunnedah Basin, New South Wales*, Geological Survey of New South Wales, Memoir Geology 12, 297-326.

Kalkreuth, W. (1982) Rank and petrographic composition of selected Jurassic - Lower Cretaceous coals of British Columbia, Canada, *Bulletin of Canadian Petroleum Geology* 30(2), 112-139.

Kalkreuth, W. and McMechan, M. (1984) Regional pattern of thermal maturation as determined from coal-rank studies, Rocky Mountain Foothills and Front Ranges North of Grande Cache, Alberta-implications for petroleum exploration, *Bulletin of Canadian Petroleum Geology* 32(3), 249-271.

Levine, J.R. (1992) Oversimplifications can lead to faulty coalbed gas reservoir analysis, *Oil and Gas Journal* Nov.23, 1992, 63-69.

Mastalerz, M., Lamberson, M. and Bustin, M. (1994) Pseudovitrinite: chemical properties and origin, *Geological Society of America* Meeting Abstract.

Mastalerz, M., Wilks, K.R. and Bustin, R.M. (1993) Variation in vitrinite chemistry as a function of associated liptinite content; a microprobe and FT-ir investigation, *Organic Geochemistry* 20(5), 555-562.

McCartney , J.T. and Teichmüller, M. (1972) Classification of coals according to degree of coalification by reflectance of the vitrinite component, *Fuel* 51, 64-68.

Mukhopadhyay, P.K. and Dow, W.G. (eds) (1994) *Vitrinite Reflectance as a Maturity Parameter: applications and limitations*, ACS Symposium Series 570, American Chemical Society, Washington, 294 pp.

Newman, J. and Newman, N.A. (1982) Reflectance anomalies in Pike River coals: evidence of variability in vitrinite type, with implications for maturation studies and "Suggate rank", *New Zealand Journal of Geology and Geophysics* 25, 233-243.

Price, L.C. and Barker, C.E. (1985) Suppression of vitrinite reflectance in amorphous rich kerogen - a major unrecognized problem, *Journal of Petroleum Geology* 8(1), 59-84.

Raymond, A.C. and Murchison, D.G. (1991) Influence of exinitic macerals on the reflectance of vitrinite in Carboniferous sediments of the Midland Valley of Scotland, *Fuel* 70, 155-161.

Stach, E., Mackowsky, M-Th., Teichmüller, M., Taylor, G.H., Chandra, D. and Teichmüller, R. (eds) (1982) *Stach's Textbok of Coal Petrology*, Gebruder Borntraeger, Berlin.

Tadros, N.Z. (ed) (1993) *The Gunnedah Basin New South Wales*, Geological Survey of New South Wales, Memoir Geology, 12, New South Wales Department of Mineral Resources, Sydney, 550 pp..

Tadros, N.Z. (1995) Gunnedah Basin, in C.R. Ward, H.J. Harrington, C.W. Mallett and J.W. Beeston (eds), *Geology of Australian Coal Basins*, Geological Society of Australia Coal Geology Group, Special Publication 1, 247-298.

Teichmüller, M. (1987) Recent advances in coalification studies and their application to geology. In Scott, A.C. and Collinson, M.E. (eds), *Coal and Coal-bearing Strata: Recent Advances*, Geological Society Special Publication 32, Blackwell Scientific Publications, Oxford, 127-169.

Teichmüller, M. and Teichmüller, R. (1982) The geological basis of coal formation, in E. Stach et al. (eds), *Stach's Textbook of Coal Petrology*, Gebruder Borntraeger Berlin, 5-87.

THE PHYSICS AND EFFICIENCY OF PETROLEUM EXPULSION FROM COAL

A quantitative evaluation of the factors which control expulsion of petroleum from coal

J.K.MICHELSEN[1] and G. KHAVARI-KHORASANI[2]

[1]*STATOIL, 4035 Stavanger, Norway;* [2]*PETEC, Prof. Olav Hanssens vei 15, P.O.Box 2503, 4004 Stavanger, Norway*

1.0 Abstract

Many coals, in particular coals of Mesozoic and Cenozoic age, generate significant quantities of "oil" constituents. At the same time few commercial oil deposits can be demonstrated to have originated from coal derived fluids. In this paper we examine mass loss and volumetric changes of coals during petroleum generation. We do not find evidence for the view that coals do not expel fluids before secondary cracking have eliminated the oil potential. Mass balance and sorption data indicate that coal constituents have lower retention capacity than e.g., classic oil source rocks. It is concluded that there must be a direct link between the generation/desorption of petroleum and the deformation of the coal. A continuum physics approach is applied to evaluate possible coal rheologies and the compaction behavior, given the state and forces which acts on the coal during petroleum generation. The coal during generation must behave in a fluid like manner. The diffusive micro scale transport of petroleum molecules in the coal matrix causes the coal to deform with a viscous rheology. The desorption of petroleum from the coal matrix instantaneously creates a mechanically unstable state, and a related fluid potential gradient in the petroleum will drive the petroleum out of the coal together with related compaction of the coal. The expulsion will be particularly efficient in normally pressured sedimentary sections, where petroleum is expelled nearly symmetrically in both directions. In sections with hard overpressure prior to generation/expulsion, there will be a stronger tendency for upward expulsion. The main driving force for expulsion is independent of the volume expansion of the organic matter. The deficiency of oil deposits from coals, must be related to factors other than the expulsion efficiency.

2.0 Introduction

Many authors have been discussing the oil generative potential of coal (Thomas, 1982; Rigby and Smith, 1982; Durand and Paratte, 1983; Shanmugan, 1985; Khavari-Khorasani, 1987; Horsfield *et al.*, 1988; Hunt, 1991; Noble *et al.*, 1989; Katz *et al.*, 1991; Khavari-Khorasani and Michelsen, 1991; Boreham and Powell, 1993; Clayton, 1993; Katz, 1994). While, there is little disagreement that many coals can generate significant quantities of "oil" constituents, there are contrasting views regarding if such coals are capable of efficiently expelling these constituents.

The main argument against coal as a source for oil have been that "liquids" generated from coal will be retained in the coals and cracked to gas during further burial, i.e., poor expulsion efficiency (e.g., Hunt, 1991; Katz, 1994). Authors, presenting this view have been referring to the adsorptive properties of coals, and/or to the petroleum molecules being trapped in the coal "micro porosity".

Durand and Paratte (1983), presented a completely opposite view. These authors argued that any micro-porosity in coals, which resembles the "common view of pores" with a rigid structure and well defined pore walls does not appear before at maturities above a reflectance of 1.5%. They further argued that a discussion of microporosity is irrelevant with respect to oil trapping, and that the relevant discussion is around the molecular mechanisms for the creation of petroleum phases from solid phases. The same authors argued that it is the high compressibility of the coal which permits transmission of the total load stress onto the petroleum fluids, and aided by the transfer of mass from the coal matrix to the petroleum fluid, creating "macro" porosity, permitting an efficient expulsion. It was also pointed out that it is likely that the coal matrix will be essentially oil wet and the oil flow monophasic (at least during early stages of expulsion), and hence relative permeability effects will not set up additional resistance to the expulsion.

Durand and Paratte (1983), also reviewed estimates of the adsorption capacity of coals and concluded that the capacity was far below what would be required to efficiently prevent petroleum expulsion (prior to secondary cracking). By comparing generation capacity with the observed retained quantities, and estimated adsorption capacity they concluded that coals are good expellers. The lack of association between coal beds and oil deposits was therefore not considered to be due to geochemical reasons, but rather geological ones.

Boreham and Powell (1993), used the mass balance scheme of Cooles *et al.* (1986), to calculate the petroleum expulsion efficiency from the Walloon and Patchaawarra Formation coals. They calculated petroleum expulsion efficiencies as high as 80-90% for a petroleum generation index (PGI) of 0.4-0.7 (PGI can be approximated to the transformation ratio). Pepper (1991), using a similar mass balance scheme concluded that the petroleum expulsion efficiency of coal can be anything from very low to very high, and basically controlled by the initial hydrogen index and maturity.

The accumulation of oil and gas is not only related to source potential and expulsion efficiency, but also on the transport efficiency of the different organic constituents in the petroleum fluids, which is closely tied to the PVT properties (Khavari-Khorasani et al., 1998).

Therefore, there are four aspects for evaluating the possibility of accumulating oils derived from coal or coaly organic matter: (a) the generated compositions and how it changes with maturity, (b) interactions between the coal matrix and the free fluids and how this determines the composition of the free fluid as a function of maturity, (c) how and why the fluids are transported out of the coal (if they are expelled), and (d) what phase relationships can we expect for the free fluids and how will these determine the transport properties of the fluids.

In this paper we will mainly discuss point (a), (b) and (c) above, since effective expulsion of the oil constituents at present in the literature, is viewed as the major obstacle for coal beds to become oil source rocks. Mass balance will be used to estimate the volumetric changes in coals during generation. A continuum physics approach is then applied to evaluate possible coal rheologies and the compaction behavior, given the state and forces which acts on the coal during petroleum generation. The predicted free petroleum compositions will be used to asses the sample losses for the measured retained petroleum quantities in the coal, while PVT properties and transport properties of the fluids will be discussed elsewhere.

3.0 Methods

On-line temperature resolved flash pyrolysis GC was used to examine the thermal evolution of coal and of asphaltenes from coal extracts. The details of the methods were given by Khavari-Khorasani et al. (1998). The thermal stability of coals and of coal asphaltenes was estimated, using multiheating rate bulk flow pyrolysis and the Lawrence Livermore Kinetics program (Braun and Burnham, 1990).

The compositional generation and expulsion model used in this paper for estimating sample losses, is based on mass balance of the composition of generated, expelled and retained petroleum. The model is based on the observation that most petroleum in source rocks is stored in sorbed states in kerogen (Young and McIver, 1977; Stainforth and Reindeers, 1990; Pepper, 1991; Sandvik et al., 1992). The calibration data are provided by the on-line, temperature-resolved flash Py-GC data from coal, the bulk composition of the retained petroleum (average group type proportions) in the coal at different levels of maturity, source rock mass balance and solubility calculations (Sandvik *et al.*, 1992). The method was described in detail by Khavari-Khorasani *et al.* (1998).

All the calculation results presented for mass balance, volumetrics and compositional predictions, were performed using the *PetroTrak PCP* (PETEC AS) software package.

4.0 Results and Discussion

4.1 THE RETENTION STATE OF PETROLEUM IN COAL

It was stated by Stainforth and Reindeers (1990) that the amount of extractable petroleum from kerogen concentrates of oil-prone source rocks is comparable to extract from powdered rocks. In consistency with this, it has been concluded by several authors, that for oil-prone source rocks, most of the retained petroleum occur in sorbed states and not as free fluids (Philippi, 1965; Young & McIver, 1977; Stainforth & Reindeers, 1990; Pepper, 1991; Sandvik *et al.*, 1992; Michelsen & Khavari-Khorasani, 1995).

While adsorption has been suggested to represent the sorption state of petroleum (Pepper, 1991), swelling experiments on kerogen indicate that a solid solution (absorption) is the most important state of petroleum for oil-prone kerogen (e.g., Sandvik *et al.*, 1992). Sandvik *et al.* (1992) also noted that the absorption decreased for kerogen with decreasing hydrogen index. It is today acknowledged that the oil generating components of coal are principally similar types of polymers as for conventional oil source rocks. In coals these aliphatic polymers are diluted in the more aromatic "humic" polymers.

The generally observed reduction of retained petroleum in coals, for maturities above a vitrinite reflectance of 1% (Figure 1) (see also Durand and Paratte, 1983) could be consistent with that also for coals, most of the retained (at surface conditions) petroleum is retained in an absorbed state. The adsorption capacity increases with increasing maturity (Khavari-Khorasani and Michelsen, this volume), while the absorption capacity decreases because the aliphatic polymers involved in absorption become more aromatic and cross-linked as petroleum is generated. Alternatively, or in addition to a change in sorption capacity, the general drop in observed retention could be related to increasing sample loss, because the petroleum products become lighter with increasing maturity (see later data and discussion).

Since coals generally have lower hydrogen indices than conventional oil source rocks, the absorption capacities for "oil" constituents is lower for coals than for conventional oil source rocks, consistent with the view and observations of Durand and Paratte (1983). The "holes" in the aliphatic polymers which holds the "dissolved" petroleum molecules, would be comparable in size with estimates of the "micro-porosity" dimensions of coals as determined from X-ray studies (e.g., Rouzaud and Oberlin 1981). It is important to emphasize that molecules in a polymer solution of this type, is not trapped in the sense discussed by Hunt (1991). The diffusion coefficients of the petroleum molecules in such a polymer solution at generation PT conditions, are in fact quite significant, as estimated by Thomas and Clouse (1990).

Adsorption, is the main sorption state which account for the methane produced during degassing of coal at low temperatures and pressures. The relative significance of absorption vs. adsorption for the pressures and temperatures of petroleum generation is discussed by Khavari-Khorasani and Michelsen (this volume).

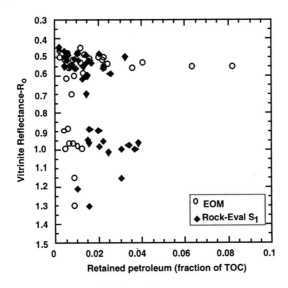

FIGURE 1. The relationship between retained petroleum and vitrinite reflectance for desmocollinite-rich coals of the Åre formation, Haltenbanken, Norway. Note the difference between the thermal extract data and the solvent extract data.

For coals at low to intermediate maturities, it is often observed that the solvent extractable bitumen, exceeds the thermal extract, while at higher maturities the thermal extract exceeds the solvent extractable bitumen (Examples in Figure 1 is a maturity series from the Åre coals. Here the reversal occurs at a vitrinite reflectance (R_o) of around 0.9%. The difference in estimated retention between these two different techniques is most likely related to that the Rock-Eval thermal extract discriminates molecules with carbon number above around C_{20} (the heavy material goes to the S_2 peak) while the solvent extractable petroleum is, due to sample work-up, mainly a measure of the C_{15+} petroleum. The two tech-

niques hence image different fractions of the retained petroleum. With increasing
maturity the amount of the heavier fractions is decreasing. Therefore, either the
heavier fractions have been thermally transformed, or the coal has been constantly expelling petroleum. (The dilution of the heavier components by progressive generation of lighter fluids does not change the amount of the material
which was already present.)

Asphaltenes represent the most thermally labile of the material present in a
petroleum fluid (Tang and Stauffer 1994; Khavari-Khorasani et al., 1998). In
Figure 2 is shown an example of the thermal decomposition of the retained
asphaltenes compared to the coal kerogen. The most thermally labile part of the
retained petroleum will probably not have undergone any significant cracking up
to the maturity level where the EOM/S_1 shift ($R_o \approx 0.9$) is observed in Figure 1.

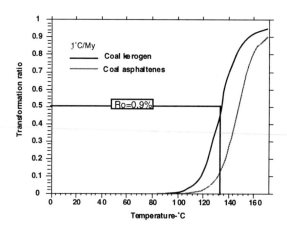

FIGURE 2. The thermal decomposition of coal kerogen vs. coal asphaltenes for an Åre coal. Notice
that coal asphaltenes display a higher thermal stability compared to the coal kerogen. The thermal
stability is estimated by multi-heating rate, bulk flow pyrolysis experiments.

Furthermore, the cracking products of the asphaltenes are not light hydrocarbon dominated. In fact, the pyrolysis products for the asphaltenes contain
higher quantities of C20+ than the primary pyrolysis products from the kerogen
(Figure 3). It is therefore indicated for the studied Åre coals, that the coals must
have been continuously expelling petroleum, including significant quantities of
C_{20+} petroleum molecules.

Free viscous petroleum, known as exudatinite, is commonly observed in
coals, both in cell lumens and in small fractures (e.g., Teichmüller and Teichmüller, 1968; Teichmüller and Ottenjann, 1977; Teichmüller and Durand,1983;
Khavari-Khorasani, 1987; Hvoslef et al., 1988; Khavari-Khorasani and Michelsen, 1991). It is important to evaluate how large mass fraction of the total
retained petroleum, this free petroleum represents. Hvoslef et al. (1988) showed

that for the Åre coals studied (Late Triassic to early Jurassic) the volumetric contribution of such free petroleum to the total retained (extractable) petroleum was
insignificant. Large errors (of the order of 4-5 times) would not change this conclusion.

There do exist in some coals, local zones of more extensive enrichment of
heavy viscous fluids. However, to our experience, even for coals with such
occurrences the exudatinites represent less than 10% of the total extractable
organic matter mass. We hence, conclude that for most coals, most of the
retained petroleum resides in sorbed states, and absorption is probably the most
significant sorption mechanism in the "oil-window" maturity range.

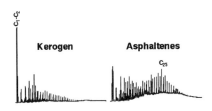

**FIGURE 3. The flash pyrolysis composition of an Åre coal kerogen at $R_o \approx 0.9$ vs. the corresponding
asphaltenes of the coal. Notice the higher concentration of "oil" constituents and low methane-
ethane yield of the asphaltenes.**

Katz (1994) used 17 coal samples with R_o=0.21-0.32, 5 coal samples with
R_o range 0.52-0.59, and an immature torbanite, to discuss the generation and
expulsion capacity of coals. The author stated that the large Rock-Eval S_1 values
in these coals (the largest values associated with the samples with $R_o \approx 0.25$-0.32)
probably represent "labile components present during deposition which have not
undergone significant geopolymerization". Nevertheless, he used the comparison between the Rock-Eval S_1 versus C_{15}+ extractable hydrocarbons from these
coals to conclude that, coals have a low expulsion efficiency. However, the
retained organic matter / hydrocarbons in *immature* to low maturity coals, cannot
be used to obtain any information regarding the expulsion efficiency of coals in
general. The expulsion efficiency can only be evaluated with a mass balance
scheme which account for and compare the quantities (and preferably compositions) of generated and retained petroleum for a maturity series.

4.2 MASS BALANCE

Many mass balance schemes have be presented for source rocks (e.g., Pelet
1985; Cooles *et al.,* 1986; Pepper 1991). The mass balance scheme presented
below, mainly deviates from previous schemes, in the manner retained and initial

petroleum quantities are accounted for. The coal will have four major physically distinct types of phases, minerals (M), kerogen (K), water (W) and one or several petroleum phases (P). We start with considering a closed, infinitively small unit mass coal volume, fixed to a point in the section:

$$m_M + m_K + m_W + m_P = 1 \qquad (1)$$

where m represent the mass fractions and the subscript refer to the phase of concern (Minerals:$_m$; Kerogen:$_K$; Water:$_W$ and Petroleum:$_P$). The kerogen mass fraction includes any petroleum molecules which are sorbed in the kerogen. We treat the mineral phases as inert, i.e., the initial mass fraction equals the mass fraction at any time:

$$^o m_M = m_M \qquad (2)$$

where $^o m_M$ represent the initial mineral mass fraction. The kerogen can both generate and store H_2O. However, we will here only be concerned with the maturity range where petroleum is generated, and most of the H_2O has already been released from the kerogen at this stage (van Krevelen, 1961). Given this simplification we write:

$$^o m_W = m_W \qquad (3)$$

The initial stage we will consider is the time immediately before the first petroleum is generated. The thermally extractable or solvent extractable organic compounds in an immature source rock, is considered a part of the kerogen, and not as an initial oil as in the mass balance scheme of Cooles *et al.* (1986). Since initially, there are no free petroleum, we have that:

$$^o m_W + {}^o m_M + {}^o m_K = 1 \qquad (4)$$

At least conceptually, the kerogen can be divided into a reactive and an inert mass fraction (Pelet, 1985; Cooles *et al.*, 1986; Patience *et al.*, 1992). Then:

$$m_K^r + m_K^i + m_K^P = m_K \qquad (5)$$

where the superscripts (r,i,P) refer to the reactive, the inert and the petroleum molecules absorbed in the kerogen. The transformation ratio of the kerogen (t) can be expressed:

$$\tau = 1 - \frac{m_K^r}{{}^o m_K^r} \qquad (6)$$

The initial mass fraction of kerogen can be found from the initial Total Organic Carbon (TOC) which represent the mass% of Carbon of the total rock:

$$^{o}m_{K} = \frac{\dfrac{^{o}TOC}{100}}{^{o}C_{K}} \qquad (7)$$

where $^{o}C_{K}$ represent the initial mass fraction of Carbon in the kerogen (≈ 0.65-0.85). The data of van Krevelen (1961) was used to derive $C_{K}(\tau)$. The initial mass fraction reactive carbon can be estimated from the hydrogen index (HI=mg Petroleum/gram Carbon):

$$^{o}m_{K}^{r} = \frac{HI\,^{o}m_{K}\,^{o}C_{K}}{1000} \qquad (8)$$

The retained petroleum in kerogen can be expressed as a function of τ and the amount of kerogen. Typical quantities of retained petroleum, can be assessed from extract or thermal extraction data, and if the cumulative retained composition can be established independently (as will be demonstrated later), the sampling losses can be estimated and the subsurface retention calculated. A function of the form:

$$F(\tau) = \left\{1 - [1-(1-\tau)]^{\wedge} E\right\} * \left[F(\tau=0) - F(\tau=1)\right] + F(\tau=1) \qquad (9)$$

has been found to be simple to fit to available data. E will determine how fast the retention level decrease as a function of τ. The retained petroleum mass in the kerogen can then be expressed:

$$m_{K}^{P} = F(\tau)\left[^{o}m_{K} - \tau\,^{o}m_{K}^{r}\right] \qquad (10)$$

The total generated petroleum mass is:

$$m_{P} + m_{K}^{P} = \tau\,^{o}m_{K}^{r} \qquad (11)$$

and the free petroleum is generated minus retained:

$$m_{P} = \tau\,^{o}m_{K}^{r} - F(\tau)\left[^{o}m_{K} - \tau\,^{o}m_{K}^{r}\right] \qquad (12)$$

providing that:

$$\tau\,^{o}m_{K}^{r} > F(\tau)\left[^{o}m_{K} - \tau\,^{o}m_{K}^{r}\right] \qquad (13)$$

If equation [13] is not satisfied, no free petroleum is present. Hence, a coal is in either of two states during generation, before or after the first free fluid appear in the coal.

4.2.1 Sample Loss

The composition of retained vs. desorbed petroleum must follow that same mass balance constraints as outlined previously. If phase fractionation between kerogen and the free fluid is ignored, the governing equation for the desorption was given by Khavari Khorasani et al. (1998):

$$d\left(yC_r^P\right) - yd\left(C_r^P\right) = f(\tau)d\tau - yd\tau \qquad (14)$$

where y represent a given petroleum component and C_r^P represent the amount of retained petroleum. The function $f(t)$ is a regression function and describe the generation of the individual petroleum components.

Figure 4 shows the evolution of retained Gas to Oil Ratio (GOR: mass/mass) at any time in the coal during generation and expulsion. The generated composition is derived from temperature resolved flash Pyrolysis Gas Chromatography as outlined by Khavari Khorasani et al. (1998), for a coal from the Åre Fm. (Norway).

If we assume that all the petroleum which is gas at surface conditions represent the sample loss (from subsurface conditions until analysis), the true subsurface retention can be estimated. It is indicated that a sample loss of around 20% is viable at low maturities, increasing to around 70% (by mass) at very high maturities. Combining this information with the Åre coal data in Figure 1, indicate a retention level of around 0.1 at early stages (mass fraction of TOC) down to around 0.05 at high maturities.

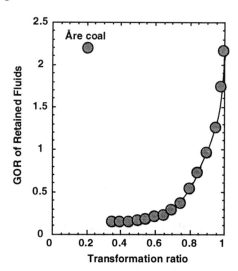

FIGURE 4. The evolution of GOR (mass/mass) of the petroleum retained in the coal as a function of the transformation ratio. The calibration data are from Åre coals of Haltenbanken (Norway) and the compositions are calculated for an initial hydrogen index of 400.

4.2.2 Desorption Efficiency

Figure 5a, shows the desorption efficiency (desorbed/generated) as a function of transformation ratio for kerogen with HI varying from 900 to 100. The black curve at the bottom of the diagram display the retention function $F(\tau)$. This function has been tuned to be consistent with observed retention levels as a function maturity, and sample losses as outlined previously.

Because the ratio between two time-temperature integrals, using a distributed activation energy model and kinetic parameters approximately in the same range, is essentially independent on heating rate (Khavari Khorasani & Michelsen 1994; Michelsen & Khavari Khorasani, 1995) a universal relationship between the transformation ratio and vitrinite reflectance can be derived providing the kinetics for petroleum generation is available for the coal of concern. In this paper we use kinetic parameters obtained from Åre coals from Haltenbanken (Norway) and the model of Burnham & Sweeney (1989) to derive such a translation. Figure 5b, shows the relationship between desorption efficiency and vitrinite reflectance for the Åre coals.

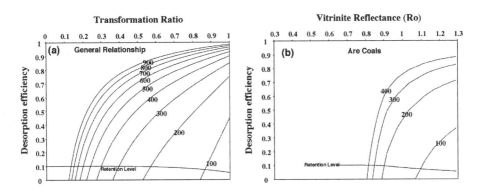

FIGURE 5. The generalized relationship between desorption efficiency and the transformation ratio (a) and between the desorption efficiency and vitrinite reflectance of the Åre coals (b), according to the mass balance formulation given previously. Each curve is labeled with the initial HI.

As pointed out previously (e.g., Pepper, 1991), the desorption efficiency is strongly controlled by the initial HI. Coals with HI generally less than 400, will have free petroleum first at much higher maturities than typical oil prone source rocks. However, coals still have the potential for high desorption efficiencies. If the free fluids represent oils, the retention time and temperatures are too small to expect significant secondary cracking of the primary fluids (see e.g., Horsfield et al., 1995 for recent estimates of the thermal stability of oils.)

Figure 6a, shows the change in HI as a function of transformation ratio with identical parameters as in Figures 5. Figure 6b shows the same relationship as a function of vitrinite reflectance.

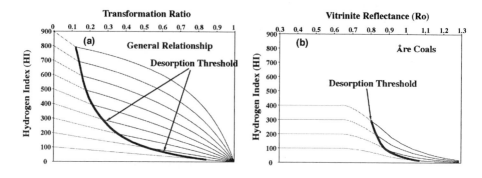

FIGURE 6. The generalized evolution of the hydrogen index as a function of the transformation ratio (a) and vs. the vitrinite reflectance of Åre coals (b), according to the mass balance formulation given previously. The thick line represent the saturation threshold.

Note that the relationship between HI and transformation ratio is linear, only in the maturity interval prior to the appearance of the first free fluids. During desorption the non-linearity increases with increasing initial HI. The desorption rate is controlled by two factors: (a) the rate of generation and (b) the rate of reduction in the kerogen mass, because less kerogen is available for sorption as kerogen in transformed into petroleum.

The kinetic parameters used to translate the transformation ratio to equivalent reflectance was obtained from a relatively hydrogen rich, immature Åre coal with a hydrogen index of 310. Examples of the measured relationship between the HI and vitrinite reflectance (R_0) is given in Figure 7.

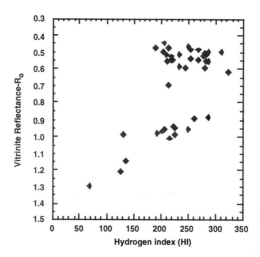

FIGURE 7. The relationship between the vitrinite reflectance and HI for Åre coals from Haltenbanken (Norway).

For the Åre coals, the drop in HI with increasing vitrinite reflectance is generally more rapid than e.g., for the HI vs. R_0 data presented by Durand and Paratte (1983) for Paleozoic coals. This difference is related to a difference in thermal stability between the Åre coals and the coals studied by these authors.

4.3 EXPULSION

There are no agreement in the literature regarding the physical and chemical processes of most importance for expulsion of petroleum from coals, or from other potential source rocks. We will therefore start with a quantitative evaluation of the main physical and chemical changes which must take place in a coal. From there, we will derive a system description, and compare this with the expulsion models which previously have been qualitatively and/or quantitatively evaluated.

4.3.1 Coals Must Expel: Volumetric Consequences of Lack of Expulsion

Consider a coal undergoing petroleum desorption in a closed system. If we assume that only very small amounts of water is generated from the kerogen in the maturity range where petroleum is generated, the volumetric changes can be evaluated, providing the density changes of the petroleum, kerogen, water and mineral matrix is known for the PT gradient of concern. The specific volume of a unit mass of coal is:

$$V = \frac{{}^o m_W}{\rho_W} + \frac{{}^o m_M}{\rho_M} + \frac{m_K}{\rho_K} + \frac{m_P}{\rho_P} \qquad (14)$$

which by the equations given for mass balance of kerogen and petroleum can be expanded to:

$$V = \frac{{}^o m_W}{\rho_W} + \frac{1 - {}^o m_K - {}^o m_W}{\rho_M} + \frac{(1-\tau){}^o m_K^r + {}^o m_K - {}^o m_K^r + F(\tau)\left[{}^o m_K - \tau {}^o m_K^r\right]}{\rho_K} + \frac{\tau {}^o m_K^r - F(\tau)\left[{}^o m_K - \tau {}^o m_K^r\right]}{\rho_P} \qquad (15)$$

The mass fraction of water can be expressed as a function of the water content at the onset of petroleum generation, assuming that all the water resides in a separate water phase:

$$^o m_W = \frac{{}^o \rho_W \, {}^o \phi_W \left[{}^o m_K \left(1 - \frac{{}^o \rho_M}{{}^o \rho_K}\right) - 1 \right]}{{}^o \phi_W \left({}^o \rho_M - {}^o \rho_W\right) - {}^o \rho_M} \qquad (16)$$

In the test scenario we will discuss, the coal is assumed to subside along a known pressure-temperature gradient, corresponding to a heating rate of 2°C per million years. The density of coal macerals were obtained using surface density measurements, compressibilities and thermal expansibilities from van Krevelen (1961). Density of free petroleum was obtained from the predicted compositions and a Peng and Robinson (1976) equation of state. Examples of the volumetric evaluations are shown in Figure 8 for four different initial hydrogen indices in the range observed in coal. The four different coal constituents (kerogen, petroleum, water and minerals) are labelled with different shades of gray.

While there are uncertainty in all assumptions, particularly the density functions applied (since the kerogen, petroleum and water densities are not linked through an equation of state to the sorption equilibria), the petroleum porosities become large particularly for the most "oil-prone" coal. For any reasonable range of properties, the free petroleum volume will exceed by far what is ever observed in a coal seam. Hence, there are little doubt that indeed petroleum continuously is being expelled from coals, in full agreement with the conclusion of e.g., Durand & Paratte (1983) and Boreham and Powell (1993).

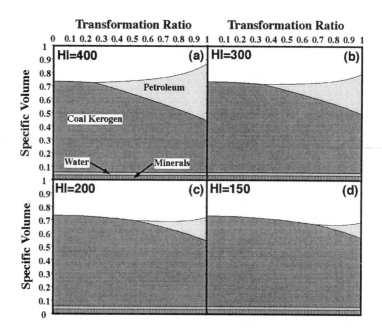

FIGURE 8. Volumetric changes in coal seams as a function of the transformation ratio. The coals have a TOC of 70%. (a) Initial HI = 400, (b) initial HI=300, (c) initial HI=200 and (d) initial HI=150.

The Figures 8a-8d show clearly, that the major change in a coal seam during generation is the mass transfer from the coal kerogen to the free petroleum. The overall density change of coal kerogen plus petroleum is small in comparison to the mass/volume transfer related to the generation/desorption. Similarly the thermal expansion of water is of subordinate character. Hence, for a coal seam, the main properties of interest for evaluating expulsion is the petroleum yield and physical properties of the organic coal constituents which dominate the rock matrix.

The initial HI is fundamental, because it determine the yield relative to the amount of sorbing polymer according to the kerogen-petroleum mass balance given previously (see also Pelet 1985; Pepper 1991). The TOC is fundamental, because it determines how much of the coal which represent "weak" organic matter.

4.3.2 Rheology and Subsurface Stress State

The stress state of the coal together with coal rheology, will determine how the coal deform as petroleum porosity is formed and the coal matrix shrinks due to the mass transfer from the kerogen to the free petroleum. Since the organic coal constituents represent the dominant part of the coal matrix (Figure 8), there can be no doubt that the stress acting on the kerogen is the total overburden stress according to the view of e.g., Palciauskas (1991).

In published quantitative expulsion models it is generally assumed that the kerogen is subjected to as stress corresponding to the fluid pressure (e.g., Ungerer et al., 1990). This is also the underlying assumption in the more geochemical expulsion evaluations (e.g., Lewan, 1994). However, in coals and source rocks in general, the kerogen makes up such a high volume fraction of the rock that it is not possible that kerogen is floating within the initial water porosity of the rock, and shielded from the overburden pressure. The models which assume that this is the case and disregards that the overburden stress is transmitted directly onto the kerogen (e.g., Lewan, 1994), will naturally lead to very different results/conclusions than what is presented in this communication.

How will the shape of the overburden stress field looks like in a generating coal seam? It is commonly assumed in the literature, that the horizontal stress represent around 70-80% of the vertical stress. However, such observations are generally from intermediate depths 1-3Km, while for the depths corresponding to the high temperatures ($> \approx 120°C$) of actively generating coals, leak-of-tests commonly indicate that the horizontal stress is comparable to the vertical stress. Hence, there are indications that at depths of petroleum generation, the sediments on a geological time scale are not deforming purely elastically, and creep rates are too high for the rocks to sustain a significant ($>$ m scale) differential stress field. While this observation applies to shale, silt and sandstones which volumetrically dominate the sedimentary sections of concern, the rheology will also play a fundamental role for expulsion of petroleum from coals.

Present day models for petroleum expulsion generally assume that the rocks have perfectly elastic rheologies, i.e. the *strain* is proportional to the applied stress (e.g., Ungerer et al., 1990). However, very simple physical arguments can be used to demonstrate that for sediments undergoing petroleum generation and desorption, viscous rheologies are much more plausible, i.e., the *strain rate* is proportional to the applied stress.

When mass is transferred from kerogen to petroleum, the kerogen volume shrinks and the petroleum phase expands. The petroleum molecules in the kerogen must be transported from their initial site in the kerogen to the site of the free petroleum. Hence petroleum molecules must diffuse through the kerogen to arrive to the site of free petroleum. The driving forces for this diffusion will be concentration gradients due to differential stress (the kerogen is a deformable · and compressible substance, and differential deformation/compression will lead to differential concentrations), and forces due to the pressure gradient itself, i.e., related to differences in the change of partial molal volumes between different molecules along a pressure gradient. Hence the change of shape of the kerogen and the petroleum porosity, are controlled by diffusive molecular transport within the kerogen. This creep process is very similar to diffusion creep in inorganic rocks as discussed by e.g., Turcotte and Schubert (1982).

The rheology will be viscous, with a viscosity having an Arrhenius type dependency on temperature and pressure and be *intimately coupled to the generation and desorption rates.* The high diffusion coefficients estimated for petroleum molecules diffusing in kerogen (Thomas and Clouse, 1990), indicates that very low viscosities can be expected. However, the absolute magnitude of the viscosity of the coal will also depend on the petroleum porosity.

The coal viscosity must depend on the petroleum porosity and petroleum viscosity. If the coal was 100% petroleum, the coal viscosity would have resembled the petroleum viscosity. Hence, if a coal generate petroleum in excess of what can be sorbed, the generated petroleum porosity will weaken the coal, and the coals total viscosity will be reduced. The relationship between the coal viscosity and petroleum porosity is, however at present poorly constrained. A total decrease in viscosity of the coal during generation and desorption is, however, only the case if the increasing condensation of the coal polymers, due to generation, does not have a larger effect on the viscosity than the increase in petroleum porosity.

4.3.3 Expulsion Dynamics

Mass balance of an expelling coal. The conservation of mass requires a direct coupling between the coal matrix and the petroleum, since neither petroleum, nor the coal matrix is individually conserved. This is stated in equations [17] and [18] which represent mass conservation equations for the system written for one spatial dimension (1-D), as opposed to the 0-D mass balance equations discussed previously. A discrete version of these equations are shown graphically in Figure 9. In these equations the coal is subdivided into two parts, (1) the coal matrix (including minerals, sorbed petroleum and all the water in the system) and (2) the free petroleum.

Equation [17] state that the change in mass concentration of free petroleum in some infinitively small volume (fixed in space), corresponds to the petroleum in-flux minus the petroleum out- flux (v represent the local petroleum velocity) plus the rate of mass transfer from the coal matrix to the petroleum (DM_m/Dt).

$$\frac{\partial}{\partial t}(\rho_p \phi) + \frac{\partial}{\partial z}(\rho_p \phi v) = \frac{D M_m}{D t} \tag{17}$$

Equation [18] state that the change in mass concentration of the coal matrix in some infinitively small volume (fixed in space), corresponds to the coal matrix in-flux (the coal matrix move through the control volume due to both subsidence and compaction: see Figure 9) minus the coal matrix out-flux (V represent the local coal matrix velocity) minus the rate of mass transfer from the coal matrix to the petroleum.

$$\frac{\partial}{\partial t}(\rho_{rm}(1-\phi)) + \frac{\partial}{\partial z}(\rho_{rm}(1-\phi)V) = -\frac{D M_m}{D t} \tag{18}$$

The rate of mass transfer is measured at a point following the motion (due to subsidence and compaction: see Figure 9) of the coal matrix:

$$\frac{DM_m}{Dt} = \frac{\partial M_m}{\partial t} + v\frac{\partial M_m}{\partial z} \tag{19}$$

The rate of mass transfer will be larger than the generation rate (providing the coal kerogen is saturated with petroleum) because the amount of the coal matrix which retain the petroleum, decreases as petroleum is generated from it. As discussed previously, during generation and desorption, the stress-strain relationship for the coal must be coupled to the generation, i.e., the coal deform on a micro scale by diffusion of petroleum in the kerogen.

The Origin of the Forces which Drive Petroleum out of the Coal. The stress which act on the coal kerogen is the overburden stress. If the coal creep is proportional to the desorption of petroleum, then the pressure of the petroleum fluid will be the overburden pressure providing that there is no change of density of the coal kerogen plus petroleum. This is illustrated in Figure 10b where the circles illustrate the petroleum "droplets". The vertical pressure gradient in the fluid will be the same as the overburden pressure gradient. If there is a volume expansion due to the petroleum desorption, and the strength of the rock and the resistance to flow is sufficiently high, the pressure could exceed the overburden pressure.

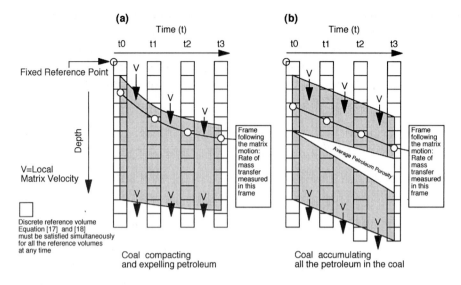

FIGURE 9. Mass balance principles for an expelling and compacting coal (a) or an accumulating non-compacting coal (b) during progressive burial and petroleum generation. (The compaction and petroleum porosity is exaggerated for illustration purposes.) Notice that the local matrix velocity (V) will display a continuous change vertically in the coal during compaction and expulsion (a) and that a coal which do not compact must accumulate a large petroleum porosity (b).

If the coal is *not* compacting during the mass transfer from kerogen to the free petroleum fluid, then the petroleum velocity relative to the rock matrix is zero and the petroleum pressure, P_p must therefore be hydrostatic (if stable):

$$\frac{dP_p}{dz} = -\rho_p g \qquad (20)$$

where ρ_p is petroleum density. The hydrostatic petroleum fluid pressure gradient is shown in Figure 10a.

The overburden pressure must everywhere support the weight of the overlying petroleum/ coal matrix mixture:

$$\frac{dP_{ob}}{dz} = -\left[(1-\phi)\rho_m + \phi\rho_p\right]g \qquad (21)$$

where z is vertical position, P_{ob} is overburden pressure, ϕ is petroleum porosity, ρ_m is matrix density and g is gravity. For [20] and [21] to be simultaneously true, the petroleum porosity must be 100% ($\phi = 1$) i.e., the coal matrix must be absent! Hence, if the coal is *not* compacting and expelling petroleum over geological time during generation and desorption, it must be unable to creep due to this mechanical instability and/or there must be infinite resistance to petroleum flow! This is in strong contrast to the required link between the coal viscosity and the petroleum generation/desorption as discussed previously. Also, as pointed out previously, the stress field in the deep parts of sedimentary basins, indicate that most sediments are creeping, and e.g., pressure solution creep commonly observed in the same depth intervals shows that even quartz rich sandstones (which are much less viscous than coal) creep efficiently at petroleum generation temperatures.

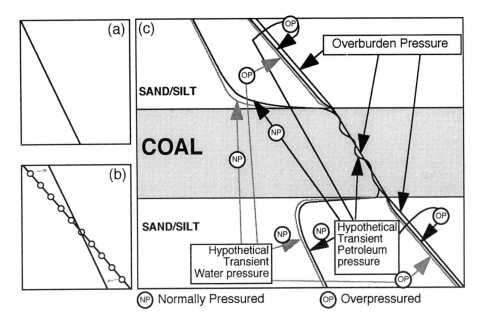

FIGURE 10. Pressure gradients of significance to petroleum expulsion. (a) hydrostatic gradient, (b) the lithostatic gradient in the coal (with gradient in initially desorbed petroleum) compared to the hydrostatic gradient. (c) The petroleum and water pressure gradients in a coal bearing section during expulsion. The gradients are indicated for normally pressured (NP) and overpressured (OP) situations.

If we accept that the coal must deform by diffusion (at least nearly) proportionally to the creation of accommodation space for the free petroleum then Equation [20] and [21] implies that it cannot do so, (if the petroleum does not have the same density as the coal matrix) without creating a mechanical unstable state. The coal must respond by compaction, and segregate the petroleum out of the coal.

In the poro-elastic models generally used to model sediment compaction and petroleum expulsion (e.g., Ungerer *et al.*, 1990), the rocks can only compact a prescribed amount for the applied stress, before the "elastic forces" balance the forces related to the mechanical instability described above (expressed as a an empiric effective stress vs. porosity function). However, for the high diffusion coefficients of petroleum in kerogen (Thomas and Clouse, 1990), the coal must creep viscously and the poro-elastic model does not apply.

The described expulsion process will operate independently of any volume expansion of the coal plus petroleum. The process can potentially expel all the free petroleum fluid in the coal. The process is driven by the density difference between the coal and the petroleum. An analogy will be the partial melting of (initially impermeable) mantle rocks where the mantle compacts and expel basaltic magmas all the way to the surface, sometimes at very low degrees of partial melting (e.g., McKenzie, 1984). The overburden gradient and petroleum hydrostatic gradient in Figure 10b, illustrate the potentials due to the different gradients. Notice that only the difference in gradients are significant, since the absolute pressures depend on many other factors.

Because the petroleum pressure in the coal is close to the overburden pressure, if the pore water pressure in the surrounding sediments is lower than the overburden pressure (which is the normal case), a large fluid potential gradient exist between the petroleum in the coal, and the pore water in the surrounding sediments. The force related to this potential gradient will normally far exceed the forces related to the density contrast between the petroleum and the coal. We will come back to this later.

A volume increase of the coal kerogen plus petroleum, will contribute to the expulsion efficiency. However, this feature by itself can only contribute to an expelled amount corresponding to the total volume increase (see Figure 8). Since actual coals are nearly void of free petroleum, this phenomenon can only be of secondary importance.

Coal Deformation. For published expulsion models, (for a review see Ungerer, 1993) the deformation of the rock matrix is modeled applying poro-elastic theory. However, this theory is only relevant to describe problems where the rock matrix is conserved (and actually behave elastically). Poro-elastic theory is not relevant to the creep process which is related to the transfer of mass from the solid to a low viscosity fluid. This is not to say that the elastic properties of the coals are unimportant. The different coal constituents have different petroleum yield and different elastic properties (e.g., van Krevelen, 1961) which certainly will influence the coal deformation. However, the point is that the dominant process in a coal seam (or any other source rock) is the mass transfer between the solid and petroleum (Figure 8), and this must lead to a related viscous creep process.

It is important to emphasize on the difference between this process and the purely diffusive expulsion model of Stainforth & Reindeer (1990). In their model, the petroleum is transported through the entire source rock by molecular diffusion in the kerogen. Hence, the diffusion length scale is the vertical thickness of the source bed. In the model presented here, the diffusion length scale in the coal is microscopic to cm scale and driven by local distortions of the stress field on that scale. We consider the transport of petroleum out of the coal to be essentially a bulk flow transport of separate petroleum phases (Michelsen *et al.*, 1997).

Durand and Paratte (1983) emphasized on the compressibility of the coal matrix as the key to expulsion. However, while we agree with most of these authors conclusions, the physical processes outlined above are not much dependent on the compressibility of the coal constituents, although the coal as a whole can be considered a very compressible fluid during petroleum expulsion. The expulsion/compaction process relies on minor differences in either compressibilities, normal and shear elastic or viscous properties between the coal constituents to create micro-scale distortions of the stress field. The general observation of micro-scale differences in deformation of coal constituents shows that this requirement is fulfilled.

Force Balance. As the petroleum porosity increases due to desorption, the viscosity of the total coal, and the resistance (permeability) to petroleum flow will decrease. Hence, during the compaction and expulsion, there must be a balance between the desorption rate, compaction rate and potentials. Since, only minute quantities of free petroleum is normally present in the coal, only a very small petroleum porosity must be required to balance the forces during petroleum flow in the system. The pressure balance between the petroleum and the coal only occur inside the coal volume occupied by petroleum.

The momentum conservation equations for the petroleum and coal matrix will be similar to the fluid/solid conservation equations described by McKenzie (1984) and shown in Figure 11.

Petroleum:

$$v - V = -\frac{k(\phi)}{\mu\phi}\frac{\partial}{\partial z}(P_p - \rho_p gz)$$

Coal Matrix:

$$v - V = -\frac{k(\phi)}{\mu\phi}\frac{\partial}{\partial z}\left((1-\phi)P_p - \left(\zeta + \frac{4}{3}\eta\right)\frac{\partial V}{\partial z}\right) - \frac{(1-\phi)}{\phi^2\mu}k(\phi)\rho_{rm}g$$

Petroleum viscosity: μ **Permeability:** k

Total Coal matrix viscosity: $\left(\zeta + \frac{4}{3}\eta\right)$

FIGURE 11. The momentum conservation equations for the petroleum expulsion process from a viscously behaving coal.

The momentum equation for the petroleum is essentially Darcy's law, while for the matrix, the petroleum pressure is only acting within the volume which is occupied by petroleum. The total viscosity of the coal matrix depend on both the bulk and shear viscosities of the matrix.

Micro-fractures. The formation of fractures have been considered by many authors to be essential for petroleum expulsion (e.g., Düppenbecker and Welte, 1991; Vernik, 1994). However, it is not important for the model presented here, whether the petroleum flows essentially in the self generated porosity or self generated fractures. There are evidence of micro-fracture formation in coals and source rocks in general. However, it should be emphasized that the horizontal micro-crack measurements of Vernik (1994) for the Bakken formation of the Williston basin (US), were not performed on a generating source rock, but rather on a shale which has been uplifted, de-stressed and cooled and presently is not generating petroleum. This is a general problem for interpretation of brittle structures in coals, since samples mostly have been recovered from coal mines (where the coals have been extensively uplifted), or violently de-stressed during sample recovery from deep petroleum wells.

The relationship between petroleum filled space and permeability will certainly be different, depending on "porosity model", but the expulsion process on a scale larger than a cm scale will probably not be very different. Because, the initial petroleum pressure in the model is close to the overburden pressure, the system could in principle instantaneously be situated on the fracture gradient. However, due to the creep behavior of the medium, it is not obvious what scale such fractures would have and when we should refer to a petroleum filled volume as a fracture or "normal" porosity. The, occurrence of exudatinite micro scale sills and dikes in the coal, however, document some brittle failure formed during generation/desorption and are probably related to heterogeneities in the material properties of the coal.

Synthesis. So far, we have described the petroleum flow in the coal as a monophasic flow, and only described the conservation of mass within the coal seam. If free water is present in the flow paths of the petroleum in the coal and/or the petroleum occur in several petroleum phases (e.g., oil and gas), relative permeability/ capillary effects will increase the resistance to the flow. However, as pointed out by Durand and Paratte (1983), these effects are probably much smaller in coal seams than in conventional source rocks. The coal seams are interbedded with clastic initially completely water wet sediments which must be in pressure continuity with the coal. How, the petroleum pressure and water pressure profiles are likely to appear within and around a coal seam expelling petroleum will be discussed below.

Figure 10c, outline hypothetical transient pressure profiles for petroleum and water in the area adjacent to the coal during expulsion. The pressure profiles are given for both a normally pressured coal bearing section, and for a highly overpressured section. The pore pressure in the clastic sediments surrounding the coal seam will have a major influence on the expulsion behavior.

If the sedimentary section is normally pressured prior to expulsion, i.e., the inorganic rock matrix is so strong that the forces required to compact the inorganic rock matrix is larger than the forces required to expel the water during compaction, there will be a large fluid pressure gradient between the free petroleum in the coal and the pore water in the surrounding sediments. This pressure difference will in this case probably dominates the petroleum flow pattern, and petroleum will be discharged out of the coals nearly symmetrical up and down. This is so because the fluid pressure gradient from the coal to the surrounding sediments can be up to several hundred bars per meter, while the density contrast between the coal and the petroleum will only give origin to a potential difference of around 0.1 bar per meter.

In the surrounding water wet sand/silt, the capillary pressure will balance the difference (saturation dependent) in water and petroleum pressures and the movement of petroleum into the surrounding sediments represent a drainage process (a non wetting fluid displacing a wetting fluid). The width of the pressure transition zone from the coal into the surrounding sediments depend on the permeability of the surrounding sediments, and the higher the resistance to flow the wider the zone in the surrounding sediments where the pore pressure is increased. The water pressure lines are not continued into the coal. This is done to simplify the graph.

If the coal is petroleum wet, (most likely) there will a rapid change of wetting fluid phase at the coal interface. Water in this case will display some drainage into the coal. If the coal is water wet, a local imbibition zone will probably exist some way into the coal.

If the pore pressure of the surrounding sediments is close to the overburden pressure (strong overpressure), i.e. the forces required to compact the inorganic rock matrix is much smaller than the forces required to expel the water, the petroleum expulsion will be more controlled by the local mechanical instability, and the petroleum will have a larger tendency to be expelled upwards (the buoyancy force due to the density contrast between the coal matrix and petroleum (around 0.1 bar per meter) always acts upwards).

Within the coal seam, it is indicated that the petroleum pressure will oscillate close to the overburden pressure. This is analogous to the porosity waves outlined by McKenzie (1987), for compaction of a viscously behaving mixture, with initial differences in porosity. In a coal seam, significant variation in HI, will create vertical variability in the desorption rates and therefore variability in petroleum porosity. The porosity waves can form, due to the coupling between coal viscosity and petroleum porosity, the dependency of resistance to flow on porosity, and the viscous forces due to the relative velocity between the coal matrix and the petroleum. The collision between such a porosity wave and a coal heterogeneity, e.g., a vitrinite layer, with aromatic polymers with appreciable elasticity, set the ideal location for the formation of a hydraulic fracture, i.e., an exudatinite vein.

5.0 Conclusions

Mass balance and observed quantities of retained petroleum in coals indicate that coal seams are effectively expelling petroleum. An analysis of the forces available in a coal seam, and the likely coal rheology during petroleum generation, strongly suggest that coal seams are fully capable of segregating the petroleum out of the coal. This expulsion is both related to the density contrast between the coal and the petroleum (gravitational segregation) and because of the potentially large fluid potential gradients between pore water in surrounding sediments and the petroleum in the coal. The stress which act on the coal kerogen is the total overburden stress, and during initial desorption, the petroleum pressure gradient will coincide with the overburden pressure gradient.

The coal will deform as a viscous fluid, and compact due to the mechanical unstable state created by petroleum desorption. The process will be particularly amplified in normally pressured sections. In such instances, the petroleum will be expelled nearly symmetrically up and down. In strongly overpressured sections, there will be a larger tendency for upwards expulsion (normal gravitational segregation). The behavior of coals with respect to expulsion, is probably very similar to the expulsion behavior of conventional source rocks as discussed by Michelsen et al. (1997).

The composition of expelled fluids of many coals contain significant quantities of potential "liquid" components. Hence the reason for the extremely few documented oil deposits sourced by coal, must be either purely geological factors as suggested by Durand and Paratte (1983) or the particular PVT properties of the coal derived fluids. This will be discussed in a later communication.

6.0 References

Boreham, C.J. and Powell, T.G. (1993) Petroleum source rock potential of coal and associated sediments: Qualitative and Quantitative aspects, in B.E. Law and D.D. Rice (eds.) *Hydrocarbons from coal.* AAPG Studies in Geology, **38**, 1993, 203-218.

Burnham, A.K., & Sweeney J.J., (1989) A chemical model of vitrinite maturation and reflectance. *Geochimica et Cosmochimica Acta*, **53**, 2649-2657.

Braun, R.L., and Burnham, A.K. (1990) *Kinetics: A computer program to analyze chemical reaction data*. Lawrence Livermore National Laboratory, December 1990. pp. 11.

Clayton, J.L. (1993) Composition of crude oils generated from coals and coaly organic matter in shales, in B.E. Law and D.D. Rice (eds.) *Hydrocarbons from coal*. AAPG Studies in Geology, **38**, 1993, 185-200.

Cooles G.P., Mackenzie A.S. and Quigley T.M. (1986). Calculation of petroleum masses generated and expelled from source rocks. *Organic geochemistry*, **10**, 235-245.

Durand, B. and Paratte, M. (1983) Oil potential of coals: a geochemical approach, in J. Brooks, ed., *Petroleum Geochemistry and Exploration of Europe*, Geological Society Special Publication, **12**: Oxford, Blackwell Scientific Publications, 255-265.

Düppenbecker, S.J. and Welte, D.H. (1991) Petroleum expulsion from source rocks - insight from geology, geochemistry and computerized numerical modelling. *Proc. of the 13th World Petroleum Congress*, Buenos Aires, October 1991, (3)2, Wiley, London.

Hvoslef, S., Larter, S.R. and Leythaeuser, D. (1988) Aspects of generation and migration of hydrocarbons from coal-bearing strata of the Hitra formation, Haltenbanken area, offshore Norway. *Organic Geochemistry*. **13**, 525-536.

Horsfield, B., Yordy, K.L. and Crelling, J.C. (1988) Determining the petroleum-generating potential of coal using organic geochemistry and organic petrology. *Organic Geochemistry*, **13**, 121-129.

Horsfield, B, Schenk, H. J., and Primio, R. (1995) Kinetics governing the conversion of oil into gas in petroleum reservoirs: *Report of findings from an industry-sponsored study*. KFA, Jülich.

Hunt, J.M. (1991) Generation of gas and oil from coal and other terrestrial organic matter. *Organic Geochemistry*, **17**, 673-680.

Katz. B.J. (1994) An alternative view on indo-Australian coals as source of petroleum. *APEA*, 256-267.

Khavari Khorasani, G. (1987). Oil-prone coals of the Walloon Coal Measures, Surat Basin, Australia, in A.C. Scott (ed.) *Coal and Coal-bearing Strata: Recent Advances*, Geological Society Special Publications, **32**, 303-310.

Khavari Khorasani G. and Michelsen J.K. (1991) Geological and laboratory evidences for early generation of large amounts of liquid hydrocarbons from suberinite and subereous components. *Organic Geochemistry*, **17**, 849-863.

Khavari-Khorasani, G., Dolson, J. and Michelsen, J.K. (1998) The factors controlling the volume and movement of heavy vs. light oils, as constrained by data from the Gulf of Suez–Part I The effect of expelled petroleum composition, PVT and the petroleum system geometry. *Organic Geochemistry* (in press)

Lewan, M.D. (1994) Assessing natural oil expulsion from source rocks by laboratory pyrolysis, in L.B. Magoon and W.G Dow (eds.) *The petroleum system from source to trap*. AAPG Memoir **60**, 201-210.

McKenzie D. P. (1984) The generation and compaction of partially molten rock. *Journal of Petrology*, **25**, (3) 713-765.

McKenzie D. P. (1987) The compaction of igneous and sedimentary rocks. *Journal of the Geological Society*, **144**, 299-307.

Michelsen, J.K. and Khavari-Khorasani, G. (1995) The kinetics of thermal degradation of individual oil-generating macerals: calibration with microscopical fluorescence spectrometry and bulk flow pyrolysis. *Organic Geochemistry*, **22**, 179-189.

Michelsen, J.K., Khavari-Khorasani, G., and Diaz, J. (1997). The physics of petroleum expulsion/source rock compaction: Solution to an old controversy in petroleum geology. 1997 *AAPG Annual Convention*, April 97, Dallas.

Noble, R.A, Alexander, R., Kagi, R.I., and Knox, J. (1986) Identification of some diterpenoid hydrocarbons in petroleum. *Organic Geochemistry*, **10**, 825-829.

Palciauskas, V.V., (1991) Primary Migration of Petroleum. In: Foster, N.H., & Beaumont, E.A. (Eds), Source and Migration Processes and Evaluation Techniques, *Treatise of Petroleum Geology*, Handbook of Petroleum Geology, The American Association of Petroleum Geologists, Tulsa, 13-22.

Patience R.L., Mann A.L. and Poplett I.J.F. (1992) Determination of molecular structure of kerogens using 13C NMR spectroscopy: III. The effects of thermal maturation on kerogens from marine sediments. *Geochimica et Cosmochimica Acta*, **56**, 2725-2742.

Pelet R. (1985) Evaluation quantitative des produits formés lors de l'évolution géochimique de la matiére organique. *Rev. Inst. Fr. Pét.* **40**, 551-562.

Peng, D.Y. and Robinson, D.B. (1976) A new two-constant equation of state. *Ind.Eng. Chem. Fundament.* **15**, 59-64.

Pepper A.S. (1991) Estimating the petroleum expulsion behavior of source rocks: a novel quantitative approach, in W.A. England and A.J. Fleet (eds.) *Petroleum Migration. Geol.Soc.London, Spec. Publ.*, **59**: 9-31.

Philippi G.T. (1965) On the depth, time and mechanism of petroleum generation. *Geochimica et Cosmochimica Acta*, **29**, 1021-1049.

Rigby, D. and Smith, J.W. (1982) A reassessment of stable carbon isotopes in hydrocarbon exploration: *Erdøl und Kohle*, 35, 415-417.

Rouzaud, J.N. and Oberlin, A. (1982) In *Proceedings of International Conference on coal science*. 7-9th Sept. 1981, Düsseldorf, Germany, D12, 663-668

Sandvik E,I., Young W.A. and Curry D.J. (1992) Expulsion from hydrocarbon sources: the role of organic absorption. *Organic Geochemistry*. **19**, (1-3), 77-87.

Shanmugan, G. (1985) Significance of coniferous rain forest and related organic matter in generating commercial quantities of oil, Gippsland basin, Australia. *AAPG Bull*, **69**, 1241-1254.

Stainforth J.G. and Reindeers J.E.A (1990) Primary migration of hydrocarbons by diffusion through organic matter networks and its effect on oil and gas generation. *Organic Geochemistry* **16**, 61-74.

Tang, Y. and Stauffer, M. (1994) Multiple cold trap pyrolysis gas chromatography: anew technique for modeling hydrocarbon generation. *Organic Geochemistry*, **22**, 863-872.

Teichmüller, M., & Teichmüller, R., (1968) Geological aspects of coal metamorphism, in D.G. Murchison and T.S. Westoll (eds.), *Coal and Coal-bearing strata*, 233-267.

Teichmüller, M and Ottenjann K. (1977) Art und diagenese von liptiniten und lipoiden stoffen in einem erdölmuttergestein auf grund fluoreszenzmikroskopischer untersuchungen. *Erdöl Kohle.* **30**, 387-398.

Teichmüller M. and Durand B. (1983) Fluorescence microscopical rank studies on liptinites and vitrinites in peats and coals, and comparison with results of the Rock-Eval pyrolysis. *Int. J. Coal Geol.* **2**, 197-230.

Thomas, B.M. (1982) Land-plant source rocks for oil and their significance in Australian basins: *APEA Journal*, 164-178.

Thomas M. T. and Clouse J.A. (1990) Primary migration by diffusion through kerogen: III. Calculation of geologic fluxes. *Geochimica et Cosmochimica Acta*, **54**, 2793-2797.

Turcotte, D.L. and Schubert, G. (1982) *Geodynamics, Application of continuum physics to geological problems.* John Wiley & Sons.

Ungerer, P., Burrus, J., Doligez, B., Chenet, P.Y. and Bessis, F. (1990) Basin evaluation by integrated two-dimensional modelling of heat transfer, fluid flow, hydrocarbon generation and migration. *AAPG Bull*, **74** (3), 309-335.

Ungerer, P., (1993) Modelling of petroleum generation and expulsion - an update to recent reviews, in Dore *et al.*, (eds.) *Basin Modelling: Advances and Applications*, Norsk Petroleumsforening / NPF Special Publication, Elsevier, Amsterdam, 3,219-232.

Vernik, L. (1994) Hydrocarbon-generation-induced micro-cracking of source rocks. *Geophysics*, **59**, (4), 555-563.

van Krevelen, D.W., (1993) *Coal.* Amsterdam: Elsevier, pp. 979.

Young W.A. and McIver R.D. (1977) Distribution of hydrocarbons between oil and associated fine-grained sedimentary rocks - physical chemistry applied to petroleum geochemistry, II *AAPG Bull.* **61**, 1401-1436.

JURASSIC COALS AND CARBONALEANS MUDSTONES: -THE OIL SOURCE IN THE JUNGGAR AND TURPAN-HAMI BASINS, CHINA

JIN KUILI, YAO SUPIG, WEI HUI AND HAO DUOHU

Beijing Graduate School, China University of Mining and Technology,
Beijing, 100083, China

Abstract

This paper using sedimentology, organic petrology and geochemistry studies the oil-related, Early and Middle Jurassic terrestrial coal measures, Junggar and Turpan-Hami Basins, NW China, shows that: (1) the oil generated in the main generating stage is earlier and generates from fluorescent desmocollinite (desmocollinite B), bituminite, cutinite and suberinite; (2) the oil-generating models of some individual macerals and oil expulsion experiment of vitrain sample are given; (3) the new organic petrological oil-source correlation techniques using laser-induced fluorescence, CLSM (Confocal Laser Scanning Microscope) and TEM (Transmission Electron Microscope) parameters for oil-source correlation are expressed and (4) the sedimentary organic facies will be founded by sedimentary, organic petrological and organic geochemical parameters, of which the running water swamp/marsh facies is best concerned with coal measures related oil.

Key Words oil-generating maceral, oil-source correlation, simulation of hydrocarbon generation, oil-expulsion experiment, organic facies, coal measures related oil

1. Introduction

The important Junggar and Turpan-Hami Basins (Figure 1) occur in Xinjiang Autonomous Region, NW China and occupy an area of 130000 km^2 and 50000 km^2 respectively. Many Chinese researchers documented that both of these basins constituted a great flood basin created due to collision of the surrounding plates during the Late Carboniferous to Early Permian. There had been marine transgression and a volcanic pyroclastic event occurring in its early history. At the beginning of the Upper Permian the basin was filled with nonmarine sediments and was separated into two basins after Late Mesozoic collision.

The Lower and Middle Jurassic coal-bearing strata (Figure 11) vary in thickness from less than 1000m up to 2500m. These strata contain mudstones, sandstones and more than forty coal seams, 100m or more in total thickness. The coal seams occur in Badaowan and Xishanyao Formations and were deposited in a lacustrine-swamp system. It is difficult to determine the sedimentary cycles on nonmarine coal measures, here people use the lake level as the indicator [1], then there are three sedimentary cycles from Badaowan Fm to Xishanyao Fm.

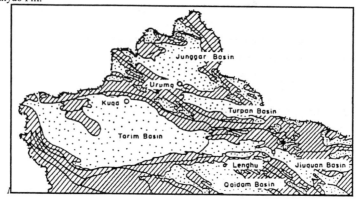

Figure 1. Location of the Junggar and Turpan-Hami Basins

2. Oil-Generating Macerals

The main macerals of coal and carbonaceous mudstone for liquid hydrocarbon generation are fluorescent desmocollinite (desmocollinite B), bituminite, cutinite and suberinite, which have generated liquid hydrocarbon in low rand coal/low maturity level carbonaceous mudstone (VRr = 0.4% - 0.6%).

This is the results of simulation of hydrocarbon generation for individual macerals at the place in question (Table 1, see their high hydrocarbon potential), and of both observations on hydrocarbon generating, macerals and organic material maturity of coal/carbonaceous samples in the locality. Under these circumstances, we study exsudintinites and their source macerals to know the hydrocarbon generating macerals, and get vitrinite's reflectance accompanied with hydrocarbon generating indications, e.g. oil drops, oil films, micrinites and also exsudintinites to indicate above mentioned maturation level.

3. Oil-Source Correlation

At first, we examined oil samples from these basins with Confocal Laser Scanning Microscope (CLSM) and Transission Electron Microscope (TEM) [2], and discovered some vitrodetrinites under CLSM. The reflectance of these vitrodetrinites (Figure 2) is similar to that of coal or carbonaceous mudstone in coal measures of the two basins, in addition, a lot of submicro-macerals and Jurassic microfossils were discovered under TEM (Figure 3). These basins' conventional oil is short of these evidences.

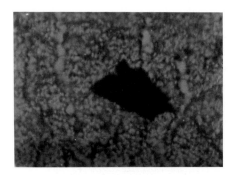

Figure 2. Relic vitrinite in oil from the

Junggar basin under CLSM, ×3300

Figure 3. A lot of coal maceral relics in oil

from the Turpan-Hami basin under TEM

1. Relic sporinite   2. Bacterium

Moreover, in order to verify oil-source correlation, the different oils and aromatic fraction from source rocks including coal and carbonaceous mudstone were used for correlation by indicators, such as fluorescence spectrum, fluorescence lifetime fingerprint, standard compound and maturity indicator, from which the so-called laser-induced fluorescence method [3] was proposed by us.

Here, we only select fluorescence spectrum and fluorescence lifetime fingerprint as parameters for correlation. On the left of the Figure 4, the coal measures related oil (2) stems from the carbonaceous mudstone (1), whereas the Figure 4 (b) shows that the same coal measures related oil (1) and (2) from different wells' reservoirs is quite distinct from the Tarim's or Junngar's conventional oils by their lifetime fingerprints. In short, regardless of the spectra or the lifetime fingerprint they are the reflecting of material's composition, in which the distinction will be only shown by spectra or by connections of every peak's fluorescence lifetime on spectra.

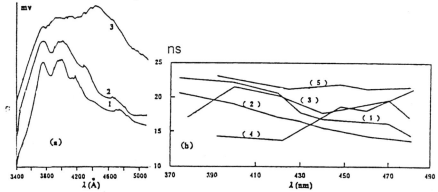

Figure 4. Oil-source correlations used in the Junggar and Turpan-Hami basins

(a) oil-source rock spectra: (1) carbonaceous mudstone; (2) coal-generating oil; (3) conventional oil.

(b) oil-oil lifetime fingerprints: (1) ,(2) coal-generating oil; (3) Junggar's conventional oil; (4), (5) Tarim's conventional oil

4. Simulation of Hydrocarbon Generation for Individual Macerals

For the sake of hydrocarbon generation study, it is necessary that using the Rock-Eval defines hydrocarbon potential (S_1+S_2) for individual macerals (Table 1), and using the pressure vessel together with quartz tube methods define the hydrocarbon generating process for individual macerals.

The pressure vessel simulation is under 6/8 temperature (150°C-330°C/150°C-360°C) conditions (see Table 2) together with quartz tube simulation under 7 temperature (200°C-400°C) conditions and analyzed by PY-GC for the former, the Micro-FT-IR as well as fluorescence microscopy for the latter. Here, we give desmocollinite B's diagrams (Figures 5-7).

TABLE 1. Macerals' hydrocarbon potential (S_1+S_2)

Maceral	VR, (%)	S_1 (mg/g)	S_2 (mg/g)	S_1+S_2 (mg/g)
Cutinite	0.5	9.91	358.23	368.14
Desmocollinite B	0.48	6.08	245.74	251.82
Suberinite	0.44	3.90	140.8	144.7
Mineral-bituminous groundmass	0.51	0.63	9.03	9.66
Bituminite	0.51	9.1	4.39	448.1

TABLE 2. Oil yields from simulation results of pressure vessel

Maceral	Temperature (°C)	VR, (%)	Gas-generating yield (ml/g)	Pyrolysis oil yield (mg/g)	Heavy oil yield (mg/g)	Total liquid hydrocarbon yield (mg/g)
	150		4.50	0.24	0.40	0.65
	180		6.20	0.34	0.27	0.61
	210	0.51	8.15	1.01	0.30	1.13
Deamocollinite	260		18.53	2.81	0.57	3.38
	290	0.87	45.00	7.40	9.10	16.50
	330		41.55	5.21	5.17	10.38
	Total		124.93	16.4	15.31	33.83
	150		3.5	0.34	0.71	1.05
	180		8.25	1.06	0.28	1.34
	210	0.51	9.80	1.37	0.21	1.58
Cutinite	260		10.95	1.58	0.85	2.45
	290	0.87	22.20	6.81	1.83	8.64
	330		38.85	20.03	5.63	25.66
	348		134.20	102.15	25.64	127.79
	360		191.20	31.28	0.81	32.09
	Total		423.95	164.60	35.96	200.60

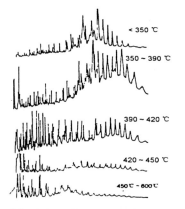

Figure 5. Pyrolysis gas chromatograms of desmocollinite B

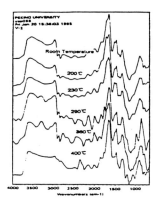

Figure 6. Micro-FT-IR spectra of desmocollinite B

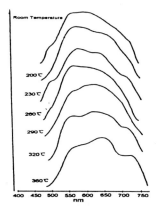

Figure 7. Fluorescence spectra of desmocollinite B

Oil-generating phase	VRₛ (%)	Alginite	Cutinite	Sporinite	Resinite	Suberinite	Bituminite	Desmocollinite	
								A	B
immature	0.0				BL				BL
	0.5	BL	BL			BL	BL		
mature				BL					
	1.0			DL		DL			
		DL	DL		DL		DL		DL
overmature	1.5								

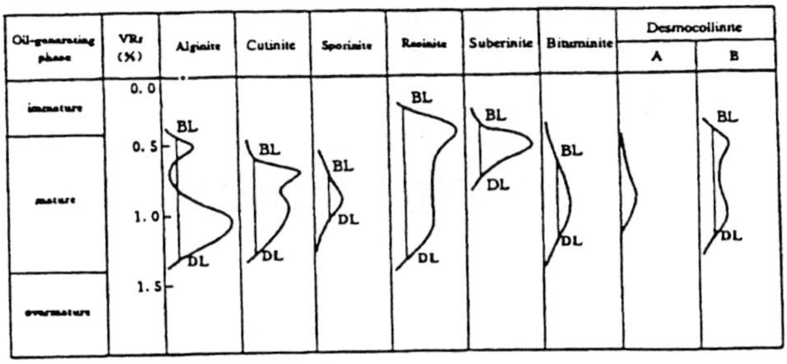

Figure 8. The oil-generating models in Jurassic coal measures

By means of these studies, the results not only confirm that the above mentioned macerals can generate oil, but also define general aspects of hydrocarbon generation. Therefore, we set up the oil-generating models for individual macerals (Figure 8), from which, it clearly manifests the birth/dead line and oil-generation window for whichever maceral.

5. Oil-Expulsion Experiments

In order to take further experiments to perfect "oil can be expelled from coal" [4], we arrange some studies as fellows:

First of all, the coal macropore volume of extracted and unextracted vitrain samples were compared based on the mercury pressure porosimetry and the SEM/TEM. The results display the total pore volume increases after extracting (Table 3) and the pore connection may be in series/parallel pattern except isolate pores. Figure 9 shows the SEM image of macrinites with intergranular/intragranular macro-pores. These indicate that oil can be expelled.

TABLE 3. Comparison of pore volumes between unextracted and extracted vitrain samples

Sample		Pore volume				Total pore
		15-2 (Å)	2-0.2 (Å)	0.2-0.02 (Å)	0.02-0.000005 (Å)	volume (Å)
1	unext.	0.0186	0.0110	0.0157	0.0453	0.0895
	ext.	0.0271	0.0148	0.0179	0.0410	0.0999
	ext.-unext.	+0.0085	+0.0038	+0.0022	-0.0043	+0.0104
2	unext.	0.0076	0.0045	0.0083	0.0286	0.0490
	ext.	0.0286	0.0096	0.0097	0.0329	0.0808
	ext.-unext.	+0.0210	+0.0051	+0.0014	+0.0043	+0.0318
3	unext.	0.0179	0.0050	0.0144	0.0514	0.0857
	ext.	0.0416	0.0144	0.0109	0.0383	0.1022
	ext.-unext.	+0.0237	+0.0046	-0.0005	-0.0131	+0.0165

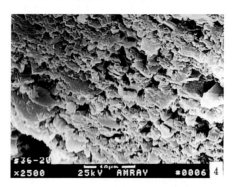

Figure 9. Macrinites with intergranular/intragranular macro-pores, 2000×, SEM

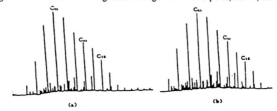

Figure 10. GC spectrum of Press-in oil (a) compared with that of the expelled (b)

Moreover, an oil-expulsion experiment, which the authors designed, is as follows:

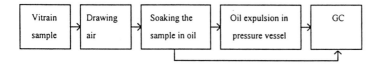

The oil expulsion is also using pressure vessel under 72 hrs., below 210 °C and 18 atm. After thorough comparison of GC spectra (Figure 10) of both soaking and expulsive oil, we find out the latter is basically the former.

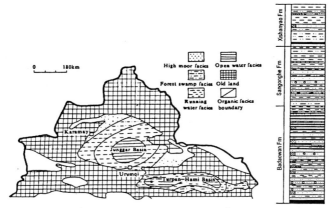

Figure 11. The sketch map of sedimentary organic facies of the Xishanyao period in the Junggar and Turpan-Hami basins, Xinjiang Autonomous Region, N.W. China

TABLE 4. The classification of the sedimentary organic facies

	Type	High moor facies	Forest swamp/ marsh facies	Running water marsh facies	Open water facies
Facies marks	Number	1	2	3	4
Organic petrologic character	V+I %	>90	70-90	40-70	<40
	E %	0-10	10-30	30-60	>60
	Main maceral	Fusinite	Telocollinite	Desmocollinite	Alginite, Cutinite
	Microlithotype	Fusite	Vitrite, Clarite	Clarite	Durite
Sedimentary character	V/I	<1	>1	>1	>1
	GI	1~2	2~50	0~50	2~10
	TPI	0~2	2~6	0~2	2~10
Organic geochemical character	H/C	<0.95	0.95-1.15	1.15-1.4	>1.4
	HI (mg/g COT)	<125	125-250	250-400	>400
	S_1+S_2 (mg/g)	<50	50-200	200-300	>300
	Organic matter type	III	II B	II A	I
Jones' organic facies		D, CD	C, BC	BC, B	B, AB, A

6. Sedimentary Organic Facies

Organic facies is first proposed by Rogers [5], and Jones [6] defined it as a mappable subdivision of a designed stratigraphic unite. Based upon this idea and especially that of Huc's [7], a key organic facies to improve quantitative petroleum evaluation, we emphasize sedimentary parameters including maceral and palynofacies analyses in addition to lithofacies study for the sake of putting the revised organic facies into paleographic map to predict and evaluate quantitatively the coal measures-related oil resources. Therefore, we call the revised organic facies as sedimentary organic facies and use the sedimentary, organic petrologic, organic geochemical parameters to determine the sedimentary organic facies. Four sedimentary organic facies of coal and carbonaceous mudstone were suggested: namely high moor, forest swamp/marsh, running water swamp/marsh and open water facies (Table 4 and Figure 11). The term of running water facies originates from C. H. Haymoba [8], and the others from M. Teichmüller [9].

In 1993, symposium on the geochemistry and petrography of kerogen/macerals [10] greatly inspired us, then, we consider that the maceral contains dual natures of both petrologic (including sedimetary) and geochemical characters, we may quickly set up sedimentary organic facies.

In addition to sedimentary and organic petrological parameters of facies samples, the organic geochemical parameters may also be determined from maceral analyses and statistics. For example, it is true that the S_1+S_2 values can be obtained from actual measurements of facies samples (Table 4), but they are given by maceral analyses and statistics from those samples based on know data (Table 1). That is to say, the sum of S_1+S_2 converted based on hydrocarbon potential of oil-generating macerals is closer to the actual measurements of Rock-Eval. Similar conclusion in dividing sedimentary organic facies has been drawn using above two methods. The running water swamp/marsh facies zone may be the best for coal measures related oil (Figure 11), because an allochthonous/hypautochthonous process as well as disintegration, which are developed within the running water swamp/marsh, which causes accumulation of macerals rich in hydrogen content, such as fluorescent desmocollinite, cutinite, bituminite and suberinite.

7. Conclusions

1. It is the results of simulation of hydrocarbon generation for individual macerals, and of studies on hydrocarbon generating phenomenon of coal/carbonaceous mudstone in the locality that they not only confirm fluorescent desmocollinite, cutinite, bituminite, suberinite may play an important role for oil generation in low rank coal or low maturity level carbonaceous mudstone, but also can be established some macerals' oil generating models.

2. The indicators from CLSM/TEM such as vitrodetrinite/submicro-macerals together with laser-induced fluorescence parameters may provide a new organic petrological technique to make oil-source correlation.

3. The mercury pressure porosimetry study shows the total pore volume increases after extracting. Moreover, oil-expulsion experiment illustrates that there is no reason to treat coal differently than other types of source rocks.

4. Sedimentary organic facies will relate directly with coal measures related oil and the running water swamp/marsh facies may be the best for it. There are sedimentary, organic petrological and organic geochemical parameters to subdivide them.

References

1. Geological Team 1, Xinjiang Bureau of Geology and Mineral Resouces (1992) *Origin and Development of Turpan-Hami Coal Basin and Accumulation for Coal Seams*, Science-Medicine Press, Urumqi, pp. 1-4. (in Chinese)

2. Yao Suping, Zhang Jingrong and Jin Kuili (1997) Organism relics or kerogens in oils as well as oil-source rock correlation indicator, *Since in China* (Series D), Vol.40 No.3., 253-258.

3. Jin Kuili, Fang Jiahu, Guo Yingting, Zhao Changyi and Qiu Nansheng (1995) The use of laser-induced fluorescence indicators in determining thermal maturation of vitrinite-free source rock and oil-source rock correlation (abs.), *Twelfth meeting of the Soc. For Org. Petrol.* 12, 12-22.

4. Durand B. and Paratte M. (1983) Oil potential of coal: a geochemical approach, in: *Petroleum Geochemistry and Exploration of Europe*, Geological Society Special Publication 12, Ed. Brooks J. Black-well Scientific Publication, Oxford, pp. 255-265.

5. Rogers M. A. (1980) Application of organic facies concepts to hydrocarbon source rock evaluation, *Proc. 10th World Petr. Cong.* 2, 23-30.

6. Jones R. W. (1987) Organic facies, in: *Advance in Petroleum Geochemistry*. Eds. Brooks J. and Welte D., Springer-Verlag, Berlin, 2, 1-90.

7. Huc A. Y. (1990) Understanding organic facies: a key to improved quantitative petroleum evaluation of sedimentary basins. In: *Organic Facies*, Ed. Huc A.Y. AAPG, Tulsa, Oklahoma 1-11.

8. Haymoba C. H. (1940) The genesis classification for coal of suburbs-Basin of Moscow, *USSR Mineral Resources Institute Bull.* 159 (in Russian)

9. Teichmüller M. (1982) Origin of macerals, in: *Stach's Textbook of Coal Petrology*, Eds. Stach E., Mackowsky M.-TH, Teichmüller M. et al., Gebrüder Borntraeger, Berlin, Stuttgart, pp. 285-290.

10. TSOP. (1994) Symposium on the geochemistry and petrography of kerogen/macerals, *Energy and Fuels*, V.8. No.6

Examples of the methane exchange between litho- and atmosphere: the coal bearing Ruhr basin, Germany .

T. THIELEMANN (1, 2) & R. LITTKE (2)

(1) *Research Centre Juelich, ICG-4, 52425 Juelich, Germany*

(2) *Aachen University of Technology, 52056 Aachen, Germany*

Abstract

In coal mining areas, the emission of coal bed methane into the atmosphere via mine shafts is a well known phenomenon. Only a few data exist about the coal bed methane release through the lithosphere/atmosphere-interface in a coal-bearing basin. Here four testfields within the german Ruhr basin are presented (Fig. 1) which differ geologically and in mining intensity. Methane fluxes across the surface of these testfields were measured and a possible correlation between mining intensity and coal bed methane emissions at the surface is discussed, as well as the consumption potential of methane oxidizing bacteria in soils.

1. Introduction

Since the beginning of the 1980s an increase of the atmospheric methane concentration has been documented. Because of methane absorbing infrared light of a wavelength different from CO_2 one single CH_4-molecule has a 21 times higher greenhouse potential than one additional CO_2-molecule in the atmosphere. Since about 1700 the atmospheric concentration of methane increased from 0.7 up to 1.7 ppmV, with a recent annual rise of 0.9 to 1.1 %. Whereas the rise in CO_2 is mainly related to fossil fuel burning, a variety of sources contribute to atmospheric methane of which the most important ones are swamps, irrigated rice fields, ruminant's digestion, the use of oil and gas as energy sources and coal mining. The latter emits 25 to 64 Tg/a into the atmosphere, or about 5 to 12 % of all emitted methane (CICERONE & OREMLAND 1988).
Emission of methane is partly balanced by its oxidation. Atmospheric methane undergoes a worldwide decay, especially by reaction with atmospheric OH-radicals and by bacterial consumption in soils. On the continents the concentration patterns of methane vary considerably, mainly influenced by local geology, climate and human activities. The quantity of methane data produced so far is poor considering methane variability and in view of its importance as a major contributor to the anthropogenically induced greenhouse effect.

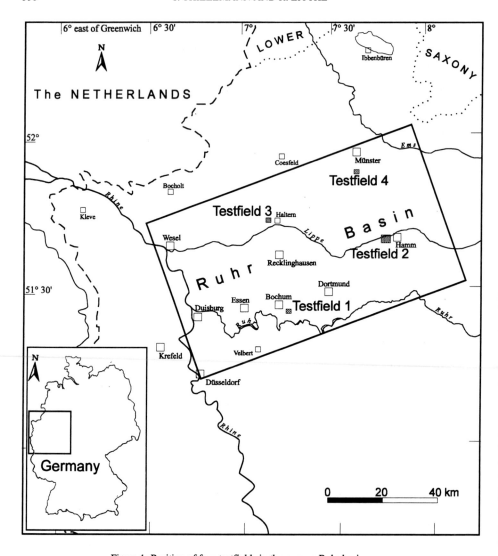

Figure 1: Position of four testfields in the german Ruhr basin

To be able to predict future methane concentrations detailed regional studies are necessary. The aim of this project is to deliver these data for the Ruhr basin as one example of coal bearing basins, with widespread and long-lasting underground mining activities.

2. Geology

The Ruhr basin is an external fold and thrust belt situated north of the Variscan orogenic belt in Western Europe. The coals are of Carboniferous (Westphalian) age and reached a rank of low to high volatile bituminous coals during the late Carboniferous or earliest Permian. Cretaceous and Quaternary sediments unconformably cover the truncated Westphalian coal measures. The gas content of the coals generally rises from west to east in the basin, ranging from less than 3 m^3 up to 12 m^3/t coal. Locally coal bed methane (CBM) emissions at the surface may lead to hazardeous gas concentrations in some buildings, especially in the eastern half of the densely populated area.

3. Methods

In this project representative testfields were chosen and methane exchange rates were measured with flux chambers. These chambers isolate the boundary layer between soil and atmosphere so that a change in methane concentration with time can be recorded within the chamber. Also soil air samples were taken at different depths to obtain profiles of methane concentrations. This technique offers clues to the sources of methane in the subsurface and to the activity of methanotrophic bacteria in the soil.

4. Results and discussion

Testfields 1 and 2 are situated in a zone of completed mining, whereas number 3 was chosen in a zone of prospected mining and coals in number 4 are under a very thick Mesozoic cover far north of the coal mining activities (Fig. 1). In testfield 1 the coal measures are covered by 15 m of Quaternary fluvial deposits and have been mined on one level at only 40 m depth, some decades ago. Within an area of 0.2 ha a slow methane consumption of -0.1 to -2.4 mg/(m^2*d) was measured, giving this field a methane consumption potential of around -2 g/d. No surface methane emissions are recorded. The coal measures in testfield 2 are 600 m below surface, unconformably overlain by Cretaceous marine sediments and by 5 m of Quaternary fluvial deposits. Low volatile bituminous coal was mined from 1965 to 1988, on up to four levels. On a meadow of 1.8 ha methane flux rates are measured on 41 locations every fortnight. At this site a very heterogenous and unique degassing pattern was observed. In and next to a pond CBM emissions were measured, within an area of 300 m^2. Gas is constantly emitted by ebullition as visible in a pond, where gas bubbles rise, preventing the water from freezing during winter times. The gas has a composition of 75 to 94 % methane, 0.7 to 3 % ethane and around 0.02 % propane and shows methane flux rates between 170 mg and 400 g/(m^2*d). Around this main emission zone a 0 to 5 m broad belt of strong methane consumption shows flux rates of -2 to -4 mg/(m^2*d), values of only -0.4 to -2 mg/(m^2*d) were measured further away. The high methane consumption around the main gas emissions indicates a higher population of methanotrophic bacteria due to a steady supply of methane in high concentrations.

This fact implies that the actual methane fluxes from the bedrocks into the soil are higher than into the atmosphere due to the bacterial consumption in the soils. Also the soil has a self cleaning potential towards methane emissions. 30% of the coal lying up to 900 m deep below the degassing center have been mined, whereas 80 m away only 12% and another 30 m away none of the coal has been mined at all. Thus there could be a correlation between mining intensity and gas emission. The difference between methane emission and bacterial consumption gives an overall balance of 5 to 10 kg/d for the 1.8 ha in testfield 2. As all the meadows in this area show a methane consumption potential of around -2 mg/(m^2*d), 250 to 500 ha are necessary to compensate these methane emissions.

In testfield 3 where the Carboniferous occurs below a Cretaceous cover of 960 m and in testfield 4 where the Carboniferous is at a depth of 1300 to 1350 m below surface no coal bed methane emissions can be observed. On the meadows, fields and forests in these areas methane consumptions of -0.3 to -2.7 mg/(m^2*d) were recorded.

Finally, the measurements in testfield 1 indicate that some decades after the last near-surface mining activities no coal bed methane emissions occur, whereas CBM can be released in high concentrations over decades in mining areas where coal gas contents have been high (7 to 10 m^3/t coal), where a caprock sealed the gas before being disrupted by mining and where a high percentage of the coal still is left. The strongest gas emissions occur at places of most intensive mining, as long as much of the total coal has been left in the district. Most of the gas at the surface is emitted alongside a NNW-SSE-trending line, indicating that degassing predominantly occurs at some cleavage planes in the Cretaceous marl below the Quaternary cover. As coal mining moves forward into testfield 3 in two years time, there might be CBM observed at the surface in future, especially when mining subsidence opens migration paths within the Cretaceous. However, with a recent gas content of only up to 3.3 m^3/t coal in the uppermost Carboniferous, the gas yield is lower than it was in testfield 2. Also the Carboniferous in testfield 3 is less intensly folded and faulted than in testfield 2. This all makes surface emissions in testfield 3 unlikely.

A dependence of methane fluxes from local climate conditions such as temperature, atmospheric pressure or precipitation was not observed. In the Ruhr basin recent local CBM emissions at the surface might be induced by underground mining. They are concentrated at some sites in the eastern part of the basin and are of minor importance for the atmospheric methane concentration. Most of the emitted CBM can be expected to be consumed because of a high bacterial methane consumption potential of dryland soils in the Ruhr basin.

References

Cicerone, R.J. and Oremland, R.S. (1988) Biogeochemical aspects of atmospheric methane, *Global Biogeochemical Cycles* **2**(4), 299-327.

DESORPTION AS A CRITERION FOR THE ESTIMATION OF METHANE CONTENT IN A COAL SEAM

J.MEDEK and Z.WEISHAUPTOVÁ

Institute of Rock Structure and Mechanics
Academy of Sciences of the Czech Republic
182 09 Prague 8, Czech Republic

1. Introduction

For the purpose of methane extraction the pressure and volume parameters of methane contained in a coal seam are usually evaluated using its high-pressure isotherm determined under laboratory conditions. Volumes interpolated on the adsorption branch of methane isotherm measured on the evacuated exploited coal is in many cases not identical with the real gas content in the seam at the same pressures. Practical arguments for this statement are results of the canister test, where the volume of the released gas is frequently more or less lower than the volume read from the adsorption isotherm. Assuming that the canister test at lossless sampling includes the total real quantity of gas contained in the coal seam, the cause of such a difference is to be looked for in the interpretation of the sorption isotherm. Although several sources quote that - during the sorption of methane on coal - the desorption branch coincides with the adsorption one [1-3], while the reversibility occurs only exceptionally [4], it cannot be precluded, that the actual methane content in coal lies rather on the desorption branch of the sorption isotherm.

The aim of this study was therefore to investigate the possibility of determination of the adsorption and desorption branches of the high-pressure methane isotherm on coal, to compare their shape, and to find an eventual application to the evaluation of potential methane content in the coal seam.

2. Experimental

The mentioned research study was carried out on samples of medium volatile bituminous coal from three representative gas-bearing seams in the Ostrava district. Coal samples were drawn from the drill cores after their degassing in canister. Individual samples differed by their sampling depth and were characterised by the content of water, ash, volatile matter and the mean light reflectance, as specified in Table 1.

The complete sorption isotherms within the pressure interval up to 11 MPa were measured on a volume apparatus [5], the pressure values were indicated by means of a digital converter with precision of ± 0.001 MPa. Isotherms were determined on dry samples with the weight of about 25 g, evacuated to 3.10^{-3} Pa. The establishment of the equilibrium was recorded when the pressure alterations remained for the time of minimum 20 min within the interval of 0.002 MPa. During the adsorption, the average time was 2 to 3 hours. During the desorption, which proceeded in all cases at a higher rate, the time was somewhat shorter. For the last point of the adsorption, the time was extended to 12 to 16 hours, in order to guarantee the equilibrium state for the subsequent desorption. All measured quantities were corrected on behaviour of real gas. The adsorption branch was constructed from standard 10 points, the desorption one from adequately lower number of points.

TABLE 1. Characteristics of coal samples

Depth of sampling	Sample	W	A^d	V^{daf}	R_o
[m]		[wt%]	[wt%]	[wt%]	-
664	a	0.7	8.9	29.1	1.10
1062	b	1.4	2.3	33.6	0.81
1240	c	0.6	6.4	30.0	1.00

W - moisture, A^d - ash in dry basis, V^{daf} - volatile matter in dry- ash- free basis, R_o - mean light reflectance

3. Results and Discussion

Not in a single case of the measured samples a reversible course could be achieved, as it is shown on 3 samples in Figures 1a,b,c. In the area of the highest pressures, the desorption branch forms first a short hysteresis loop, which can be explained by the strongly bound portion of methane dissolved in the coal matrix [6]. At the upper half of the pressure range, the desorption branch intersects the adsorption branch and then it runs below it till the point of the origin.

As illustrated in Figures 2a,b,c, both isotherms branches could very well be rectified in Langmuir coordinates with a high correlation coefficient, the desorption branch proving a more satisfactory agreement than the adsorption one. Table 2 depicts characteristic parameters of Langmuir isotherm equation.

As recently documented by the authors of this study [5], methane is not present in coal in just a single form of sorbed gas, but above critical temperature, there exist in principle four forms which differ by the bond intensity of the gas to the solid phase. The strongest bond occurs in micropores, it decreases on the surface of meso- and macropores, and the least strong is the bond of methane dissolved in the present pore water. Not interaction exists between the free moving gas in the pore voids of coal. Individual forms can be determined separately within different pressure intervals;

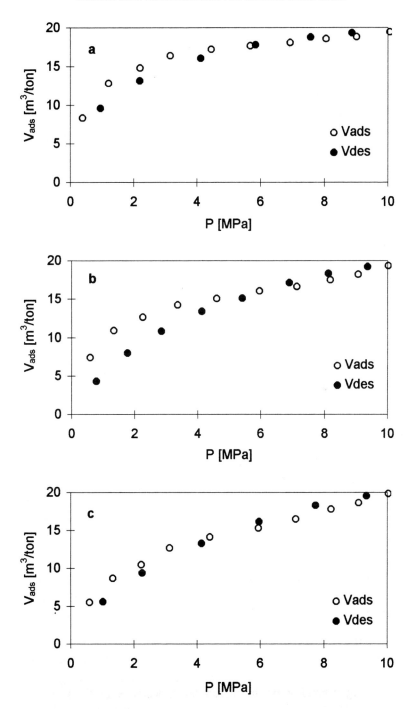

Figure 1. Adsorption and desorption branches of methane isotherm on samples a, b, c

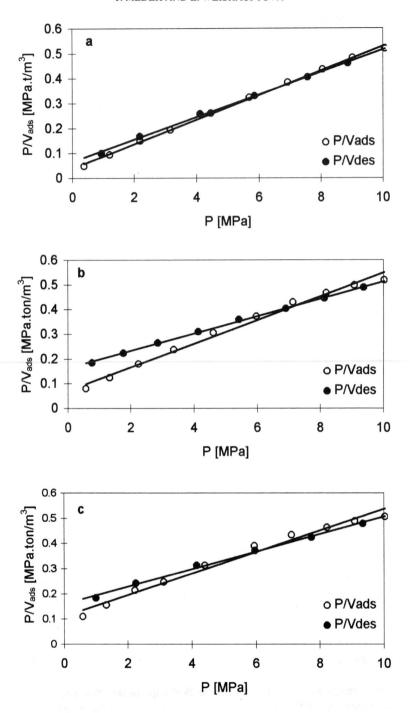

Figure 2. High pressure methane isotherms in Langmuir coordinates, samples a, b, c

the high-pressure isotherm includes all these bond forms [5], the most important being the sorption in micropores by the mechanism of volume filling according to Dubinin's theory [7].

A significant part of methane present in coal at higher pressures is represented by the free moving gas, whose molecules are not bonded by adsorption forces as considered by Langmuir equation. Because this equation has been derived according to principles of adsorption in a monolayer, only the analytical function of this equation has been used in this case, by which both the adsorption and desorption branches, which does not at all agree with the postulates of monolayer adsorption, can be formally converted into the linear form. In so doing, the constants in the Langmuir equation lose here their primary physical significance and become only adjustable parameters with formal possibility of assuming the dimensions of these constants.

TABLE 2. Parameters of high-pressure sorption according to Langmuir equation

Parameter	Sample	Adsorption	Desorption
Capacity in monolayer	a	20.4	22.3
V_m [m³(STP)/ton]	b	20.9	28.5
	c	23.5	28.8
Surface area	a	93	102
S [m²/g] *)	b	96	130
	c	107	131
Sorption coefficient	a	1.32	0.70
k [1/MPa]	b	0.69	0.22
	c	0.39	0.22
Correlation coefficient	a	0.999	0.999
	b	0.995	0.999
	c	0.991	0.997

*) contact area of methane molecule in monolayer $\sigma = 0.17$ nm²

Obviously, the unequal course of both sorption branches is affected by the specific mechanism of sorption in connection with individual bonding forms. The difference between the rates of adsorption and desorption can be explained by the different behaviour of the free moving gas during the two sorption activities. In the first case, the macroscopic equilibrium is maintained by the gradual pressure increasing, so that the free moving gas content continuously increases. Unlike that, during the desorption, the equilibrium connected with the reduction of external pressure is permanently disturbed by the quick escape of the free gas, which equalizes its pressure with that of the neighbourhood, this loss being accompanied with release of the adsorbed gas. It could be confirmed, by the course of degassing curves, carried out till the equilibrium state during the canister test of the drill cores of studied coals that the sequence of methane releasing is governed by its bonding forms, whose strength has been derived from the desorption carried out under laboratory conditions.

4. Conclusion

The application of the desorption branch of the sorption isotherm for the estimation of the quantity of methane in coal can be justified by the relation between the momentary state of gas in coal and the character of isotherm. The adsorption branch measured in laboratory on extracted and for the experiment urgently evacuated coal represents an ideal state of maximum saturation of coal with methane, which includes all bond forms, as mentioned above [5], without differentiation. The gas quantity in coal read from the adsorption branch can be interpreted as its gas-bearing capacity, i.e. the maximum potential gas volume contained in coal, which under the given temperature and pressure conditions without action of external effects would exist in an equilibrium state. On the contrary, the desorption branch corresponds to a dynamic state caused by a long term degasification of the seam by disturbing its primary adsorption equilibrium roughly corresponding at the same pressure with the equilibrium on experimental adsorption branch. The gas quantity thus read may be designated as gas-yielding capacity, i.e. the momentary maximum volume of gas, which would be released from coal by an artificial degasification. It was suggested [8] that the primary methane capacity of 100 to 200 m^3/ton formed by bituminous coals during the metamorphism process was reduced, step to step, to the actual 50 - 60 m^3/ton by the diffusion degassing into the seam neighbourhood, as it may be supported from the presence of methane in porous layers of overlying rocks.

Acknowledgements

The authors are very grateful to J.Němec DSc, General Director of EUROGAS Ltd, for his encouraging interest in this work and many helpful suggestions.

5. References

1. van der Sommen, J., Zwietering, P., Eillebrecht, B.J.M. and van Krevelen, D.W. (1955) Chemical Structure and Properties of Coal XII-Sorption Capacity for Methane, *Fuel*, **35**, 444-448
2. Gunther, J. (1965) Etude de la Liaison Gaz-Charbon, *Revue de l'Industrie Minerale*, **47**, 693-708
3. Mavor, M.J., Owen, L.B. and Pratt, T.J. (1990) Measurement and Evaluation of Coal Sorption Isotherm Data, SPE 20728, *SPE 65th Annual Technical Conf. and Exhib.* New Orleans, La
4. Bell, G.J. and Rakop, K.C. (1990) Hysteresis of Methane/Coal Sorption Isotherms, SPE 15454, *SPE 61st Annual Technical Conf. and Exhib.*, New Orleans, La
5. Weishauptová, Z. and Medek, J. (1998) Bound forms of methane in the porous system of coal, *Fuel*, **77**, 71-76
6. Ceglarska-Stefanska, G. (1995) Sorption investigation of coal on methane, in J.A.Pajares and J.M.D.Tascón (eds.), *Coal Science*, Elsevier, Vol.I., pp. 27-30.
7. Dubinin, M.M. (1958) The Porous Structure and Adsorption Properties of Active Carbon, *Industrial Carbon and Graphite*, Soc. Chem. Industry, London, pp. 219-230
8. Skotchinskij,A.A. and Chodot,V.V. (1958) *Metan v ugolnych plastach*, Ugletechizdat, Moscow

GRADING OF RESERVES AND RESOURCES OF COALBED GAS IN CHINA

Sun Wanlu*, Ying Wenmin*, Fan Mingzhu*, Wang Shuhua*
* North China Petroleum Bureau, Zhengzhou, Henan450006

Abstract

The grades of coalbed gas reserves and resources extents are mainly determined at the exploration stage and by a wildcat area. According to the numbers of development wells, the number of assessment parameters of reserves and resources extents, China's coalbed gas could be classified into two kinds of resources extents and four types of geological reserves. Based on incomplete statistics, the total coalbed gas resources buried shallower than 2000m is about 326366×10^8 m^3, within which the resources buried between 1000-2000m is about 215666×10^8 m^3 and the resources shallower than 1000m is about 110700×10^8 m^3.

1. General Aspects

Regional divisions of coalbed gas resources include national large scale division and local second division. Depending on geotectonic framework as well as factors of economic development, coalbed gas resources in China were divided into regions (Figure 1). Based on incomplete statistics(1-4 regions), there are more than 60 coalbed gas bearing basins whose area is larger than 55×10^4 KM^2. By calculations, coalbed gas perspective resources(potential resources) buried shallower than 2000m are about $326,388\times10^8$ m^3, within which the resources buried between 1000-2000m are about 215666×10^8 m^3(take 66%) and the resources shallower than 1000m are about 110700×10^8 m^3(take 34%).

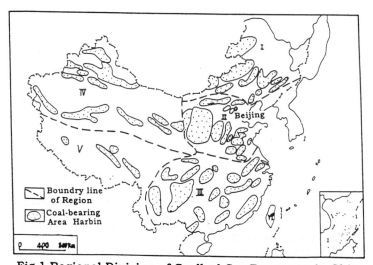

Fig.1 Regional Division of Coalbed Gas Resources in China

According to coal rank, coalbed gas resources distribution is: within low rank coal about $189161 \times 10^8 \, m^3$ (about 57.96%), in medium rank coal about $47127 \times 10^8 \, m^3$(take about 14.44%), and in high rank coal is about $90078 \times 10^8 \, m^3$(take about 27.60%).

Primary coal-generating phase is distinctly different in different region, so the resources vary with change of the regions/geological age as follows: Region 1; coal of Late Jurassic to Early Cretaceous, the resources are about $24784 \times 10^8 \, m^3$ (about 7.6%); Region 2; coal of Carboniferous, Permian and Early-middle Jurassic, the resources are about $201254 \times 10^8 \, m^3$(take about 61.7%); Region 3: Late Permian coal, the resources are about $24784 \times 10^8 \, m^3$ (about 7.6%); Region 4: Early-middle Jurassic coal, the resources are about $50636 \times 10^8 \, m^3$ (about 15.5%). The experience of coalbed gas exploration and development in USA showed that coalbed gas sourced from medium rank coalseams, buried at shallower than 1000m depth has the largest commercial development value.

2. Grading

Gas field development usually experiences two periods of pre-prospecting assessment; exploration assessment, and development assessement. Based on the different exploration periods, these can be further divided into inferred resources, and potential resources; Reserves could be further classified into three grades which are; forecasting reserves, controllable reserves and proven reserves. Proven reserves also could be classified into three types, namely of developed, undeveloped and basic proved reserves.

Coalbed gas and conventional natural gas are significantly different in their reservoir qualities, and therefore the formulation of their reserves' standard will also differ.

(1) Coalbed gas exploration is usually conducted after coal field exploration, so exploration methods such as geophysical and geochemical ones are seldom used, the main exploration method used for coalbed gas exploration in China is drilling . So, depth of the coalbed, the gas content, coal properties, desorption factor and permeability are used as parameters for reserves classification.

(2)After a coalbed well was completed, usually it needs a length time for draining from the coalbeds after being gas - desorbed, with production rate of gas gradually reaching peak time of gas production, so then the assessment to the well can be made. However, the productivity of the field can not be proven by only one successfully draining well. There is need to do draining tests of numerous small wells and obtain their drainage patterns.

(3) Coalbed gas reservoir in which the boundary of gas and water changes gradually is different from a conventional gas reservoir which exists in a gas-bearing trap where the contact of water and gas (in conventional gas field) is distinct. In some cases, the boundary of gas and water is an incomplete geological boundary. That shows that there may be a large area in gas reservoirs which may include many independent gas reservoirs in one coabed gas carrying basin.

Table 1 Gradation of Reserves and Resources of Coalbed Gas in China

Period	Gradation			Exploration Level
Development	Geological reserves	Proved	Developed	All development wells designed by simulation and production tests of small well pattern of the whole block were completed, essential assessment data was obtained, feasibility plan program was made for development production.
			Undeveloped	Essential appraisal which control development block were drilled, production tests in several well pattern were completed, reservoir simulation analysis was conducted, essential assessment data was obtained, feasibility assessment for development was made.
		Basic proved		Several appraisal wells were drilled in development block, production tests in one group of wells and reservoir simulation analysis were completed, essential assessment data was obtained, feasibility assessment for on-going exploration and development was made.
Exploration	Control reserves			Several exploration wells and at least one appraisal well was drilled, production test for single well was conducted, essential assessment data was obtained, control reserves was calculated for control block, feasibility assessment for development was made.
	Forecasting reserves			Exploration and development data of coal-bearing basin was integrated, at least one exploration well was drilled, essential data was obtained for coalbed gas assessment, forecasting reserves of control block was calculated, exploration assessment was made.
Preprospecting	Prospective resources	Potential		Exploration and development data of coal field in coal-bearing basin was integrated, at least one coalbed gas data well was drilled, essential data was obtained for coalbed gas resources assessment, potential resources within working depth was estimated, feasibility assessment for exploration was made.
		Inferred		Exploration and development data of coal fields in one or more than one coal-bearing basin was integrated, coalbed gas resources of coal-bearing rock series was inferred, preprospecting assessment was made.

(4) Currently, the exploration depth of a coalbed gas well usually is shallower than 2000m. So, when the potential resources value is calculated and the grade of reserves is classified, the resources and reserves at different grades, buried at less than 2000m productive depth are generally considered.

Referring to national natural gas (conventional) reserves standard, we here provide a preliminary method of grading of coalbed gas reserves and prospective resources based on the above features of coalbed gas.

Prospective resources are classified into inferred resources and potential resources which correspond with the inferred resources and reserves in conventional natural gas standards. The difference between coalbed gas and conventional natural gas is that some additional necessary well data need to be incorporated, so that potential resources can be calculated during the late phase of pre-exploration period. However, the data used for potential resources calculation are obtained from coalbeds buried shallower than 2000m which is the usual depth of coalbed gas exploration wells. The data used for inferred resources calculation which is not limited by the exploration depth includes all the data obtained from coal-bearing rock series in a coal-bearing basin.

The difference of forecasting reserves and controllable reserves during exploration period with proven reserves during the development period is that during the exploration period just exploration wells and a few assessment wells are drilled, whereas during 'late period' many more assessment wells are drilled, and at the latest phase of the development period, a well pattern of the whole development block should be completed. Simultaneously production tests in more than one small well are needed for obtaining their behavioural pattern.

Development period can not reach an advanced phase from a low phase of production test if small well behavioural patterns are unsuccessful. Therefore a high grade reserves report could not be provided.

The difference of basic proved reserves, undeveloped reserves and developed reserves during development period is that, basic proved reserves can be calculated after some assessment wells were drilled and production test at least in one well pattern was accomplished, and simultaneously reservoir simulation analysis was conducted. Developed reserves can be obtained after development well patterns of the whole development block designed by reservoir simulation were accomplished, and production tests of more than one small well patterns obtained and found to be a major control of he whole block with regional distribution.

With respect to assessment contents and geological understanding degree of different grades of prospective resources and geological reserves, their difference is distinct.

Assessment from prospective resources to reserves is a development process from low grade to high grade. Each exploration or development phase must depend on the assessment of resources and reserves of the proceeding phase. And each phase should have corresponding resources and reserves appraisal so that scientific evaluation could be made.

References

1. Cheng Yuqi, "An Introduction to Geology in China" Beijing, Geological Publishing House, 1994, 17.

2. Boyer C M. Coallbed methane-state of the industry Review of Methane form Coal Seams Technology, 1993,11(1):1~52.

3. Sichuan Petroleum Administrative Bureau, CNPC, "Nature Gas Reserves Standard" (GB 270-88), Beijing: Chinese Standard Publish House,1988, 2-4.

Acknowledgement

The manuscript was reviewed and revised by M.Glikson

ANHYDRIDE THEORY

A NEW THEORY OF PETROLEUM AND COAL GENERATION

C. WARREN HUNT
ANHYDRIDE OIL CORPORATION
1119 Sydenham Road SW, Calgary, Alberta, Canada T2T 0T5
Tel. 403-244-3004, Fax 403-244-2834, E-mail <archeanc@telusplanet.net>

Abstract

The theory presented in this paper is a synthesis of the observeable geology of petroleum occurrences and new information on the ability of microfauna to generate petroleum from methane. Three paradigms of petroleum generation, anhydride theory, conventional diagenesis, and cosmic or inner earth abiogenesis, are compared as to their relevant geology.

The author calls attention to the well-known fact that methane effuses from earth's interior and to varying degrees pervades all crustal terranes, crystalline, volcanic, and sedimentary. He points out that the energy from this methane can be utilized by hyperthermophyllic bacteria and archaea, which obtain it by stripping away its hydrogen. Dehydrogenated methane molecules can be defined as *anhydrides,* and their recombinations as petroleum. Anhydrides of this origin are *biologically-derived through dehydrogenation of methane*, and thus, are products of *biogenesis by living, microbial organisms rather than biogenesis of fossil biomass* (kerogen). Treating coal as the *"terminal anhydride"* classifies coalification also as a process of *biogenesis by living organisms.*

Petroleum in anhydride theory may thus be generated either in association with *source rocks* or in their absence. Coal may result from the coalification of peat by addition of externally-derived carbon, or it may be deposited in veins as asphaltite absent any peat. Oil in igneous host rocks and asphaltite in non-sedimentary terranes attest to the validity of *anhydride theory.*

Anhydride theory is a paradigm shift that portends serious implications for petroleum discovery and recoverability from terranes thought heretofore to be barren. It also implies the possibility for rejuvenation of producing or depleted resources, and thus challenges industry to note reservoir conditions suggesting rejuvenation, which may be occurring in a producing oil field or may be in progress in a previously depleted reserve.

Methane Effusion from Earth's Interior

This phenomenon is widely recognized as a worldwide phenomenon. Hardrock miners and geologists cope with it; petroleum and coal geologists exploit it. Nevertheless finding it in previously unsuspected places still evokes expressions that verge on wonder. For example, in a recent article on drilling of Indian Ocean crust, William C. Evans of the USGS[1] describes *"a 0.5 km core of oceanic crustal layer 3, consisting of gabbroic rocks that underlie mid-ocean ridge basalts (MORB) and extend down several kilometres to the upper mantle. Surprisingly, CH_4 is an abundant volatile species in fluid inclusions throughout the length of the core, and in some cases the only major volatile."*

Methane is also abundant in other terranes. The conventional sources in sedimentary rocks need no elaboration. Methane encapsulated by seawater, as *"hydrates"* in the sediments on and immediately below the ocean floors in many places worldwide is generally recognized to exceed in volume the proven gas reserves in conventional fields. Methane influx into hardrock mines in continental shields is widespread and a problem for the miners. Quebec and Ontario mines as well as the Bushveld and Witwatersrand mines are typical in this respect.

A well drilled by the author at Fort McMurray, Alberta, Canada penetrated one thousand metres of granite beneath the famous bituminous sands of the area. Gas chromatography on the mud throughout the drilling recorded pervasive hydrogen and carbon dioxide. Intermittent methane and petroleum were also recorded as the drill bit pulverized the rock. Quite evidently, the presence of methane is a general condition of all crustal rocks, not just sedimentary rocks.

The Conventional View of Petroleum Generation

Fossil biomass leaves kerogen, an insoluble residue from decayed life forms. Burial then with optimum temperature and pressure conditions known as an *"oil window"* causes *"diagenesis"* of the organic molecules in the kerogen to *"mature"* into petroleum. The concept fails to provide any source for additional hydrogen or energy, which the reduction of the partially oxidized molecules requires. Neither can it be invoked to explain the source of methane or petroleum found in crystalline and volcanic terranes beneath or absent sedimentary cover, where there is no fossil kerogen.

Anhydride Theory

The new theory starts with the proposition that the carbon forms comprising kerogen and peat are partially oxidized, and hence carry a slight positive charge, whereas the carbon in methane from earth's highly reduced interior environment carries a negative charge. The opposite charges attract each other, causing methane to be drawn into the peat or kerogen. Its presence there gives the misleading impression that it originated there.

Microorganisms and methane

Microorganisms, which pervade the shallow crust and all fossil biomass, are demonstrably able to strip hydrogen progressively from the abundant methane that effuses from earth's interior and thus to provide the partial molecules that can recombine to form longer-chain hydrocarbons.[2] The end of the process is terminal carbon, either in the form of coal or asphaltite.

Microbiological transformation of methane to petroleum as an alternative to biogenesis has not been recognized by geologists as feasible despite industrial production of petroleum from coal or methane by the Fischer Tropsch and Sasol processes, which use metal catalysts (instead of microbiota) to bring about dissociation of the methane. Industrially, temperatures of $>600^{\circ}C$ are involved.

Microbiota require more energy than they can get from either kerogen or peat. Hydrogen provides that energy to them when they strip it from methane. The stripping produces fractional molecules that by their nature are relatively unstable and should combine with each other to give larger alkane and other hydrocarbon molecules. Carried to conclusion, dehydrogenation yields pure carbon. Microbiological research in the last few years[2] has led to the recognition that hyperthermophillic bacteria and archaea actually do this. They are able to *"dehydrogenate alkane hydrocarbons up to [carbon number] C20 anaerobically"* and to produce *"molecular hydrogen ... in situ ... microbiologically"* at temperatures less than $110^{\circ}C$. This relatively low temperature is the tolerance level of hyperthermophillic bacteria and archaea, and it occurs at earth depths near 6,500m (21,000') in most regions. One must wonder whether it could be coincidental that the depth of the deepest known productive oil reservoirs coincides with the temperature toleration of hyperthermophyllic bacteria and archaea. Such a coincidence suggests, rather, that hyperthermophillic bacteria and archaea are the agents that generate the petroleum.

Methane from earth's interior infused into a terrane rich in kerogen or peat is a setup for synergistic conversion of fossil biomass and immigrant methane. Conversion renders their carbon into petroleum and coal. Neither kerogen nor peat is necessary, however, for the production of petroleum and coal. Hyperthermophillic bacteria and archaea can do it all. This fact is abundantly illustrated by the occurrence of petroleum and veins of asphaltite[3] in igneous and otherwise organically barren rocks worldwide.

Because orthodoxy has not embraced the idea of generation of higher carbon numbered hydrocarbons from methane in nature or the concept of carbon addition to peat in coalification, let us review the evolution of thinking that has led to the present stasis.

The History of the *"Fossil Fuel"* Theory

In antiquity the Greeks named the oil they found in rocks petroleum, "rock oil," to distinguish it from oil obtained by compressing olives or rendering animal products. To them rock oil was self-evidently *"abiogenic,"* and coal was *"the stone that burns,"* a simple observable fact - or so it seemed.

19th century theorists, understanding fossils and knowing one could not compress or render rock to obtain oil, found it easy to assume that oil originated from the only organic source they could see, fossil biomass. Thus originated the idea of *"biogenesis."* There was no evident alternative hypothesis for the origin of petroleum, and the idea of *"fossil fuel"* was born by default, another seemingly simple, observed fact.

In the 20th century kerogen was recognized as the fossilized, insoluble organic residue found in sedimentary rocks, and porphyrins were observed in petroleum and coal as complex, organic metal-bearing compounds having clear biological origins. These discoveries solidified the support for the *"fossil fuel"* concept. In the case of coal, the presence of plant macerals had much earlier led easily into the idea that coal is merely compressed plant material, another *"fossil fuel."* The kerogen, porphyrin, and maceral observations gave the *"fossil fuel"* interpretation the appearance of truth despite alternative possible interpretations and inconvenient conflicting data.

Thomas Gold in his 1987 book, POWER FROM THE EARTH (out-of-print), pointed out the now well known fact that hydrocarbons are present in meteorites and abundant throughout the solar system, apparently in habitats where life as we see it on earth is unable to exist. He reasoned that petroleum must be producible without the help of living organisms, i.e. *"abiogenically,"* another seemingly simple, observable fact.

Since 1987, mounting evidence from the space and ocean exploration programs and deep continental drilling are finding life in many habitats that were previously considered hostile. Thus, kerogen recently reported in a martian meteorite is interpreted by the authors as support for abiogenesis.[4,5] The choices, life in space or inorganic kerogen, are both ineluctable contradictions of *"fossil fuel"* theory.

The initial dismissal of Gold's thesis (often derisive) by most petroleum geoscientists that followed is more muted today. The dismissal derived first from the fact that petroleum has almost exclusively been found in sedimentary rocks, and secondly, that the ubiquitous *"porphyrins"* and the kerogen residues are seen to constitute irrefutable evidence for biological origins. As fossils have been the only biologically derived material recognized to be associated with petroleum, the *fossil fuel* theory prevailed at first; Gold lost. Today, the tables are turning.

This author contributed to the turning in his 1992 book, EXPANDING GEOSPHERES[6] with support for Gold's abiogenesis concept on the basis that petroleum occurs in igneous rocks. In that book he also posed the concept that methane is necessary for the creation of coal from peat. He did not link coal with petroleum generation.

Anhydride theory, born in 1996, reverses that support on the basis that *living microbial forms*, as opposed to *fossilized forms*, better explain observable facts of the creation, topology, and trace element composition of petroleum and coal.   Creation of hydrocarbons and coal may well have been assisted by the presence of kerogen and coal macerals, but dependence for the creation on these components is not shown by available evidence.   The work of Karl Stetter and associates demonstrating the role of *living* microorganisms to produce alkane hydrocarbons and hydrogen from methane by hyperthermophyllic bacteria and archaea is a milestone in organic science.   It is a small step from their demonstration to anticipate that such microorganisms should occur in different assemblages in all crustal rocks with appropriate temperatures for their survival.

Anhydride theory explains porphyrins as the remains of *living,* as opposed to fossilized, biomass. In this context the porphyrins in petroleum represent organisms that lived beneath the earth's surface rather than on it.

Resources of petroleum in igneous terranes are becoming well known and imply a much larger domain for biogenesis than has previously been recognized. It is worth observing in this context that most oil and gas has been found in sedimentary terranes up to now because first of all, these are the terranes where drilling has occurred, and second, that sedimentary rocks provide entrapment structures far more readily than igneous rocks. Oil fields illustrative of resources in crystalline rocks are typified by the White Tiger field offshore Vietnam. Discovered in the 1970s, this field produces 120,000 barrels per day from 20 wells that penetrate the granite surface at about 1,000 metres and produce from multiple fractured zones down to 5,000 metres. The original field pressure has been maintained sufficiently that a resupply process appears implicit.   The feature of rejuvenation of depleting reserves is a feature not seen only in the White Tiger field. Some oil fields that produce from sedimentary rocks with no apparent proximal *"source rocks"* are the Saudi Arabian fields.   Producing from fractured carbonates, they are experiencing little drawdown from virgin pressures while indicating huge net increases in reserves despite decades of high capacity withdrawals[7].

The Topology of Petroleum

Topology of petroleum resources represents this author's original argument against fossil biomass as the sole source of petroleum. The argument has two divisions: (1) depth ranges of petroleum occurrences, and (2) quantitative distribution with respect to reservoir rocks.

The depth ranges aspect is highlighted by the fact that kerogen transforming progressively into bitumens and thence to higher gravity oils at increasing depths and pressures should have resulted in prominent resources of black oils and bitumens *at generation depths*. This is not what exploration has found. Worldwide, black oils and bitumens are mostly near or at the surface, well above the supposed generation depths; few resources of black oils and bitumens occur today *at generation depths*.

The quantitative distribution aspect is highlighted by the example of western Canada basin reserves. Where one or two hundred billion barrels of medium and high gravity oil have been discovered in this basin, a belt some three hundred miles (500 km) in length and thirty miles (50 km) width on the northeast edge of the basin contains perhaps four trillion barrels of bituminous oil. These black oils are all in shallow, Cretaceous reservoirs; the light and medium oils are variously in Devonian to Cretaceous reservoirs. Whereas the Cretaceous host rocks are about the same thickness and presumably more or less the same kerogen content over the whole area, some mechanism has concentrated 95% of all the oil in the basin into 10% of the land area and perhaps 5% of the stratigraphic column. Whereas migration can be invoked for some relocation of petroleum, it is not sufficient to explain the great quantity anomaly. Relative flatness of the strata at the time of migration in the late Cretaceous and high viscosity of the black oils mitigate against enormous relocation. Lateral migration in nearly-flat, discontinuously-porous strata is unreasonable as an explanation for the observed distribution of petroleum in the western Canada basin..

The topology of petroleum occurrences also can be raised in opposition to the Gold view of abiogenesis. That scenario predicts that petroleum created from inner earth methane by pressure and temperature at mantle depths migrates upward from deeper levels. If true, the ratio of gas to oil should increase with the assent to the surface, because gas is far more mobile than oil. And there should be no upward decrease of the ratio once it reaches a given level, because gas should always outpace migrating oil in the system.

Thus, Gold's theory should give the result that gas would predominate at shallower depths, while oil should lag at deeper levels. Needless to say, observed conditions are precisely opposite. Petroleum geologists know that there is no significant oil found below about six km, that gas is more abundant than oil at four to five km depths, and black oils occur at shallow levels or at the surface.

Coal Seam Methane

Methane in coalbeds has been regarded as largely an evolutionary product from coalified peat, and thus biogenic. The possibility that much methane could result by infusion into the peat and its antecedent coal from external sources has not been given much credence despite some compelling arguments. Let us consider the evidence for infusion.

1. The ash from coal, which is predominantly silica, alumina, oxides of calcium and magnesium, and lesser amounts of water-soluble metal oxides, when compared with

the carbon in the coal is often much less than in the supposed precedent vegetation. Where plant ash is normally 8-15% of dry plant weight in the writer's experience, individual coal beds often have only 3% or even as little as 1% ash. Implicitly, the overabundance of carbon in coal compared with the precedent plants suggests that carbon has been concentrated or added from an external source. The addition of carbon as coal or as methane within the coal explains the increase of thermal value in the transition from peat to coal.

2. Trace metals such as gold, silver, copper, nickel, etc., that are foreign to the vegetation before coalification, vary widely among various coals without relation to former subcrop grades, and often at levels far above the levels in present-day vegetation. Metals of this kind could be introduced as metal hydrides accompanying methane and deposit as native or oxidized mineral forms or metabolize as organometallic complexes in microbiota. In any case, they must be regarded as additions in the coalification process.

3. Most coal measures have no fossilized regolith below them that would represent the soil horizon in which the original peat-contributing vegetation was rooted. This suggests that the carbon present volume is far greater than in the original peat or that the final depositional locality of a coal may differ from the site where the precedent plant life grew.

4. Peat and kerogen are made up of partially oxidized organic molecules, that is to say, molecules from which electrons have been lost. They are, thus, partly ionized and positively charged. Further oxidation releases methane from peat along with carbon dioxide, but it does not generate petroleum; and oxidation would have precluded coalification. Thus, further oxidation is not what has happened. Instead, reduction has led to coalification and, because coal represents more carbon than its peat predecessor, the methane found in many coal seams must be mainly from an external source. Calorific value is enhanced by the reduction process, and would be an impossibility in an oxidizing regime.

Metals in Petroleum and Coal

Metal contents of petroleum and coal are particularly problematical to explain within existing theories. For example, the Firebag coal deposit of northeastern Alberta is laterally contiguous with the Cretaceous Athabasca bituminous sands, which contain over two trillion barrels of bitumens in the immediate area. The 900 million ton coal deposit is a shoreline replacement of the estuarine, now-bituminous sands, which interfinger northeastward into it and overlap it. Paleozoic carbonates that underlie the sand/coal horizon are not known to be mineralized. Thus, in the Firebag coal and its associated petroleum metals not normally present in either oil or coal beyond trace amounts occur abundantly. Analyses run up to levels such as the following: (in ppm) Au 1.04, Ag 0.9, Cu 34, Pb 9, Zn 15, Co 35, Cr 398.

Plant uptake of water that has passed through metal ores can give such impregnations in vegetation, but such uptake was not a precedent during Firebag peat deposition, so far as known. A better explanation for the metals is that they arrived as hydrides along with the methane that promoted coalification. Metal hydrides are volatile and would be stable in the reducing environment of effusing methane, but oxidized so as to deposit native metals, oxides, or sulphides when they entered a peat bed in the process of coalification.

Metal content of oil in the Athabasca bituminous sands adjoining the coal is likewise elevated far above the metal content found in Cretaceous oils farther southwest in the western Canada basin. The enhancements of particular metals in the Syncrude bitumens mine some thirty miles to the west of the Firebag site are shown in the following table:[8]

Metal	Alberta plains	Ft. McMurray	Enhancement
Na	0.32 ppm	40.33 ppm.	126 x
Fe	0.50 "	141.7 " 254 "	283 x 508 x
Mn	Na	3.853 "	
Ni	0.80 ppm	74.11 " 71.88 "	93 x 90 x
Cr	Na	1014.0 ppb 1682.0 "	
Co	1.25 ppb	1349.0 " 1998.0 "	1079 x 1598 x
As	4.17 "	400.3 " 320.9 "	96 x 77 x
Se	Na	190.7 "	
Cs	0.36 "	25.9 " 68.53 "	72 x 190 x
V	1.107 ppm	176.5 ppm	159 x
Eu	.091 ppb	9.0 ppb	99 x
Au	Na	1.316 "	
Ga	Na	315 ppm 267 "	

Conclusions

It seems clear to this author from the foregoing that an external agency capable of supplying carbon, hydrogen, energy, and metals has been at work in the generation of the world's coal deposits.

Anhydride theory, thus supports the proposition that the carbon forms comprising peat and kerogen are partially oxidized, and hence carry a slight positive charge, whereas the carbon in methane from earth's highly reduced interior environment carries a negative charge, that methane has been drawn into peat or kerogen and once there, has given the misleading impression that it originated there, that microorganisms, which pervade the shallow crust and all fossil biomass, are demonstrably able to strip hydrogen progressively from the abundant methane that effuses from earth's interior and thus to provide the partial molecules that can recombine to form longer-chain hydrocarbons,[2] and that the end of the process is terminal carbon, either in the form of coal or asphaltite.

In conclusion it can be said that Gold's abiogenesis and diagenetic biogenesis are parlous theories that fail to fit the facts. Only Anhydride Theory accommodates the observed data on coal, with respect to excess carbon, metal content, and incremental energy above the energy of peat. Only anhydride theory adequately explains the geological distribution in igneous rocks as well as sedimentary rocks of methane, high gravity oils, black oils, and their contained porphyrins, ash and metals. Only Anhydride theory allows for rejuvenation of depleting and depleted oil and gas fields, which is apparently a real phenomenon in our time and should be monitored.[9]

Observation therefore leads to the some permissible deductions:

1. Petroleum produced by dehydrogenation of methane is a *"mixture of anhydrides,"* and hence, constitutes a *biogenic, renewable fuel, not a "fossil fuel."*

2. Porphyrins and metals that occur in petroleum and coal are residues from living microbial metabolisms that produced the petroleum and coal, and not fossilized residues.

3. Coal in varying degrees comprises a mixture of peat macerals and terminally dehydrogenated methane. Coal, like petroleum, should be considered a *biogenic,* renewable resource, not a *"fossil fuel."*

4. *"Age"* assignment to oil or coal should be regarded as the age of its microbiotic generation, not the age of the host rock, which may be much younger.

5. *"Typing"* of oil based on the residues from an originating microbiota cannot be taken to demonstrate an originating *"source rock,"* but only the spectrum of microbiotas that acted on it during migration.

6. Peat and kerogen may or may not have been involved in the origin of coal and petroleum, which may have been generated independently from methane.

[1] Evans, Wm. C. 1996, A gold mine of methane, *Nature*, v 381 p.114-115

[2] Stetter, K.O., R. Huber, E. Blochl, M. Kurr, R.D. Eden, M. Flelder, H. Cash, I. Vance, 1993, Hyperthermophyllic archaea are thriving in deep North Sea and Alaskan oil reservoirs, *Nature*, v.365, p743-745.

[3] Asphaltite also goes by other names: uintaite, grahamite, gilsonite, wurtzilite

[4] Clemett, S.J. and R.N. Zare, 1996 Science, v..274, 2122-2123

[5] Anders, E. 1996, Science, v.274, 2119-2121

[6] Hunt, C.W., 1992, *Expanding Geospheres*, Polar Publishing, Calgary

[7] Mahfoud, R.F., and J.N. Beck, 1995, Why the middle east oilfields may produce forever, *Offshore Magazine*, April, 1995 p.56-106

[8] Hitchon, B, and Filby, R.H., Alberta Crude Oils, OFR 1983-02, Alberta Research Council

[9] Indications of natural pressure maintenance may demonstrate rejuvenation that counteracts expected depletion should not be dismissed immediately as indicative of error in earlier estimation of reserves.

Looking Back on Development History of Coalbed Methane in China

Zhang Sui An

(China United Coalbed Methane Co. Ltd.)

Abstract

China is a large country on coalbed methane resources. The total volume of CBM resources in China is about 30~35 × $10^{10}m^3$.[1] So she has greatness development potential of coalbed methane in China. The developments of CBM in China go through three major stages. This paper presents the history of Chinese development coal seams gas, including in-mine gas drainage in the mining areas, Gob well drainage and vertical well development.

The development program of coal seams gas in China can be divided into three distinct stages – non-industrialized development stage, an early exploration stage and a more recent experimental stage.

1 Non-industrialized Development

This stage is a stage of in-mine gas drainage. China's coal seams gas drainage history dates back almost one-half century to the 1950's. Beginning at the Longfeng mine of the Fushun mining administration, gas drainage techniques and technology were developed primarily for improved mine safety. Subsequently, these procedures were expanded to all of the high gassy mines in China.

At present, approximately 130 underground coal mines had in-mine gas drainage systems in operation.[2,3] During the last 50 years, in-mine gas drainage has changed from being only mine safety related to mine safety and energy related. Notwithstanding it's potential energy value, in-mine drainage remains one of the major methods to reduce emissions and avoid gas outbursts in the coal mines of China.

The selection of the appropriate in-mine gas drainage method depends on many variables, including in-mine gas (CBM) reservoir properties, coal seam conditions, and mining operations (mining rate, type, etc.). Dependent upon the source of the CBM in the mining operation, in-mine drainage can be divided into (1) in-seam drainage; (2) adjacent seam drainage; (3) gob drainage; and (4) adjacent strata drainage. Drainage can be further divided, based on the time of drainage, into pre-mining drainage, post-mining drainage, and artificial stress relief drainage; based on the applied techniques, the drainage can be divided into hole drainage (drainage through drilled holes) or entry drainage (drainage through mine entries); Based on the relationship between mining and the drainage, it can be divided into pre-mining, with mining, and post-mining drainage. Within China, the primary division of drainage type is associated with gas source.

Techniques for In-Seam Drainage -- In-seam drainage technique refers to the use of holes (or entries) within the seam being mined and this technique is widely applied in China. Currently, the adopted practice is to employ holes drilled from the seam being mined into the unmined areas or from a different level into the seam being mined.

The use of pre-mining boreholes is a simple operation with low cost and with favorable mine safety results and is adopted in most mines. This method drains the CBM under virgin seam conditions and is of two general types - one along the dip of the seam and the other across strata to the mined seam. For example, in the Huainan Administration, Liozhi Administration, and the Luling mine of the Huaibei Administration, in-seam gas drainage is achieved through the drilling of boreholes in the gate road entries with a drainage time often of one year. The Songzao sub-project of the "Development of Coalbed Methane Resources of China" project is testing the use of long horizontal boreholes, another form of in-seam drainage. The use of boreholes across the strata (cross-measure) to drain CBM from the mined seam is employed in the Fushun, Zhongliangshan, and Fengfeng Administrations and in the Xieer mine of the of the Huainan Administration and the Luling mine of the Huaibei Administration. In some areas with favorable coal seam permeability, the technique which employs sealed mine entries to drain the gas is used, such as at the Fushun Administration.

Techniques for Adjacent Seam Drainage -- Adjacent seam drainage techniques are employed when over- and under-lying coal seams are the source of the CBM and affect the mine workings due to natural fracturing caused by the mining operation (this could be considered a type of gob well drainage). Two methods are commonly employed, hole drainage and entry drainage with the former the most widely used.

The key for success of adjacent seam drainage is that the borehole should be drilled into the adjacent relaxed zone and avoid the collapsed and highly fractured zone. This technique has been adopted by the Yangquan, Baotou, Nantong, and Tianfu Administrations. Recently, at the Daxing mine of the Tiefa Administration, three experimental long sub-horizontal to horizontal boreholes were drilled 10 to 20 meters above the seam being mined (#7 coal). This is a new approach and is still in the testing phase but confirmed results.

The effect of entry drainage is characterized by large amounts of drained gas, high efficiency, large areas effectively drained, and effective removal of gas from the working face. Adjacent seam drainage through entries is effective at the Didao mine of the Jixi Administration and the Xiaoming mine of the Tiefa Administration. The optimum drainage method for the Yangquan Administration is also using the high-efficiency entry system to remove gas from neighboring gob areas.

The drilling of large diameter boreholes is currently replacing the driving of entries for adjacent seam drainage. For example, the Yangquan Administration in conjunction with the Xi'an Branch - CCRI is conducting tests in the No. 3 and No. 5 mines at this administration.

Technique for Gas Drainage of Gob Areas -- During mining operations, especially under multiple-coal seam conditions, CBM flows into the gob area from adjacent seams, surrounding strata, coal pillars, and coal abandoned at the working face.   Therefore, large amounts of CBM can be stored in old or new gob areas.   Gas drainage of these gob areas is therefore not only possible but also quite effective.

Gas drainage in gob areas can operate either with the longwall mining operation or can effectively drain the gas once the gob area is sealed (either totally or partially sealed).   One advantage of gob well drainage is that it is applicable to most any geologic or mining condition and does not require a large amount preparatory work.   However, the gas drained in this manner is often of low methane concentration and can be hazardous due to spontaneous combustion in the gob area.   Because of this, progress in the research and development of in-mine gob well drainage has been slow, but recently favorable results have been obtained.   Based on an investigation of 60 mines, about half of the mines have 25 to 30 percent of the extracted gas coming from the gob drainage program.   In some mines, this can   be as high as 40 to 50 percent.

2   Early Exploration on Coal Seams Gas Development

During the 1970's to 1980's, about twenty-two wells were drilled for mine safety, utilizing hydraulic fracturing or cavitation completion to the coal seams as pay bed.   These wells were tested for gas production at the Fushun administration (Liaoning Province), Yangquan administration (Shanxi Province) and the Jiaozuo administration (Henan province).   Because of numerous reasons, all of these tests resulted in poor production performance.

In the mid-1980's, to look for gas reservoir of the coal-formed gas, few vertical wells were drilled and tested by the Ministry of Geology and Mineral Resources at Kailuan administration (Hebei province) and Yangquan administration (Shanxi province) and another administration, but again these were unsuccessful.

At same time and for sameness purpose, Ministry of petroleum industry drilled few wells in Zhongyuan Oil Field and Chinese South Sea and so on, a part of the wells are producing still until present.

3 Development Test Stage

The Xi'an Branch of the China Coal Research Institute has completed a national key project of the Seven-Five-Year Plan of China, "Assessment of Coalbed Methane Resources in China", this project is a first study project on coalbed methane in China. This report caused Chinese government and overseas companies, for example ICF Resources Corporation. The some overseas companies have begun to come in Chinese mark of CBM development. The first national coalbed methane study workshop was held by the former Ministry of Energy in Shenyang city in December 1989. Beginning with that workshop a new era in CBM development in China began with a new level of activity in CBM exploration and development. This led to the initiation of exploration tests by the various Chinese agencies, mining administrations, local governments, and joint-venture companies.

Beginning in May 1990, a pilot project, approved by the National Planning Committee and funded by the National Energy Investment Corporation was started at the Tangshan mine of the Kailuan administration.[4] The ICF Resources Corporation (USA) provide technical service. Two assessment wells were drilled during to execute this project. Concurrent with this, eight CBM assessment wells and a production test well were drilled by the Shenyang Coal Gas Company, also with assistance from ICF Resources.

During the Eighth-Five-Year Plan of China, two national key project, "Area Identification for Coalbed Methane Exploration and Development and Related Techniques" and "Research on the Development of Coalbed Adsorbed Gas in Favorable Areas", were completed and over ten CBM wells were drilled for assessment or produce test at Anyang (Henan Province), Huainan (Anhui Province), and Dacheng (Hebei Province). One of the wells, Dacan No. 1 achieved commercial gas production rates of up to 6,000 m³ per day.[5,6] China National Petroleum Corporation has conducted exploration efforts at Xialiaohe (Liaoning Province), Fengfeng (Hebei Province), Qinshui (Shanxi Province), Xinyang (Henan Province), Fengcheng (Jiangxi Province), and Hongshandian and Lengshuijiang (Hunan Province).[7]

The United Nations Development Program (UNDP) has provided a large amount of support to China' CBM development program. In 1992, UNDP with the former Ministry of Energy (now the Ministry of Coal Industry) initiated the project "DEVELOPMENT OF COALBED METHANE RESOURCES IN CHINA." As part of this, the "National Coalbed Methane Resource Assessment" sub-project tested ten assessment wells in eight coal administration areas (Tiefa, Anyang, Hebi, Jiaozuo, Jincheng, Pingdingshan, Huaibei, and Huainan). Seventeen target coal seams in these wells were cored and tested, resulting in the collection of a large database of reservoir properties.

In 1993, the UNDP and the China International Economic Technology Exchange Center of the Ministry of Economy and Trade initiated a project "Coalbed Methane Exploration in Deeper Strata." The Ministry of Geology and Natural Resources, the project's executing agency, drilled six CBM production test wells, which installed the first pilot test field of China, production rates as high as 7,200 m³ per day have been achieved.

Jincheng administration joined with the Sino-American Energy Company (USA) and has drilled one assessment and three production test wells at Panzhuang (Shanxi Province). The three production test wells have been completed and produced, with one well in production since September 1994. Average production rate of CBM from a well has been approximately 2,000~5,000 m³ per day. It is the second successful CBM development pilot test field of China. [8]

At last years, the China National Administration of Coal Geology tested three CBM exploration wells at Tiefa (Liaoning Province), Yangquan (Shanxi Province), and Hancheng (Shaanxi Province). The well at Hancheng tested at 3,000 to 4,000 m³ per day during the early stage.

This favorable trend of CBM development by vertical wells has led to the interest and attention of many foreign companies, leading to cooperative ventures. Enron (USA) in cooperation with the Ministry of Coal Industry tested ten assessment and production test wells, four at Huainan (Anhui Province), five at Sanjiao (Shanxi Province), and one at Pingdingshan (Henan Province). Subsequent to this, Enron and the North China Bureau of Petroleum Geology and the Ministry of Geology and Natural Resources tested two wells at Shilou (Shanxi Province). CBM Energy Associates (USA) has cooperated with the Committee of Fuxin City (Liaoning Province) and the Energy Group Corporation (Shanxi Province). Four wells have been drilled at Linxian and Xinxian (Shanxi Province) as a result of this venture. In addition to their cooperation with the Jincheng administration (Shanxi province), the Sino-American Energy Company has also cooperated in a CBM venture with the Natural Gas Company of Zhengzhou city (Henan Province), with a well drilled at Xinyang. Lowell (Australia) and Amoco (USA) have both cooperated with the North China Bureau of Petroleum Geology, with wells drilled by each venture at Liulin and Puxien (respectively).

In May 1996, the State Council along with the Ministry of Coal Industry, Ministry of Geology and Natural Resources, and the China National Petroleum Corporation, approved the formation of a joint company, the China United Coalbed Methane Co., Ltd. This organization is an important enterprise responsible for the overall coordination of CBM exploration, development, transportation, marketing, and utilization. This organization has specific rights to conduct business with foreign enterprises interested in CBM in China. The establishment of this organization will push the development of CBM forward and is indicative that CBM development and utilization is entering a next new phase.

References

2. Zhang Xinmin, Zhang Suian et al, "Coalbed Methane Resources in China", Shaanxi Technology Publishing House, 1991.

4. Li Zhongqi, "Special Issue of Coal Gas, Thermal Force and Gas Resource Utilization," Coal Gas and Thermal Force magazine, 1994, P. 128-136.

5. Coalbed Methane Information Center of the Ministry of Coal Industry and the U.S. Environmental Protection Agency, "United Report on Prospective of China's Coalbed Methane Development," USEPA 1995, p. 36-49.

6. L. D. Sheehy, S. Stevens, C. M. Boyer II, and Lu Fumin, "Summary of CBM Development in Shenyang of China," Coalbed Methane 1, 1994, P.43-49.

7. Coalbed Methane Information Center of the Ministry of Coal Industry and the U.S. Environmental Protection Agency, "United Report on Prospective of China's Coalbed Methane Development," 1995, USEPA, p. 56-60.

8. Bai Qingzhao, "Evaluation of Optimum Areas for Development in China," in Coalbed Gas in China, Vol. 1, 1995, p. 18-20.

9. Zhang Jianbo, Li Anqi, "Study on Increasing Gas Production and Field Testing," in Coalbed Gas in China, Vol. 1, 1995, p. 50-53.

10. Jin Anxin, Li Guofu, et al, "Elementary Inquiring of Transforming Permeability of Anthracite Coal Seam in Jincheng by Hydraulic Fracturing Method," Proceedings of International Conference on Coalbed Methane Development and Utilization, 1995, p. 315.

The Study of the Influence of Pressure on Coalbed Permeability

Zhang Jianbo,   Li Jingming,   Wan Yujin ,Wang Hongyan
Langfang Branch Institute, RIPED
P.O. Box 44, Langfang City, Hebei Province
People's Republic of China

1. Introduction

The permeability of coalbed is not only controlled by geological conditions, but it is also influenced by later production process (Zhang xinmin, 1991), especially the influence of producing pressure on the permeability of coal bed near wellbore is very significant. It has less reported about the study of the influence of production status on the permeability in literature for coalbed well. For the producing well, the producing pressure drop should be properly controlled during drainage in order to prevent coal fine from moving which can decrease coal permeability. That is extremely important for the success of experimental well and the evaluation of the previous experiment process in the key stage of exploration and development of coalbed methane. The correlation between the producing pressure drop and the permeability of coalbed have been established by using the experimental technology, the results of this paper may play an instructing role for coalbed methane production. Here are major elements of this study:

(1) Pore structure features and its influence on permeability of coal has been studied.
(2) The permeability at different confining pressure has been measured and the influence of confining pressure on coal permeability has been studied.
(3) The influence of upstream pressures on coal permeability has been studied; the range of wellhead pressure correspondence to the biggest permeability has been gotten.

2. Experimental Procedures

Two kinds of samples were selected to do the experiments, the pore size distribution is measured by using liquid nitrogen adsorption method, the apparatus used here is ASAP200 which can measure the specific area and the micro pore size distribution.

The core sample used to measure the permeability is about 3-4 inches long and 2 inches in diameter, The sample was putted into hydrostatic core holder, the helium permeability was measured at different confining pressure and at different upstream pressure.

3. Experimental Results

This paper shows a series of experiment results of two kinds of samples(figure 1 to figure 6).

Figure 1 and figure 2 are pore size distribution diagrams of two coal samples. Sample No.1 shows a kind of samples with more large pores (20-100nm), while sample No.2 represents samples with more small pores (5 ~ 20nm).

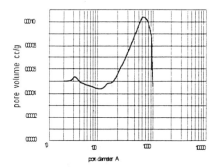 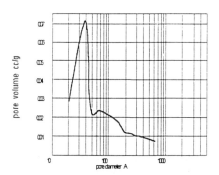

Fig.1 Pore size distribution   **Fig.2 Pore size distribution**
diagram of Sample No.1   **diagram of Sample No.2**

Figure 3 and figure 4 are diagrams of permeability variation with confining pressure. With the increasing confining pressure, coal permeability goes down, but the permeability of sample with more large pores decreases rapidly.

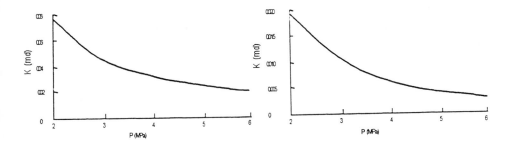

Fig.3 The affection of confining
pressure on permeability of
Sample No.1

Fig.4 The affection of confining
pressure on permeability of
Sample No.2

Figure 5 and figure 6 show the variation of permeability with upstream pressure. The variation of sample No.1 is less than that of sample No.2. With the increase of upstream pressure, the coal permeability drops down first and then goes up later, there is a transparent spot from decreasing to increasing.

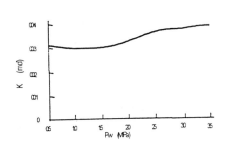

Fig.5 The affection of upstream
pressure on permeability
of Sample No.1

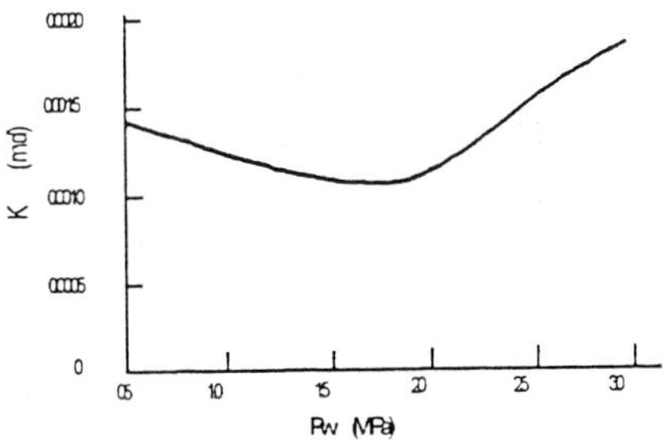

Fig.6 The affection of upstream
pressure on permeability
of Sample No.pressure2

4. Analysis and Discussion

The experiment shows that the coal permeability directly depends on the pore size distribution. For the coal, the bigger the proportion of its large pores, the higher the permeability will be.

Due to coal's compressibility is bigger and coal has both micro-pores and fracture pores, the permeability varies from high to low and then from low to high with the change of pressure drop. The magnitude of permeability variation is affected by the development of micro pores and fracture pores. Sample No.1 has more large pores, so the magnitude of permeability variation is less than that of sample No.2, which has small pores, with the variation of upstream pressure. The reason of the permeability changes less is that the fluid can flow through the coal sample when the large pores were compressed, permeability decrease a lot when the small pores were compressed. To different coal bed, the permeability is different with the variation of production pressure and will have the lowest spot (value) .

During producing, the experimental to the coal samples should be done. According to the experiments results we can select properly production status, in order to avoid producing within the lower permeability range.

5.   Conclusion

(1) The confining pressure affected the coal permeability obviously. With the pressure increasing, the coal permeability decreased rapidly.

(2) To the coal, the bigger the proportion of its large pores, the higher the permeability will be. With the variation of upstream pressure, the ranges of the permeability variation are less than that of coal samples having larger of small pores.

(3) The coal permeability varies with its upstream pressure. There is a lowest permeability spot in range of   0-3.5MPa, the spot depends on different coal porous structure.

Note: $1mD=9.8 \times 10^{-4} \mu m^2$
                 $1A=0.1nm$

References:

1. Zhang xinmin, The coalbed methane in China, Xi-an, Science and Technology Publishing House , Shan-xi Province,1991.
2. Qian-kai, Exploration Theory and Development Test Technology on the Coalbed Methane, Beijing, Petroleum Publishing House, 1996.